Neuromethods

Series Editor
Wolfgang Walz
University of Sasatchewan
Saskatoon, SK, Canada

For further volumes:
http://www.springer.com/series/7657

Neuromethods publishes cutting-edge methods and protocols in all areas of neuroscience as well as translational neurological and mental research. Each volume in the series offers tested laboratory protocols, step-by-step methods for reproducible lab experiments and addresses methodological controversies and pitfalls in order to aid neuroscientists in experimentation. *Neuromethods* focuses on traditional and emerging topics with wide-ranging implications to brain function, such as electrophysiology, neuroimaging, behavioral analysis, genomics, neurodegeneration, translational research and clinical trials. *Neuromethods* provides investigators and trainees with highly useful compendiums of key strategies and approaches for successful research in animal and human brain function including translational "bench to bedside" approaches to mental and neurological diseases.

Stroke Biomarkers

Edited by

Philip V. Peplow

Department of Anatomy, University of Otago, Dunedin, New Zealand

Bridget Martinez

Physical Chemistry and Applied Spectroscopy, Chemistry Division, Los Alamos National Lab, Los Alamos, NM, USA

Svetlana A. Dambinova

Brain Biomarkers Research Lab, Emory Decatur Hospital, Emory Healthcare, Decatur, GA, USA

Institute of Pharmacy, I.M. Sechenov First Moscow State Medical University, Moscow, Russia

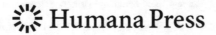 Humana Press

Editors
Philip V. Peplow
Department of Anatomy
University of Otago
Dunedin, New Zealand

Bridget Martinez
Physical Chemistry and Applied Spectroscopy
Chemistry Division
Los Alamos National Lab
Los Alamos, NM, USA

Svetlana A. Dambinova
Brain Biomarkers Research Lab
Emory Decatur Hospital
Emory Healthcare
Decatur, GA, USA

Institute of Pharmacy
I.M. Sechenov First Moscow
State Medical University
Moscow, Russia

ISSN 0893-2336   ISSN 1940-6045   (electronic)
Neuromethods
ISBN 978-1-4939-9684-1   ISBN 978-1-4939-9682-7   (eBook)
https://doi.org/10.1007/978-1-4939-9682-7

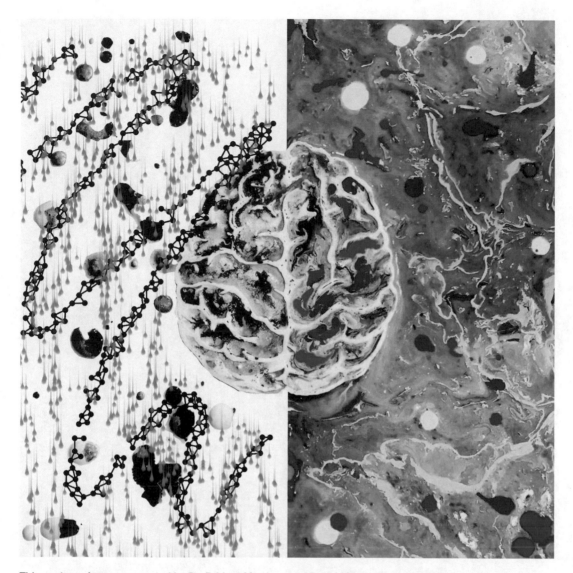

This custom piece was created by Dr. Bridget Martinez and Donald Mario Robert Harker; using acrylic paint on canvas with the addition of graphic design

This artistic rendition encapsulates the hemispherical dominances in a visual capacity. Whereas one represents an objective, linear, logical, analytical and sequential school of thought, the other reveals itself to be less reasonable and dabbles in the idiosyncrasies of subjectivity by manifesting itself as a holistic network invoking an emotional, more intuitive and creative thought. Yet whether it is the poet or the accountant which dominates in each of us, it has become increasingly clear that the complete absence of either, leaves one in a universe devoid of both rhyme and reason.

Dedication

This book is dedicated to our universe which inspires, challenges, creates, and molds beauty, even from the most unexpected corners and circumstances. The pursuit of knowledge is a great adventure. In loving memory of my father Elmer Martinez. Endless gratitude to my mother Eda Noguera, my sister Karla "Zeo" Sanchez, and my brother Elmer Martinez.

Bridget Martinez

The world is still a beautiful place and full of kindness and love. I wish to dedicate this book to Bridget who is a shining light and inspiration in my life and also to my family, friends, and colleagues for their support and encouragement.

Philip V. Peplow

In memory of Robert Rose Rhinehart, a strong believer in a great future of brain biomarkers and constant support.

Svetlana A. Dambinova

Preface to the Series

Experimental life sciences have two basic foundations: concepts and tools. The Neuro-methods series focuses on the tools and techniques unique to the investigation of the nervous system and excitable cells. It will not, however, shortchange the concept side of things as care has been taken to integrate these tools within the context of the concepts and questions under investigation. In this way, the series is unique in that it not only collects protocols but also includes theoretical background information and critiques which led to the methods and their development. Thus, it gives the reader a better understanding of the origin of the techniques and their potential future development. The Neuromethods publishing program strikes a balance between recent and exciting developments like those concerning new animal models of disease, imaging, in vivo methods, and more established techniques, including, immunocytochemistry and electrophysiological technologies. New trainees in neurosciences still need a sound footing in these older methods in order to apply a critical approach to their results.

Under the guidance of its founders, Alan Boulton and Glen Baker, the Neuromethods series has been a success since its first volume published through Humana Press in 1985. The series continues to flourish through many changes over the years. It is now published under the umbrella of Springer Protocols. While methods involving brain research have changed a lot since the series started, the publishing environment and technology have changed even more radically. Neuromethods has the distinct layout and style of the Springer Protocols program, designed specifically for readability and ease of reference in a laboratory setting.

The careful application of methods is potentially the most important step in the process of scientific inquiry. In the past, new methodologies led the way in developing new disciplines in the biological and medical sciences. For example, Physiology emerged out of Anatomy in the nineteenth century by harnessing new methods based on the newly discovered phenomenon of electricity. Nowadays, the relationships between disciplines and methods are more complex. Methods are now widely shared between disciplines and research areas. New developments in electronic publishing make it possible for scientists who encounter new methods to quickly find sources of information electronically. The design of individual volumes and chapters in this series takes this new access technology into account. Springer Protocols makes it possible to download single protocols separately. In addition, Springer makes its print-on-demand technology available globally. A print copy can therefore be acquired quickly and for a competitive price anywhere in the world.

Saskatoon, SK, Canada   *Wolfgang Walz*

Preface

There is an increasing incidence of stroke in adults, and this trend is expected to continue to increase due to changing demographics. This increased incidence in stroke is further exacerbated by improved medical care and as health provision leads to extended life expectancy in developed and developing countries. Increasing effort is being given to develop reliable biomarkers for brain disorders, including stroke, and this will dramatically accelerate research on the etiology, pathophysiology, and disease progression of a number of prevalent and devastating nervous system diseases.

It is the goal of this book to provide a forum for international experts in the field of stroke research, both experimental and clinical, to present recent data on the latest achievements in new and emerging technologies for stroke biomarkers and innovations in stroke assessment. Theoretical background together with tested protocols reproducing experimental, clinical laboratory and instrumental methods for educational purposes are presented. It is hoped that the topics covered herein will extend the knowledge on the role of the upcoming biomarkers in different types of stroke and that will lead to a more effective approach to clinical management and benefit to patient care.

We wish to express our deep appreciation to each of the chapter authors for the time and effort spent on writing informative reviews on their respective areas of clinical and research interest. Also, we wish to thank Professor Wolfgang Walz, Series Editor, Springer Neuromethods series, and Anna Rakovsky, Assistant Editor, Springer Protocols, for their help and advice in putting together this book.

Dunedin, New Zealand   *Philip V. Peplow*
Los Alamos, NM, USA   *Bridget Martinez*
Decatur, GA, USA   *Svetlana A. Dambinova*

The original version of this chapter was revised. The correction to this chapter is available at https://doi.org/10.1007/978-1-4939-9682-7_21

Contents

Dedication . . . . *vii*
Preface to the Series . . . . *ix*
Preface . . . . *xi*
Contributors. . . . . *xv*

PART I INTRODUCTION

1   Integrative Biomarkers in Stroke   3
    Svetlana A. Dambinova, Bridget Martinez, and Philip V. Peplow

PART II RESEARCH METHODS

2   Vascular-Related Biomarkers of Ischemic Stroke   9
    Daniel Arteaga and Bradford B. Worrall

3   Generation and Applicability of Genetic Risk Scores
    (GRS) in Stroke 23
    Natalia Cullell, Jonathan González-Sánchez,
    Israel Fernández-Cadenas, and Jerzy Krupinski

4   Neuroplasticity Biomarkers in Experimental Stroke Recovery. 35
    Philip V. Peplow, Bridget Martinez, D. Mascareñas,
    and Svetlana A. Dambinova

5   Methods of Mitochondrial and Redox Measurements
    in Ischemic Stroke 61
    Oiva Arvola, Anand Rao, and Creed M. Stary

6   Blood Biomarkers for Stroke Differentiation 79
    Deepti Vibha and Shubham Misra

7   Laser-Capture Microdissection for Measurement of Angiogenesis
    After Stroke 113
    Mark Slevin, Xenia Sawkulycz, Laura Combes, Baoqiang Guo,
    Wen-Hui Fang, Yasmin Zeinolabediny, Donghui Liu,
    Glenn Ferris, and Anna Ludlaim

PART III TRANSLATIONAL INNOVATIVE METHODS

8   Blood-Borne Biomarkers of Hypertension Predicting Hemorrhagic
    and Ischemic Stroke. 125
    Alina González-Quevedo, Marisol Peña Sánchez,
    Sergio González García, María Caridad Menéndez Saínz,
    and Marianela Arteche Prior

9   RNA Gene Expression to Identify the Etiology of Acute
     Ischemic Stroke: The Biomarkers of Acute Stroke Etiology
     (BASE) Study 157
     *Edward C. Jauch, W. Frank Peacock IV, Judy Morgan, Jeff June,
     and James Ireland*

10   Neurotoxicity Biomarker Assay Development 171
     *Galina A. Izykenova, German A. Khunteev, Ivan I. Krasnjuk,
     Vladimir L. Beloborodov, and Svetlana A. Dambinova*

11   Glutamate Receptor Peptides as Potential Neurovascular Biomarkers
     of Acute Stroke 195
     *Svetlana A. Dambinova, J. D. Mullins, J. D. Weissman,
     and A. A. Potapov*

12   Antibodies to NMDA Receptors in Cerebral and Spinal Cord
     Infarctions 225
     *G. V. Ponomarev, E. V. Alexandrova, Svetlana A. Dambinova,
     D. S. Asyutin, N. A. Konovalov, and A. A. Skoromets*

13   Impaired Retinal Vasoreactivity as an Early Marker of Stroke
     Risk in Diabetes 245
     Kerstin Bettermann and Kusum Sinha

PART IV   CLINICAL METHODS

14   Imaging Biomarkers: Keys to Decision-Making in Stroke 259
     J. D. Weissman, J. C. Boiser, C. Krebs, and G. V. Ponomarev

15   Neuroimaging Methods for Acute Stroke Diagnosis and Treatment 297
     Mathew Elameer and Christopher I. Price

16   Ultrasound Assessments of Risk for TIA and Stroke in Vascular Surgery 335
     Melvinder Basra and Robert E. Brightwell

17   Preoperative and Intraoperative Markers of Cerebral Ischemia 349
     *V. A. Lukshin, D. Yu Usachev, A. V. Shmigelsky, A. A. Shulgina,
     and A. A. Ogurtsova*

18   Time is Brain: The Prehospital Phase and the Mobile Stroke Unit 371
     Shrey Mathur and Klaus Fassbender

19   Acute Clinical Intervention and Chronic Management of Cerebral
     Vascular Accident 397
     D. M. R. Harker

PART V   CONCLUSION

20   Trends in Biomarkers Development for Stroke 419
     Philip V. Peplow, Bridget Martinez, and Svetlana A. Dambinova

Correction to: Laser-Capture Microdissection for Measurement
of Angiogenesis After Stroke C1

Index *423*

Contributors

E. V. ALEXANDROVA • *Burdenko National Science and Practical Centre for Neurosurgery, Moscow, Russia*

DANIEL ARTEAGA • *Department of Neurology, University of Virginia, Charlottesville, VA, USA*

MARIANELA ARTECHE PRIOR • *Institute of Neurology and Neurosurgery, Havana, Cuba*

OIVA ARVOLA • *Department of Anesthesiology, Perioperative and Pain Medicine, Stanford University School of Medicine, Stanford, CA, USA*

D. S. ASYUTIN • *Burdenko National Science and Practical Centre for Neurosurgery, Moscow, Russia*

MELVINDER BASRA • *Vascular Surgery, Norfolk & Norwich University, NHS Foundation Trust, Norwich, UK*

VLADIMIR L. BELOBORODOV • *Institute Pharmacy, I.M. Sechenov First Moscow State Medical University, Moscow, Russia*

KERSTIN BETTERMANN • *Department of Neurology, Penn State College of Medicine, Hershey, PA, USA*

J. C. BOISER • *Emory at Decatur Neurology, Emory at Decatur Hospital, Emory Healthcare, Decatur, GA, USA*

ROBERT E. BRIGHTWELL • *Vascular Surgery, Norfolk & Norwich University, NHS Foundation Trust, Norwich, UK; Imperial College London, London, UK*

LAURA COMBES • *School of Healthcare Science, Manchester Metropolitan University, Manchester, UK*

NATALIA CULLELL • *Stroke Pharmacogenomics and Genetics, Fundació Docència i Recerca MútuaTerrassa, Hospital Mútua de Terrassa, Barcelona, Spain; Stroke Pharmacogenomics and Genetics, Institut de Recerca Hospital de la Santa Creu i Sant Pau, Barcelona, Spain; Department of Neurology, Hospital Universitari Mútua Terrassa, Barcelona, Spain; Facultat de Medicina, Universitat de Barcelona, Barcelona, Spain*

SVETLANA A. DAMBINOVA • *Brain Biomarkers Research Lab, Emory Decatur Hospital, Emory Healthcare, Decatur, GA, USA; Institute of Pharmacy, I.M. Sechenov First Moscow State Medical University, Moscow, Russia*

MATHEW ELAMEER • *Institute of Neuroscience, Newcastle University, Newcastle Upon Tyne, UK*

WEN-HUI FANG • *School of Healthcare Science, Manchester Metropolitan University, Manchester, UK*

KLAUS FASSBENDER • *Department of Neurology, Saarland University Medical Center, Homburg, Germany*

ISRAEL FERNÁNDEZ-CADENAS • *Stroke Pharmacogenomics and Genetics, Fundació Docència i Recerca MútuaTerrassa, Hospital Mútua de Terrassa, Barcelona, Spain; Stroke Pharmacogenomics and Genetics, Institut de Recerca Hospital de la Santa Creu i Sant Pau, Barcelona, Spain; Department of Neurology, Hospital Universitari Mútua Terrassa, Barcelona, Spain*

GLENN FERRIS • *School of Healthcare Science, Manchester Metropolitan University, Manchester, UK*

SERGIO GONZÁLEZ GARCÍA • *Institute of Neurology and Neurosurgery, Havana, Cuba*

ALINA GONZÁLEZ-QUEVEDO • *Institute of Neurology and Neurosurgery, Havana, Cuba*

JONATHAN GONZÁLEZ-SÁNCHEZ • *Stroke Pharmacogenomics and Genetics, Fundació Docència i Recerca MútuaTerrassa, Hospital Mútua de Terrassa, Barcelona, Spain; Stroke Pharmacogenomics and Genetics, Institut de Recerca Hospital de la Santa Creu i Sant Pau, Barcelona, Spain; Department of Neurology, Hospital Universitari Mútua Terrassa, Barcelona, Spain; Centre of Bioscience, School of HealthCare Sciences, Manchester Metropolitan, Manchester, UK*

BAOQIANG GUO • *School of Healthcare Science, Manchester Metropolitan University, Manchester, UK*

D. M. R. HARKER • *Department of Medicine, St. Georges Medical University, Grenada, West Indies*

JAMES IRELAND • *Ischemia Care, Oxford, OH, USA*

GALINA A. IZYKENOVA • *CIS Biotech, Inc., Atlanta, GA, USA; Department of Pharmaceutical Technology, I.M. Sechenov First Moscow State Medical University, Moscow, Russia*

EDWARD C. JAUCH • *Mission Research Institute/Mission Health, Asheville, NC, USA*

JEFF JUNE • *Ischemia Care, Oxford, OH, USA*

GERMAN A. KHUNTEEV • *CIS Biotech, Inc., Atlanta, GA, USA*

N. A. KONOVALOV • *Burdenko National Science and Practical Centre for Neurosurgery, Moscow, Russia*

IVAN I. KRASNJUK • *Department of Pharmaceutical Technology, I.M. Sechenov First Moscow State Medical University, Moscow, Russia*

C. KREBS • *Emory at Decatur Neurology, Emory at Decatur Hospital, Emory Healthcare, Decatur, GA, USA*

JERZY KRUPINSKI • *Stroke Pharmacogenomics and Genetics, Fundació Docència i Recerca MútuaTerrassa, Hospital Mútua de Terrassa, Barcelona, Spain; Department of Neurology, Hospital Universitari Mútua Terrassa, Barcelona, Spain; Centre of Bioscience, School of HealthCare Sciences, Manchester Metropolitan, Manchester, UK*

DONGHUI LIU • *School of Healthcare Science, Manchester Metropolitan University, Manchester, UK*

ANNA LUDLAIM • *Department of Life Sciences, Manchester Metropolitan University, Manchester, UK*

V. A. LUKSHIN • *N.N. Burdenko National Medical Research Center of Neurosurgery, Moscow, Russia*

BRIDGET MARTINEZ • *Physical Chemistry and Applied Spectroscopy, Chemistry Division, Los Alamos National Lab, Los Alamos, NM, USA*

D. MASCAREÑAS • *Engineering Institute, Los Alamos National Laboratory, Los Alamos, NM, USA*

SHREY MATHUR • *Department of Neurology, Saarland University Medical Center, Homburg, Germany*

MARÍA CARIDAD MENÉNDEZ SAÍNZ • *Institute of Neurology and Neurosurgery, Havana, Cuba*

SHUBHAM MISRA • *Department of Neurology, All India Institute of Medical Sciences, New Delhi, India*

JUDY MORGAN • *Ischemia Care, Oxford, OH, USA*

J. D. MULLINS • *Department of Surgery, Piedmont Hospital, Atlanta, GA, USA*

A. A. OGURTSOVA • *N.N. Burdenko National Medical Research Center of Neurosurgery, Moscow, Russia*

MARISOL PEÑA SÁNCHEZ • *Institute of Neurology and Neurosurgery, Havana, Cuba*

W. FRANK PEACOCK IV • *Department of Emergency Medicine, Baylor College of Medicine, Dallas, TX, USA*

PHILIP V. PEPLOW • *Department of Anatomy, University of Otago, Dunedin, New Zealand*

G. V. PONOMAREV • *Pavlov's First St. Petersburg State Medical University, St. Petersburg, Russia*

A. A. POTAPOV • *Burdenko National Science and Practical Centre for Neurosurgery, Moscow, Russia*

CHRISTOPHER I. PRICE • *Institute of Neuroscience, Newcastle University, Newcastle Upon Tyne, UK*

ANAND RAO • *Department of Anesthesiology, Perioperative and Pain Medicine, Stanford University School of Medicine, Stanford, CA, USA*

XENIA SAWKULYCZ • *School of Healthcare Science, Manchester Metropolitan University, Manchester, UK*

A. V. SHMIGELSKY • *N.N. Burdenko National Medical Research Center of Neurosurgery, Moscow, Russia*

A. A. SHULGINA • *N.N. Burdenko National Medical Research Center of Neurosurgery, Moscow, Russia*

KUSUM SINHA • *Department of Neurology, Penn State College of Medicine, Hershey, PA, USA*

A. A. SKOROMETS • *Pavlov First Saint Petersburg State Medical University, St. Petersburg, Russia*

MARK SLEVIN • *School of Healthcare Science, Manchester Metropolitan University, Manchester, UK*

CREED M. STARY • *Department of Anesthesiology, Perioperative and Pain Medicine, Stanford University School of Medicine, Stanford, CA, USA*

D. YU USACHEV • *N.N. Burdenko National Medical Research Center of Neurosurgery, Moscow, Russia*

DEEPTI VIBHA • *Department of Neurology, All India Institute of Medical Sciences, New Delhi, India*

J. D. WEISSMAN • *Emory Decatur Hospital, Emory Decatur Hospital, Emory Healthcare, Decatur, GA, USA; Emory at Decatur Neurology, Emory at Decatur Hospital, Emory Healthcare, Decatur, GA, USA*

BRADFORD B. WORRALL • *Department of Neurology, University of Virginia, Charlottesville, VA, USA; Department of Public Health Sciences, University of Virginia, Charlottesville, VA, USA*

YASMIN ZEINOLABEDINY • *School of Healthcare Science, Manchester Metropolitan University, Manchester, UK*

Part I

Introduction

Chapter 1

Integrative Biomarkers in Stroke

Svetlana A. Dambinova, Bridget Martinez, and Philip V. Peplow

Abstract

Stroke assessment is critical, complex, and should include integrative techniques combining clinical/functional, neurostructural, and biochemical markers. Accurate assessment of stroke and its appropriate timely management are essential to ensure a favorable outcome avoiding the risk of recurrent events. There is still an unmet medical need in search of key indicators that will be associated with the complexity of structural/sub-structural lesions, reflect insufficiency in neurovascular territories, and respond to immunity due to brain dysfunction. The book unveils current and upcoming methods in stroke biomarker research, translational innovative approaches to diagnosis, and advances in medical technologies.

Key words Brain, Stroke, Neurovascular territories, Integrative, Functional, Neurostructural, Biochemical, Biomarkers

1 Introduction

Stroke or "brain attack" could be considered an emergent multi-systemic disorder involving cerebrovascular insufficiency of thrombotic or embolic origin, causing neuronal impairment, and immune system activation. Stroke causes a major disability worldwide and has the second leading rate in mortality. To improve diagnostic certainty of probable stroke onset it would be plausible to add affordable blood test detecting brain biomarkers to standard clinical imaging protocol. There is still an unmet medical need in search of key indicators that will be associated with the complexity of cerebral structural/sub-structural connectivity, reflect insufficiency in different neurovascular territories, and respond to cellular immunity activation due to brain lesions. Besides, stroke is subdivided into three major subtypes: the transient ischemic attack (TIA), cerebral ischemia, and hemorrhage that are conventionally presented as separate disorders. Ideally a multi-biomarker panel should demonstrate a diagnostic performance efficacy similar to that of troponin which is a "perfect" heart failure biomarker.

Philip V. Peplow et al. (eds.), *Stroke Biomarkers*, Neuromethods, vol. 147, https://doi.org/10.1007/978-1-4939-9682-7_1,
© Springer Science+Business Media, LLC, part of Springer Nature 2020

Ischemic stroke represents the most common (75–80%) type and is subclassified as thrombotic when a clot forms directly in an artery supplying the brain and accounts for approximately 50% of all strokes, while embolic stroke depends on a microemboli formed due to cardiac conditions and then transported through the bloodstream to the brain. Cortical and subcortical lesions usually occur approximately in 80%, vertebral basilar damage occurs in about 8%, and up to 12% are found in the brainstem area.

Hemorrhagic strokes account for 20% of all strokes with mortality rate up to 50% and are divided into categories depending on the site and cause of bleeding. In *intracerebral hemorrhage* (ICH), bleeding occurs from a ruptured blood vessel within the brain. A *subarachnoid hemorrhage* (SAH) involves bleeding from a damaged blood vessel that accumulates at the surface of the brain.

Transient ischemic attacks (TIAs) are defined as brief episodes of neurologic dysfunction caused by ischemia or inadequate blood flow when clinical symptoms typically last less than 1 h without evidence of acute infarction. They are increasingly underrecognized, underreported, and undertreated. After a first TIA, 10–20% of patients are likely to have a stroke within the next 90 days, and in 50% of these patients the stroke occurs within the first 2 days after a TIA.

Furthermore, up to 20% of patients presenting at Emergency Departments have no cerebral ischemia but present with the so-called "stroke mimics," including palsies, complicated migraine, postictal paresis, psychological disturbance, and other etiologies. These conditions should be ruled out from strokes. Another key question is whether a patient has suffered a TIA, or acute ischemic or hemorrhagic stroke. The third major issue is to promptly assess the location of vascular territory of lesion(s) that are connected to severity of the brain altered state and outcome.

Stroke assessment is critical, complex, and should include integrative techniques combining clinical/functional (stroke scales), structural (neuroimaging), and biochemical (single or panel of brain indicators) markers. There are current neuroimaging biomarker approaches to stroke diagnosis that are available in clinical practice. However, taken alone they often do not reach sufficient accuracy, particularly in the diagnosis of small vessel strokes. That might be achieved when neuroimaging is integrated with rapid point-of-care blood testing additional to clinical observations.

Accurate assessment and appropriate timely management of stroke are essential to ensure a favorable outcome avoiding the risk of recurrent events. The emergency pathway for suspected stroke is initiated by ambulance personnel, but early recognition is challenging due to heterogeneous clinical presentations of stroke and an absence of portable diagnostic technology. An integrative biomarkers approach could rapidly stratify patients for personalized therapy avoiding treatment-related risks without benefits.

Development of reliable biomarkers for brain disorders would dramatically accelerate research on the etiology, pathophysiology, and disease progression of a number of prevalent and devastating nervous system diseases.

The present issue is devoted to upcoming methods in stroke biomarker research, translational innovative approaches to diagnosis, and current techniques utilized in clinical practice. International experts in the field of stroke research both experimental and clinical have contributed chapters to this volume that present the latest achievements in cutting-edge and well-established medical technologies for stroke.

Part II

Research Methods

Vascular-Related Biomarkers of Ischemic Stroke

Daniel Arteaga and Bradford B. Worrall

Abstract

Stroke remains a leading cause of death and disability worldwide, with ischemic stroke accounting for the vast majority of all stroke cases. Despite advances in the treatment of acute stroke with tissue-plasminogen activator and more recently with mechanical thrombectomy, limitations in the use and availability of these advanced treatment options have excluded most stroke patients from benefit. Moreover, the rapid diagnosis of acute stroke remains challenging due to "stroke mimics" (other entities with similar presentations) and the lack of a rapid, easily accessible means to rule in cerebral ischemia with imaging modalities such as magnetic resonance imaging or computed tomography perfusion in the acute or hyperacute setting. While biomarkers in other areas of medicine (e.g., troponin in myocardial injury) are well-established and commonly used in clinical practice, the identification and implementation of stroke-related biomarkers has proven to be challenging. This chapter provides a brief overview of several key diagnostic vascular biomarkers in ischemic stroke, highlighting the current development status of these biomarkers, limitations in their use, and their potential application in and implications for stroke care.

Key words Ischemic stroke, Vascular, Biomarker, Molecule, Inflammation

1 Introduction

As the second leading cause of death and the major cause of adult disability worldwide, stroke has a substantial public health and economic impact across the globe [1]. Clinical classification into ischemic and hemorrhagic subtypes reflects the underlying mechanism, with ischemic stroke accounting for more than 85% of all strokes. The most common etiologies of ischemic strokes include large-vessel arterial atherosclerosis, occlusion by an in situ thrombus or embolus, small-vessel occlusion by degenerative vessel wall changes, and cerebrovascular hypoperfusion, although up to 40% of ischemic strokes are cryptogenic [2]. This chapter focuses on the ischemic stroke subtype.

Treatment for ischemic stroke has rapidly evolved over the last two decades, with the introduction of both tissue-plasminogen activator (tPA) and the release of landmark mechanical thrombectomy trials demonstrating substantial benefit of intracranial clot

Philip V. Peplow et al. (eds.), *Stroke Biomarkers*, Neuromethods, vol. 147, https://doi.org/10.1007/978-1-4939-9682-7_2,
© Springer Science+Business Media, LLC, part of Springer Nature 2020

retrieval in the acute setting. These treatments have drastically changed the triage and administration of acute stroke care, transforming entire hospital systems to ensure timely delivery of care. However, although recent trials have extended the window for thrombectomy to 24 h [3, 4], the requirement for a proximal large vessel occlusion excludes all but a small proportion of stroke patients. Moreover, tPA also has strict inclusion and exclusion criteria that permit its use only within a 4.5-h window from last known normal, which also significantly limits the proportion of patients eligible for this treatment. Other treatments including newer generation thrombolytics [5], other revascularization strategies [6], and prehospital initiated therapy [7] remain under investigation.

With regard to the diagnosis of acute ischemic stroke, the advent of magnetic resonance imaging (MRI), and specifically diffusion-weighted imaging (DWI), has revolutionized the ability of clinicians to reliably detect cerebrovascular infarcts. Several recent major clinical trials have validated the use of computed tomography (CT) perfusion imaging for both the detection of infarct core (i.e., irreversibly damaged tissue) and penumbra (tissue that is at risk for irreversible damage). The validation and adoption of impactful therapeutic approaches in recent clinical trials [3, 4, 8] has created the potential for a more widespread implementation of such advanced imaging. Unfortunately, the cost and upkeep of MRI and CT perfusion imaging often prohibits their use in many medical centers across the world, and the required staffing to provide these services can prohibit their use 24 h a day even in advanced countries. Thus, there remains a need for rapid and cost-effective serum testing for the timely diagnosis of acute ischemic stroke.

Numerous identified blood-based vascular biomarkers in ischemic stroke could allow rapid and economical diagnosis patients with stroke. While the implementation of biomarkers in other fields of medicine such as cardiology has vastly changed clinical diagnosis and triage, few well-validated vascular biomarkers for ischemic stroke have achieved wide use in the clinical setting for the reasons outlined below. Serum testing would certainly increase the number of patients correctly diagnosed with stroke. This in turn would serve to improve patient outcomes regardless of successful use of therapeutic interventions such as tPA or thrombectomy because it would allow for enhanced patient education and more reliable execution of preventive management strategies. Moreover, vascular biomarkers could guide the administration of thrombolysis, helping to determine which patients would most likely benefit from tPA and in which patients the risk of hemorrhage might outweigh the benefits. The following sections will review the data supporting some of these biomarkers.

2 Blood–Brain Barrier Function

The blood–brain barrier (BBB) serves an important function in maintaining a strict extracellular environment, forming a tight barrier to allow the entry of molecules critical for normal neuronal function and to restrict the influx of neurotoxic substances. Under normal physiologic conditions, endothelial tight junctions of the BBB prevent many peripherally circulating molecules from entering the central nervous system. However, cerebrovascular ischemia causes bi-directional breakdown of the BBB and the release of signaling molecules from the CNS into the peripheral circulation. This consequently allows for the recruitment of immune cells into the brain to repair cellular injury.

Matrix metalloproteinases (MMPs) are molecules capable of degrading key structural elements that normally serve to maintain the structure and integrity of the BBB. Tissue damage leads to activation of MMPs, causing the cerebrovascular endothelium to become more permeable and less selective. Among the several different MMP subtypes, MMP-2 and MMP-9 play particularly important roles in cerebrovascular ischemia [9]. Notably, circulating MMP-9 levels have potential as a biomarker of both cerebral ischemia [10] and stroke severity [11]. Supporting these data, MMP-9 knockout mice have significantly reduced BBB disruption after ischemic stroke, with smaller final infarct volume sizes [12]. Individuals with ischemic stroke have higher concentrations of circulating MMP-9 and MMP-2 compared to controls [13]. Interestingly, MMPs may also play a role in recovery after ischemic stroke, stimulating the formation and growth of new neurons as well as assisting in the process of vascular remodeling [14].

Increased levels of MMP-9 and MMP-2 result in fragmentation of the tight junction proteins occludin and the claudins [15], with serum levels of these two endothelial protein classes closely associated with BBB disruption after ischemic stroke. Specifically, levels of occludin markedly increase at 4.5 h after middle cerebral artery occlusion, suggesting that the release of occludin in ischemic cerebral microvessels may occur after a threshold of damage in the BBB is reached [16]. Moreover, MMP-2 mediates occludin degradation in the early stages of ischemic stroke, further leading to BBB disruption [17]. Of the claudin protein subtypes, the BBB has claudin-5 in greatest abundance, and claudin-5 plays a particularly important role in the induction and maintenance of tight-junction tightness [18]. Similar to MMP-9 knockout mice, claudin-5 knockout mice tend not have generalized breakdown of tight junctions, but there is selective opening for small molecules <800 Da [19]. Both of these tight junction proteins also appear closely associated with the risk of hemorrhagic transformation discussed later in this chapter.

3   Inflammatory Cells

Inflammation plays an important, and predominantly harmful, role in acute ischemic stroke. Much of the damage caused by the inflammatory cascade in response to ischemia occurs gradually over the course of hours to days [20–22]. Leukocytes play a central role in this process. Once activated by platelets at the site of occlusion, they release inflammatory cytokines and express antigens which promote adherence to the endothelium. This in turn leads to further ischemic injury through release of reactive oxygen species, proteases, and other inflammatory mediators in the vicinity of the ischemic territory.

Throughout the various stages of ischemic stroke, expression of numerous leukocyte–platelet–endothelial adhesion antigens occurs as part of the inflammatory cascade. These antigens include the β2-integrins macrophage antigen-1 (Mac-1) and lymphocyte function-associated antigen-1 (LFA-1), the selectin family of adhesion molecules including endothelial selectin (E-selectin) and leukocyte selectin (L-selectin), and members of the immunoglobulin family including intercellular adhesion molecule 1 (ICAM-1) and vascular cell adhesion molecule 1 (VCAM-1). The recruitment of lymphocytes involves at least two different stages: (1) an initial low affinity binding via rolling interactions by leukocytes with the endothelium and (2) later high affinity interactions leading to stronger adhesion. The rolling of leukocytes during the early stages of ischemic stroke occurs after the release of cytokines from damaged neurons. This in turn leads to activation of L-selectin on leukocytes and E-selectin on endothelial cells. Over the course of hours, leukocytes form even stronger adhesive bonds with the vascular endothelium via the expression of adhesion molecules, notably Mac-1, LFA-1, ICAM-1, and VCAM-1 [23, 24].

Our understanding of the pathophysiology of these adhesion molecules opened up numerous therapeutic possibilities targeting these molecules and their inflammatory pathways. Compared to wild-type mice, ICAM-1 knockout mice have reduced leukocyte adhesion, smaller infarcts, and lower mortality after cerebral ischemia [25]. In humans, co-administration of anti-ICAM-1 antibodies with tPA may improve neurological outcomes and enhance the efficacy of thrombolytic therapy [26]. Similarly, mutant mice lacking either the LFA-1 or Mac-1 molecule had smaller infarcts and improved neurological outcomes [27]. Unfortunately, clinical trials employing anti-adhesion strategies based upon these initial findings have failed to deliver on their initial promise. Phase III clinical trials examining anti-ICAM antibodies within 6 h of symptom onset actually led to increased mortality, larger infarct volumes, and side effects relative to controls [28]. This unexpected negative finding may reflect an immune response triggered by administration of murine antibodies. Trials of anti-Mac-1 antibodies have also led to negative outcomes [29, 30].

Confounding factors such as age, sex, and the presence of comorbid conditions with associated underlying inflammatory states may explain these apparently contradictory findings. Such factors may lead to anomalous levels of inflammatory markers during ischemic stroke relative to individuals without these comorbid conditions. In addition, the heterogeneity of ischemic stroke, including the duration and magnitude of the degree of ischemia as well as the underlying mechanism, may also require closer examination. Regardless, there remain promising aspects of leukocyte adhesion-directed therapies yet to be fully explored, such as the combination of anti-adhesion molecules with thrombolytic agents. With the push to expand treatment windows for ischemic stroke patients, the continued effort to identify the role of these adhesion molecules during cerebral ischemia remains justified.

4   Key Coagulation Proteins

This section will focus on two key coagulation proteins (subsequent chapters will focus on other important coagulation proteins). Fibrinogen serves an important role in hemodynamic homeostasis, contributing both to platelet aggregation and leukocyte-endothelial cell interactions. Endothelial injury initiates recruitment of fibrinogen, which in turn leads to an inflammatory response and an increase in hepatic fibrinogen release. Thrombin then cleaves the fibrinogen to form insoluble fibrin that forms a mesh network and stabilizes the thrombus allowing formation of a definitive homeostatic plug. Studies examining fibrinogen in acute ischemic stroke have shown greater severity and worsening stroke prognosis with higher levels of fibrinogen [31–33]. Fibrinogen may have a stronger predictive effect for ischemic stroke than for myocardial infarction [34]. Elevated fibrinogen also increases the risk for recurrent stroke [35]. Interestingly, animal studies suggest that fibrinogen has intrinsic CNS toxicity and can promote both apoptosis and neurodegeneration [36, 37]. Ultimately, while fibrinogen concentrations may represent more of an epiphenomenon in response to ischemia, its presence may nonetheless provide an important marker of stroke outcome.

Von Willebrand factor (vWF) also plays an important role in platelet adhesion and aggregation at sites of vascular injury. While it normally serves an important role in maintaining homoeostasis, its release from endothelial cells in the setting of atherosclerotic damage can lead to sudden thrombus formation. Specifically, fibrillar collagen Type I and III exposed at the site of damage immobilizes vWF, which subsequently promotes binding to the GPIbα receptor on platelets. In addition, vWF promotes various inflammatory processes, including supporting leukocyte tethering and rolling under high shear stress [38]. Clinically, elevation in levels of vWF increases the risk of first-ever and recurrent stroke and is an independent risk factor for mortality after stroke [39–41]. Supporting these findings, mice deficient in vWF are protected against brain ischemia and reperfusion injury [42]. Monoclonal antibodies that can block both vWF-collagen and vWF-GPIbα binding are currently in preclinical and clinical phases of testing [43]. When used in combination with tPA, these vWF antagonists have the additive potential to prevent ongoing thrombus formation and promote spontaneous thrombolysis.

5   Hemorrhagic Transformation Associated with Thrombolysis

Thrombolysis with intravenous-tPA, while ultimately more beneficial than harmful in select patients with acute ischemic stroke, still carries a 6% risk of hemorrhagic transformation [44]. Specific biomarkers that may predict the occurrence of secondary intracranial hemorrhage with thrombolytic treatment have thus gained attention in the last few decades. Such biomarkers have the potential to allow exclusion of patients deemed at highest risk to avoid this feared complication of tPA administration. Currently, however, no such biomarkers have sufficient evidence to support withholding IV t-PA and none are in regular clinical practice. Several such biomarkers discussed below hold promise and warrant formal testing.

Both hepatocytes and cerebrovascular endothelia produce fibronectin, a ubiquitous glycoprotein that serves important roles in wound healing, hemostasis, and platelet aggregation. Fibronectin contains functional domains that interact with a variety of other molecules involved in the coagulation cascade and inflammatory response, including fibrin and fibrinogen as well as circulating leukocytes. Plasma levels of fibronectin strongly correlate with vascular damage [45, 46]. Notably, Castellenos et al. demonstrated that levels of fibronectin >3.6 mg/mL predicted the risk of parenchymal hemorrhage associated with tpA administration with 100% sensitivity and 60% specificity, generating enthusiasm for its potential as an initial screening tool. While further work will need to determine the reliability of this biomarker to predict hemorrhage after thrombolytic therapy and identify target thresholds, these preliminary findings are very encouraging.

Lower levels of plasminogen activator inhibitor-1 (PAI-1) and higher levels of thrombin-activated fibrinolysis inhibitor (TAFI) have also been associated with hemorrhagic transformation after tPA [47]. The combination of these two biomarkers produces the best testing characteristics, with PAI-1 levels >180% and TAFI levels <21.4 ng/mL predicting hemorrhagic conversion after tpA with a sensitivity of 75% and a specificity of 97%. As mentioned in the next section, both of these molecules play an important role in fibrinolysis, with PAI-1 serving as the main inhibitor of tPA and TAFI serving as an inhibitor of clot lysis.

Matrix metalloproteinase-9 (MMP-9) levels may also predict hemorrhage after thrombolysis. As previously discussed, MMPs play a significant role in the maintenance and integrity of the blood–brain barrier. tPA itself activates MMP-9. Unsurprisingly, therefore, elevated MMP levels predict tPA-associated hemorrhage well. In fact, baseline levels of MMP-9 prior to the administration of tPA strongly predict hemorrhagic transformation [48]. Moreover, MMP-9 levels above 140 ng/mL in ischemic stroke patients not treated with tPA predicts intracranial bleeding with a 92% sensitivity and 74% specificity [45].

6 Recanalization Status

Currently, vascular imaging offers the only clinically available modality for determining recanalization status after ischemic stroke. Importantly, no diagnostic test used in clinical practice can predict the probability of successful reperfusion after thrombolytic therapy. However, several recently identified and important protein biomarkers have the potential to enable serum-based testing of revascularization. These biomarkers may not only identify successful revascularization after thrombolysis, but may also predict the likelihood that reperfusion is effective. The fact that endogenous factors involved in fibrinolysis and coagulation likely modulate physiological responses to thrombolysis makes the principle underlying this test more easily understood. The clinical implications of such testing include improved selection of patients for thrombolytic therapy and prioritization of patients who would otherwise not be good candidates for thrombolysis alone for transfer to comprehensive stroke centers with endovascular treatment capabilities.

Promising biomarkers evaluating reperfusion include a disintegrin and metalloproteinase with a thrombospondin type 1 motif, member 13 (ADAMTS13), soluble endothelial protein C receptor (sEPCR), soluble thrombomodulin (sTM), and α2-antiplasmin. Specifically, elevated levels of ADAMTS13 and reduced levels of sEPCR, sTM, and α2-antiplasmin have been associated with an increased likelihood of achieving recanalization after tpA [49–51]. As a metalloproteinase, ADAMTS13 (also known as von Willebrand factor-cleaving protease, VWFCP) cleaves and down-regulates vWF released by endothelium in the setting of shear stress, thereby decreasing thrombogenic activity. Low levels of ADAMTS13 indicate a prothrombotic state associated with acute myocardial infarction and ischemic stroke [52, 53]. While the significance of a low ADAMST13 after unsuccessful recanalization with tPA is not completely understood, Bustamante et al. [49] hypothesized that reduced levels may either serve as a trigger for ischemia and/or promote thrombus extension.

Contrary to ADAMTS13, low levels of α2-antiplasmin indicate an anti-thrombotic state. α2-antiplasmin is the most powerful inhibitor of plasmin, which in turn is responsible for degrading fibrin clots. A level of α2-antiplasmin >85% of normal plasma predicts recanalization with a sensitivity of 25% and a specificity of 85% [50]. Finally, sEPCR and sTM are both members of the protein C pathway which serves to slow down clot formation. Endothelial cells that are not damaged tend to have higher expression of sEPCR and sTM (with consequently low serum levels), protecting the endothelium against thrombotic events. In contrast, loss of sEPCR and sTM from the endothelial surface (leading to higher serum levels) may suggest endothelial dysfunction and favor thrombosis.

Other potential biomarkers of revascularization status include follistatin-like 1 (FSTL1), small platelet microparticles, and thrombin–antithrombin complex (TAT). FSTL1 is a cytokine released by muscle shown to stimulate revascularization in rats following ischemic injury via activation of nitric-oxide synthase-dependent signaling within endothelial cells [54]. More recently, small platelet microparticles released from activated platelets have been identified as a potential biomarker for cerebrovascular recanalization [55]. In this study published by Bivard et al., ischemic stroke patients treated with intravenous-tPA demonstrated a significant increase in CD41+ microparticle levels. These microparticles are thought to be shed by platelets after restoration of blood flow. Not surprisingly, patients with elevated CD41+ microparticle levels also tended to have improved long-term outcomes. Lastly, the protease thrombin, a main enhancer of the coagulation cascade, functions to cleave fibrinogen to fibrin, activate platelets, activate Factor XIII which in turn stabilizes fibrin/clots, and activate protein C. Thus, low levels of thrombin, specifically TAT, may lead to improved outcome after ischemic stroke. Fernandez et al. confirmed this by determining that pretreatment levels of TAT complex correlated with recanalization status after tPA treatment [56]. While pretreatment TAT levels in this study did not correlate with mortality rates, other studies examining TAT levels at 24 h after tPA administration (which also had increased statistical power) have indeed demonstrated lower mortality with lower TAT levels [57].

Finally, it is worth mentioning that polymorphisms influencing the activity rate of two fibrinolysis inhibitors, plasminogen activator inhibitor-1 (PAI-1) and thrombin-activatable fibrinolysis inhibitor (TAFI), also influence the rate of successful recanalization of tPA [58]. PAI-1, the predominant inhibitor of tPA, prevents plasminogen activation by binding and forming a complex with tPA. Levels of PAI-1 < 34 ng/mL have been shown to predict poor response to thrombolysis [59]. TAFI is a proenzyme that removes lysine residues from fibrin, in effect inhibiting further fibrinolysis. Maximal activation of TAFI requires up to eightfold higher t-PA concentrations to achieve similar clot lysis times [60]. Co-administration of an inhibitor of TAFI decreases the time to reperfusion by threefold [61]. These data suggest that administration of these fibrinolysis inhibitors along with tPA could increase the efficacy of thrombolysis and thus improve functional outcomes.

7   Ischemic Stroke Prevention

In addition to being used for diagnosis and prognosis, vascular biomarkers may also allow better understanding of treatments for ischemic stroke. Therapy with aspirin plus extended release dipyridamole (Aggrenox®) or aspirin alone has been shown to lead to a rapid recovery of plasma endothelial nitric oxide synthase three levels [62]. The predominant functions of nitric oxide include maintenance of vascular tone, reduction of the inflammatory response, and balance of coagulation homeostasis. Endothelial nitric oxide synthase may have a neuroprotective role in ischemic stroke. Aspirin plus extended release dipyridamole alone has also been associated with oxidized low-density lipoprotein (LDL) levels [62]. Oxidized LDL can stimulate peripheral leukocyte chemotaxis, platelet adhesion, and development of atherosclerotic plaques [63]. Oxidized LDL has been found to be associated with high risk of death and poor functional outcome after ischemic stroke [64].

Thromboxane A2 production and its stable byproduct 11-dehyroxythromboxane B2 may reflect aspirin resistance [65, 66]. Identifying patients using these two biomarkers may help select patients who might benefit from aspirin and patients who may need an alternative antiplatelet therapy. The thromboxane A2 receptor is expressed in circulating platelets. Its activation mediates platelet activation and aggregation, and increased synthesis has been associated with ischemic stroke [67]. Notably, thromboxane A2 receptor antagonists inhibit the ischemia-induced inflammatory response and reduce reperfusion injury in a mouse model [67]. Levels of 11-dehyroxythromboxane B2, a stable product of nonenzymatic conversion of thromboxane A2 and reliable indicator of thromboxane A2 production, are elevated in patients with ischemic stroke relative to controls [68].

The use of P2Y12 platelet function testing and single nucleotide polymorphisms (SNP) testing is already commonly used in clinical practice to determine clopidogrel resistance [69, 70], although proof of their clinical utility remains lacking or incomplete [70, 71]. While point of care testing has also been developed to detect aspirin resistance, with assays including the PFA-100 system, modified thromboelastrograph system, and Rapid Platelet

Function Analyzer system, agreement in results among these assays can be quite variable [72]. One possible explanation for this is that as these tests examine platelet aggregation, it is possible that other factors aside from antiplatelet agents, such as levels of vWF, serve as the primary drivers for platelet aggregation.

8 Conclusion

In conclusion, blood-based biomarkers can provide useful information regarding the pathophysiological events underlying acute ischemic stroke. While numerous promising biomarkers have been discovered, their translation to clinical applications has proven to be challenging. Consequently, there remains an ongoing need for validation of comprehensive biomarker panels in large cohorts of ischemic stroke patients. Such biomarker panels may eventually not only facilitate diagnosis and treatment stratification but could also allow for improved patient education and elucidate better strategies for stroke prevention.

References

1. Benjamin EJ et al (2017) Heart disease and stroke statistics—2017 update: a report from the American Heart Association. Circulation 135(10):e146–e603. https://doi.org/10.1161/CIR.0000000000000485

2. Kolominsky-Rabas PL et al (2001) Epidemiology of ischemic stroke subtypes according to TOAST criteria: incidence, recurrence, and long-term survival in ischemic stroke subtypes: a population-based study. Stroke 32 (12):2735–2740

3. Nogueira RG et al (2018) Thrombectomy 6 to 24 hours after stroke with a mismatch between deficit and infarct. N Engl J Med 378 (1):11–21. https://doi.org/10.1056/NEJMoa1706442

4. Albers GW et al (2018) Thrombectomy for stroke at 6 to 16 hours with selection by perfusion imaging. N Engl J Med 378(8):708–718. https://doi.org/10.1056/NEJMoa1713973

5. Campbell BCV et al (2018) Tenecteplase versus Alteplase before thrombectomy for ischemic stroke. N Engl J Med 378(17):1573–1582. https://doi.org/10.1056/NEJMoa1716405

6. Nacu A et al (2017) NOR-SASS (Norwegian Sonothrombolysis in Acute Stroke Study): randomized controlled contrast-enhanced sonothrombolysis in an unselected acute ischemic stroke population. Stroke 48(2):335–341. https://doi.org/10.1161/STROKEAHA.116.014644

7. Shkirkova K et al (2018) Frequency, predictors, and outcomes of prehospital and early postarrival neurological deterioration in acute stroke: exploratory analysis of the FAST-MAG randomized clinical trial. JAMA Neurol. https://doi.org/10.1001/jamaneurol.2018.1893

8. Thomalla G et al (2018) MRI-guided thrombolysis for stroke with unknown time of onset. N Engl J Med 379(7):611–622. https://doi.org/10.1056/NEJMoa1804355

9. Morancho A et al (2010) Metalloproteinase and stroke infarct size: role for anti-inflammatory treatment? Ann N Y Acad Sci 1207:123–133. https://doi.org/10.1111/j.1749-6632.2010.05734.x

10. Reynolds MA et al (2003) Early biomarkers of stroke. Clin Chem 49(10):1733–1739

11. Montaner J et al (2001) Matrix metalloproteinase expression after human cardioembolic stroke: temporal profile and relation to neurological impairment. Stroke 32(8):1759–1766

12. Asahi M et al (2001) Effects of matrix metalloproteinase-9 gene knock-out on the proteolysis of blood-brain barrier and white matter components after cerebral ischemia. J Neurosci 21(19):7724–7732

13. Horstmann S et al (2003) Profiles of matrix metalloproteinases, their inhibitors, and laminin in stroke patients: influence of different therapies. Stroke 34(9):2165–2170. https://

doi.org/10.1161/01.STR.0000088062. 86084.F2

14. Bergers G et al (2000) Matrix metalloproteinase-9 triggers the angiogenic switch during carcinogenesis. Nat Cell Biol 2 (10):737–744. https://doi.org/10.1038/ 35036374

15. Yang Y et al (2007) Matrix metalloproteinase-mediated disruption of tight junction proteins in cerebral vessels is reversed by synthetic matrix metalloproteinase inhibitor in focal ischemia in rat. J Cereb Blood Flow Metab 27 (4):697–709. https://doi.org/10.1038/sj. jcbfm.9600375

16. Pan R et al (2017) Blood occludin level as a potential biomarker for early blood brain barrier damage following ischemic stroke. Sci Rep 7:40331. https://doi.org/10.1038/ srep40331

17. Liu J et al (2012) Matrix metalloproteinase-2-mediated occludin degradation and caveolin-1-mediated claudin-5 redistribution contribute to blood-brain barrier damage in early ischemic stroke stage. J Neurosci 32(9):3044–3057. https://doi.org/10.1523/JNEUROSCI. 6409-11.2012

18. Lv J et al (2018) Focusing on claudin-5: a promising candidate in the regulation of BBB to treat ischemic stroke. Prog Neurobiol 161:79–96. https://doi.org/10.1016/j. pneurobio.2017.12.001

19. Nitta T et al (2003) Size-selective loosening of the blood-brain barrier in claudin-5-deficient mice. J Cell Biol 161(3):653–660. https:// doi.org/10.1083/jcb.200302070

20. Kawabori M, Yenari MA (2015) Inflammatory responses in brain ischemia. Curr Med Chem 22(10):1258–1277

21. Vogelgesang A, Becker KJ, Dressel A (2014) Immunological consequences of ischemic stroke. Acta Neurol Scand 129(1):1–12. https://doi.org/10.1111/ane.12165

22. Iadecola C, Anrather J (2011) The immunology of stroke: from mechanisms to translation. Nat Med 17(7):796–808. https://doi.org/10. 1038/nm.2399

23. Tsai NW et al (2009) The value of leukocyte adhesion molecules in patients after ischemic stroke. J Neurol 256(8):1296–1302. https:// doi.org/10.1007/s00415-009-5117-3

24. Simundic AM et al (2004) Soluble adhesion molecules in acute ischemic stroke. Clin Invest Med 27(2):86–92

25. Ishikawa M et al (2003) Molecular determinants of the prothrombogenic and inflammatory phenotype assumed by the postischemic cerebral microcirculation. Stroke 34

(7):1777–1782. https://doi.org/10.1161/ 01.STR.0000074921.17767.F2

26. Bowes MP et al (1995) Monoclonal antibodies preventing leukocyte activation reduce experimental neurologic injury and enhance efficacy of thrombolytic therapy. Neurology 45 (4):815–819

27. Arumugam TV et al (2004) Contributions of LFA-1 and Mac-1 to brain injury and microvascular dysfunction induced by transient middle cerebral artery occlusion. Am J Physiol Heart Circ Physiol 287(6):H2555–H2560. https://doi.org/10.1152/ajpheart.00588. 2004

28. Enlimomab Acute Stroke Trial Investigators (2001) Use of anti-ICAM-1 therapy in ischemic stroke: results of the Enlimomab Acute Stroke Trial. Neurology 57(8):1428–1434

29. Becker KJ (2002) Anti-leukocyte antibodies: LeukArrest (Hu23F2G) and Enlimomab (R6.5) in acute stroke. Curr Med Res Opin 18(Suppl 2):s18–s22

30. Sughrue ME et al (2004) Anti-adhesion molecule strategies as potential neuroprotective agents in cerebral ischemia: a critical review of the literature. Inflamm Res 53(10):497–508. https://doi.org/10.1007/s00011-004-1282- 0

31. del Zoppo GJ et al (2009) Hyperfibrinogenemia and functional outcome from acute ischemic stroke. Stroke 40(5):1687–1691. https:// doi.org/10.1161/STROKEAHA.108. 527804

32. Hennerici MG et al (2006) Intravenous ancrod for acute ischaemic stroke in the European Stroke Treatment with Ancrod Trial: a randomised controlled trial. Lancet 368 (9550):1871–1878. https://doi.org/10. 1016/S0140-6736(06)69776-6

33. Di Napoli M, Papa F (2006) Should neurologists measure fibrinogen concentrations? J Neurol Sci 246(1–2):5–9. https://doi.org/ 10.1016/j.jns.2006.03.005

34. Siegerink B, Rosendaal FR, Algra A (2009) Genetic variation in fibrinogen; its relationship to fibrinogen levels and the risk of myocardial infarction and ischemic stroke. J Thromb Haemost 7(3):385–390. https://doi.org/10. 1111/j.1538-7836.2008.03266.x

35. Williams SR et al (2016) Shared genetic susceptibility of vascular-related biomarkers with ischemic and recurrent stroke. Neurology 86 (4):351–359. https://doi.org/10.1212/ WNL.0000000000002319

36. Ryu JK, McLarnon JG (2008) VEGF receptor antagonist Cyclo-VEGI reduces inflammatory reactivity and vascular leakiness and is

neuroprotective against acute excitotoxic striatal insult. J Neuroinflamm 5:18. https://doi.org/10.1186/1742-2094-5-18

37. Ryu JK, McLarnon JG (2009) A leaky blood-brain barrier, fibrinogen infiltration and microglial reactivity in inflamed Alzheimer's disease brain. J Cell Mol Med 13(9A):2911–2925. https://doi.org/10.1111/j.1582-4934.2008.00434.x

38. Bernardo A et al (2005) Platelets adhered to endothelial cell-bound ultra-large von Willebrand factor strings support leukocyte tethering and rolling under high shear stress. J Thromb Haemost 3(3):562–570. https://doi.org/10.1111/j.1538-7836.2005.01122.x

39. Williams SR et al (2017) Genetic drivers of von Willebrand factor levels in an ischemic stroke population and association with risk for recurrent stroke. Stroke 48(6):1444–1450. https://doi.org/10.1161/STROKEAHA.116.015677

40. Wieberdink RG et al (2010) High von Willebrand factor levels increase the risk of stroke: the Rotterdam study. Stroke 41(10):2151–2156. https://doi.org/10.1161/STROKEAHA.110.586289

41. Carter AM et al (2007) Predictive variables for mortality after acute ischemic stroke. Stroke 38(6):1873–1880. https://doi.org/10.1161/STROKEAHA.106.474569

42. Kleinschnitz C et al (2009) Deficiency of von Willebrand factor protects mice from ischemic stroke. Blood 113(15):3600–3603. https://doi.org/10.1182/blood-2008-09-180695

43. Staelens S et al (2006) Humanization by variable domain resurfacing and grafting on a human IgG4, using a new approach for determination of non-human like surface accessible framework residues based on homology modelling of variable domains. Mol Immunol 43(8):1243–1257. https://doi.org/10.1016/j.molimm.2005.07.018

44. Miller DJ, Simpson JR, Silver B (2011) Safety of thrombolysis in acute ischemic stroke: a review of complications, risk factors, and newer technologies. Neurohospitalist 1(3):138–147. https://doi.org/10.1177/1941875211408731

45. Castellanos M et al (2004) Plasma cellular-fibronectin concentration predicts hemorrhagic transformation after thrombolytic therapy in acute ischemic stroke. Stroke 35(7):1671–1676. https://doi.org/10.1161/01.STR.0000131656.47979.39

46. Castellanos M et al (2007) Serum cellular fibronectin and matrix metalloproteinase-9 as screening biomarkers for the prediction of parenchymal hematoma after thrombolytic therapy in acute ischemic stroke: a multicenter confirmatory study. Stroke 38(6):1855–1859. https://doi.org/10.1161/STROKEAHA.106.481556

47. Ribo M et al (2004) Admission fibrinolytic profile is associated with symptomatic hemorrhagic transformation in stroke patients treated with tissue plasminogen activator. Stroke 35(9):2123–2127. https://doi.org/10.1161/01.STR.0000137608.73660.4c

48. Montaner J et al (2003) Matrix metalloproteinase-9 pretreatment level predicts intracranial hemorrhagic complications after thrombolysis in human stroke. Circulation 107(4):598–603

49. Bustamante A et al (2018) Usefulness of ADAMTS13 to predict response to recanalization therapies in acute ischemic stroke. Neurology 90(12):e995–e1004. https://doi.org/10.1212/WNL.0000000000005162

50. Marti-Fabregas J et al (2005) Hemostatic markers of recanalization in patients with ischemic stroke treated with rt-PA. Neurology 65(3):366–370. https://doi.org/10.1212/01.wnl.0000171704.50395.ba

51. Faille D et al (2014) Endothelial markers are associated with thrombolysis resistance in acute stroke patients. Eur J Neurol 21(4):643–647. https://doi.org/10.1111/ene.12369

52. Kaikita K et al (2006) Reduced von Willebrand factor-cleaving protease (ADAMTS13) activity in acute myocardial infarction. J Thromb Haemost 4(11):2490–2493. https://doi.org/10.1111/j.1538-7836.2006.02161.x

53. McCabe DJ et al (2015) Relationship between ADAMTS13 activity, von Willebrand factor antigen levels and platelet function in the early and late phases after TIA or ischaemic stroke. J Neurol Sci 348(1–2):35–40. https://doi.org/10.1016/j.jns.2014.10.035

54. Ouchi N et al (2008) Follistatin-like 1, a secreted muscle protein, promotes endothelial cell function and revascularization in ischemic tissue through a nitric-oxide synthase-dependent mechanism. J Biol Chem 283(47):32802–32811. https://doi.org/10.1074/jbc.M803440200

55. Bivard A et al (2017) Platelet microparticles: a biomarker for recanalization in rtPA-treated ischemic stroke patients. Ann Clin Transl Neurol 4(3):175–179. https://doi.org/10.1002/acn3.392

56. Fernandez-Cadenas I et al (2009) Lower concentrations of thrombin-antithrombin complex (TAT) correlate to higher recanalisation rates among ischaemic stroke patients treated

with t-PA. Thromb Haemost 102(4):759–764. https://doi.org/10.1160/TH08-06-0398

57. Tanne D et al (2006) Hemostatic activation and outcome after recombinant tissue plasminogen activator therapy for acute ischemic stroke. Stroke 37(7):1798–1804. https://doi.org/10.1161/01.STR.0000226897.43749.27

58. Fernandez-Cadenas I et al (2007) Influence of thrombin-activatable fibrinolysis inhibitor and plasminogen activator inhibitor-1 gene polymorphisms on tissue-type plasminogen activator-induced recanalization in ischemic stroke patients. J Thromb Haemost 5 (9):1862–1868. https://doi.org/10.1111/j.1538-7836.2007.02665.x

59. Ribo M et al (2004) Admission fibrinolytic profile predicts clot lysis resistance in stroke patients treated with tissue plasminogen activator. Thromb Haemost 91(6):1146–1151. https://doi.org/10.1160/TH04-02-0097

60. Sakharov DV, Plow EF, Rijken DC (1997) On the mechanism of the antifibrinolytic activity of plasma carboxypeptidase B. J Biol Chem 272 (22):14477–14482

61. Philippou H (2014) Thrombin activatable fibrinolysis inhibitor (TAFI): more complex when it meets the clot. Thromb Res 133 (1):1–2. https://doi.org/10.1016/j.thromres.2013.10.034

62. Serebruany V et al (2011) Effects of Aggrenox and aspirin on plasma endothelial nitric oxide synthase and oxidised low-density lipoproteins in patients after ischaemic stroke. The AGgrenox versus aspirin therapy evaluation (AGATE) biomarker substudy. Thromb Haemost 105 (1):81–87. https://doi.org/10.1160/TH10-05-0316

63. Nishi K et al (2002) Oxidized LDL in carotid plaques and plasma associates with plaque instability. Arterioscler Thromb Vasc Biol 22 (10):1649–1654

64. Wang A et al (2017) Association of oxidized low-density lipoprotein with prognosis of stroke and stroke subtypes. Stroke 48 (1):91–97. https://doi.org/10.1161/STROKEAHA.116.014816

65. Dobaczewski M et al (2008) Targeting the urine and plasma determinants of thromboxane A2 metabolism in detection of aspirin effectiveness. Blood Coagul Fibrinolysis 19 (5):421–428. https://doi.org/10.1097/MBC.0b013e3283049686

66. Sharp FR et al (2000) Multiple molecular penumbras after focal cerebral ischemia. J Cereb Blood Flow Metab 20(7):1011–1032. https://doi.org/10.1097/00004647-200007000-00001

67. Yan A et al (2016) Thromboxane A2 receptor antagonist SQ29548 reduces ischemic stroke-induced microglia/macrophages activation and enrichment, and ameliorates brain injury. Sci Rep 6:35885. https://doi.org/10.1038/srep35885

68. Dharmasaroja PA, Sae-Lim S (2014) Comparison of aspirin response measured by urinary 11-dehydrothromboxane B2 and VerifyNow aspirin assay in patients with ischemic stroke. J Stroke Cerebrovasc Dis 23(5):953–957. https://doi.org/10.1016/j.jstrokecerebrovasdis.2013.08.001

69. Nordeen JD et al (2013) Clopidogrel resistance by P2Y12 platelet function testing in patients undergoing neuroendovascular procedures: incidence of ischemic and hemorrhagic complications. J Vasc Interv Neurol 6 (1):26–34

70. Pare G et al (2010) Effects of CYP2C19 genotype on outcomes of clopidogrel treatment. N Engl J Med 363(18):1704–1714. https://doi.org/10.1056/NEJMoa1008410

71. Kass-Hout T et al (2015) Neurointerventional stenting and antiplatelet function testing: to do or not to do? Interv Neurol 3(3–4):184–189. https://doi.org/10.1159/000431261

72. Harrison P et al (2005) Screening for aspirin responsiveness after transient ischemic attack and stroke: comparison of 2 point-of-care platelet function tests with optical aggregometry. Stroke 36(5):1001–1005. https://doi.org/10.1161/01.STR.0000162719.11058.bd

Chapter 3

Generation and Applicability of Genetic Risk Scores (GRS) in Stroke

Natalia Cullell, Jonathan González-Sánchez, Israel Fernández-Cadenas, and Jerzy Krupinski

Abstract

In this chapter genetic risk scores (GRS), also called polygenic risk scores (PRS), are described as feasible biomarkers to predict stroke risk. GRS are based on genome-wide association studies (GWAs) results. Use of GWAs has allowed the discovery of multiple variants associated with stroke which has led to an increase in stroke pathophysiology knowledge. However, it would be necessary to consider the cumulative risk of all associated variants to make GWAs results applicable in clinical practice. GRS summarize information from significant genetic variants into a single score. The most frequent way for GRS construction is based on the selection of the most significant single nucleotide polymorphisms (SNPs) from the largest GWAs for a trait. Then, allele dosage for each SNP is weighted using estimated effect sizes for each variant. Next, global fit, discrimination capacity, calibration, and reclassification of GRS have to be evaluated. Furthermore, replication in independent cohorts is necessary. Stroke is a disease with complex genetic architecture. Multiple loci with small effect sizes have been found to increase stroke risk. In this case, GRS could enable the translation of genetics into clinical practice in different ways, such as stroke risk prediction or differentiation of stroke subtypes to give specific secondary prevention drugs depending on the type of ischemic stroke.

Key words Genetic risk scores, GRS, PRS, GWAs, Biomarkers, Stroke

1 Introduction

In complex diseases such as stroke, many variants with small effect sizes are involved in their physiopathology. As an initial step for the discovery of the mechanisms uncovered in the disease, GWAs are an efficient way to discover new biomarkers of susceptibility, prognosis, and treatment (pharmacogenetics) [1]. GWAs are based on an agnostic approach to evaluate genetic variants along the genome. They have allowed for the discovery of over a thousand variants associated with susceptibility of complex diseases [2].

Systematic approaches are necessary to use genetics as a biomarker for stroke. These approaches have to be able to

Philip V. Peplow et al. (eds.), *Stroke Biomarkers*, Neuromethods, vol. 147, https://doi.org/10.1007/978-1-4939-9682-7_3,
© Springer Science+Business Media, LLC, part of Springer Nature 2020

accommodate more than one genetic result and have to be flexible to be modified when new genetic biomarkers are found.

Genetic risk scores (GRS) have been developed for many disease like schizophrenia, multiple sclerosis, cardiovascular disease, and also stroke [3–5]. GRS are a summary of risk associated variation from multiple SNPs. Thus, GRS are an efficient and effective approach to construct genome-wide risk measurements from GWAs [6]. They are a systematic and replicable way to select SNPs from different GWAs results to be included in a single score. Implementation of GRS into clinical practice is important to allow genetics to be relevant in clinician decisions.

This chapter describes methodology for GRS construction and gives some examples of its applicability in stroke.

2 Methods

2.1 Genome-Wide Association Studies

GWAs are an agnostic and cost-effective approach to analyze SNPs along the genome using array technology. SNPs are variant positions in the genome with normally two different possible alleles [7]. The frequency of the less common allele in a specific population is called minor allele frequency (MAF) [1]. GWAs usually analyze common variants in the genome with MAF higher than 1%. Normally about 300,000–1 millions of variants are detected by current chips but using imputation is possible to infer information about 40 million variants [1, 8].

For the correct analysis of GWAs results it is important to follow a strict methodology protocol before the performance of the GWAs. The main steps of GWAs analysis are described below [1, 7, 9, 10]:

2.1.1 Trait Selection

For the study of a disease it is possible to analyze the trait as a categorical variable (cases and controls) or to use intermediate phenotypes which can be categorical or continuous variables. Continuous variables tend to increase the power of the analysis [1]. The type of variable will determine the statistical test for final GWAs analysis.

2.1.2 Subjects Selection

Depending on the study design, cohorts used in GWAs analysis can be prospective, retrospective, and family-based cohorts. Family-based studies are more useful in low-frequency pathologies with large effects of a gene while it is more powerful to use unrelated cohorts for diseases with complex genetic architecture [1].

2.1.3 Imputation

Genotype of untyped variants can be assessed using imputation. This technique is able to increase the power of GWAs analysis, facilitating comparison and joining of different studies genotyped with different arrays [8]. It uses haplotype information from a reference panel to infer genetic information in the same haplotypes from subjects in the study [11]. The most used reference panels for imputation are obtained from HapMap Project and The 1000 Genomes Project [12–14]. However, larger imputation panels have been developed, for example by the Haplotype Reference Consortium (HRC) to increase accuracy and allowing for imputation of low-frequency variants [8]. It is important to correctly select the imputation panel according to the study population. The frequency of genetic variants varies among populations. For that reason, it is important to impute genetic variants in each specific community [15]. These projects include genotypes of individuals distributed in about 25 different locations to be used as reference [11]. IMPUTE, MACH and BEAGLE are some of the software used for imputation [1].

2.1.4 Quality Controls (QC)

Frequently genotyping process produce missing in genotype calling in some variants or samples. In order to reduce the number of false-positive or false-negative results it is important to perform quality controls before GWAs analysis. The main software used for QC are PLINK and R statistical package [9]. QC can be classified in two classes [1, 9]:

- Sample QC: it consists of the removal of individuals in the study applying different filters:

  - Discordant sex: it uses information from X-chromosome homozygosity to identify individual sex. The genotyped sex is compared with sex annotated in the clinical database. If there is discordance in sex it can indicate some error in sex annotation or some mix-up in sample preparation. Individuals with sex discordance have to be removed from the study if discrepancy could not be solved.

  - Samples with low genotyping call rate: different reasons could affect the genotyping rate. For example, low concentration or quality of DNA. Commonly, individuals with >5% of missing genotypes are removed.

  - Identity-by-descent (IBD): it consists of the identification of related or duplicated subjects. In studies of unrelated subjects is important to identify samples from familiars or duplicates introduced by error. IBD calculates from autosome SNPs the proportion of genome sharing among two individuals. Two samples with the same genotype will have an IBD of 1 while siblings or parents-offspring relationships will have an IBD of 0.5. Normally, individuals with IBD > 0.1875 are removed because it indicates duplication or mixing of samples. In case of sample mixing, they probably would be removed in heterozygosity QC.

- Principal component analysis (PCA): genetics is very ancestry-dependent. As geographically closest individuals are, easier is the identification of genetic variants associated with a trait. Identifying population subgroups, it is important to avoid the finding of associations not related with the studied phenotype. PCA is very useful to detect population stratifications through normalization and linear transformation of input SNPs. Different principal components are calculated, being the first one the most explicative. The threshold used for individual removal is 6 standard deviations away from the mean of the top principal component.

- Variant QC: it consists of the removal of genetic variants for different reasons:

  - SNPs with low MAF: normally SNPs with MAF < 1% are removed from GWAs. GWAs are usually prepared to analyze common SNPs. Thereby, variants with low MAF are considered to be at risk of genotyping error.
  - SNPs with high missing genotype ratio: some SNPs are difficult to genotype and then missing genotype is detected in many individuals. The conventional threshold to remove polymorphisms is 5%.
  - SNPs out of Hardy–Weinberg equilibrium (HWE) [9]: it is expected that variants from a large and homogeneous population with random mating follow HWE law. Deviations from HWE are calculated in controls using chi-square goodness-of-fit test and these SNPs are removed from the analysis.

2.1.5 Association Tests

Before deciding the correct statistical test to be used for association analysis it is important to select the appropriate genetic model: dominant model assumes that the presence of the risk allele increase risk with independence of the allele dosage. In contrast, the additive model will weight the risk depending on the allele dosage. The recessive model will consider only risk when the two risk alleles are present.

For analysis of categorical phenotypes a different test can be used depending on the chosen model. The most common ones are chi-squared test, odds ratio (OR) test, Fisher's exact test and Armitage's trend test. In the case of continuous phenotypes, ANOVA and t-test can be used.

Usually in GWAs analysis it is necessary to take into account confounding variables such as age and gender. Furthermore, GWAs tend to be adjusted by PCA to avoid effect of population stratification. In this case, the test used is the generalized linear model (GLM). This model uses logistic regression for categorical variables and linear regression for continuous variables [1].

Multiple test correction is necessary in GWAs to select the significance p-value threshold [16]. Different methods, such as Bonferroni correction, have been developed. SNPs with p-values bellow 5×10^{-8} are usually considered genome-wide significant in GWAs analysis [1, 17, 18].

2.1.6 Replication

Results from GWAs have to be replicated in independent cohorts. It is important to select replication cohorts with enough sample size to be able to replicate results. Consequently, a sample size analysis using the data obtained in the GWAs is necessary. Furthermore, replication cohorts have to be similar to the discovery set and the phenotype has to be described using the same parameters [1].

2.2 GWAs in Stroke

Different GWAs have analyzed the association of common genetic variants with stroke, ischemic stroke and stroke subtypes risk. A total of 32 loci have been found to be associated with all stroke or ischemic stroke subtypes [19–21]. Two loci were found to be associated specifically with hemorrhagic stroke [22, 23].

The MEGASTROKE study is the largest and last GWAs performed in stroke. It included 521,612 individuals (67,162 cases and 454,450 controls) and identified 22 new signals of the total 32 loci associated with all stroke, ischemic stroke and specific stroke subtypes. Moreover, these results showed a correlation of genetics among different vascular traits and stroke [21]. In the specific case of ischemic stroke, GWAs results allowed for the identification of mechanisms underlying stroke that were not known before [21].

For intracerebral hemorrhage (ICH), one meta-analysis including six GWAs in patients from European ancestry (1,545 cases and 1,481 controls) found two loci associated with this pathology. However, in the multiethnic replication, only one of the loci was significantly associated with ICH [22]. Another GWAs in 2,189 ICH cases and 4,041 controls found APOE alleles ε2/ε4 to be genome-wide significantly associated with this disease [23].

2.3 Development of GRS

2.3.1 GWAs Selection

The first step prior the construction of a GRS is to select the proper source of genetic data. Different GWAs can be published for the same trait. Normally, the strategy followed to select GWAs is to focus on meta-analysis of different GWAs. However, meta-analysis is not always available for a trait. In this case, the GWAs with larger sample size are the best option to select polymorphisms for the GRS. Furthermore, if diverse GWAs are available for a trait, it is important to select the one performed in a population from the same ethnicity than the cohort used to design the GRS [6].

Different sources are available to find GWAs results [6]:

- GWAs databases.
- Web-based whole-genome analysis tools.
- Own genome-wide data.

2.3.2 Polymorphisms Selection

The second step for the construction of a GRS is selection of polymorphisms. Normally, SNPs are selected depending on their p-value. However, there is other information which could be useful to select polymorphisms: linkage disequilibrium (LD), imputation quality, and imputation type [3].

Currently, the most common ways to select SNPs are the "top hits" and the "best guess" approach. The first one is based on the selection of SNPs with lower p-values from published GWAs. The second approach consists of the selection of nominal SNPs located in genes with relevant function in the disease.

Nevertheless, it is important to take into account LD of SNPs. Variants without LD, which are independently inherited are incorporated in GRS in order to not have repetitive information in GRS [3].

In the case that multiple GWAs are published for a particular trait, another strategy is plausible to find SNPs to be included in GRS [6]:

- Extraction: SNPs below a threshold are selected from the different GWAs.

- Clustering: LD is considered to cluster variants extracted.

- Selection: for each block, the SNP with the lower p-value and replicated in a higher number of studies is selected.

2.3.3 GRS Construction

Depending on the weight assigned to each genetic variant included in a GRS it is possible to classify GRS in different types [3, 5, 6] (*see* **Note 1**):

- Equal weighting: is the simplest approach for GRS construction. It consists of summing the allele dosage of the risk allele for each polymorphism. The allele dosage will be 0, 1, or 2.

$$\text{GRS}_{\text{unweighted}} = \sum D_i.$$

In this formula, "D_i" is the dosage for each allele. In the case of imputed SNPs, allele dosage will be a continuous number from 0 to 2 accounting for the probability of having the allele. Example: we consider a GRS with three SNPs. For an individual with SNP1 = AA (risk allele A), allele dosage = 2; SNP2 = TC (risk allele C), allele dosage = 1 and SNP3 = CC (risk allele T), allele dosage = 0, if we consider a GRS with equal weighting, the final score will be

$$\text{GRS}_{\text{unweighted}} = D_{\text{allele1}} + D_{\text{allele2}} + D_{\text{allele3}} = 2 + 1 + 0 = 3.$$

However, assuming equal relevance for each SNP in the disease could not be realistic, including a bias in your analysis (all the SNPs have the same importance, and this is not true).

- Weighted GRS: in this case, GRS are constructed assuming different weight for each specific polymorphism included in the score. Normally, alleles are weighted according to effect sizes (estimated log odds ratio) for the risk allele in the outcome (*see* **Note 2**). Each log odds ratio will be multiplied by the number of risk alleles:

$$\text{GRS}_{\text{weighted}} = \sum \beta_i \times D_i.$$

  "β_i" corresponds to the effect size (β value). For an example considering the three previous different SNPs with effect sizes of SNP1 = 2.2, SNP2 = −2.9, and SNP3 = 4, the GRS will have a score of

$$\begin{aligned}\text{GRS}_{\text{weighted}} &= \beta_{\text{allele1}} \times D_{\text{allele1}} + \beta_{\text{allele2}} \times D_{\text{allele2}} + \beta_{\text{allele3}} \times D_{\text{allele3}} \\ &= 1.3 \times 2 - 2.9 \times 1 + 4 \times 0 = 2.6 - 2.9 + 0 = -0.3.\end{aligned}$$

  Weights for each allele can be provided also from hazard ratio or relative risk from one or more cohort studies [3].

  Using this method, the contribution of each allele on the score will be different, thus being a more realistic method highlighting the specific importance of each polymorphism. In the example, we can observe as despite SNP1 has higher allele dose, the weight of SNP2 is higher because of high OR.

- Explained variance weighted GRS [24]: this method includes two different information for each polymorphism: effect size and MAF. It is a way to increase the information regarding the variance explained by each variant:

$$\text{GRS}_{\text{explained\_variance}} = \sum \beta_i \times D_i \times \text{MAF}_i$$

  where MAF_i refers to the MAF for each allele.

  If we consider the information from the previous example SNPs adding a MAF for SNP1 of 0.8, SNP2 0.1 and SNP3 0.5, we will have the next result:

$$\begin{aligned}\text{GRS}_{\text{explained\_variance}} &= \beta_{\text{allele1}} \times D_{\text{allele1}} \times \text{MAF}_{\text{allele1}} + \beta_{\text{allele2}} \\ &\quad \times D_{\text{allele2}} \times \text{MAF}_{\text{allele2}} + \beta_{\text{allele3}} \times D_{\text{allele3}} \\ &\quad \times \text{MAF}_{\text{allele3}} \\ &= 1.3 \times 2 \times 0.8 - 2.9 \times 1 \times 0.1 + 4 \times 0 \times 0.5 \\ &= 0.4 = 2.08 - 0.29 + 0 = 1.79\end{aligned}$$

  In the example we can observe as MAF is important for weighting polymorphisms. In the weighted GRS (without considering MAF), SNP2 had the higher weight. However, as it has a very small MAF, its final contribution in the GRS and thereby in the disease is smaller than for SNP1.

- Combined GRS: it is also possible to combine GRS with classical clinical risk scores including demographic and clinical variables to increase the predictive value of the score.

Values of GRS have not meaning per itself. For that reason, it is recommended to standardize results to a mean of 0 and a variance of 1 to be easier to interpret and more comparable [3].

2.4 Evaluation of GRS

After GRS construction, they have to be evaluated in an independent cohort (target sample) in relation to the trait studied. Linear and logistic regression models are used depending on the variable type [6].

Sometimes, SNPs used in the creation of GRS are not genotyped in the target samples. Some methods have been developed to solve this problem [25]. Usually these SNPs are imputed, omitted or replaced by proximal SNPs. Imputation of SNPs is not always feasible. For that reason, omission and replacement by proximal SNPs are alternative options. Comparing omission and replacement of SNPs, the removal of the missing SNPs supposes a great decrease in the statistical power and predictive ability [25].

GRS are usually evaluated by comparing them with previous models for predicting the specific disease risk. The main objective will be to develop a model able to predict better than previous biomarkers the risk of a disease and which will have enough power to alter or support clinician decisions. Furthermore, it is expected that the model gives a result which will be close to the real risk of the patient. The statistical measures that are important to take into account for this comparison are [26]:

- Measures of global fit: how GRS is fitting data.
- Discrimination: the capacity to correctly select individuals who will develop the disease.
- Calibration: the comparison of the predicted risk with the observed risk.
- Reclassification: the ability of changing risk of subjects depending on the model.

2.4.1 Measures of Global Fit

These measures are used to determine how the model is expected to explain data. The main measures used are R-squared, Akaike Information Criterion, and Bayes Information Criterion [26, 27]. They are based on the number of variables included in the GRS. As fewer variables are included, better will be the predictive model.

2.4.2 Discrimination

This measure explains how GRS is able to discriminate patients who will develop the disease from patients who will do not. Given a random selection of a subject from a control group and a random selection of a subject from case group, it is expected that the subject from control group had a lower predicted risk than the patient. If this situation happens, the GRS will be good on discrimination.

The main method to evaluate the capacity of discrimination of the model is C statistic or area under receiver operator characteristic curve (AUC). An AUC with value 1 will represent perfect discrimination, while a value of 0 will mean non-ability to discriminate.

An alternative method for discrimination accounting is Integrated Discrimination Improvement (IDI) [26, 28]. It considers the difference between predicted risks for the group of people who will develop the disease and the group who will do not. It is expected that the difference in predicted risk average between both groups would be as higher as possible to be a good discrimination model.

2.4.3 Calibration

This measure is useful to know if the new model is better than previous ones to predict risk of a disease. The main method used to measure calibration of a model is Hosmer–Lemeshow [26]. It compares the predicted number of subjects in each risk group with the observation risk. Thereby, this test will be affected by how groups are formed (deciles, quantiles, etc.).

2.4.4 Reclassification

It is important that a model has the ability of changing persons from a group risk to another or to change the way subjects will be followed up or treated for a disease.

The Net Reclassification Improvement (NRI) is the method used to measure reclassification taking into account the proportion of subjects with events and without events who will increase or decrease the risk category where they are classified [26, 29].

2.5 GRS Studies in Stroke

Stroke is a multifactorial complex disease for which genetics is also relevant. Different GRS have been generated in stroke with the objective to predict stroke risk in general population improving current non-genetic scores.

One study included two significant ($p < 5 \times 10^{-8}$) SNPs associated with ischemic stroke and replicated in an independent cohort. Furthermore, they analyzed SNPs associated with nine risk factors for stroke (blood pressure, atherosclerosis, arrhythmia, etc.), including different traits for each risk factor. Finally, 34 traits were analyzed. The most significant SNP for a locus was selected from the largest meta-analysis. For each trait a risk score was constructed by weighting allele dosage with log odds ratio for each SNP. GRS was applied to 2,047 patients with stroke and 22,720 controls from CHARGE consortium, with European ancestry. The score was analyzed in stroke patients alone or together with clinical variants to compare ROC curves. Furthermore, NRI and IDI were calculated to assess the clinical added value. GRS was replicated in an external cohort to study its predictive power. They concluded that GRS provided an improvement in the risk of subjects for future stroke. Furthermore, AUC improved for GRS comparing with the score including only clinical variables [5].

Another study developed a GRS including SNPs from modifiable risk factors for stroke obtained from GWAs catalogue (hypertension, blood pressure, smoking, type 1 diabetes mellitus, etc.).

For GRS construction, variants were weighted according to the estimated effect sizes. GRS was derived in a sample of over 3,000 cases and 6,000 controls from the Wellcome Trust Case Control Consortium 2 [30]. GLM was used to asses association of GRS with ischemic stroke risk. The score was replicated in a cohort of more than 3,000 cases and 4,000 controls from METASTROKE consortium. The incident risk of stroke was measured in an independent prospective cohort using Cox regression model and R-squared to calculate the variance explained by the model. They found their GRS was able to predict risk of ischemic stroke. However, no improvement was observed in net reclassification [4].

A GRS constructed with the purpose of evaluating its association with stroke subtypes was developed including 934 SNPs nominally associated with atrial fibrillation. SNPs included had a p-value $<10^{-4}$, were LD independent ($r^2 > 0.5\%$), had good imputation information (>0.8) and MAF $> 1\%$. To test the association of the GRS with different stroke subtypes, they used logistic regression and then calculated the variance explained by the model. This GRS was significantly associated with cardioembolic stroke subtype and stroke of undetermined cause. These results could be very useful for the identification of stroke patients initially classified as undetermined which in fact are cardioembolic stroke. In this case, this type of patient could have benefit of anticoagulant treatment instead of antiplatelet drugs, which is less effective in cardioembolic stroke patients [31].

2.6 Advantages of GRS Comparing Single Variant Results

GRS are robust to imperfect linkage disequilibrium because they include several variants. It is common that variants found in GWAs associated with a disease are not really the causative variant but are in LD with real causative variants. Due to large sample sizes used in GWAs, associations are found despite correlation between the found variant and the real causative variant had a correlation <1. However, when these results are translated to smaller sample sizes, the effects of variants not perfectly correlated are lost. GRS help in solving this problem. In the same way, GRS are less sensitive to other effects like SNP MAF [6].

GRS can address the problem that in some cases GWAs for the same disease provide different results. Different phenotyping, chips, genotyping quality, and analysis of data could lead to different top variants for a locus associated with the trait or indeed a different associated locus. GRS could solve this problem including variants from different GWAs into the same score [6]. Thus, GRS could be very helpful tools to unify information provided from different polymorphisms and GWAs with the main objective to translate genetics into clinical practice [6].

3 Notes

1. Different methods for GRS construction could make their comparison and interpretation difficult. Thus, unifying methodologies is important.

2. Most articles are published including GRS weighted using effect sizes. However, other information such as LD or imputation parameters is also important and could modify the predictable value of GRS.

References

1. Wang MH, Cordell HJ, Van Steen K (2019) Statistical methods for genome-wide association studies. Semin Cancer Biol 55:53–60

2. Pasaniuc B, Rohland N, McLaren PJ et al (2012) Extremely low-coverage sequencing and imputation increases power for genome-wide association studies. Nat Genet 44:631–635. https://doi.org/10.1038/ng.2283

3. Goldstein BA, Yang L, Salfati E, Assimes TL (2015) Contemporary considerations for constructing a genetic risk score: an empirical approach. Genet Epidemiol 39:439–445. https://doi.org/10.1002/gepi.21912

4. Malik R, Bevan S, Nalls MA et al (2014) Multi-locus genetic risk score associates with ischemic stroke in case-control and prospective cohort studies. Stroke 45:394–402. https://doi.org/10.1161/STROKEAHA.113.002938

5. Ibrahim-Verbaas CA, Fornage M, Bis JC et al (2014) Predicting stroke through genetic risk functions the CHARGE risk score project. Stroke 45:403–412. https://doi.org/10.1161/STROKEAHA.113.003044

6. Belsky DW, Moffitt TE, Sugden K et al (2013) Development and evaluation of a genetic risk score for obesity. Biodemogr Soc Biol 59:85–100. https://doi.org/10.1080/19485565.2013.774628

7. Bush WS, Moore JH (2012) Genome-wide association studies. PLoS Comput Biol 8(12): e1002822. https://doi.org/10.1371/journal.pcbi.1002822

8. Das S, Forer L, Schönherr S et al (2016) Next-generation genotype imputation service and methods. Nat Genet 48:1284–1287. https://doi.org/10.1038/ng.3656

9. Laurie CC, Doheny KF, Mirel DB et al (2010) Quality control and quality assurance in genotypic data for genome-wide association studies. Genet Epidemiol 34:591–602. https://doi.org/10.1002/gepi.20516

10. Dehghan A (2018) Genome-wide association studies. Methods Mol Biol 1793:37–49. https://doi.org/10.1007/978-1-4939-7868-7_4

11. Howie B, Marchini J, Stephens M (2011) Genotype imputation with thousands of genomes. G3 1:457–470. https://doi.org/10.1534/g3.111.001198

12. Belmont JW, Hardenbol P, Willis TD et al (2003) The international HapMap project. Nature 426:789–796. https://doi.org/10.1038/nature02168

13. Auton A, Abecasis GR, Altshuler DM et al (2015) A global reference for human genetic variation. Nature 526(7571):68–74. https://doi.org/10.1038/nature15393

14. Sudmant PH, Rausch T, Gardner EJ et al (2015) An integrated map of structural variation in 2,504 human genomes. Nature 526:75–81. https://doi.org/10.1038/nature15394

15. Rosenberg NA, Huang L, Jewett EM et al (2010) Genome-wide association studies in diverse populations. Nat Rev Genet 11:356–366. https://doi.org/10.1038/nrg2760

16. Johnson RC, Nelson GW, Troyer JL et al (2010) Accounting for multiple comparisons in a genome-wide association study (GWAS). BMC Genomics 22:724. https://doi.org/10.1186/1471-2164-11-724

17. Fadista J, Manning AK, Florez JC, Groop L (2016) The (in)famous GWAS P-value threshold revisited and updated for low-frequency variants. Eur J Hum Genet 24:1202–1205. https://doi.org/10.1038/ejhg.2015.269

18. Pe'er I, Yelensky R, Altshuler D, Daly MJ (2008) Estimation of the multiple testing burden for genomewide association studies of nearly all common variants. Genet Epidemiol 32:381–385. https://doi.org/10.1002/gepi.20303

19. Traylor M, Farrall M, Holliday EG et al (2012) Genetic risk factors for ischaemic stroke and its subtypes (the METASTROKE Collaboration): a meta-analysis of genome-wide association studies. Lancet Neurol 11:951–962. https://doi.org/10.1016/S1474-4422(12)70234-X

20. NINDS Stroke Genetics Network (SiGN); International Stroke Genetics Consortium (ISGC) (2015) Loci associated with ischaemic stroke and its subtypes (SiGN): a genome-wide association study. Lancet Neurol 15:174–184

21. Malik R, Chauhan G, Traylor M et al (2018) Multiancestry genome-wide association study of 520,000 subjects identifies 32 loci associated with stroke and stroke subtypes. Nat Genet 50:524–537. https://doi.org/10.1038/s41588-018-0058-3

22. Woo D, Falcone GJ, Devan WJ et al (2014) Meta-analysis of genome-wide association studies identifies 1q22 as a susceptibility locus for intracerebral hemorrhage. Am J Hum Genet 94:511–521. https://doi.org/10.1016/j.ajhg.2014.02.012

23. Biffi A, Sonni A, Anderson CD et al (2010) Variants at APOE influence risk of deep and lobar intracerebral hemorrhage. Ann Neurol 68:934–943. https://doi.org/10.1002/ana.22134

24. Che R, Motsinger-Reif AA (2012) A new explained-variance based genetic risk score for predictive modeling of disease risk. Stat Appl Genet Mol Biol 11:15. https://doi.org/10.1515/1544-6115.1796

25. Chagnon M, O'Loughlin J, Engert JC et al (2018) Missing single nucleotide polymorphisms in Genetic Risk Scores: a simulation study. PLoS One 13:e0200630. https://doi.org/10.1371/journal.pone.0200630

26. McGeechan K, Macaskill P, Irwig L et al (2008) Assessing new biomarkers and predictive models for use in clinical practice: a clinician's guide. Arch Intern Med 168 (21):2304–2310. https://doi.org/10.1001/archinte.168.21.2304

27. Vrieze SI (2012) Model selection and psychological theory: a discussion of the differences between the Akaike information criterion (AIC) and the Bayesian information criterion (BIC). Psychol Methods 17:228–243. https://doi.org/10.1037/a0027127

28. Pencina MJ, D'Agostino RB, D'Agostino RB, Vasan RS (2008) Evaluating the added predictive ability of a new marker: from area under the ROC curve to reclassification and beyond. Stat Med 27:157–172. https://doi.org/10.1002/sim.2929

29. Jewell ES, Maile MD, Engoren M, Elliott M (2016) Net reclassification Improvement. Anesth Analg 122:818–824. https://doi.org/10.1213/ANE.0000000000001141

30. Bellenguez C, Bevan S, Gschwendtner A et al (2012) Genome-wide association study identifies a variant in HDAC9 associated with large vessel ischemic stroke. Nat Genet 44:328–333. https://doi.org/10.1038/ng.1081

31. Pulit SL, Weng L-C, McArdle PF et al (2018) Atrial fibrillation genetic risk differentiates cardioembolic stroke from other stroke subtypes. Neurol Genet 4:e293. https://doi.org/10.1212/NXG.0000000000000293

Chapter 4

Neuroplasticity Biomarkers in Experimental Stroke Recovery

Philip V. Peplow, Bridget Martinez, D. Mascareñas, and Svetlana A. Dambinova

Abstract

This chapter focuses on the main classes of plasticity biomarkers consisting various neurovascular specific regulators, glial/neuronal proteins, and peptide growth factor profiles related to selective types of experimental stroke recovery. Radiological methods supporting the progress of spontaneous, therapy-induced restoration, and cell-induced replacement recovery are presented.

Key words Neuroplasticity biomarkers, Experimental stroke recovery, Ischemic stroke, Glial proteins, Neuronal proteins, Peptide growth factors, Radiology

1 Introduction

Neuroplasticity is a unique ability of the central nervous system (CNS) to adapt to a changing environment through reorganization and compensation to structural and functional impairments following injuries. Complete or partial recovery of brain functions being a prolonged process it is dependent on the type of injury and location of lesion(s) after stroke rather than the size of damage [1].

Neuroplasticity as a part of spontaneous recovery after stroke might extend up to 3 months in animal models and up to 6 months in humans. Functional restoration of CNS is associated with reorganization of microvascular and neuronal networks, glial expansion, and synaptic contact growth [2]. In recent studies devoted to recovery after experimental stroke, the conscientious attention to biomarkers of neuroplasticity capable of examining the progress of cerebral blood flow and nervous cell restorations is still unfolding [3].

The distinct endogenous biomarkers revealing restorative patterns after ischemic or hemorrhagic stroke of different severity are expected to be recognized [3]. Additionally, trafficking of these

Philip V. Peplow et al. (eds.), *Stroke Biomarkers*, Neuromethods, vol. 147, https://doi.org/10.1007/978-1-4939-9682-7_4,
© Springer Science+Business Media, LLC, part of Springer Nature 2020

biomarkers from the CNS into biological fluids should relate to dynamic processes of angiogenesis, neurogenesis, and gliogenesis [4–6]. Comparable research of neuroplasticity biomarkers in respect to human and animal recovery after cerebrovascular accident might assist in understanding mechanisms of neuroregeneration and in the search of personalized rehabilitative therapy.

There is a medical need for specific biomarkers capable to navigate therapy-induced restoration and stem cell transplantation, and prognosticate recovery after stroke. Combining neuroplasticity biomarkers or their respective panels with up-to-date neuroimaging might confer invaluable data concerning monitoring of neuroregeneration.

A window of opportunity in brain research emerges with the development of in vivo research methods like optical imaging and 7.0 or 9.4 T magnetic resonance imaging (MRI) with capabilities to depict soft tissue properties with 50–100 μm resolution have been extensively used in experimental stroke [7]. Several integrative biomarkers of neuroplasticity in conjunction with advanced radiological methods are explored.

2   Substrates of Neurorestorative Processes in Cortical Ischemic Stroke

In cortical ischemic stroke, the most frequent type of onset in humans, spontaneous angiogenesis and neurogenesis are initiated within the first few weeks in a time-dependent manner. As time passes, delayed neuroinflammation reactions occur following the disruption of the blood–brain barrier (BBB) and the systemic immune responses also could contribute to recuperation [8].

Spontaneous recovery after induced cortical ischemic stroke yielding relatively mild to moderate lesions often occurs [9]. The restoration in the gray matter network is a lengthy process and proceeds by reorganization of short axonal contacts in white matter and recuperation of the neurovascular unit (NVU) [10]. NVU includes pericytes, microglia, astrocytes, neurons, endothelial cells, and basal lamina [11].

In experimental stroke, systemic inflammation at the time of middle cerebral artery occlusion (MCAO) in rats induced the development of a deleterious autoimmune response to myelin basic protein (MBP) after 1 month [12]. Furthermore, in a similar experimental model, impairment of neurological deficits associated with a Type 1 T helper (Th1) cells response to MBP in the spleen was reported as soon as 48 h after induction of ischemia [13].

Angiogenesis is the growth of blood vessels from the existing vasculature. Stimulation of angiogenesis is important after cerebral ischemia that depends on endogenous hemodynamic critical survival factors of vascular networks and structural adaptations of vessel walls. This process can be complete or partial depending on the severity of injury and extent of consequences.

Microvasculature plays a key role in stroke pathophysiology both during initial damage and extended neural repair. Moreover, angiogenesis seems to be a promising target for future neurorestorative therapies. However, dynamic changes of microvessels after stroke still remain unclear and might be explained through additional research of biomarkers involved in functions of vasoconstriction and vasodilatation during stroke recovery.

Gliogenesis is the developmental process by which glial cells—astrocytes, oligodendrocytes, Schwann cells, and microglia—are generated. It includes the production of glial progenitor cells and their differentiation into mature glia. Oligodendrocytes influence neuronal conductance by forming insulating myelin sheets, which increase the speed of neural processing [14].

Neurogenesis is the growth and development of nervous tissue that promotes functional recovery. The ability of nervous tissue to undergo reorganization by forming new neural connections in response to ischemic damage represents neuronal plasticity and remodeling. Post-ischemic endogenous plasticity partly compensates for the loss of axons in target structures of damaged fiber bundles at various levels of the brain [15, 16].

Adult neurogenesis mainly occurs at the subventricular zone (SVZ) on the walls of the lateral ventricle and the subgranular zone (SGZ) of the dentate gyrus (DG) [17, 18]. These areas of the brain retain the ability to generate neuronal progenitor cells in adult life. The survival rate of adult neural precursor cells in the brain parenchyma is poor and only a small percentage of the cells differentiate into neurons [19].

Stem cell (SC) transplantation is a potential option to expedite the recovery that has been explored in various models of stroke [20]. It is expected that transplanted SCs should facilitate long-term functional recovery by migrating to the damaged brain areas augmenting endogenous repair processes.

Intracerebrally injected neural stem cells (NSCs) differentiating in specific locations (dentate gyrus, ventricles, striatum, neocortex, substantia nigra, and spinal cord) and mesenchymal stem cells (MSCs) had better utilization of treating induced transient cerebral ischemia [21, 22]. At the same time, effects of neural progenitors (NPs), induced pluripotent stem cell (iPSCs), and multilineage-differentiating stress-enduring cells (MUSECs) had lesser utilization for treating experimental ischemia and hemorrhage [20].

Generally, SC transplantation studies while demonstrating neurotrophic, neuroprotective and anti-inflammatory properties, suffer from a non-systematic approach: different cell types are used, transplantations are performed at various time intervals after induced stroke, multiple behavioral tests to assess the transplant efficacy are employed [23]. The optimal time for in vitro-cultured SC transplantation was day 3 after 30-min MCAO in terms of preventing host cell apoptosis, attenuation of ischemic zone expansion, and better preserved neurological performance [24].

3 Prospective Biomarkers of Spontaneous Recovery After Experimental Hemorrhage

Microhemorrhagic transformations are coincident with compromised BBB permeability and microbleeding into nervous tissue associated with lacunar stroke [9]. Microbleeding in subcortical areas activates mechanisms of spontaneous recovery of white matter occurring for small lesions due to the gliogenesis [17, 25]. Intracerebral hemorrhage (ICH) causes disturbances in neuronal connectivity [11]. The recovery in this case is related to the resolution and absorption of hematomas, decrease in swelling, return electrolyte and neurochemical balance partially due to immune system responses that during vascular rapture become activated to generate antibodies to nervous tissue antigens [26].

Hemorrhagic microbleeding triggers an initial immune response leading to intracellular changes within the nervous system [27]. The neuroinflammation after induced ICH activates an innate immune response with immunocompetent reaction intrinsic to the CNS. Adaptive response with external reaction of the immune system is delayed and extended up to several months with levels of autoantibodies to neuronal antigens maintaining steadily elevated levels compared to ischemic stroke [28].

New neurons, oligodendrocytes, and astrocytes can be produced in the adult brain throughout the entire life. Under certain pathological conditions such as cerebral hemorrhage, the brain is able to produce additional glial cells theoretically aimed to regenerate the damaged region within a certain period of time. However, slow replacement of damaged neurons as well as weakness of electrical signal transductions delay the improvement of neurological functions [29, 30].

Experimental models of stroke can cover only specific aspects of this complex disease. It should be recognized that complete spontaneous recovery may occur in rodent models while even partial restoration of motor functions occurs less often in humans after stroke [31].

There are some immune active recombinant biomarkers of neuroplasticity found to produce antibodies as humoral immune response to stroke [32, 33]. It seems plausible such antibodies to neuroplasticity biomarkers may be used as tracers of spontaneous recovery.

Ultralow doses of antibodies to S-100 protein increased rat survival, reduced neurological deficit, eliminated myorelaxation, and improved movement coordination and cognitive functions in rats with experimental hemorrhagic stroke; the efficiency of the preparation was not inferior to that of nimodipine [34]. In contrast to nimodipine, ultralow doses of antibodies to S-100 protein exhibited pronounced anxiolytic properties.

Increased glial fibrillary acidic protein (GFAP) autoantibodies in cerebrospinal fluid (CSF) and peripheral blood are presumed to be connected to brain hemorrhagic transformation due to microvessel rapture and compromised BBB [35]. Abnormally high GFAP antibodies were identied for ischemic lesions as well. However, it is still unclear if serum anti-GFAP would be useful in assessment of hemorrhagic stroke. Further investigations could demonstrate tendecy of autoantibodies to protect nervous tissue and prevent it from further deterioration.

4 Potential Biomarkers of Angiogenesis

Brain derived neurotrophic factor (BDNF), a member of the neurotrophin family, has been suggested to be involved in angiogenesis, synaptogenesis and functional recovery after experimental stroke [36]. Intracerebroventricular (ICV) injection of collagen-binding BDNF promoted angiogenesis, reduced cell apoptosis, and improved functional recovery after MCAO in rats [37]. To prove the direct involvement of BDNF in the pro-angiogenic effect on cerebral microvascular endothelial cells, authors [36] knocked down BDNF by 70% in rat brains 14 days before 90-min MCAO. Then candesartan, an angiotensin type 1 receptor blocker, at the time of reperfusion has been intravenously (IV) administered. The rodent group that received candesartan showed better functional outcome as well as increased vascular density and synaptogenesis compared to saline treatment. BDNF is directly involved in candesartan-mediated angiogenesis and synaptogenesis as well as functional recovery [36].

A recent study described the microvascular plasticity after transient cerebral ischemia at molecular and structural levels observed in vivo by MRI [38]. The results at 1–3 days (acute stage) are characterized by high levels of angiopoietin 2 (Ang 2), vascular endothelial growth factor receptor 2 (VEGFR-2), and endothelial nitric oxide synthase that are associated with compromised BBB and vasodilation. The "subacute stage" takes place after day 7 and is characterized by high levels of Ang1, Ang2, vascular endothelial growth factor (VEGF), VEGFR-1, and transforming growth factor-$\beta1$ that may lead to stabilization and maturation of vessels. Imaging studies highlighted that multiparametric MRI (T1-weighted, T2-weighted, and gradient-echo) is useful to assess post-stroke angiogenesis, and could be a biomarker of the neurorestorative process [38].

Quenault et al. [39] hypothesized that endothelial activation is a hallmark of cerebrovascular events, especially transient ischemic attack (TIA). Transcriptional and immunohistological analyses for

adhesion molecules depicted inflammatory responses of the endothelium suggesting that brain endothelial P-selectin might be a
potential biomarker for TIA. The conventional 7.0 T MRI
(T2-weighted, 2D time-of-flight, and diffusion-weighted images)
showed a robust decrease of cerebral blood velocity by 75% after
MCAO and the absence of brain lesions similar to clinical presentation of TIA. Ultra-sensitive molecular MRI based on P-selectin
antibody-bound to iron microparticles unmasked activated endothelial cells and allowed after 15 min MCAO (TIA model) distinguishing of mice with kainite-induced seizures (epilepsy model)
from those with nitroglycerine-induced migraine [39]. Epilepsy
and migraine are two major mimic disorders to TIA that often
simultaneously present at emergency rooms.

5   Neuronal Biomarkers in Neurogenesis After Stroke

Among neuronal biomarkers the αII-spectrin breakdown products
(SBDPs) could be used to evaluate the neurogenesis activated by
experimental stroke. SBDPs were detected in CSF of some rats
subjected to 2 h MCAO [4, 6]. Another potential biomarker,
ubiquitin carboxy-terminal hydrolase L1 (UCH-L1), which is a
highly abundant protein in the neuronal cell body, was significantly
elevated in CSF from 6 to 72 h after 30 min MCAO and from 6 to
120 h after 2 h MCAO [4, 5]. Evidence from preclinical animal
models has clearly established the role of cathepsin cysteine proteases in the development and progression of vascular lesions [40].

Attempts to increase the survival rate of pyramidal tract axons
after cerebral ischemia by use of growth factors and SC delivery
have been undertaken in rodents [15, 41]. It has been estimated
that use of SCs in experimental models of stroke produced 24–44%
improvement in the functional outcome [41].

Granulocyte-colony stimulating factor (G-CSF) is a hemato-
poietic growth factor that controls proliferation and differentiation
of neural SCs. Recently comparative studies applying 45 min of
MCAO to three groups of mice showed that ligand binding densities of alpha-amino-3-hydroxy-5-methyl-4-isoxazole propionic
acid (AMPA) receptor in G-CSF-deficient mice were substantially
enhanced compared to wild type mice [42]. G-CSF administration
in mice lacking G-CSF largely reversed this effect. At the same time
infarct volumes did not significantly differ between the experimental groups. The latter demonstrated neuroprotective properties of
endogenous G-CSF after ischemic stroke.

It was shown that utilization of VEGF promotes neurogenesis
of axonal fibers from intact contralateral motor cortex to dener-
vated neurons in lesioned structures [15, 16].

In the living mouse, substantial loss of yellow fluorescent protein (YFP)-labeled axonal and dendritic structures as well as the

formation of abnormal dendritic bulbs were detected in the ische-mic boundary regions 24 h after stroke compared with that 1 h after stroke by confocal microscopy imaging [43]. Dynamic measure-ments of neuronal responses after stroke exhibit significant increases in spine density in the ischemic boundary region indicat-ing dendritic plasticity during stroke recovery in living animals.

6 Prospective Biomarkers of Gliogenesis

Numerous astrocyte-specific proteins including GFAP were found in perihematomal regions of animals within 24 h after 10 min collagenase-induced ICH [44]. GFAP expression was downregu-lated in normobaric rats and coincided with reduced infarction at 24–48 h in an experimental model of focal cerebral ischemia [45].

Studies focused on ischemia-related alterations of the cytoskel-eton with a special focus on microtubule-associated proteins, neu-rofilament light and heavy chains (NF-L, NFH), microtubule-associated protein (T-tau), and MBP have been performed. Down-regulation in the most abundant component of neurofilament (NF68) and class III β-tubulin (Tuj1) was detected in white matter of experimental animals within 4 days of induced ischemia [46]. At 1 day after induced thromboembolic stroke in rats, immunoreac-tivity of T-tau and microtubule-associated protein-2 (MAP2) were reduced in the ischemia-affected striatum and the neocortex [47]. At the same time, ischemic areas displayed local agglomera-tions of NF-L-immunoreactivity that might be related to degraded cell bodies and neocortical pyramidal cells. It has been shown that NFH(+) axons and MBP(+) oligodendrocytes were substantially increased in the peri-infarct area during reperfusion after perma-nent MCAO demonstrating axonal outgrowth and myelination in the rodent ischemic brain [48].

Ganglioside GM1, which is particularly abundant in the CNS, penetrates the BBB, and is closely associated with neurotrophic effects [49]. Intraperitoneal GM1 (50 mg/kg) treatment after 2-h MCAO significantly reduced the autophagy activation up to 60%, improved modified Garcia neurobehavioral test results by 28%, and decreased infarction volume (from 26.3% to 19.5%) with-out causing significant adverse side effects [50].

Intravenous administration of allogenic bone marrow mesen-chymal stem cells (BM-MSCs) and adipose-derived mesenchymal stem cells (AD-MSCs) in animals with permanent MCAO (pMCAO) infarct was associated with good functional recovery despite the lack of a reduction in infarct volume [51]. At 14 days after MSCs administration, significantly increased levels of VEGF, synaptophysin (SYP), oligodendrocyte-2 (Olig-2) and neurofila-ment (NF) protein were registered while GFAP was reduced that pointed to activation of cell proliferation and decreased cell death.

In rats with subcortical stroke to white matter induced by injection of endothelin-1 the AD-MSC transplantation yielded better functional recovery and smaller lesion size than the control group [9]. AD-MSC-treated rodents showed significantly higher numbers of proliferating cells (including oligodendrocyte progenitors) and significantly less cell death at the lesion region compared to the control group. The group treated with AD-MSCs also had higher levels of white matter-associated markers (NF, MBP and Olig-2) than the control group, suggesting that AD-MSCs administration induced repair of white matter fiber tracts.

The spatial and temporal dynamics of the nestin biophotonic signal induction following ischemic injury was visualized using in vivo biophotonic/bioluminescence imaging [52]. Ischemic brain injury as well as controlled innate immune challenge significantly upregulated the nestin in vivo signals due to acute and chronic neuroinflammation. Authors also noted that in physiological conditions and in the adult brain, nestin expression is restricted to neural progenitors. However, the neuroinflammatory conditions following stroke are associated with marked induction of nestin in astrocytes and activated microglia/macrophages suggesting that nestin may serve as a biomarker of inflammatory response in these cells.

In the in vivo study in rats of intracerebral hemorrhage (ICH), induced in the striatum by collagenase or autologous blood, the temporal evolution from the hyperacute (within 3 h) to acute stages (3–24 h) of ICH were characterized [44]. It was demonstrated that astrocytes act as the major cell population within the first 24 h of ICH rather than microglia or macrophages. Abundant astrocytes were observed using GFAP immunohistochemistry in hyperacute and acute ICH. MRI assessments (T2-weighted and DWI modalities) at 21 h after ICH indicated the progression of hemoresolution and compromised BBB integrity. Upon suppression of astrocyte activity, ICH rats exhibited decreased size of hematoma expansion, less BBB destruction, reduced astrocyte accumulation in perihematomal regions, postponed course of hemoresolution and achieved better outcomes [44].

7   Pharmacologically Induced Recovery After Experimental Stroke

There is emerging research depicting effects of endogenous factors in spontaneous restoration after stroke in animal models and humans. Small peptides analogs of endogenous neurotrophic factors are considered as part of the therapeutic strategy for neurological disorders like dementia, stroke, and traumatic brain injury.

The heptapeptide Semax (Met-Glu-His-Phe-Pro-Gly-Pro) is an analog of the adrenocorticotropin fragment (ACTH 4–10). Administration of Semax (50 µg/kg) affects cognitive brain functions by modulating the expression and activation of the hippocampal BDNF/trkB system [53]. The peptide affects angiogenesis by increasing mRNA level of Vegf-b gene at 3 h after pMCAO [54]. Maximal neuroprotective effect of Semax has been observed in the hippocampus 12 h after occlusion [55].

Cortexin, the endogenous mixture of small peptides isolated and purified from bovine brain [56], reduced neuronal apoptosis [57], increased expression of BDNF and a nerve growth factor (NGF), and the adenosine triphosphate (ATP) content in neuronal mitochondria after exposure to toxic concentrations of glutamate, activating spontaneous regeneration [58]. The recorded mitochondrial potentials of the neurons showed that cortexin significantly downregulated the development of calcium deregulation during glutamate overload. Cortexin administration substantially reduces oxidative stress after acoustic-induced cerebral hemorrhage downregulating the formation of hemoglobin nitrozyl (Hb-NO) complexes [59].

The exposure of primary cortical neuronal culture to tetrapeptide L-Ala-L-Glu-Asp-L-Pro (Cortagen, 0.5–1000 ng/mL) significantly increased the growth of short axons and fibroblast-like cells, as well as neuroglial migration to the periphery [60].

The administration of metallopeptide Gafargin (Gly, Arg, Gln complex with Fe^{3+}) to rats with white matter hemorrhages downregulated microbleeding, edema, and minimized inflammation consequences. Gafargin also affects metalloproteases and protects subcortical areas from further microlesion development [11].

The derivative of double glutamate peptide, Glyzargin, has been synthesized and tested in preclinical studies [61]. After intraventricular injections Glyzargin improves the learning ability (28–30%) of rats and reduces neurotoxicity after cortical-induced injury. Without altering cerebral blood flow or physiological parameters, it downregulated GluR1 peptide (40–45%) and antibody (50%) levels.

Cerebrolysin is a neuropeptide mixture that mimics the effects of endogenous neurotrophic factors on brain protection and repair [62]. Experimental studies in stroke models have shown that cerebrolysin stabilizes the structural integrity of cells by inhibition of calpain and reduces the number of apoptotic cells after ischemic lesion. It induces restorative processes, decreases infarct volume/edema formation, and promotes neurogenesis in the subventricular areas.

Semax, Cortexin, Cortagen, and cerebrolysin have passed preclinical research, clinical trials, and are approved for nootropic use in Europe and former Soviet Union (FSU) countries. Open clinical trials have been initiated for some of them recently [63].

8 Conclusion

It is known that reduction in cerebral blood flow and blood supply to microvessels due to ischemia primarily affects the metabolic demands in cortical areas that are responsible for brain executive functions (motor, learning capabilities, behavioral regulations) [64]. Therefore, vascular aspects of cerebral ischemia represent the forefront of spontaneous recovery [65].

Additionally, the search of specific biomarkers that would be capable of evaluating the efficacy of spontaneous, pharmacologically induced, or cell-replacement recovery after stroke will contribute to the choice of treatment. The final lesion size and the neurological outcome depend on a number of factors including the extent and severity of ischemic and hemorrhagic events, the preexisting conditions (diabetes, arteriosclerosis, hypertension), etiology and localization of the small cerebral vessel rapture but also on age, gender, and genetic background [66].

Neuroplasticity biomarkers that distinctly differentiate angiogenesis, gliogenesis, and neurogenesis could aid in selecting an optimal therapy for CNS regeneration and rehabilitation. In this matter biomarkers detected in biological fluids, comprising high performance characteristics, and capable of differentiating certain clinically relevant indications could be employed to tremendously hasten the recovery. The decision-making process will be upgraded by an integrative approach, involving assessment of neuroplasticity biomarkers in conjunction with advanced neuroimaging. Future studies of neuroplasticity biomarkers' potential in recovery after stroke could optimize the human therapy by inducing personalized management and benefiting patient healthcare.

9   Methods

9.1   Methods Assessing Neuroplasticity Biomarkers

9.1.1   SBDPs Immunoblotting of CSF and Cortical Tissue

Homogenates of cortical tissue or CSF samples were prepared for electrophoresis as described [6]. Twenty micrograms of protein per lane was routinely resolved by sodium dodecyl sulfate–polyacrylamide gel electrophoresis (SDS-PAGE) on 6.5% or 10–20% Tris-–glycine gels for 1–2 h [5]. After electrophoresis, separated proteins were laterally transferred to polyvinylidene fluoride (PVDF) membranes in a transfer buffer for 1–2 h at ambient temperature in a semidry transfer unit. Blots were blocked for nonspecific binding and were incubated with the primary polyclonal anti-αII-spectrin antibody (1:10,000) overnight, followed by three washes, an incubation with a biotinylated secondary antibody and a 30-min incubation with Streptavidin-conjugated alkaline phosphatase. Semiquantitative evaluation of intact protein levels was performed via computer-assisted densitometric scanning and image analysis with ImageJ software (NIH, USA).

9.1.2   UCH-L1 and SBDP Sandwich ELISA

For UCH-L1 and SBDP sandwich enzyme-linked immunoassay (ELISA) 96-well plates were coated with corresponding capture antibody. Plates were then incubated overnight at 4 °C, emptied and blocked [5]. This was followed by either the addition of 100 μL/well respective antigen standard (0.05–50 ng/well) or diluted CSF samples. The plate was incubated for 2 h at room temperature, then washed. Detection secondary antibodies conjugated with horseradish peroxidase (HRP) in blocking buffer was then added at 100 μL/well and incubated for 1.5 h at room temperature followed by washing. The wells were developed with chemiluminescent substrate solution and the signal was detected using a chemiluminescence microplate reader.

9.1.3 GFAP Immunofluorescence Staining

At time intervals established by experiment after induced ICH [44], animals were perfused with saline followed by 4% paraformaldehyde. The brains were removed and embedded in paraffin. Five-micron-thick brain sections were subjected to immunofluorescence staining of GFAP (1:2000) and fluorescent-tagged secondary antibody, nuclei were counterstained with 4',6-diamidino-2-phenylindole (DAPI). Immunostaining was analyzed with a fluorescence microscope interfaced with a digital charge-coupled device camera and an image analysis system. Representative images with lesions (at ×40 magnification) were selected and the coverage areas and cell numbers were analyzed using ImageJ software (NIH, USA).

9.1.4 Primary Cortical Neurons

Briefly, the dissociated cerebral cortical cells were cultured in neurobasal medium with 2% B27 which includes serum albumin, corticosterone, insulin, and progesterone [67]. To kill astrocytes, uridine and 5-fluorodeoxyuridine were added for 2 days. At 14 days in vitro, media was changed to Ca^{2+}-free and Mg^{2+}-free Hanks balanced salt solution, and neurons were subjected to oxygen-glucose deprivation (OGD) for 3 h. To separate axons from neuronal soma, a microfluidic chamber (Xona Microfluidics) was employed [68]. The cells were cultured in the same medium with daily changes. At 7 days in vitro, neurons were subjected to OGD for 3 h, morphologically analyzed at 24 and 96 h after OGD using time lapse microscopy, and then prepared for immunocytochemistry.

9.1.5 Co-culture of Neurons with Oligodendrocyte Cells

N20.1 cells were differentiated into mature oligodendrocytes [69, 70]. Concurrently, neurons cultured in the microfluidic chamber were subjected to OGD for 3 h. Immediately after OGD, harvested differentiated N20.1 cells were placed at a density of 3×10^4 cells/chamber in the axonal compartment of microfluidic chambers with the neurobasal medium for 96 h.

9.1.6 NFH and MBP Immunochemistry and Images Acquisition/ Quantification

The primary antibodies against phosphorylated and nonphosphorylated NFH (anti-SMI31 and anti-SMI32) and rabbit anti-MBP were used in the present study. Nuclei were counterstained with 4',6-diamidino-2-phenylindole. Areas of pNFH+ were binarized and digitally level-adjusted with intensity threshold setting (pixel 22—black, pixel 255—white), where black pixels represent pNFH immunoreactive areas. The pNFH+-MBP+ areas were split into green- and red-channel images, and the image showing common positive areas in both images was obtained and adjusted as above. The area of black pixels was divided by the total area in each image to estimate the profile of axonal outgrowth after stroke [48].

9.1.7 Isolation of Mononuclear Cells and ELISPOT Assay

Mononuclear cells (MNCs) were isolated from the brain and spleen using previously described methods [71]. MNCs were cultured (10^5 cells/well) for 48 h in 96-well plates (MultiScreens-IP; Millipore) in media alone or in media supplemented with bovine myelin basic protein (MBP) (25 mg/mL). Experiments were performed in triplicate. Interferon-γ or TGF-β1 capture antibodies were used and the reaction product developed with alkaline phosphatase. Spots were independently counted under a dissecting microscope by two individuals blinded to treatment status. The difference between the number of MBP-stimulated and unstimulated spots was considered indicative of an antigen-specific response. Results are expressed as the number of MBP-specific cells per 10^5 total MNCs. If the ratio of MBP-specific secreting cells was greater than two, the animal was considered sensitized to MBP.

9.2 Methods for Assessment of Neurotropic Factors

9.2.1 Brain Tracer Injections

Brain tracer injections are used to evaluate pyramidal tract plasticity in the motor cortex. The cascade blue-labeled dextran amine (CB) for ipsilateral and biotinylated dextran amine (BDA) for contralateral to the stroke are micro injected (angles of 45°, 90° and 135° against the midline at a depth of 1.5 mm) 6 weeks after MCAO are made at 0.5 mm rostral and 2.5 mm lateral to the bregma [15]. Ten days after the tracer injection, mice are perfused, brains removed and post-fixed in paraformaldehyde, and cryoprotected. The tissue then is frozen and cut into 20 and 40 μm thick coronal cryostat sections for conventional and tract tracing histochemistry.

9.2.2 Immunohistological Stainings

Brain sections from the areas of interest are fixed, pretreated for antigen retrieval, and immersed in 10% normal donkey serum. Brain sections then are incubated with corresponding monoclonal antibodies: mouse anti-NeuN, anti-CD31, anti-GFAP, anti-CD45, and anti-ionized calcium binding adaptor protein (Iba)-1 that were detected with Cy3 or Cy2 conjugated secondary antibodies [15]. Sections were counterstained with 4'-6-diamidino-2-phenylindole (DAPI). In some experiments (CD45, Iba1) biotinylated secondary antibodies are used in avidin–biotin kits. Sections are evaluated under a fluorescence microscope connected to a camera. Surviving neurons (NeuN+), microvascular profiles (CD31+), reactive astroglia (GFAP+), leukocytes (CD45+), and microglia (Iba1+) are analyzed by counting numbers of cells or profiles in six defined regions of interest. In case of GFAP staining, the overall area of scar tissue is outlined. The degree of post-ischemic atrophy

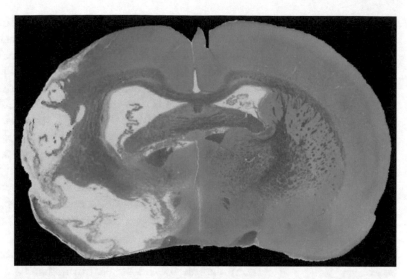

Fig. 1 With silver stain and Luxol fast blue staining, section stained at 6 weeks after stroke. Reproduced from Ding et al. 2008. This article was published in J Cereb Blood Flow Metab 28, Ding G, Jiang Q, Li L, Zhang L, Zhang ZG, Ledbetter KA, Panda S, Davarani SP, Athiraman H, Li Q, Ewing JR, Chopp M, Magnetic resonance imaging investigation of axonal remodeling and angiogenesis after embolic stroke in sildenafil-treated rats, 1440–1448, Copyright SAGE, 2008

of the striatum and corpus callosum are assessed using silver staining [72] (*see* Fig. 1).

9.2.3
Immunohistochemistry for CB and BDA

Brain sections are rinsed in buffer and immersed with polyclonal rabbit anti-Cascade Blue antibody (for CB detection), followed by incubation with a horseradish peroxidase (HRP)-labeled secondary anti-rabbit antibody. For detection of BDA, sections were incubated with avidin–biotin–peroxidase complex and staining is revealed with 3,3′-diaminobenzidine (DAB).

9.2.4 Brain Samples Preparations for RT-PCR

For gene expression studies, rodents are sacrificed at different time points after stroke by perfusion, brains immediately removed, and dissected on dry ice from 2 mm rostral to 2 mm caudal to the bregma. Brain samples are collected from regions of interest, tissue samples are homogenized, and total RNA is isolated. Then tRNA is converted to cDNA (1 μg from each 2 μg), and used for RT-PCR with predesigned TaqMan low density arrays (TLDA) [73]. Briefly, for each TLDA, there are eight separate loading ports that distributed the cDNA into a total of 48 wells, for a total of 384 different wells per card. Each well contains a specific primer and probe, capable of detecting one single gene (up to 46 different genes total) together with two housekeeping genes, glyceraldehyde-3-phosphate dehydrogenase (GAPDH) and 18S ribosomal RNA.

9.2.5   *RT-PCR*

RT-PCR is processed using a real time PCR machine with samples obtained from individual animals from each of the six regions of interest, each sample containing 10 μL cDNA (1 μg). Gene cards are analyzed using the threshold cycle (CT) relative quantification method. CT values are normalized for endogenous GAPDH reference [$\Delta CT = CT$ (target gene) $- CT$ (GAPDH)] and compared with a calibrator using the $\Delta\Delta CT$ formula [$\Delta\Delta CT = \Delta CT$ (sample) $- \Delta CT$ (calibrator)]. As calibrator sample, a brain obtained from an untreated rodent of the same age, gender and strain is utilized. Data are presented using the logarithmic transformation of fold induction (F.I.) ratios between ischemic vehicle- and non-ischemic vehicle-treated mice (MCAO effect) and of ratios between ischemic Epo- and ischemic vehicle-treated mice (Epo effect).

9.2.6   *Western Blot Analysis*

Tissue samples are harvested from the motor cortex ipsilateral and contralateral to the stroke immediately adjacent to the level at which cryostat sections were taken (0–2 mm caudal to bregma) from the brains used for immunohistochemistry. Tissue samples belonging to the same group are pooled, homogenized, sonicated and treated with protease inhibitor cocktail and phosphatase inhibitor cocktail [74]. Equal amounts of these samples after protein evaluations are subjected to SDS-PAGE, followed by protein transfer onto a PVDF membranes as described for SBDPs. Membranes after blocking nonspecific binding are incubated with mouse monoclonal SPRR1A antibody, and further incubated in peroxidase conjugated goat anti-mouse antibody. Blots are revealed using a chemiluminescence kit according to the manufacturer's protocol. Protein loading is controlled with β-actin antibody. Protein bands are evaluated by Image J Program.

9.3   *Stem Cells Transplantation Methods*

9.3.1   *Isolation of NSCs*

NSCs are isolated from the ganglion eminences dissected from E14 (14-day-old) embryos of Sprague-Dawley rats [24]. Briefly, the heads of the embryos are removed and the brain tissue is dissected into separate cortices, midbrain and stria. The dissected tissue is transferred to the NSC culture media and mechanically dissociated to reach uniform cell suspension. The cells at density 50,000 cells/mL are placed in a culture incubator. Neurospheres are detected by day 5. To identify NSCs, Nestin and CD133 immunocytochemistry is performed.

To promote NSC differentiation into neurons, oligodendrocytes, and astrocytes, single-cell NSC suspension of passage #4 is treated with 0.05% trypsin, then cells are cultured on polyornithine-coated plates for 2 days with following 0.5% fetal bovine serum (FBS) addition for 3 days [24]. To confirm the differentiation of the NSCs, immunocytochemistry is performed on NSCs mounted onto glass slides and fixed in 4% paraformaldehyde for β-tubulin III (neuron marker), glial fibrillary acidic protein (GFAP; an astrocyte marker) and oligodendrocyte marker Olig2 [75].

9.3.2 Cell Transplantation After MCAO

Thirty minutes after MCAO, rodents are divided into sham group (did not receive any treatment), control group (received 200 µL buffer), and experimental groups with cell transplantation carried out at 1 h to 7 days after MCAO. NSCs (200,000 cells suspended in 200 µL buffer) to anesthetized animals are injected into right lateral ventricle [24].

The brain slides (10 µm) are stained with Hematoxylin & Eosin to visualize the architecture and TUNEL assay is performed to determine the number of apoptotic cells in the damaged area on day 28 of respective treatment.

9.3.3 Isolation of Mesenchymal Stem Cells

Cultures are made of BM-MSC obtained from the tibia and femur of adult rodents by incubation in alcohol and Hank's ($1\times$) balanced salt solution [76]. The bone marrow is removed and placed in DMEM (1X) solution containing penicillin–streptomycin. Collected cells are washed, centrifuged, and placed in culture incubator for 3–4 weeks with medium replacement every 3 days. Cells are treated with trypsin–EDTA as soon as reaching 80–90% confluence and expanded in another flask. On the third pass cells again are trypsinized and counted before being administered to the experimental animals.

Lipoaspirates from adult female rats are washed with sterile buffer and digested with an equal volume of type I collagenase. The filtered AD-MSC cells are centrifuged and contaminating erythrocytes are removed to isolate the stromal vascular fraction (SVF). On the third pass cells are trypsinized and counted before being administered to the experimental animals [76].

At the time the cells are obtained, the cultures should be characterized to confirm the presence or absence of MSC surface markers using the flow cytometric technique and analyzed with fluorescence-activated cell sorting (FACS). The cells are incubated with the following antibodies: CD90-fluorescein isothiocyanate (FITC) (AbD Serotec, Oxford, UK), CD29-Phycoerythrin (PE) (AbD Serotec), CD45-PE (AbD Serotec), and CD11b-PE (AbD Serotec).

9.3.4 Cell Administration

Intravenous injections of 2×10^6 MSC in 650 µL saline are administered over 4 min through the femoral vein. Infarct animals underwent cerebral ischemia as in the treated animals but received only a saline infusion through the femoral vein. Sham-operated animals received the saline infusion through the femoral vein but did not undergo cerebral ischemia. The sham-operated and infarct groups both received a single 650 µL saline infusion without MSCs over 4 min. Either the saline or MSC solution is administered in the acute phase 30 min after pMCAO reperfusion [76].

9.4 Imaging Modalities in Cerebral Ischemia

Various modalities like structural MRI, magnetic resonance angiography (MRA), perfusion and diffusion MRIs, functional MRI, and diffusion tensor imaging (DTI), separately or merged in one session are carried out in experimental stroke.

9.4.1 Conventional MRI

All imaging sessions are performed using 7.0 T (Bruker, Germany) imaging system to assess infarction (T2-weighted MRI), time-of-flight (2D-TOF) angiograms for vascular permeability, T2 maps imaging for hemorrhage, and diffusion-weighted images (DWI) to evaluate penumbra area with parameters described in Table 1a–c (*see* Fig. 2).

9.4.2 Diffusion Tensor (DTI) Imaging

Diffusion tensor (DTI) imaging is used to define longitudinal imaging of axonal degeneration using 9.4 T Bruker BioSpin imaging system. Sham animals or animals subjected to cerebral hypoxia–ischemia were anesthetized (1.5% to 2% isoflurane) and DTI images in addition to anatomical scans are acquired at 3 h, 1 day,

Table 1

Magnetic resonance imaging (MRI) modalities in experimental stroke (rodents)

MRI modality	Repetition/echo time ms	Resolution, um	Reference
(a) MRI modalities assessing brain lesions (T2-weighted) and vascular permeability (2D TOF)			
T2-weighted	2500/51	$70 \times 70 \times 500$	[39]
	2500/60	$234 \times 234 \times 1000$	[38]
	36/4.2	$256 \times 196 \times 500$	[79]
	4560/60	$125 \times 125 \times 500$	[80]
	3000/70	$256 \times 128 \times 1000$	[44]
	10/10	$128 \times 128 \times 550$	[77]
2D TOF	12/7	$70 \times 70 \times 500$	[39]
	15.6/2.3	$234 \times 234 \times 1000$	[38]

MRI modality	Repetition, ms/echo time	Reference
(b) T2 maps modality evaluating cerebral hemorrhage		
T2 maps	500/7.7	[39]
	1000/[4,7,10,13]	[81]
	1800/12	[80]

MRI modality	Repetition, ms/echo time, ms	Matrix size, mm	Resolution, μm	*b*-Factor, s/mm^2	Reference
(c) DWI to estimate stroke penumbra					
DWI	2500/29	192×192	$100 \times 78 \times 750$	1000	[39]
	3700/40	–	–	[5,300,800,1200]	[81]
	2600/35	128×128	$256 \times 128 \times 1000$	0, 1100	[44]
	6500/35	128×128	$128 \times 128 \times 550$	0, 1000	[77]

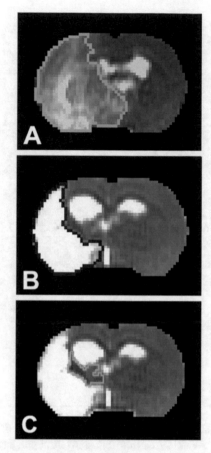

Fig. 2 T_2 map acquired at 24 h after stroke (**a**) demarcated acute ischemic lesion area with values above mean plus two SD of contralateral measurements. (**b**) Final infarction area was determined by T_2 map acquired at 6 weeks after stroke. (**c**) The difference was referred to recovery area after stroke. Reproduced from Ding et al. 2008. This article was published in J Cereb Blood Flow Metab 28, Ding G, Jiang Q, Li L, Zhang L, Zhang ZG, Ledbetter KA, Panda S, Davarani SP, Athiraman H, Li Q, Ewing JR, Chopp M, Magnetic resonance imaging investigation of axonal remodeling and angiogenesis after embolic stroke in sildenafil-treated rats, 1440–1448, Copyright SAGE, 2008

2 days, 1 week, or 4 weeks post-insult [77]. Images were acquired using a 3.5 cm diameter quadrature volume coil for radiofrequency transmission and reception. Depending on the age of the animal, each MR imaging scan consists of 25–30 slices of 0.5–0.55 mm thickness covering the cerebrum and medulla, a 2 × 2 cm or 2.5 × 2.5 cm field of view and a data matrix size of 128 × 128. For DTI, a four-shot echo-planar imaging sequence was used to acquire four averages of DWI sets (Table 1). DTI Image acquisition time was approximately 1 h (*see* Fig. 3).

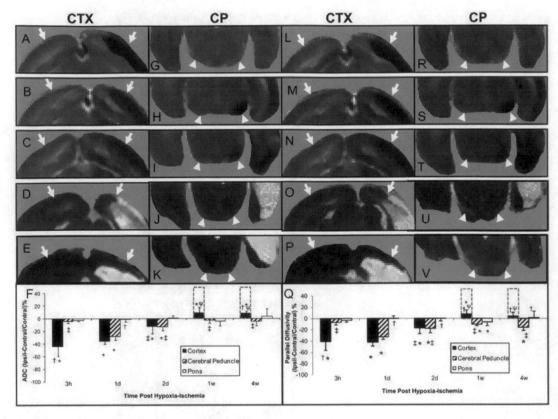

Fig. 3 Magnetic resonance imaging maps of the apparent diffusion coefficient of water (ADC) (A–K) and its parallel diffusivity (L–V) analyzed from diffusion tensor imaging scans of neonatal rat brains at 3 h (top row, A, G, L, R), 1 day (second row, B, H, M, S), 2 days (third row, C, I, N, T), 1 week (fourth row, D, J, O, U), or 4 weeks (fifth row, E, K, P, V) after transient hypoxia right cerebral ischemia. Shown are representative maps of slices containing the cerebral cortex (A–E, L–P, arrows) and the posterior cerebral peduncle (G–K, R–V, arrowheads). Mean (±SD) values of ipsilateral–contralateral differences presented as a percentage of contralateral are shown in F and Q. Development of cyst is shown as off-scale measures in a dashed line bar above pericyst cortical measures. $^*P < 0.05$; ipsilateral different from contralateral; $^\dagger P < 0.05$, cortex or pons different from cerebral peduncle within the time point; $^\ddagger P < 0.05$, different from corresponding region at 1 day; $^\Psi P < 0.05$ different from corresponding region at 3 h, 1 day, and 2 days (Two Way repeated ANOVA, Holm–Sidak multiple comparisons). Reproduced from Tuor et al. 2014. This article was published in Neuroimage Clin 6, Tuor UI, Morgunov M, Sule M, Qiao M, Clark D, Rushforth D, Foniok T, Kirton A, Cellular correlates of longitudinal diffusion tensor imaging of axonal degeneration following hypoxic-ischemic cerebral infarction in neonatal rats, 32–42, Copyright Elsevier, 2014

9.4.3 Molecular MRI
Antibody-Based
Microparticles of Iron Oxide

Cerebral blood volume (CBV) and VSI (perfused vessel caliber) imaging are performed using a steady state approach [38]. Briefly, a multigradient-echo spin-echo sequence (TR = 4000 ms, spin-echo = 40 ms; seven gradient-echoes from 2.3 to 15.6 ms; voxel size: 234 × 234 × 1000 mm; 7 slices) is applied, 2 min before and after intravenous (IV) injection of an intravascular ultrasmall super-paramagnetic iron particles (Roissy, France; 200 µmol iron/kg body weight). BBB leakage was assessed using T1 weighted images (TR/TE = 300/4.8 ms) acquired 3 min before and after IV injection of 0.2 mmol/kg Gd-DOTA.

9.4.4 Bioluminescence Imaging

Optical microscopy techniques, such as laser-scanning confocal microscopy, have been widely used for detailed postmortem assessment of changes in neurovascular anatomy after ischemic injury. For in vivo optical imaging studies in rats and mice, brain tissue can be directly exposed through a cranial window, or the skull can be thinned.

9.4.5 Preparation of Mice for In Vivo Imaging

After anesthetizing the mice with an intraperitoneal injection of chloral hydrate (400 mg/kg body weight), cerebral blood vessels were imaged through a craniotomy window centered at stereotactic coordinates 2.5 mm caudal to bregma, 2.5 mm lateral to midline [78]. The area to be observed is peri-infarcted zone in MCAO mice. The dura was removed and a metal frame of diameter 10.0 mm with a removable cover glass lid (diameter 6.0 mm) was glued to the skull to cover the craniotomy window. The space between the exposed brain surface and the cover glass was filled with 1.5% (w/v) low melting point agarose in an artificial cerebrospinal fluid. A bolus of 5 mg/kg Texas Red dextran (70 kDa, Carlsbad, California) in 0.9% NaCl was injected into the tail vein before TPLSM imaging to outline the blood vessels (*see* Fig. 4).

Vessel imaging and analysis is performed using an Olympus Fluoview1000 two-photon microscope (Tokyo, Japan) with an excitation source of a Spectra-Physics MaiTai HP DeepSee femtosecond Ti:Sa laser [78]. To acquire images (either stacks or single focal planes) from green fluorescent protein (GFP)-positive vessels, a long-working-distance (2 mm) water-immersion objective (25×, NA 1.05) is used for line-scan measurements. The images are taken

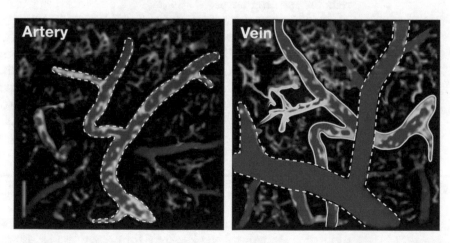

Fig. 4 TPLSM in conjunction with labeling of the blood plasma with Texas-dextran is used for mapping the angioarchitecture as well as quantify the transport of individual RBC in arteries and veins. Bar = 100 μm. Modified from Huang et al. 2014. This article was published in CNS Neurosci Ther 20, Huang JY, Li LT, Wang H, Liu SS, Lu YM, Liao MH, Tao RR, Hong LJ, Fukunaga K, Chen Z, Wilcox CS, Lai EY, Han F, In vivo two-photon fluorescence microscopy reveals disturbed cerebral capillary blood flow and increased susceptibility to ischemic insults in diabetic mice, 816–822, Copyright Wiley, 2014

at 12-bit depth with resolution of 1024×1024 pixels and a scanning rate of 10 μs/pixel and 2000 lines in total to follow the velocity of RBCs. In vivo vessel diameters are measured manually using ImageJ. RBC velocity and flux are calculated with an automated image-processing algorithm using MATLAB software (Natick, Massachusetts).

9.4.6 Laser-Scanning Fluorescence Microscopy

Multiphoton fluorescence microscopy is an advanced combination of laser-scanning microscopy with pulsed long-wavelength multi-photon fluorescence excitation, which offers high-resolution 3-dimensional images of fluorophore-labeled living tissue up to a depth of ≈ 1 mm.

Images are gathered using the IVIS 200 Imaging System (Hopkinton, Massachusetts). The luciferase substrate D-luciferin is injected intraperitoneally (i.p.) 20 min prior to the imaging session [52]. The mice are anesthetized, placed in the heated lightproof imaging chamber, and maintained anesthetized by constant delivery of the 2% isoflurane–oxygen mixture. To obtain baseline expression measurements, all animals were imaged before and on 24 h, 72 h, and 5, 7, 10, and 14 days following the MCAO [52]. The light output is quantified by determining the total number of photons emitted per second (p/s) using the Living Image 4.0 acquisition and imaging software. Regions of interest on the images are used to convert surface radiance ($p/s/cm^2/sr$) to source flux or total flux of photons expressed in photons per seconds. The data are represented as pseudo color images indicating light intensity (red and yellow, 5 most intense), which are superimposed over gray scale reference photographs.

For the acquisition of three-dimensional (3D) images, gray scale photographs are acquired and structured light images followed by a series of bioluminescent images using different wavelengths (560–660 nm). 3D images are created using diffuse light imaging tomographic (DLIT) algorithms to reconstruct for the position, geometry, and strength of the internal light sources. The modifiable parameters are analyzed across the wavelengths, source spectrum, and tissue properties (Living Image 4.0 3D analysis software).

Acknowledgements

We wish to thank R. McPhee, Department of Anatomy, University of Otago, Dunedin, New Zealand, for preparing Fig. 4.

References

1. Murphy TH, Corbett D (2009) Plasticity during stroke recovery: from synapse to behaviour. Nat Rev Neurosci 10(12):861–872. https://doi.org/10.1038/nrn2735

2. Hara Y (2015) Brain plasticity and rehabilitation in stroke patients. J Nippon Med Sch 82(1):4–13. https://doi.org/10.1272/jnms.82.4

3. Gandolfi M, Smania N, Vella A, Picelli A, Chirumbolo S (2017) Assessed and emerging biomarkers in stroke and training-mediated stroke recovery: state of the art. Neural Plast 2017:1389475. https://doi.org/10.1155/2017/1389475

4. Ren C, Zoltewicz S, Guingab-Cagmat J, Anagli J, Gao M, Hafeez A, Li N, Cao J, Geng X, Kobeissy F, Mondello S, Larner SF, Hayes RL, Ji X, Ding Y (2013) Different expression of ubiquitin C-terminal hydrolase-L1 and αII-spectrin in ischemic and hemorrhagic stroke: potential biomarkers in diagnosis. Brain Res 1540:84–91. https://doi.org/10.1016/j.brainres.2013.09.051

5. Liu MC, Akinyi L, Scharf D, Mo J, Larner SF, Muller U, Oli MW, Zheng W, Kobeissy F, Papa L, Lu XC, Dave JR, Tortella FC, Hayes RL, Wang KK (2010) Ubiquitin C-terminal hydrolase-L1 as a biomarker for ischemic and traumatic brain injury in rats. Eur J Neurosci 31(4):722–732. https://doi.org/10.1111/j.1460-9568.2010.07097.x

6. Pike BR, Flint J, Dave JR, Lu XC, Wang KK, Tortella FC, Hayes RL (2004) Accumulation of calpain and caspase-3 proteolytic fragments of brain-derived alphaII-spectrin in cerebral spinal fluid after middle cerebral artery occlusion in rats. J Cereb Blood Flow Metab 24(1):98–106. PMID:14688621

7. Yanev P, Dijkhuizen RM (2012) In vivo imaging of neurovascular remodeling after stroke. Stroke 43(12):3436–3441. https://doi.org/10.1161/STROKEAHA.111.642686

8. Famakin BM (2014) The immune response to acute focal cerebral ischemia and associated post-stroke immunodepression: a focused review. Aging Dis 5(5):307–326. https://doi.org/10.14336/AD.2014.0500307

9. Otero-Ortega L, Gutiérrez-Fernández M, Ramos-Cejudo J, Rodríguez-Frutos B, Fuentes B, Sobrino T, Hernanz TN, Campos F, López JA, Cerdán S, Vázquez J, Díez-Tejedor E (2015) White matter injury restoration after stem cell administration in subcortical ischemic stroke. Stem Cell Res Ther 6(1):121. https://doi.org/10.1186/s13287-015-0111-4

10. Hermann DM, Chopp M (2012) Promoting brain remodelling and plasticity for stroke recovery: therapeutic promise and potential pitfalls of clinical translation. Lancet Neurol 11(4):369–380. https://doi.org/10.1016/S1474-4422(12)70039-X

11. Gennarelli TA, Dambinova SA, Weissman JD (2018) Advances in diagnostics and treatment of neurotoxicity after sport-related injuries. In: Peplow PV, Dambinova SA, Gennarelli TA, Martinez B (eds) Acute brain impairment: scientific discoveries and translational research. Royal Society of Chemistry, London, pp 141–161

12. Becker KJ, Kindrick DL, Lester MP, Shea C, Ye ZC (2005) Sensitization to brain antigens after stroke is augmented by lipopolysaccharide. J Cereb Blood Flow Metab 25(12):1634–1644. https://doi.org/10.1038/sj.jcbfm.9600160

13. Zierath D, Kunze A, Fecteau L, Becker K (2015) Promiscuity of autoimmune responses to MBP after stroke. J Neuroimmunol 285:101–105. https://doi.org/10.1016/j.jneuroim.2015.05.024

14. Frisen J (2016) Neurogenesis and gliogenesis in nervous system plasticity and repair. Annu Rev Cell Dev Biol 32:127–141. https://doi.org/10.1146/annurev-cellbio-111315-124953

15. Reitmeir R, Kilic E, Kilic U, Bacigaluppi M, ElAli A, Salani G, Pluchino S, Gassmann M, Hermann DM (2011) Post-acute delivery of erythropoietin induces stroke recovery by promoting perilesional tissue remodeling and contralesional pyramidal tract plasticity. Brain 134. (Pt 1:84–99. https://doi.org/10.1093/brain/awq344

16. Reitmeir R, Kilic E, Reinboth BS, Guo Z, ElAli A, Zechariah A, Kilic U, Hermann DM (2012) Vascular endothelial growth factor induces contralesional corticobulbar plasticity and functional neurological recovery in the ischemic brain. Acta Neuropathol 123(2):273–284. https://doi.org/10.1007/s00401-011-0914-z

17. Rusznák Z, Henskens W, Schofield E, Kim WS, Fu Y (2016) Adult neurogenesis and gliogenesis: Possible mechanisms for neurorestoration. Exp Neurobiol 25(3):103–112. https://doi.org/10.5607/en.2016.25.3.103

18. Palma-Totosa S, Garcia-Culebras A, Moraga AG, Hurtado O, Perez-Ruiz A, Duran-Laforet V, Parra JD, Cuartero MI, Pradillo JM, Moro MA, Lizasoain I (2017) Specific features of SVZ neurogenesis after cortical ischemia: a longitudinal study. Sci Rep 7

(1):16343.   https://doi.org/10.1038/
s41598-017-16109-7

19. Zhang RL, Chopp M, Roberts C, Jia L, Wei M, Lu M, Wang X, Pourabdollah S, Zhang ZG (2011) Ascl1 lineage cells contribute to ischemia-induced neurogenesis and oligodendrogenesis. J Cereb Blood Flow Metab 31 (2):614–625.   https://doi.org/10.1038/ jcbfm.2010.134

20. Gorodinsky A (2018) Advanced approaches in stem cell therapy for stroke and traumatic brain injury. In: Peplow PV, Dambinova SA, Gennarelli TA, Martinez B (eds) Acute brain impairment: Scientific discoveries and translational research. Royal Society of Chemistry, London, pp 214–241

21. Ottoboni L, Merlini A, Martino G (2017) Neural stem cell plasticity: advantages in therapy for the injured central nervous system. Front Cell Dev Biol 5:52. https://doi.org/ 10.3389/fcell.2017.00052

22. Stonesifer C, Corey S, Ghanekar S, Diamandis Z, Acosta SA, Borlongan CV (2017) Stem cell therapy for abrogating stroke-induced neuroinflammation and relevant secondary cell death mechanisms. Prog Neurobiol 158:94–131. https://doi.org/10. 1016/j.pneurobio.2017.07.004

23. Horie N, Hiu T, Nagata I (2015) Stem cell transplantation enhances endogenous brain repair after experimental stroke. Neurol Med Chir (Tokyo) 55(Suppl 1):107–112. PMID:26236795

24. Ziaee SM, Tabeshmehr P, Haider KH, Farrokhi M, Shariat A, Amiri A, Hosseini SM (2017) Optimization of time for neural stem cells transplantation for brain stroke in rats. Stem Cell Investig 4:29. https://doi.org/10. 21037/sci.2017.03.10

25. Hakon J, Quattromani MJ, Sjölund C, Tomasevic G, Carey L, Lee JM, Ruscher K, Wieloch T, Bauer AQ (2017) Multisensory stimulation improves functional recovery and resting-state functional connectivity in the mouse brain after stroke. Neuroimage Clin 17:717–730. https://doi.org/10.1016/j.nicl. 2017.11.022

26. Dambinova SA (2012) Neurodegradomics: the source of biomarkers for mild traumatic brain injury. In: Dambinova SA, Hayes RL, Wang KW (eds) Biomarkers for TBI: RSC drug discovery. Royal Society of Chemistry, London, pp 66–86

27. Mracsko E, Veltkamp R (2014) Neuroinflammation after intracerebral hemorrhage. Front Cell Neurosci 8:388. https://doi.org/10. 3389/fncel.2014.00388

28. Miró-Mur F, Urra X, Gallizioli M, Chamorro A, Planas AM (2016) Antigen presentation after stroke. Neurotherapeutics 13 (4):719–728.   https://doi.org/10.1007/ s13311-016-0469-8

29. Lindvall O, Kokaia Z (2015) Neurogenesis following stroke affecting the adult brain. Cold Spring Harb Perspect Biol 7(11):pii: a019034. https://doi.org/10.1101/cshperspect. a019034

30. Koh SH, Park HH (2017) Neurogenesis in stroke recovery. Transl Stroke Res 8(1):3–13. https://doi.org/10.1007/s12975-016-0460-z

31. Sommer CJ (2017) Ischemic stroke: experimental models and reality. Acta Neuropathol 133(2):245–261. https://doi.org/10.1007/ s00401-017-1667-0

32. Jauch EC, Lindsell C, Broderick J, Fagan SC, Tilley BC, Levine SR, NINDS rt-PA Stroke Study Group (2006) Association of serial biochemical markers with acute ischemic stroke: the National Institute of Neurological Disorders and Stroke recombinant tissue plasminogen activator Stroke Study. Stroke 37 (10):2508–2513

33. Becker KJ, Kalil AJ, Tanzi P, Zierath DK, Savos AV, Gee JM, Hadwin J, Carter KT, Shibata D, Cain KC (2011) Autoimmune responses to the brain after stroke are associated with worse outcome. Stroke 42(10):2763–2769. https:// doi.org/10.1161/STROKEAHA.111. 619593

34. Voronina TA, Kheyfets IA, Dugina YL, Sergeeva SA, Epshtein OI (2009) Study of the effects of preparation containing ultralow doses of antibodies to S-100 protein in experimental hemorrhagic stroke. Bull Exp Biol Med 148(3):530–532

35. Kamchatnov PR, Chugunov AV, Ruieva NY, Dugin SF, Basse DA, Abusueva BA, Buriachkovskaya LI, Gusev EI (2010) Autoantibodies to GFAP (glial fibrillary acidic protein) and to dopamine in patients with acute and chronic cerebrovascular disorders. Health 2 (12):1366–1371. https://doi.org/10.4236/ health.2010.212202

36. Fouda AY, Alhusban A, Ishrat T, Pillai B, Eldahshan W, Waller JL, Ergul A, Fagan SC (2017) Brain-derived neurotrophic factor knockdown blocks the angiogenic and protective effects of angiotensin modulation after experimental stroke. Mol Neurobiol 54 (1):661–670.   https://doi.org/10.1007/ s12035-015-9675-3

37. Guan J, Tong W, Ding W, Du S, Xiao Z, Han Q, Zhu Z, Bao X, Shi X, Wu C, Cao J, Yang Y, Ma W, Li G, Yao Y, Gao J, Wei J, Dai J, Wang R (2012) Neuronal regeneration and

protection by collagen-binding BDNF in the rat middle cerebral artery occlusion model. Biomaterials 33(5):1386–1395. https://doi.org/10.1016/j.biomaterials.2011.10.073

38. Moisan A, Favre IM, Rome C, Grillon E, Naegele B, Barbieux M, De Fraipont F, Richard MJ, Barbier EL, Rémy C, Detante O (2014) Microvascular plasticity after experimental stroke: a molecular and MRI study. Cerebrovasc Dis 38(5):344–353

39. Quenault A, Martinez de Lizarrondo S, Etard O, Gauberti M, Orset C, Haelewyn B, Segal HC, Rothwell PM, Vivien D, Touzé E, Ali C (2017) Molecular magnetic resonance imaging discloses endothelial activation after transient ischemic attack. Brain 140 (1):146–157. https://doi.org/10.1093/brain/aww260

40. Weiss-Sadan T, Gotsman I, Blum G (2017) Cysteine proteases in atherosclerosis. FEBS J 284(10):1455–1472. https://doi.org/10.1111/febs.14043

41. Leong WK, Lewis MD, Koblar SA (2013) Concise review: preclinical studies on human cell-based therapy in rodent ischemic stroke models: where are we now after a decade? Stem Cells 31(6):1040–1043. https://doi.org/10.1002/stem.1348

42. Mammele S, Frauenknecht K, Sevimli S, Diederich K, Bauer H, Grimm C, Minnerup J, Schabitz WR, Sommer CJ (2016) Prevention of an increase in cortical ligand binding to AMPA receptors may represent a novel mechanism of endogenous brain protection by G-CSF after ischemic stroke. Restor Neurol Neurosci 34(4):665–675. https://doi.org/10.3233/RNN-150543

43. Zhang ZG, Zhang L, Ding G, Jiang Q, Zhang RL, Zhang X, Gan WB, Chopp M (2005) A model of mini-embolic stroke offers measurements of the neurovascular unit response in the living mouse. Stroke 36(12):2701–2704

44. Chiu CD, Yao NW, Guo JH, Shen CC, Lee HT, Chiu YP, Ji HR, Chen X, Chen CC, Chang C (2017) Inhibition of astrocytic activity alleviates sequela in acute stages of intracerebral hemorrhage. Oncotarget 8 (55):94850–94861. https://doi.org/10.18632/oncotarget.22022

45. Esposito E, Mandeville ET, Hayakawa K, Singhal AB, Lo EH (2013) Effects of normobaric oxygen on the progression of focal cerebral ischemia in rats. Exp Neurol 249:33–38. https://doi.org/10.1016/j.expneurol.2013.08.005

46. Song M, Woodbury A, Yu SP (2014) White matter injury and potential treatment in ischemic stroke. In: Baltan S, Carmichael ST, Matute C, Xi G, Zhang JH (eds) White matter injury in stroke and CNS disease. Springer, New York, NY, pp 39–52. https://doi.org/10.1007/978-1-4614-9123-1_2

47. Härtig W, Krueger M, Hofmann S, Preißler H, Märkel M, Frydrychowicz C, Mueller WC, Bechmann I, Michalski D (2016) Up-regulation of neurofilament light chains is associated with diminished immunoreactivities for MAP2 and tau after ischemic stroke in rodents and in a human case. J Chem Neuroanat 78:140–148. https://doi.org/10.1016/j.jchemneu.2016.09.004

48. Ueno Y, Chopp M, Zhang L, Buller B, Liu Z, Lehman NL, Liu XS, Zhang Y, Roberts C, Zhang ZG (2012) Axonal outgrowth and dendritic plasticity in the cortical peri-infarct area after experimental stroke. Stroke 43 (8):2221–2228. https://doi.org/10.1161/STROKEAHA.111.646224

49. Gao L, Jiang T, Guo J, Liu Y, Cui G, Gu L, Su L, Zhang Y (2012) Inhibition of autophagy contributes to ischemic postconditioning-induced neuroprotection against focal cerebral ischemia in rats. PLoS One 7:e46092. https://doi.org/10.1371/journal.pone.0046092

50. Li L, Tian J, Long MK, Chen Y, Lu J, Zhou C, Wang T (2016) Protection against experimental stroke by ganglioside GM1 is associated with the inhibition of autophagy. PLoS One 11(1):e0144219. 20.1271/journal.pone.0144219

51. Gutiérrez-Fernández M, Rodríguez-Frutos B, Ramos-Cejudo J, Teresa Vallejo-Cremades M, Fuentes B, Cerdán S, Díez-Tejedor E (2013) Effects of intravenous administration of allogenic bone marrow- and adipose tissue-derived mesenchymal stem cells on functional recovery and brain repair markers in experimental ischemic stroke. Stem Cell Res Ther 4(1):11. https://doi.org/10.1186/scrt159

52. Krishnasamy SK, Weng YC, Thammisetty SS, Phaneuf DJ, Lalancette-Hebert M, Kritz JS (2017) Molecular imaging of nestin in neuroinflammatory conditions reveals marked signal induction in activated microglia. J Neuroinflammation 14:45. https://doi.org/10.1186/s12974-017-0816-7

53. Dolotov OV, Karpenko EA, Inozemtseva LS, Seredenina TS, Levitskaya NG, Rozyczka J, Dubynina EV, Novosadova EV, Andreeva LA, Alfeeva LY, Kamensky AA, Grivennikov IA, Myasoedov NF, Engele J (2006) Semax, an analog of ACTH(4–10) with cognitive effects, regulates BDNF and trkB expression in the rat hippocampus. Brain Res 1117(1):54–60. https://doi.org/10.1016/j.brainres.2006.07.108

54. Medvedeva EV, Dmitrieva VG, Povarova OV, Limborska SA, Skvortsova VI, Myasoedov NF, Dergunova LV (2013) Effect of Semax and its C-terminal fragment Pro-Gly-Pro on the expression of VEGF family genes and their receptors in experimental focal ischemia of the rat brain. J Mol Neurosci 49(2):328–333. https://doi.org/10.1007/s12031-012-9853-y

55. Stavchanskii VV, Tvorogova TV, Botsina AI, Skvortsova VI, Limborskaia SA, Miasoedov NF, Dergunova LV (2011) The effect of Semax and its C-end peptide PGP on expression of the neurotrophins and their receptors in the rat brain during incomplete global ischemia. Mol Biol (Mosk) 45(6):1026–1035. Russian

56. Cortexin. https://topbrainboosters.com/cortexin/. Accessed 1 Jan 2019

57. Pinelis VG, Storozhevykh TP, Surin AM, Senilova YE, Persiyantzeva NF, Tukhmatova GR, Andreeva LA, Myasoedov NF, Granstrem O (2008) Neuroprotective effects of cortagen, cortexin and semax on glutamate neurotoxicity. 30th European peptide symposium (30EPS), Helsinki, 30 Aug–5 Sep

58. Sorokina EG, Reutov VP, Senilova YE, Khodorov BI, Pinelis VG (2007) Changes in ATP content in cerebellar granule cells during hyperstimulation of glutamate receptors: possible involvement of NO and nitrite ions. Bull Exp Biol Med 143(4):442–445

59. Reutov VP, Baĭder LM, Kuropteva ZV, Krushinskiĭ AL, Kuzenkov VS, Moldaliev ZT, Granstrem OK (2011) Experimental hemorrhagic stroke: the effect of the peptide preparation cortexin in the formation of Hb-NO-complexes and other blood paramagnetic centers. Zh Nevrol Psikhiatr Im S S Korsakova 111(8 Pt 2):56–61. Russian

60. Khavinson VK, Morozov VG, Malinin VV, Grigoriev EI (2007) Tetrapeptide stimulating functional activity of neurons pharmacological agent based thereon and method of use thereof patent (Patent #7189701) Publ. US 7189701

61. Danilenko UI, Khunteev GA, Bagumyan A, Izykenova GA (2012) Neurotoxicity biomarkers in experimental acute and chronic brain injury. In: Dambinova SA, Hayes RL, Wang KW (eds) Biomarkers for TBI: RSC drug discovery. Royal Society of Chemistry, London, pp 87–105

62. Masliah E, Díez-Tejedor E (2012) The pharmacology of neurotrophic treatment with cerebrolysin: brain protection and repair to counteract pathologies of acute and chronic neurological disorders. Drugs Today (Barc) 48(Suppl A):3–24. https://doi.org/10.1358/dot.2012.48(Suppl.A).1739716

63. Mashin VV, Belova LA, Chaplanova OI, Khusnullina AF, Manasyan AM (2016) An open clinical trial of cortexin in cerebral ischemia. Neurosci Behav Physiol 46(4):390–393. https://link.springer.com/article/10.1007/s11055-016-0247-4

64. Hattori Y, Enmi J, Iguchi S, Saito S, Yamamoto Y, Nagatsuka K, Iida H, Ihara M (2016) Substantial reduction of parenchymal cerebral blood flow in mice with bilateral common carotid artery stenosis. Sci Rep 6:32179. https://doi.org/10.1038/srep32179

65. Adamczak JM, Schneider G, Nelles M, Que I, Suidgeest E, van der Weerd L, Löwik C, Hoehn M (2014) In vivo bioluminescence imaging of vascular remodeling after stroke. Front Cell Neurosci 8:274. https://doi.org/10.3389/fncel.2014.00274

66. Min H, Hong J, Cho IH, Jang YH, Lee H, Kim D, Yu SW, Lee S, Lee SJ (2015) TLR2-induced astrocyte MMP9 activation compromises the blood brain barrier and exacerbates intracerebral hemorrhage in animal models. Mol Brain 8:23. https://doi.org/10.1186/s13041-015-0116-z

67. Buller B, Liu X, Wang X, Zhang RL, Zhang L, Hozeska-Solgot A, Chopp M, Zhang ZG (2010) MicroRNA-21 protects neurons from ischemic death. FEBS J 277(20):4299–4307. https://doi.org/10.1111/j.1742-4658.2010.07818.x

68. Taylor AM, Blurton-Jones M, Rhee SW, Cribbs DH, Cotman CW, Jeon NL (2005) A microfluidic culture platform for CNS axonal injury, regeneration and transport. Nat Methods 2(8):599–605

69. Paez PM, García CI, Davio C, Campagnoni AT, Soto EF, Pasquini JM (2004) Apotransferrin promotes the differentiation of two oligodendroglial cell lines. Glia 46(2):207–217

70. Paez PM, García CI, Campagnoni AT, Soto EF, Pasquini JM (2005) Overexpression of human transferrin in two oligodendroglial cell lines enhances their differentiation. Glia 52(1):1–15

71. Becker K, Kindrick D, McCarron R, Hallenbeck J, Winn R (2003) Adoptive transfer of myelin basic protein-tolerized splenocytes to naive animals reduces infarct size. A role for lymphocytes in ischemic brain injury. Stroke 34:1809–1815

72. Ding G, Jiang Q, Li L, Zhang L, Zhang ZG, Ledbetter KA, Panda S, Davarani SP, Athiraman H, Li Q, Ewing JR, Chopp M (2008) Magnetic resonance imaging

investigation of axonal remodeling and angiogenesis after embolic stroke in sildenafil-treated rats. J Cereb Blood Flow Metab 28 (8):1440–1448

73. Pluchino S, Muzio L, Imitola J, Deleidi M, Alfaro-Cervello C, Salani G, Porcheri C, Brambilla E, Cavasinni F, Bergamaschi A, Garcia-Verdugo JM, Comi G, Khoury SJ, Martino G (2008) Persistent inflammation alters the function of the endogenous brain stem cell compartment. Brain 131(10):2564–2578. https://doi.org/10.1093/brain/awn198

74. Kilic E, ElAli A, Kilic U, Guo Z, Ugur M, Uslu U, Bassetti CL, Schwab ME, Hermann DM (2010) Role of Nogo-A in neuronal survival in the reperfused ischemic brain. J Cereb Blood Flow Metab 30(5):969–984. https://doi.org/10.1038/jcbfm.2009.268

75. Hosseini SM, Talaei-Khozani T, Sani M, Owrangi B (2014) Differentiation of human breast-milk stem cells to neural stem cells and neurons. Neurol Res Int 2014:807896. https://doi.org/10.1155/2014/807896

76. Gutiérrez-Fernández M, Rodríguez-Frutos B, Alvarez-Grech J, Vallejo-Cremades MT, Expósito-Alcaide M, Merino J, Roda JM, Díez-Tejedor E (2011) Functional recovery after hematic administration of allogenic mesenchymal stem cells in acute ischemic stroke in rats. Neuroscience 175:394–405. https://doi.org/10.1016/j.neuroscience.2010.11.054

77. Tuor UI, Morgunov M, Sule M, Qiao M, Clark D, Rushforth D, Foniok T, Kirton A (2014) Cellular correlates of longitudinal diffusion tensor imaging of axonal degeneration following hypoxic-ischemic cerebral infarction in neonatal rats. Neuroimage Clin 6:32–42. https://doi.org/10.1016/j.nicl.2014.08.003

78. Huang JY, Li LT, Wang H, Liu SS, Lu YM, Liao MH, Tao RR, Hong LJ, Fukunaga K, Chen Z, Wilcox CS, Lai EY, Han F (2014) In vivo two-photon fluorescence microscopy reveals disturbed cerebral capillary blood flow and increased susceptibility to ischemic insults in diabetic mice. CNS Neurosci Ther 20 (9):816–822. https://doi.org/10.1111/cns.12268

79. Koch S, Mueller S, Foddis M, Bienert T, von Elverfeldt D, Knab F, Farr TD, Bernard R, Dopatka M, Rex A, Dirnagl U, Harms C, Boehm-Sturm P (2017) Atlas registration for edema-corrected MRI lesion volume in mouse stroke models. J Cereb Blood Flow Metab. https://doi.org/10.1177/0271678X17726635

80. Jimenez-Xarrie E, Davila M, Candiota AP, Delgado-Mederos R, Ortega-Martorell S, Julià-Sapé M, Arús C, Martí-Fàbregas J (2017) Brain metabolic pattern analysis using a magnetic resonance spectra classification software in experimental stroke. BMC Neurosci 18 (1):13. https://doi.org/10.1186/s12868-016-0328-x

81. Shim WH, Suh JY, Kim JK, Jeong J, Kim YR (2017) Enhanced thalamic functional connectivity with no fMRI responses to affected forelimb stimulation in stroke-recovered rats. Front Neural Circuits 10:113. https://doi.org/10.3389/fncir.2016.00113

Chapter 5

Methods of Mitochondrial and Redox Measurements in Ischemic Stroke

Oiva Arvola, Anand Rao, and Creed M. Stary

Abstract

Mitochondria are critical for maintenance of normal, physiologic cellular function by playing a central role in ATP production, apoptosis, metabolism, and numerous other key cellular processes. Mitochondria maintain the dynamic demand for ATP by coupling the electrochemical gradient to the reduction of molecular oxygen, a process maintained by the shuttling of electrons from carrier to receptor, quantified as redox state. Accurately determining mitochondrial function and redox state has therefore been a central focus for targeted stroke therapies, where even transient disruptions in O_2 can reduce cellular ATP availability and trigger neuronal cell death and dysfunction. Here, we describe both in vivo and in vitro approaches to more accurately assess mitochondrial function and redox signaling in the setting of cerebral ischemia. The current lack of alternative therapies for stroke apart from early reperfusion mandates validation of preclinical data by congruent, parallel outcome measures, to better ensure reproducibility and appropriate interpretation of data.

Key words Bioenergetics, ATP, NADH, NAD$^+$, Respiration, Respirometry, Oxidative phosphorylation, Fluorometry, Fluorescence

1 Introduction

Adenosine triphosphate, *ATP*, is a high-energy molecule central for maintainance of cellular function. Maintaining adequate ATP availability is paramount for cell survival during both normal physiological states and in response to cerebral ischemia. Cellular respiration of mitochondria relies on redox reactions coupling glycolysis, the tricarboxylic acid cycle (TCA, AKA citric acid/Kreb's cycle) and the electron transport chain (ETC) to drive oxidative phosphorylation via the electron carriers nicotinamide adenine dinucleotide (NADH) and flavin adenine dinucleotide (FADH$_2$, Fig. 1). Reduction–oxidation (redox) reactions are described by electron transfer from one chemical moiety (an atom, ion or molecule) to another: a chemical species losing electrons is oxidized, and the receiving species is reduced. The cellular "redox state" therefore generally

Philip V. Peplow et al. (eds.), *Stroke Biomarkers*, Neuromethods, vol. 147, https://doi.org/10.1007/978-1-4939-9682-7_5,
© Springer Science+Business Media, LLC, part of Springer Nature 2020

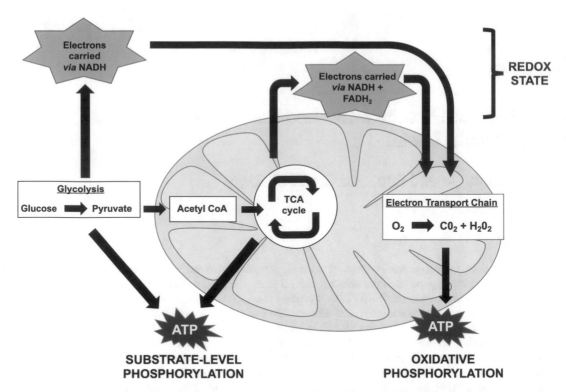

Fig. 1 Bioenergetics overview. Glycolysis and the tricarboxylic acid cycle (TCA) together generate adenosine triphosphate (ATP, via substrate-level phosphorylation) and provide electrons to power ATP formation from the electron transport chain (ECT) via reduction of nicotinamide adenine dinucleotide (NADH) and flavin adenine dinucleotide ($FADH_2$). Through chemiosmotic coupling, the ECT drives oxidative phosphorylation and ATP synthesis resulting in the reduction of molecular oxygen (O_2) to CO_2 and H_2O_2. The cellular "redox state" is the relative balance of the reduced *versus* oxidized forms of electron-transporting molecules, such as NADH and $FADH_2$. The NADH/$FADH_2$ redox state therefore determines the rate of ATP generation

describes the overall relative balance between oxidized or reduced molecular states. Energy released in the TCA and ECT powers the generation of a proton gradient across the intermembrane space, which is subsequently utilized to phosphorylate ADP to ATP. Under physiological conditions 0.2–2% of O_2 molecules are imperfectly reduced in the mitochondria, leading to single electron transfer from the ETC [1]. Reactive oxygen species (ROS) are formed as a normal byproduct from the reduction of molecular oxygen, however failure or dysfunction of ETC complexes can lead to excessive ROS production and subsequent oxidative damage. Disruptions in cerebral blood flow results in ischemia-derived neuronal oxidative stress, as perturbations occur in the ETC at the molecular level. This results in increased levels of oxidized genomic material such as nuclear DNA, mitochondrial DNA, and RNAs, as well as oxidized phospholipids, and proteins. Neurons possess multiple endogenous mechanisms to counter oxidants and repair oxidative damage independently but can also receive protection from local glia cells. These

A

<u>REDOX Equations</u>

Electron transport chain

Complex I: $NADH \rightarrow NAD^+ + 2e^- + Q \rightarrow QH_2$

Complex II: $FADH_2 \rightarrow FAD + 2e^- + Q \rightarrow QH_2$

Complex III: $QH_2 + 2\ Cyt\ c\ (Fe^{3+}) \rightarrow Q + Cyt\ c\ (Fe^{2+})$

Complex IV: $Cyt\ c\ (Fe^{2+}) + O^2 \rightarrow H_2O + 2\ Cyt\ c$ (Fe^{3+})

Citric acid cycle

$Pyruvate + 4\ NAD^+ + FAD + GDP + P_i + 2\ H_2O \rightarrow$

$\quad Coenzyme\ A + 4\ NADH + FADH_2 + GTP + 4H^+$ $\quad + CO_2$

Reactive oxygen species

$O_2 + e^- \rightarrow {}^\bullet O_2^-$

$2\ H^+ + {}^\bullet O_2^- + {}^\bullet O_2^- \rightarrow H_2O + O_2$

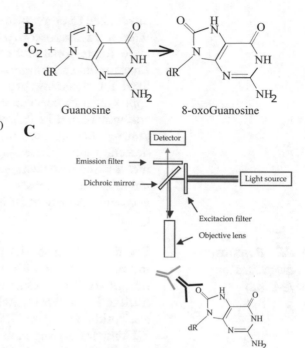

Guanosine   8-oxoGuanosine

Fig. 2 Reduction–oxidation (redox) reactions and fluorescent indicators. (**a**) Redox reactions characterize the transfer of an electron from a donor moiety to an acceptor moiety. Transfer of electrons powers oxidative phosphorylation which results in generation of reactive oxygen species as a byproduct of reduction of molecular oxygen. (**b**) 8-Hydroxydeoxyguanosine (8-OHdG) is a stable byproduct of ROS generation. (**c**) 8-OHdG can be visualized by fluorescent immunohistochemistry

counter-measures can be visualized fluorescently using appropriate immunohistochemical markers, and/or quantified using enzyme-linked immunosorbent assays. A summary of redox reactions and ROS generation are listed in Fig. 2a.

2 Immunohistochemical Markers for Oxidative Stress and Redox-Signaling

2.1 Immunohisto-chemistry

The detection of proteins in fixed cells/tissues via antibodies conjugated to fluorophores or enzymes is the basic principle for immunohistochemistry (IHC). There are numerous working protocols for different type of tissues and antibodies. Generally, after fixation (with paraffin or paraformaldehyde) the tissue is sectioned using a microtome to a desired thickness optimized for microscopy (typically 15–50 μm), followed by permeabilization, quenching of endogenous peroxidase activity and blocking for nonspecific immunostaining with an antibody of interest. One or more (if not cross-reactive) primary antibodies can then be utilized to bind protein (s) of interest with high affinity, followed by washing steps. Secondary antibodies tagged with fluorophores or stains are then introduced to target primary antibodies, which can be imaged with fluorescent microscopy or bright field microscopy techniques to allow for regional and cellular localization of protein expression. Enzyme linked immunosorbent assays (ELISAs) are typically utilized for detecting proteins in a liquid sample using a similar approach with specific antibodies. In principle, premanufactured antibody coated ELISA plates bind individual proteins from a fluid sample, then an enzyme-linked secondary antibody is added for detection after processing with substrates to allow for measurements based on enzyme-converted colors or fluorescent dye. Signal intensity measurement is therefore directly proportional to presence and quantity of antigen, which allows for a degree of protein quantification.

2.2 8-Hydroxydeoxyguanosine (8-OHdG)

Reactive oxygen species (ROS) are chemical species derived from molecular oxygen with one unpaired electron. Excessive production of ROS can result in oxidation of mitochondrial DNA and nuclear DNA during ischemic conditions [2]. ROS can oxidize nucleotide bases in DNA, forming oxidized species including 7,8-dihydro-8-oxoguanine (8-oxoguanine), which can cause transversion mutations [3, 4]. Cells have rapid repair mechanisms to maintain their genomic integrity, and newly formed 8-oxoguanine in nuclei and in mitochondria is recognized by OGG1 gene-encoded 8-oxoguanine DNA glycosylase for targeted removal [5]. The relatively stable "end-product," 8-hydroxydeoxyguanosine (8-OHdG, Fig. 2b), is transported out of the cell and renally cleared, allowing for DNA repair [6, 7]. Levels of 8-OHdG can therefore be measured via fluorescent automated cell sorting (FACS) of dissociated cells or ELISA-based analysis of blood or urine. Moreover, 8-OHdG can be used as a proxy for localized cellular production of ROS in fixed tissue via fluorescent microscopy (Fig. 2c), as 8-OHdG is considered a stable marker for oxidative stress [8].

2.3 Nuclear Factor Erythroid 2-Related Factor 2 (Nrf2)

Nrf2 is member of the leucine zipper protein family characterized as one of the major cellular defense mechanisms against oxidative stress [9]. Under physiological conditions, Nrf2 is bound to Keap1 (Kelch ECH (erythroid cell-derived protein with CNC (Drosophila transcription factor Cap 'n' collar) homology) associating protein 1), and to cytoskeletal elements within the cell [10]. With insignificant oxidative stress levels, the Nrf2-Keap1 complex is targeted for degradation by the proteasome through ubiquitination (Fig. 3) [11]. Under oxidative stress, the Nrf2-keap1 interaction is disrupted and Nrf2 translocates to the nucleus where it forms heterodimers with the Maf family proteins in the nucleus [12]. This new complex binds to antioxidant response elements (ARE) on DNA, promoting the transcriptional upregulation of antioxidant and detoxifying phase 2 enzymes, which include

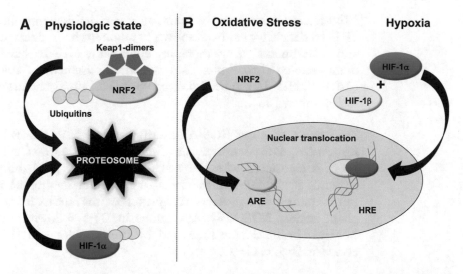

Fig. 3 Cytoprotective nuclear factor erythroid 2-related factor 2 (Nrf2) and hypoxia-inducible factor-1 (HIF-1) subunit-α. (**a**) Nrf2 and HIF-1α are transcription factors poised to respond to cellular stressors such as excessive oxidant production or decreased oxygen availability, respectively. In the resting state, Nrf2 is bound to Keap1 dimers and ubiquitins, which co-direct the complex towards proteosomal degradation. In parallel, HIF-1α subunit alone is ubiquitin-bound, resulting in targeted proteosomal degradation. (**b**) In response to oxidative stress, Keap1 and ubiquitin do not bind Nrf2, allowing for nuclear translocation and binding to DNA antioxidant response element (ARE) to drive transcription of cytoprotective genes. Similarly, HIF-1α in the presence of hypoxia does not bind ubiquitins, instead binding cofactor HIF-1β to allow translocation to the nucleus, resulting in cytoprotective transcriptional activity

protective proteins to respond to oxidative stress [13, 14]. Protein kinase C (PKC), imparts both a direct influence on Keap1 and indirectly via Nrf2 by phosphorylation events [15]. Levels of Nrf2 can be detected via ELISA, and also stained via IHC, allowing cellular and regional localization of nuclear Nrf2 levels after injury [16].

2.4 HIF-1-Alpha

A central cellular hypoxia-sensing and redox-regulating mechanism is Hypoxia-inducible factor 1 (HIF-1) [17], which is a heterodimer complex composed of two subunits, a 120 kDa α subunit and a 91–94 kDa β subunit [18]. HIF-1α is constitutively translated but subsequently destroyed by the ubiquitin-proteasome pathway (Fig. 3). In response to hypoxia, HIF-1α ubiquitination is significantly decreased, allowing cellular levels of HIF-1-α to accumulate [19]. Oxidative stress also contributes to HIF-1α stabilization under hypoxic conditions [20]. Stabilized HIF-1-α binds to other subunits and regulates multiple hypoxia-induced target genes, such as erythropoietin, glucose transporters, and vascular endothelial growth factor (VEGF) to counter both cellular and systemic hypoxia and oxidative stress [21–23]. However, hypoxia also serves as a negative feedback loop for HIF-1α as it is negatively regulated by prolyl-4-hydroxylases [24]. HIF-1α contributes to mitigating

ROS-related damage in the brain, after cerebral ischemic insult, as HIF-1α deficiency has been shown to augment brain damage [25], and experimental drug therapies employing HIF-1α have been demonstrated to increase cerebral ischemic tolerance [26, 27]. HIF-1α can be visualized with IHC after ischemia–reperfusion injury [28].

2.5 DJ-1

DJ-1 is a ubiquitous ROS-countering protein [29] that is H_2O_2-responsive, redox-sensitive, and functions as an antioxidant [30]. DJ-1 is associated with the cell death cascade: [31] during an ischemic insult DJ-1 translocates to the inner membrane of the mitochondria and protects complex I of the electron transport chain against ROS [29]. Alterations in DJ-1 following cerebral ischemic insult can be measured by immunoblot and [32] and also visualized via IHC.

2.6 Pimonidazole

Pimonidazole hydrochloride (Hypoxyprobe™) is a commercially available hypoxia indicator which penetrates the blood-brain-barrier to irreversibly bind thiol-containing proteins in hypoxic tissues. Due to pharmacokinetics in mice, it can be used for acute ischemia models such as transient middle cerebral artery occlusion (MCAO), when it is intravenously injected prior injury [33]. With a plasma half-life of only ~20 min, pimonidazole can be reliably assessed after a few half-lives after injection without notable background from tissue hypoxia caused by sample collection. Our laboratory has developed a fluorescent IHC approach to independently assess ROS production in ischemic core versus penumbra after MCAO in mice using the ROS indicator 8-OHdG complexed with the hypoxia indicator pimonidazole (Fig. 4). A protocol for pimonidazole hydrochloride 8-OHdG complexing to localize ischemic regions and ROS production after MCAO is below:

2.6.1 Procedure: Pimonidazole Hydrochloride/8-OHdG Complexing

1. Hypoxyprobe is resuspended at 30 mg/ml in 0.9% sterile saline.

2. Mice are weighed, anesthetized, and intravenously injected with pimonidazole hydrochloride from Hypoxyprobe™ (cat# HP1-1000) at 60 mg/kg body weight, 5 min prior to MCAO.

3. Pimonidazole is circulated in vivo completely for 5 min, followed by MCAO of 60 min, and a follow-up period of 30 min before the mice are euthanized according to IACUC approved protocols. Brains are then perfusion-fixed with 4% paraformaldehyde (PFA).

4. Saline flushed and PFA perfusion fixed brains are removed and set for further fixing in PFA for 48 h in 4 °C.

5. Brains are sectioned using a vibrotome/microtome to 50 μm thickness.

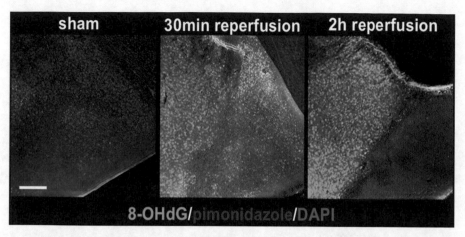

Fig. 4 Post-MCAO reactive oxygen species (ROS) generation in ischemic core and penumbra. Following an IV injection of the hypoxia indicator pimonidazole (red) animals were subjected to sham or 1-h middle cerebral artery occlusion (MCAO) and sacrificed at 30 min or 2 hours reperfusion. Fixed brains were co-labeled with the ROS indicator 8-hydroxy-2′-deoxyguanosine (8-OHdG, green) and the nuclear dye 4′,6-diamidino-2-pheny-lindole (DAPI). Substantial elevation in ROS generation is sustained in penumbra at both 30 min and 2 h reperfusion. Co-labeling of pimonidazole and 8-OHdG in ischemic core is evident at 30 min but not at 2 h. Scale bar = 100 μm

6. Sections are processed for fluorescent immunohistochemistry (IHC):

(a) Wash and permeabilize sections in phosphate-buffered saline (PBS-t) twice for 5 min.

(b) Block in IHC blocking buffer for 1 h at room temperature, or overnight at 4 °C.

(c) Block for endogenous auto-fluorescent molecules with H_2O_2 for 30 min to 2 h.

(d) Wash with PBS-t three times 5 min each.

(e) Add primary antibodies 1:200, 4 °C, overnight, gentle agitation (Pimonidazole mouse-AB, 8-OHdG goat-AB).

(f) Wash with PBS-t three times 5 min each.

(g) Add secondary antibodies 1:200, and DAPI 1:5000 4 °C, overnight, gentle agitation (donkey anti-goat, donkey anti-mouse).

(h) Wash three times with PBS-t, mount on glass slides.

(i) Image on upright microscope equipped for fluorescent capture at 350 nm (blue), 488 nm (green) and 594 nm (red).

Reagents:

1. Hypoxyprobe Kit (100 mg pimonidazole–HCl plus 1.0 ml of 4.3.11.3 mouse MAb) (Hypoxyprobe™, HP1-100Kit).

2. PBS (Sigma-Aldrich, catalog number: P-4417).

3. Goat Anti-8-OHdG antibody (Abcam, ab10802).

4. Donkey Anti-Goat IgG H&L (Alexa Fluor® 488; Abcam, ab150129).

5. Donkey Anti-Mouse IgG H&L (Alexa Fluor® 594; Abcam ab150108).

6. Triton X-100 (Sigma-Aldrich, T8787).

7. ProLong™ Glass Antifade Mountant (Invitrogen, P36980).

8. DAPI (4′,6-Diamidino-2-Phenylindole, Dihydrochloride) (Invitrogen, D1306).

9. Normal goat serum (Invitrogen, 10000C).

10. Slides (Thermo Fisher Scientific, catalog number: 12-550-15).

11. Micro cover glasses (VWR International, catalog number: 48393-106).

Solutions:

1. IHC blocking buffer (5% Normal Goat Serum in PBS) 500 μl of NGS in 10 ml PBS.

2. PBS-t (0.1% Triton X-100 in PBS) 100 μl of Triton X-100 in 100 ml of PBS.

3. 0.9% saline: Mix 0.9 g of sodium chloride (NaCl) in 100 ml milli Q H_2O.

2.7 Oxidized Phospholipids

Not all reactive oxidant species can oxidize lipids, but it is known that free radicals play a role in lipid peroxidation in ischemia–reperfusion injury [34]. Oxidized lipids are, however, not ideal for direct measurement of antioxidant levels or redox activity. Levels of many of the oxidized lipid products are affected by age and other factors and are therefore generally not regarded as suitable markers to measure redox signaling [35].

3 In Vivo Approaches for Measuring Tissue Redox State

3.1 Intravital Imaging

Cerebral perfusion, cellular function and redox signaling can be directly assessed by intravital visualization. The basic approach is to create an imaging window through skull and dura to visualize tissue using fluorescent indicators while maintaining the research animal under anesthesia. Intravital microscopy techniques have been used to study the vasculature of cerebral cortex (Fig. 5a), but also protein markers [36] and by-products of redox signaling [37]. Direct tissue visualization via fluorescent antibodies, dyes, or auto-fluorescent molecules [38] enable assessment of the regional, and cell-specific response of intact, living tissue to injury, for example Rhodamine 6G [39] for visualizing adherent and

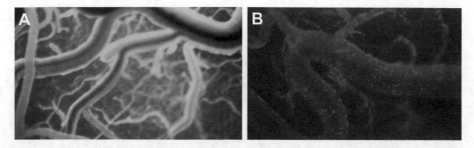

Fig. 5 (**a**) Intravenous fluorescein (green) injection allowing visualization of cerebral vasculature. Cardiac rhythm causes subtle brain movement (and blurred images) which should be considered. (**b**) Rhodamine 6G (red) labeled adherent and rolling leukocytes visualized in cerebral arteries. Consideration should be taken with intravital imaging for phototoxicity, photobleaching and any requirement for serial, invasive cranial surgeries

rolling leukocytes (Fig. 5b). *Important considerations*: real-time in vivo microscopic imaging of central nervous system at single-cell resolution represents a challenge in studying complex biological pathways in living systems. Serial measurements are limited by the requirement for an imaging window to be created in the skull [40], phototoxicity [41], and cerebral autoregulatory compensation mechanisms. Notably, intravital imaging of the brain can be limited by subtle tissue movement from the cardiac and respiratory cycles which must be considered when interpreting data.

3.2 NAD$^+$/NADH

Reduced NADH is a natural intracellular fluorophore. A major proportion of NAD$^+$ is sequestered in mitochondria behind impermeable membranes to NAD$^+$ and NADH. The NAD$^+$/NADH ratio under physiological conditions in mitochondria is under tight control [42]. Ischemia shifts the NAD$^+$/NADH redox state shifts towards NADH, contrary to the high NAD$^+$/NADH ratio during normal homeostasis. Light at a wavelength of 330 nm results in a fluorescence emission peak at 440–460 nm for NADH but NAD$^+$ does not absorb light above 300 nm [43, 44] allowing for fluorescent assessment of the NAD$^+$/NADH ratio. Fluoroscopic assessment of the NAD$^+$/NADH ratio as a measure of redox state can therefore be achieved intravitally, in situ, or in vitro.

4 In Vitro Approaches to Assess Mitochondrial Function and Redox State

4.1 High Resolution Respirometry

ATP synthesis occurs on the inner mitochondrial membrane as protons flow down the electrochemical gradient of ETC and the energy released rotates ATP-synthase comprising oxidative phosphorylation (OXPHOS, Fig. 6). Failure in any of the first four complexes of the ETC disrupts electron transport flow causing respiratory malfunction. Cerebral ischemia results in mitochondrial

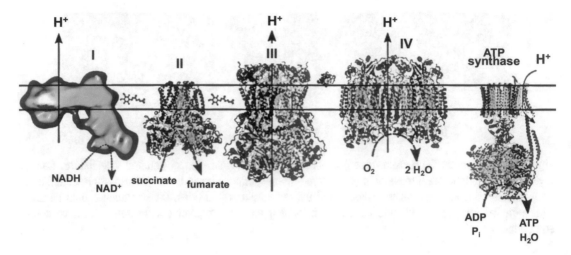

Fig. 6 Mitochondrial electron transport chain (ETC). NADH and FADH$_2$ produced in the TCA carry electrons to ETC complex I and II, respectively. Complexes I, III and IV translate the energy to electromotive force driving protons across the inner mitochondrial membrane to generate a proton-motive gradient. The proton-motive force then drives the chemiosmotic enzyme ATP synthase to phosphorylate ADP to ATP as the fundamental energy source for all eukaryotic cells

dysfunction leading to neuronal cell death, and approaches that can more accurately define the mechanisms leading to mitochondrial dysfunction can therefore be used to develop mitochondrial-targeted post-stroke therapies. The Oroboros™ O2k consists of two temperature controlled 2 ml closed chambers utilizing polarographic oxygen sensors for real time oxygen concentration measurements [45]. Both cell cultures and dissociated cells from tissue samples can be analyzed. Cells are typically permeabilized to allow exogenous ADP, substrates, and complex inhibitors to enter the cells. A substrate-uncoupler-inhibitor-titration (SUIT) protocol is then employed to assess non-phosphorylating leak-respiration, oxidative phosphorylation capacity, and electron transport chain (ETC) capacity (Fig. 7). ADP and succinate are saturated to study the effect of electron input through complexes I and II, quantifying OXPHOS capacity. Complex uncouplers (e.g., p-trifluoromethoxy carbonyl cyanide phenyl hydrazone (FCCP) or carbonyl cyanide m-chloro phenyl hydrazine (CCCP)) are used to determine maximal electron transport chain capacity. Non-respiratory O$_2$ consumption is defined by inhibiting complexes I and III with rotenone and antimycin A, blocking NADH dehydrogenase (complex I) and preventing NADH, and succinate oxidation at complex III, which results in complete inhibition of respiration. *Important considerations*: oxygen consumption rates must be normalized to tissue volume or cell count (assessed by cytometry).

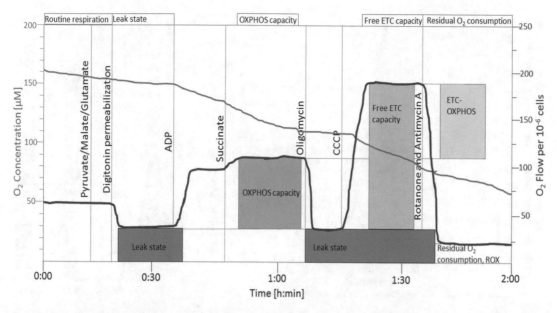

Fig. 7 Illustration of respiratory states measured by Oroboros O2k respirometer. Before any additional substrates, routine respiration is recorded. After saturating complex I with substrates and permeabilizing the cells, the non-phosporylating resting state (leak state without adenylates) is induced. Adding ADP and succinate, the stimulated oxidative phosphorylation capacity is measured (OXPHOS, green box). Inhibition of the phosphorylation system at ATP synthase with oligomycin produces another LEAK state (leak state with adenylates). During leak states (pink boxes), the oxygen flux is minimized. Stimulating with established protonophores carbonyl cyanide m-chlorophenyl hydrazone (CCCP) or carbonyl cyanide-4-(trifluoromethoxy) phenylhydrazone (FCCP) a reference state of free electron transport chain capacity can be measured (orange and blue boxes)

4.2 Respirometry/ Fluorometry Complexing

Traditional methods to assess OXPHOS have been technically limited in resolution. Direct assessment of O_2 consumption and ATP concentration have historically required separate, independent measurements, introducing inter-sample variability, while assessment of O_2 consumption at a known level of ADP concentration as substrate for OXPHOS is limited in dynamic range and does not account for other sources of ADP rephosphorylation (e.g., from phosphocreatine/creatine kinase). However, recent advances in fluorescent imaging techniques have dramatically improved the ability to simultaneously observe intracellular biochemical processes in real time by complexing measures of O_2 consumption with other biological processes. One advantage of the O2k for stroke research is that dissociated cells from ipsilateral and contralateral cortex can be used to assess and compare the disruption in mitochondrial function that occurs after MCAO (Fig. 8). Another advantage is the ability to complex ROS measurements by simultaneous fluorometry (Fig. 8). Hydroethidine (HEt), or the mitochondria-targeted analog Mito-SOX can be used to detect real time intracellular superoxide formation. Despite a disadvantage in detecting intracellular superoxide formation in tissue samples via

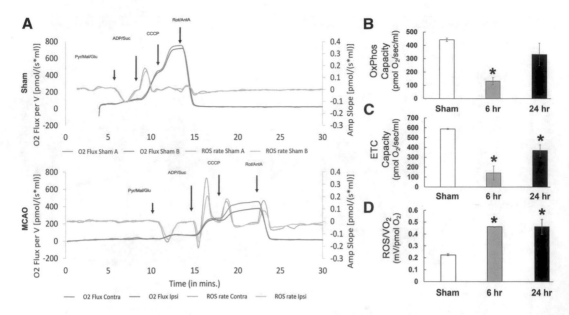

Fig. 8 High resolution respirometry/fluorometry complexing to assess mitochondrial function and reactive oxygen species (ROS) generation after middle cerebral artery occlusion (MCAO). (**a**) Tracings of rates of O_2 consumption (O_2 flux/VO_2, pmol O_2/s/ml) and rates of ROS production (hydroethidine fluorescence, V/s/ml) in dissociated cells from cortex of sham animals (above) and ipsilateral and contralateral cortex 24 h post-MCAO (below) during substrate-uncoupler-inhibitor-titration: complex I substrate (Pyr/Mal/Glu) to determine proton leak; complex II substrate (ADP/Suc) to determine oxidative phosphorylation (OXPHOS) capacity, complex uncoupler (CCCP) to determine maximal electron transport chain (ETC) capacity; complex I and III inhibitors (Rot/AntA) to determine non-respiratory O_2 consumption. (**b**) Quantification of OXPHOS capacity. (**c**) Quantification of maximal ETC flux. (**d**) Quantification of ROS production rate per unit O_2 consumption. *Asterisk* significant ($p < 0.05$) difference vs. sham. *Pyr* pyruvate, *Mal* malate, *Glu* glutamate, *CCCP* carbonyl cyanide m-chlorophenyl hydrazine, *Rot* rotenone, *AntA* antimycin A

fluorescence microscopy alone, Mito-SOX is the most commonly used ROS-detecting probe for quantitative intracellular measurements of oxidants [46]. Another system that permits complexing of respirometry and fluorometry is the Seahorse FX™ Extracellular Flux analyzer. Compared to the operator driven 2-chamber O2k, the automated Seahorse XF Extracellular Flux Analyzer can perform high-throughput assays with 96-well plates. Both systems can assess real-time mitochondrial respiratory data in isolated mitochondria and cultured cells. *Important considerations:* the limitations of the Seahorse XF are the high cost of optimizing, performing the assays, and the cost of fluorescent plates, as well as the potential interference with injectable compounds and sensor fluorescence [47, 48].

A more recent complexing approach to assess OXPHOS efficiency has been developed [49] in permeabilized skeletal muscle fibers utilizing an enzymatically coupled approach. Enzymatically coupling ATP as substrate for $NADP^+$, and simultaneously assessing reduced $NADP^+$ (NADPH) bioluminescence and molecular

O_2 consumption results in a measure of OXPHOS affinity for ADP during physiologic steady-state conditions. This approach, in which ADP and ATP concentrations are clamped, provide a high-resolution measure of OXPHOS efficiency by: (1) reducing inter-sample variability with simultaneous measurements of O_2 consumption and NADPH; and (2) increasing the fidelity of measurement with repeated measures within a single experiment. This approach could also theoretically be applied to primary brain cell cultures.

4.3 Bioluminescence Assessment of ATP

An alternative indirect technique for characterizing mitochondria function is the assessment of ATP production using bioluminescence. Luciferase catalyzes a two-step reaction that oxidizes luciferin in an ATP-dependent reaction that generates a light signal proportional to ATP concentration [50, 51]. Mitochondria isolated from homogenized tissue respire and generate ATP when provided with the appropriate substrates. When combined with a recombinant-luciferase based commercially available assay this reaction enables measurement of rates of ATP synthesis using a microplate luminometer/plate reader [51]. ATP synthesis rates are calculated from the repeated measurements taken, where the slope represents rate and concentration is represented as a function of time. *Important considerations*: it is important to normalize data to mitochondrial content or variations in inherent mitochondrial properties. Data is most commonly normalized to mitochondrial protein content, mitochondrial DNA copy number, or citrate synthase activity [51]. Improper homogenization can damage mitochondria and affect capacity for respiration.

5 Live-Cell Imaging

Mitochondria are highly dynamic organelles with variable morphologies. Moreover, mitochondria undergo fission, fusion and are actively transported throughout the cell. Thus, the visualization of mitochondria and related-processes is a valuable source of information.

5.1 Mitochondrial Membrane Potential

Fluorescent microscopy is a powerful tool for the assessment of mitochondrial membrane potential, and thus assessing the effects of variables reflective of toxicity, disease and injury that affect mitochondrial function. Dyes and reagents such as Rhodamine 123 and tetramethylrhodamine methyl ester (TMRE) are used for the visualization of mitochondria and are useful for assaying mitochondrial membrane potential (Fig. 9). They are reversible probes that label mitochondria with an intensity that varies depending on mitochondrial membrane potential. Other labels such as MitoTracker™ incorporate and accumulate in active mitochondria

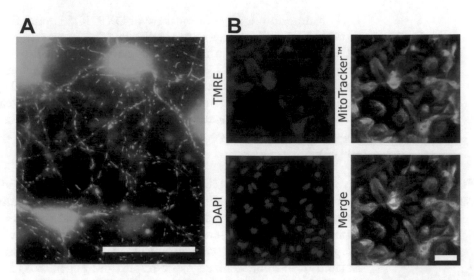

Fig. 9 Live cell fluorescent imaging of mitochondrial function. (**a**) Primary cortical neuronal cultures stained with MitoTracker Green™. (**b**) Primary cortical astrocyte cultures stained with the mitochondrial membrane potential indicator tetramethylrhodamine ethyl ester (TMRE, red). The nuclear indicator DAPI (blue) provides for assessment of cell count to standardize the TMRE signal. MitoTracker Green™ similarly allows for standardization of mitochondrial density. Merged images can also be used to assess and compare cell-specific mitochondrial membrane potential. Bar = 15 μm

secondary to the transmembrane membrane potential. Potential insensitive dyes such as CellLight™ can be combined with potential-dependent dyes for dual emission fluorescence measurements that help investigators account for variability in morphology. Dual emission probes such as JC-1™ have different fluorescence emission dependent on membrane potential within individual mitochondria. At low membrane potential, JC-1 exists as a monomer that fluoresces in the green emission wavelength. When potential increases, JC-1 forms red-fluorescent "J-aggregates" and the ratio between green to red JC-1 can be used to measure membrane potential independent of mitochondrial density, length, and morphology, factors that might affect the aforementioned single emission probes [52]. Alternatively, single emission probes can be used together. One example of such application is co-staining with tetramethylrhodamine ethyl ester (TMRE) and MitoTracker™ dyes. This combination allows the assessment of mitochondrial potential with standardization by mitochondrial density. *Important considerations:* Intensity of fluorescence with single-wavelength dyes may be affected by dye concentration, photobleaching, and movement. When utilizing two or more single emission probes serial staining (i.e., first staining with the first dye, washing out the excess, then staining with the next dye) may be necessary to prevent competitive binding or other off-target effects affecting fluorescence signal intensity [53].

5.2 Mitochondrial Calcium

Intracellular calcium levels are integral to cell functioning and mitochondria play a central role in buffering calcium to maintain homeostasis. Mitochondrial buffering capacity can be measured with the rhodamine derivative RHOD-2™, a positively charged calcium indicator capable of crossing the cell membrane and labeling mitochondria in living cells. When using RHOD-2- mitochondria are typically only visible after calcium uptake. This feature, when coupled with a potential-independent mitochondrial marker, can provide valuable information regarding the mitochondrial location and its spatial relationship to aspects of intracellular calcium release and sequestration. *Important considerations:* care must be taken if co-labeling with other mitochondrial indicators to utilize a serial-incubation approach in order to minimize off potential target effects from competitive binding.

6   Conclusions

There remains a critical need for the development of novel alternative treatments for stroke. Advances in live-cell imaging techniques, real-time assessment of OXPHOS, and IHC complexing provide a novel platform for cell- and organ-specific mitochondrial-targeted approaches. Layering these techniques with growing innovations in real-time imaging, such as two-photon intravital microscopy, magnetic resonance spectroscopy or optical O_2 measurement via phosphorescence decay will serve to enhance the fidelity of measurement and advance our understanding of the fundamental mechanics of mitochondrial bioenergetics. Multimodal complexing is one approach to provide parallel outcome measures for preclinical testing of novel drugs, a growing necessity in this age of translational failure for novel stroke therapies. Advancing the development and application of these techniques to other types of living cells will further help to identify universal, cell- and organ-specific pathways regulating mitochondrial function and redox state, thereby clarifying the role mitochondria play in both normal physiological function and in the evolution of injury following stroke.

Acknowledgment

Funding: Supported by Finnish Cultural Foundation Grant #00171200 to O.A. and American Heart Association Grant 14FTF19970029 to C.M.S.

References

1. Madamanchi NR, Runge MS (2007) Mitochondrial dysfunction in atherosclerosis. Circ Res 100(4):460–473

2. Turrens JF (2003) Mitochondrial formation of reactive oxygen species. J Physiol 552 (Pt 2):335–344. https://doi.org/10.1113/jphysiol.2003.049478

3. Greco NJ, Sinkeldam RW, Tor Y (2009) An emissive C analog distinguishes between G, 8-oxoG, and T. Org Lett 11(5):1115–1118. https://doi.org/10.1021/ol802656n

4. Kubo N, Morita M, Nakashima Y et al (2014) Oxidative DNA damage in human esophageal cancer: clinicopathological analysis of 8-hydroxydeoxyguanosine and its repair enzyme. Dis Esophagus 27(3):285–293. https://doi.org/10.1111/dote.12107

5. Iida T, Furuta A, Nakabeppu Y et al (2004) Defense mechanism to oxidative DNA damage in glial cells. Neuropathology 24(2):125–130

6. Ba X, Aguilera-Aguirre L, Rashid QT et al (2014) The role of 8-oxoguanine DNA glycosylase-1 in inflammation. Int J Mol Sci 15(9):16975–16997. https://doi.org/10.3390/ijms150916975

7. Mazurek A, Berardini M, Fishel R (2002) Activation of human MutS homologs by 8-oxo-guanine DNA damage. J Biol Chem 277 (10):8260–8266. https://doi.org/10.1074/jbc.M111269200

8. Chiou CC, Chang PY, Chan EC et al (2003) Urinary 8-hydroxydeoxyguanosine and its analogs as DNA marker of oxidative stress: development of an ELISA and measurement in both bladder and prostate cancers. Clin Chim Acta 334(1–2):87–94. S0009898103001918 [pii]

9. Liu Y, Zhang L, Liang J (2015) Activation of the Nrf2 defense pathway contributes to neuroprotective effects of phloretin on oxidative stress injury after cerebral ischemia/reperfusion in rats. J Neurol Sci 351(1–2):88–92. S0022-510X(15)00125-2 [pii]

10. Itoh K, Wakabayashi N, Katoh Y et al (1999) Keap1 represses nuclear activation of antioxidant responsive elements by Nrf2 through binding to the amino-terminal Neh2 domain. Genes Dev 13(1):76–86

11. Nguyen T, Sherratt PJ, Huang HC et al (2003) Increased protein stability as a mechanism that enhances Nrf2-mediated transcriptional activation of the antioxidant response element. Degradation of Nrf2 by the 26 S proteasome. J Biol Chem 278(7):4536–4541. https://doi.org/10.1074/jbc.M207293200

12. Toki T, Itoh J, Kitazawa J et al (1997) Human small Maf proteins form heterodimers with CNC family transcription factors and recognize the NF-E2 motif. Oncogene 14 (16):1901–1910. https://doi.org/10.1038/sj.onc.1201024

13. Itoh K, Wakabayashi N, Katoh Y et al (2003) Keap1 regulates both cytoplasmic-nuclear shuttling and degradation of Nrf2 in response to electrophiles. Genes Cells 8(4):379–391. 640 [pii]

14. Rushmore TH, Morton MR, Pickett CB (1991) The antioxidant responsive element. Activation by oxidative stress and identification of the DNA consensus sequence required for functional activity. J Biol Chem 266 (18):11632–11639

15. Huang HC, Nguyen T, Pickett CB (2000) Regulation of the antioxidant response element by protein kinase C-mediated phosphorylation of NF-E2-related factor 2. Proc Natl Acad Sci U S A 97(23):12475–12480. https://doi.org/10.1073/pnas.220418997

16. Srivastava S, Alfieri A, Siow RC et al (2013) Temporal and spatial distribution of Nrf2 in rat brain following stroke: quantification of nuclear to cytoplasmic Nrf2 content using a novel immunohistochemical technique. J Physiol 591(14):3525–3538. https://doi.org/10.1113/jphysiol.2013.257964

17. Semenza GL (2012) Hypoxia-inducible factors in physiology and medicine. Cell 148 (3):399–408. https://doi.org/10.1016/j.cell.2012.01.021

18. Wang GL, Semenza GL (1995) Purification and characterization of hypoxia-inducible factor 1. J Biol Chem 270(3):1230–1237

19. Kallio PJ, Wilson WJ, O'Brien S et al (1999) Regulation of the hypoxia-inducible transcription factor 1alpha by the ubiquitin-proteasome pathway. J Biol Chem 274(10):6519–6525

20. Brunelle JK, Bell EL, Quesada NM et al (2005) Oxygen sensing requires mitochondrial ROS but not oxidative phosphorylation. Cell Metab 1(6):409–414. S1550-4131(05)00140-3 [pii]

21. Semenza GL, Roth PH, Fang HM et al (1994) Transcriptional regulation of genes encoding glycolytic enzymes by hypoxia-inducible factor 1. J Biol Chem 269(38):23757–23763

22. Ebert BL, Firth JD, Ratcliffe PJ (1995) Hypoxia and mitochondrial inhibitors regulate expression of glucose transporter-1 via distinct Cis-acting sequences. J Biol Chem 270 (49):29083–29089

23. Bernaudin M, Tang Y, Reilly M et al (2002) Brain genomic response following hypoxia and re-oxygenation in the neonatal rat.

Identification of genes that might contribute to hypoxia-induced ischemic tolerance. J Biol Chem 277(42):39728–39738. https://doi.org/10.1074/jbc.M204619200

24. Gunter J, Ruiz-Serrano A, Pickel C et al (2017) The functional interplay between the HIF pathway and the ubiquitin system – more than a one-way road. Exp Cell Res 356(2):152–159. S0014-4827(17)30131-3 [pii]

25. Sheldon RA, Osredkar D, Lee CL et al (2009) HIF-1 alpha-deficient mice have increased brain injury after neonatal hypoxia-ischemia. Dev Neurosci 31(5):452–458. https://doi.org/10.1159/000232563

26. Li L, Yin X, Ma N et al (2014) Desferrioxamine regulates HIF-1 alpha expression in neonatal rat brain after hypoxia-ischemia. Am J Transl Res 6(4):377–383

27. Davis CK, Nampoothiri SS, Rajanikant GK (2018) Folic acid exerts post-ischemic neuroprotection in vitro through HIF-1alpha stabilization. Mol Neurobiol. https://doi.org/10.1007/s12035-018-0982-3

28. Yang ML, Tao T, Xu J et al (2017) Antiapoptotic effect of gene therapy with recombinant adenovirus vector containing hypoxia-inducible factor-1alpha after cerebral ischemia and reperfusion in rats. Chin Med J 130 (14):1700–1706. https://doi.org/10.4103/0366-6999.209909

29. Pantcheva P, Elias M, Duncan K et al (2014) The role of DJ-1 in the oxidative stress cell death cascade after stroke. Neural Regen Res 9(15):1430–1433. https://doi.org/10.4103/1673-5374.139458

30. Taira T, Saito Y, Niki T et al (2004) DJ-1 has a role in antioxidative stress to prevent cell death. EMBO Rep 5(2):213–218. https://doi.org/10.1038/sj.embor.7400074

31. Canet-Aviles RM, Wilson MA, Miller DW et al (2004) The Parkinson's disease protein DJ-1 is neuroprotective due to cysteine-sulfinic acid-driven mitochondrial localization. Proc Natl Acad Sci U S A 101(24):9103–9108. https://doi.org/10.1073/pnas.0402959101

32. Yang RX, Lei J, Wang BD et al (2017) Pretreatment with sodium phenylbutyrate alleviates cerebral ischemia/reperfusion injury by upregulating DJ-1 protein. Front Neurol 8(256). https://doi.org/10.3389/fneur.2017.00256

33. Aguilera KY, Brekken RA (2014) Hypoxia studies with pimonidazole in vivo. Bio Protoc 4(19):e1254 [pii]

34. Halliwell B, Whiteman M (2004) Measuring reactive species and oxidative damage in vivo and in cell culture: how should you do it and what do the results mean? Br J Pharmacol 142

(2):231–255. https://doi.org/10.1038/sj.bjp.0705776

35. Lee CY, Seet RC, Huang SH et al (2009) Different patterns of oxidized lipid products in plasma and urine of dengue fever, stroke, and Parkinson's disease patients: cautions in the use of biomarkers of oxidative stress. Antioxid Redox Signal 11(3):407–420. https://doi.org/10.1089/ARS.2008.2179

36. de la Rosa X, Santalucia T, Fortin PY et al (2013) In vivo imaging of induction of heat-shock protein-70 gene expression with fluorescence reflectance imaging and intravital confocal microscopy following brain ischaemia in reporter mice. Eur J Nucl Med Mol Imaging 40(3):426–438. https://doi.org/10.1007/s00259-012-2277-7

37. Yannopoulos FS, Arvola O, Haapanen H et al (2014) Leg ischaemia before circulatory arrest alters brain leucocyte count and respiratory chain redox state. Interact Cardiovasc Thorac Surg 18(3):272–277. https://doi.org/10.1093/icvts/ivt415

38. Ricard C, Arroyo ED, He CX et al (2018) Two-photon probes for in vivo multicolor microscopy of the structure and signals of brain cells. Brain Struct Funct. https://doi.org/10.1007/s00429-018-1678-1

39. Respicio NC, Heitz JR (1981) Comparative toxicity of rhodamine B and rhodamine 6G to the house fly (Musca domestica l.). Bull Environ Contam Toxicol 27(2):274–281

40. Shih AY, Driscoll JD, Drew PJ et al (2012) Two-photon microscopy as a tool to study blood flow and neurovascular coupling in the rodent brain. J Cereb Blood Flow Metab 32 (7):1277–1309. https://doi.org/10.1038/jcbfm.2011.196

41. Klinger A, Krapf L, Orzekowsky-Schroeder R et al (2015) Intravital autofluorescence 2-photon microscopy of murine intestinal mucosa with ultra-broadband femtosecond laser pulse excitation: image quality, photodamage, and inflammation. J Biomed Opt 20 (11):116001. https://doi.org/10.1117/1.JBO.20.11.116001

42. Ying W (2006) NAD+ and NADH in cellular functions and cell death. Front Biosci 11:3129–3148. [2038 pii]

43. Chance B, Schoener B, Oshino R et al (1979) Oxidation-reduction ratio studies of mitochondria in freeze-trapped samples. NADH and flavoprotein fluorescence signals. J Biol Chem 254(11):4764–4771

44. Mayevsky A, Rogatsky GG (2007) Mitochondrial function in vivo evaluated by NADH fluorescence: from animal models to human

studies. Am J Physiol Cell Physiol 292(2): C615–C640. 00249.2006 [pii]

45. Djafarzadeh S, Jakob SM (2017) High-resolution respirometry to assess mitochondrial function in permeabilized and intact cells. J Vis Exp 120. https://doi.org/10.3791/54985

46. Zielonka J, Kalyanaraman B (2010) Hydro-ethidine- and MitoSOX-derived red fluorescence is not a reliable indicator of intracellular superoxide formation: another inconvenient truth. Free Radic Biol Med 48(8):983–1001. https://doi.org/10.1016/j.freeradbiomed. 2010.01.028

47. Sauerbeck A, Pandya J, Singh I et al (2011) Analysis of regional brain mitochondrial bioenergetics and susceptibility to mitochondrial inhibition utilizing a microplate based system. J Neurosci Methods 198(1):36–43. https://doi.org/10.1016/j.jneumeth.2011.03.007

48. Horan MP, Pichaud N, Ballard JW (2012) Review: quantifying mitochondrial dysfunction in complex diseases of aging. J Gerontol A Biol Sci Med Sci 67(10):1022–1035. https://doi.org/10.1093/gerona/glr263

49. Gouspillou G, Rouland R, Calmettes G et al (2011) Accurate determination of the oxidative phosphorylation affinity for ADP in isolated mitochondria. PLoS One 6(6):e20709. https://doi.org/10.1371/journal.pone. 0020709

50. DeLuca M, McElroy WD (1974) Kinetics of the firefly luciferase catalyzed reactions. Biochemistry 13(5):921–925

51. Lanza IR, Nair KS (2009) Functional assessment of isolated mitochondria in vitro. Methods Enzymol 457:349–372. https://doi.org/10.1016/S0076-6879(09)05020-4

52. Distelmaier F, Koopman WJ, Testa ER et al (2008) Life cell quantification of mitochondrial membrane potential at the single organelle level. Cytometry A 73(2):129–138. https://doi.org/10.1002/cyto.a.20503

53. Mitra K, Lippincott-Schwartz J (2010) Analysis of mitochondrial dynamics and functions using imaging approaches. Curr Protoc Cell Biol 4:25.1–25.21. 10.1002/0471143030. cb0425s46

Chapter 6

Blood Biomarkers for Stroke Differentiation

Deepti Vibha and Shubham Misra

Abstract

Acute stroke is a neurological emergency. Time sensitive treatment decisions depend on the correct diagnosis of stroke and its subtype: ischemic or hemorrhagic. An early diagnosis of ischemic stroke is required for prompt decision making and administration of thrombolytic therapy within the time frame. Non-contrast computed tomography (CT) scan is currently used for the diagnosis of stroke. However, CT scan facility is not widely available in resource limited settings. Therefore, there is an urgent need to identify blood-based biomarkers for rapid diagnosis and differentiation of stroke in the acute stages for better treatment and management strategies. A blood test that can rapidly identify the correct stroke type could help the practitioners in deciding stroke type specific treatment, thereby improving the management of stroke. This chapter gives an update on all the diagnostic test studies conducted to determine potential blood-based protein biomarkers to diagnose and differentiate ischemic stroke from hemorrhagic stroke, stroke mimics, transient ischemic attack, and mixed cases of ischemic stroke with hemorrhagic transformation.

Key words Ischemic stroke, Hemorrhagic stroke, Transient ischemic attack, Blood biomarkers, Stroke differentiation, Stroke mimics, Hemorrhagic transformation

1 Introduction

Stroke has emerged as the major global health problem [1] and the second most common cause of mortality all over the world [2]. Currently a non-contrast computed tomography (NCCT) head scan is routinely used for confirming the diagnosis of stroke and differentiating an ischemic stroke (IS) from hemorrhagic stroke (HS). Although reliable, CT scan is an expensive technique and is not available easily across all the hospitals, especially in the remote areas of low–middle income countries. It also has the disadvantage of exposing patients to the radiation. It is largely used to diagnose or exclude an HS but has very poor sensitivity to diagnose IS.

In over one-third of cases, IS is not diagnosed correctly during the early phase of onset [3]. In a study conducted by Hand et al. a group of physicians evaluated 350 patients with suspected acute stroke out of which 109 (31%) patients initially diagnosed with

Philip V. Peplow et al. (eds.), *Stroke Biomarkers*, Neuromethods, vol. 147, https://doi.org/10.1007/978-1-4939-9682-7_6,
© Springer Science+Business Media, LLC, part of Springer Nature 2020

acute IS received a final diagnosis of mimic [4]. Thus, stroke mimics such as seizures, systemic infection, brain tumors, and toxic metabolic syndromes, also need to be timely diagnosed to avoid wrong management [5]. Along with mimics, early diagnosis of patients with a risk of developing hemorrhagic transformation (HT) is extremely crucial, to triage these patients to specialized stroke units for closer monitoring, thereby decreasing the incidence of this complication.

Therefore, despite advances in the field of neuroimaging, potential limitations like their time-consuming nature, expensive equipment, less availability, and variation in the analyses of radiological images have hampered its use in stroke diagnosis and differentiation [6]. Thus, a rapid blood-based biomarker test which can diagnose and differentiate stroke in acute stages at prehospital settings is required for effective treatment strategies to be implemented and to prevent adverse outcomes. Blood tests may reduce the cost of diagnostic procedures significantly and may be used in home or prehospital settings. Blood biomarkers may offer a reliable, rapid, and a cost-effective way of differentiating IS from potential similar conditions.

This chapter gives an update on the diagnostic test studies conducted to date to determine potential blood-based protein biomarkers to diagnose and differentiate ischemic stroke from hemorrhagic stroke, stroke mimics, transient ischemic attack, and mixed cases of ischemic stroke with hemorrhagic transformation for better management of stroke.

2 Blood Biomarkers for the Differentiation of Stroke

2.1 Blood Biomarkers for the Differentiation of Ischemic Stroke from Hemorrhagic Stroke

2.1.1 Individual Biomarkers

A study published by Allard et al. described the potential role of apo C-I and apo C-III proteins in distinguishing IS and HS [7]. The levels of apo C-I and Apo C-III were expressed in relative fluorescence units (RFU). Apo C-I was reported to have a sensitivity of 94% and specificity of 73% with the cutoff point at 60 RFU, while apo C-III was reported to have a sensitivity of 94% and specificity of 87% with a cutoff point at 36 RFU in distinguishing between 15 IS and 16 HS patients. The levels of both Apo C-I and Apo C-III were found to be raised in IS as compared to HS within 6 h of symptom onset. Several years later, Lopez et al. [8] assayed a panel of nine apolipoproteins (apo A-I, apo A-II, apo B, apo C-I, apo C-II, apo C-III, apo D, apo E, apo H) by using a selective reaction monitoring (SRM) based assay for distinguishing IS and HS patients in the first week of symptom onset. They observed that out of the nine apolipoproteins assayed, apo C-III (individually) differentiated between IS and HS with an area under the curve (AUC) of 0.85.

A recent study by Walsh et al. [9] measured the levels of matrix metalloprotease-3 (MMP-3), MMP-9, paraoxonase-1, apo A-I, apo C-I, and apo C-III and observed that apo A-I and paraoxonase-I levels significantly differentiated between IS and HS in a small cohort of patients within 4.5 h of symptom onset. Both the Apo A-I (140 in IS vs. 180 mg/dl in HS) and Paraoxonase levels (250,500 in IS vs. 366,000 mg/dl in HS) were found to be significantly lower in IS as compared to HS. However, unlike the previous two studies [7, 8] they did not find Apo C-III as a potential biomarker to differentiate the two stroke types.

A panel of four biomarkers (BNP, D-dimer, MMP-9, and S100B) was studied by Kim et al. but they observed only a single biomarker (i.e., BNP) whose levels were significantly higher in IS patients (90.8 ± 156.4) as compared to HS patients (16.3 ± 10.8) and it differentiated the two stroke types with an AUC of 0.61 [10].

Several studies have demonstrated that the *GFAP* levels are significantly raised in intracerebral hemorrhage (ICH) as compared to IS within 2–6 h of symptom onset [11–20]. Foerch et al. performed a pilot study on 93 IS and 42 ICH patients and observed that serum GFAP can reliably detect ICH from IS within first 6 h with a sensitivity of 79% and a specificity of 98% at a cutoff point of 2.9 ng/l [13]. Six years later the same author performed a larger study on 163 IS and 39 HS patients and again demonstrated that GFAP can efficiently differentiate ICH from IS with a diagnostic sensitivity and specificity of 84.2% and 96.3% respectively at a cutoff point of 0.29 µg/l within a much smaller time window of 4.5 h [14]. Another study by Dvorak et al. followed the same trend and found GFAP to be significantly differentiating ICH from IS with a sensitivity of 70% and specificity of 100% within a varying period from 2 to 48 h at a cutoff level of 0.04 ng/ml [15]. Unden et al. [16] analyzed a panel of biomarkers, S100B, neuron-specific enolase (NSE), GFAP, and activated Protein C–protein C inhibitor complex (APC-PCI) in 83 IS and 14 HS samples and found only GFAP and APC-PCI which significantly distinguished ICH from IS. The APC-PCI levels varied more in IS patients (0.04–3.55 µg/l) than in the ICH patients (0.19–0.49 µg/l), and at a cutoff value of <0.35 µg/l, it differentiated ICH from IS with a sensitivity of 96% and a specificity of 42%. GFAP levels were found to be significantly higher in ICH patients (40–160 ng/l) as compared to IS patients (<30–70 ng/l); and at a cutoff value of $\geq$40 ng/l and a sensitivity and specificity of 79% and 64%, it differentiated ICH from IS within 24-h onset [16]. Another study by Xiong et al. confirmed the potential role of serum GFAP in distinguishing IS and HS with a sensitivity and specificity of 86% and 76.9% respectively at a cutoff point of 0.7 ng/ml [17]. Recently Ren et al. further validated these results by demonstrating a sensitivity of 61% and a specificity of 96% of GFAP in distinguishing IS and ICH at a cutoff point of 0.34 ng/ml [18]. Llombart et al., found that RBP4 > 48.75 µg/ml (sensitivity 68.4%, specificity 84%) and GFAP < 0.07 ng/ml (sensitivity 32%, specificity 100%)

differentiated IS and HS [19]. A latest study by Luger et al. recruited 146 IS and 46 ICH patients and observed a sensitivity of 77.8% and a specificity of 94.2% of serum GFAP in distinguishing ICH from IS under 6 h at a cutoff value of 0.03 µg/l [20]. Katsanos et al. also confirmed the role of plasma GFAP in differentiating 34 ICH from 121 IS patients with a sensitivity of 91% and specificity of 97% under 6 h of symptom onset at a cutoff value of 0.43 ng/ml [11]. Another study conducted by Rozanski et al. assessed the potential of plasma GFAP in differentiating ICH from IS in pre-hospital settings. In 49 IS and 25 ICH patients, GFAP >0.29 ng/ml had a sensitivity of 36% and a specificity of 100% to differentiate ICH from IS [12].

A recent study by Zhou et al. confirmed the role of S100B within first 6 h as an individual biomarker in differentiating the two stroke types with a sensitivity of 95.7% and a specificity of 70.4% at a cutoff value of 67 pg/ml in 71 IS and 46 ICH cases [21].

Only a single study conducted on 32 stroke patients by Roudbary et al. assessed the concentration of *hs-CRP* in serum of patients with IS and HS in the first 24 h of symptom onset [22]. They found the levels of hs-CRP to be increased in IS patients (18.92 ± 11.28) as compared to HS patients (2.65 ± 1.7).

2.1.2 Panel of Biomarkers

Montaner et al. used a panel of blood biomarkers including S100B, MMP-9, sRAGE, CRP, D-dimer, BNP, NT-3, caspase-3, chimerin-II, secretagogin, cerebellin, and NPY to examine their predictive value in differentiating IS from ICH [23]. They observed that high S100B (107.58 in ICH vs. 58.70 ng/ml in IS, p value < 0.001) and low sRAGE levels (0.77 in ICH vs. 1.02 ng/ml in IS, p value = 0.009) in combination could differentiate between the two stroke types with a sensitivity of 22.7% and specificity of 80.2% within 6 and 3 h of symptom onset. The role of S100B as a marker of stroke differentiation was earlier ascertained by Kavalci et al. using a panel of biomarkers (S100B, MMP-9, D-dimer and BNP) to differentiate IS and HS within 24 h of stroke [24]. The levels of most of the biomarkers assessed in the panel were found to be increased in HS as compared to IS: BNP (172 in HS vs. 76.7 pg/ml in IS, p value < 0.001, sensitivity-65.5%, specificity-60.6); D-dimer (1780 in HS vs. 574.1 ng/ml in IS, p value<0.001, sensitivity-58.6%, specificity-59.2%); MMP-9 (445 in HS vs. 170 ng/ml in IS, p value<0.001, sensitivity-65.5%, specificity-66.2%); S100B (100 in HS vs. 100 pg/ml in IS, p value>0.05, sensitivity-13.8%, specificity-98.6%).

Llombart et al. found a panel of Retinol binding protein 4 (RBP4) (>61 µg/ml) and GFAP (<0.07 ng/ml) differentiated ICH from IS with a specificity of 100% [19]. Unden et al. observed that APC-PCI (cutoff < 0.35 µg/l) and GFAP (cutoff ≥ 40 ng/l) had a sensitivity of 71% and a specificity of 73% when used in combination for differentiating the two stroke types [16]. After applying a multiple marker algorithm approach, Lopez et al. observed that a combination of Apo C-III and Apo A-I efficiently distinguished between IS and HS with an AUC of 0.92 [8].

The stroke chip study by Bustamante et al. assayed a panel of 21 biomarkers and observed NT-pro BNP > 4.9 (sensitivity: 44.8%; specificity: 74.9%) and Endostatin > 4.9 (sensitivity: 18.8%; specificity: 90.8%) along with age, sex, blood pressure, stroke severity, atrial fibrillation, and hypertension had a predictive accuracy of 80.6% to differentiate 941 IS patients from 174 HS patients within 6 h of symptom onset [25].

A panel of S100 (cutoff > 0.250 μg/l) and Plasma DNA (cutoff > 2500 kilogenome/equivalents/l) was found to diagnose HS from IS with a sensitivity of 47% (for S100) and 31% (for Plasma DNA) and specificity of 81% (for S100) and 83% (for plasma DNA), respectively, in a study by Rainer et al. [26]. Sharma et al. developed a multivariate model of a panel of five biomarkers including eotaxin, EGFR, S100A12, and prolactin which differentiated ICH from IS with a discriminating capacity of $C = 0.082$ [27].

Table 1 lists all the studies with biomarkers which have been discovered for differentiation of IS from HS and gives a brief account of the methodology adopted in assaying the biomarker levels.

2.2 Blood Biomarkers for the Differentiation of Ischemic Stroke from Stroke Mimics

2.2.1 Individual Biomarkers

In a small sample of 34 IS and 29 mimics, Glickmann et al. in 2011 [28] analyzed five serum based biomarkers, namely, CRP, MMP-9, S100B, BNP, and D-dimer for their potential to diagnose IS from stroke mimics. They observed a model of C-Reactive Protein and National Institutes of Health Stroke Scale (CRP and NIHSS) to be highly predictive of IS with a discrimination capacity of $C = 0.95$. The levels of CRP biomarker were significantly higher in IS patients as compared to stroke mimics (IS: 37.6 ± 33.1 vs. mimics: 9.7 ± 11, $p < 0.001$). MMP-9 and S100B were found to be moderately predictive of IS when combined with NIHSS in a bivariate model with a discrimination capacity of $C = 0.92$ for MMP-9 and $C = 0.87$ for S100B. The median time from symptom onset to blood sampling was 5 h (interquartile range: 3.5–8 h). The levels of S100B were also assessed by Gonzalez-Garcia et al. in 2012 [29] along with neuron-specific enolase (NSE) to diagnose total stroke from mimics + transient ischemic attack (TIA) within 48 h of symptom onset. In a sample of 61 stroke (IS = 44, HS = 17) and 11 TIA + mimics, NSE had a sensitivity of 53% and specificity of 64% at a cutoff value of 14 μg/l, while S100B had a sensitivity of 55% and specificity of 64% at a cutoff value of 130 ng/l to differentiate total stroke from TIA + mimics. They concluded that neither NSE nor S100B improved the diagnosis of acute stroke. Another study by An et al. in 2013 [30] assayed the levels of MMP-9 and S100B along with NSE, VSNL-1, hFABP, and Ngb, GFAP, MMP-9, IL-6 and TNF-α to diagnose IS within 24 h of symptom

Table 1
List of studies including blood based protein biomarkers for differentiating ischemic and hemorrhagic stroke

S. no	Study	Sample size	Protein profile assayed	Methodology	Duration of stroke	Conclusion	Limitations of study	Sensitivity	Specificity
1.	Allard (2004) (Switzerland) [7]	26—IS 19—HS *Validated* 11—IS 10—HS	Plasma proteins	SELDI, LC-ESI-MS, ELISA	6 h (range 40 min to 3 days)	Apo C-I and Apo C-III, first plasmatic biomarkers to distinguish IS and HS in a small no. of patients	Small sample size in discovery and validation phases	*Apo C-I* 94% *Apo C-III* 94%	*Apo C-I* 73% *Apo C-III* 87%
2.	Montaner (2012) (Spain) [23]	776—IS 139—ICH	CRP, D-dimer, sRAGE, MMP9, S100B, BNP, NT-3, caspase-3, chimerin-II, secretagogin, Cerebellin, NPY (Plasma)	ELISA	Within 24 h; (<6 h; <3 h)	S100B and sRAGE as a rapid blood test might help to distinguish IS and HS	Validation not performed and low sensitivity of biomarkers reported, clinically relevant control group not taken	22.7%	80.2%
3.	Sharma (2014) (USA) [27]	57—IS 32—ICH	262 Serum Biomarkers	ELISA	Within 24 h	5 biomarkers (cotaxin, EGFR, S100A12, TIMP-4, and prolactin) distinguished IS and ICH (C = 0.82)	Sample size not large enough to perform external validation, sensitivity and specificity of the panel to distinguish IS and HS not reported	–	–
4.	Walsh (2016) (USA) [9]	14—IS 23—HS	Apo A-I, Apo C-I, Apo C-III, MMP-3, MMP-9, and paraoxonase-1 (Plasma)	Multiplex Assays, ELISA	<12 h	Apo A-I and paraoxonase-1 levels differed between IS and HS cases	Small sample size, validation phase not performed	–	–

#	Author (year) (country) [ref]	Biomarker	Method	Sample	Time window	Conclusion	Limitations	Sensitivity	Specificity
5.	Lopez (2012) (USA) [8]	9 Serum Apolipoproteins	Multiplex SRM Assay (MS)	54—IS 26—HS	<1 week	apo C-III and apo A-I differentiated IS and HS	Validation required in large samples, samples in first few hours required	–	–
6.	Kavalci (2011) (Turkey) [24]	4 Plasma biomarkers	Triage stroke panel—a biochemical multimarker assay	71—IS 29—HS *Validated* 100—Stroke	Within 24 h	A combination of BNP, D-dimer, MMP9, and S100b plasma biomarkers can differ IS and HS	Study not adequately powered, sensitivity and specificity of the panel not high, larger validation cohort required	*BNP* 65.5% *D-dimer* 58.6% *MMP-9* 65.5% *S100B* 13.8%	*BNP* 60.6% *D-dimer* 59.2% *MMP-9* 66.2% *S100B* 98.6%
7.	Roudbary (2011) (Iran) [22]	hs-CRP (serum)	Immunonephelometric method	16—IS 16—HS	Within 24 h	hs-CRP level is increased in patients with IS but not in HS	Less sample size, Validation not done, clinically relevant control group not taken	–	–
8.	Llombart (2016) (Spain) [19]	RBP4, GFAP (Plasma)	Multiplex Sandwich ELISA	*Discovery* 36—IS 10—HS *First rep.* 16—IS 16—HS *Second rep.* 38—IS 28—HS	<6 h	RBP4 and GFAP useful as diagnostic biomarkers to differentiate IS and ICH	Results obtained in the discovery phase were not corrected for multiple testing; replication cohorts are small	*RBP4* 68.4% *GFAP* 32%	*RBP4* 84% *GFAP* 100%
9.	Foerch (2012) (Germany) [14]	Plasma GFAP	Electrochemiluminometric immunoassay	163—IS 39—HS	<4.5 h	GFAP test performed within 4.5 h of symptom onset is a reliable tool for the	Validation phase not performed, only a phase-1 study, less number of HS patients recruited	84.2%	96.3%

(continued)

Table 1
(continued)

S. no	Study	Sample size	Protein profile assayed	Methodology	Duration of stroke	Conclusion	Limitations of study	Sensitivity	Specificity
						differentiation between ICH and IS			
10.	Foerch (2006) (Germany) [13]	93—IS 42—HS	Serum GFAP	Elecsys	<6 h	GFAP raised in ICH as compared to IS in first 6 h	Determined GFAP at time point of hospital admission, not in prehospital setting; validation with same cutoff required in other populations	79%	98%
11.	Dvorak (2009) (Germany) [15]	45—IS 18—HS	Serum GFAP	ELISA	2–48 h	GFAP was reported to be higher in ICH than in IS with best time window between 2 and 6 h	Small sample size; clinically relevant control group not taken	70%	100%
12.	Unden (2009) (Sweden) [16]	83—IS 14—HS	S100B, NSE, GFAP, APC-PCI (Serum)	ELISA	<24 h	Admission GFAP and APC-PCI levels may rule out ICH	Small sample size; validation with same cutoff required in other populations; clinically relevant control group not taken, less number of HS patients	*APC-PCI 96% GFAP 79% Panel (GFAP, APC-PCI) 71%*	*APC-PCI 42% GFAP 64% Panel (GFAP, APC-PCI) 73%*

No.	Study	Sample size	Biomarker	Assay	Time	Finding	Limitation		
13.	Xiong (2015) (China) [17]	65—IS 43—HS	Serum GFAP	ELISA	~2–6 h	GFAP test within 2–6 h after stroke onset could be used to differentiate ICH and IS	Small Sample size; clinically relevant control group not taken; influence of other parameters on GFAP level not considered	86%	76.9%
14.	Ren (2016) (China) [18]	79—IS 45—ICH	Serum UCH-L1 and GFAP	Sandwich ELISA	Within 4.5 h	GFAP differentiated between IS and ICH	Modest sample size; validation phase not performed	GFAP 61%	GFAP 96%
15.	Rainer (2007) (China) [26]	118—IS 35—HS	Serum S100 and Plasma DNA	qRT-PCR, ELISA	<24 h	A combination of S100 and Plasma DNA differentiated ICH from IS	Less sample size, validation phase not performed	S100 47% Plasma DNA 31%	S100 81% Plasma DNA 83%
16.	Kim (2010) (South Korea) [10]	89—IS 11—HS	BNP, D-dimer, MMP-9, S100B (Plasma)	Fluorescence Immunoassay	6 h (range: 0–120 h)	Only BNP distinguished between AIS and HS	Less sample size in HS, validation not performed	–	–
17.	Luger (2017) (Germany) [20]	146—IS 45—ICH	Serum GFAP	Electrochemiluminometric immunoassay	Within 6 h	GFAP differentiated ICH from IS	Less number of ICH cases recruited, no validation phase	77.8%	94.2%
18.	Zhou (2016) (China) [21]	71—IS 46—ICH	Plasma S100B	Electrochemiluminescence immunoassay	Within 6 h	S100B distinguished between IS and ICH	Small sample size, no validation phase performed	95.7%	70.4%
19.	Katsanos (2017) (Greece) [11]	121—IS 34—ICH	Plasma GFAP	ELISA	Within 6 h	Plasma GFAP distinguished ICH from IS with optimum	Small sample size, no validation phase	91%	97%

(continued)

Table 1
(continued)

S. no	Study	Sample size	Protein profile assayed	Methodology	Duration of stroke	Conclusion	Limitations of study	Sensitivity	Specificity
						diagnostic yield			
20.	Rozanski (2017) (Germany) [12]	49—IS 25—ICH	Plasma GFAP	ELISA	62.5 min (36–139 min)	GFAP levels >0.29 ng/ml were seen only in ICH, thus confirming the diagnosis of ICH during prehospital care	Small sample size, IS patients significantly older than ICH patients, validation phase not performed, poor sensitivity	36%	100%
21.	Bustamante (2017) (Spain) [25]	941—IS 174—HS	21 protein biomarkers	ELISA	Within 6 h	Only NT-proBNP and endostatin differentiated IS from HS with moderate predictive accuracy of 80.6%	Low reproducibility between interim and final analysis, poor sensitivity	*NT-pro BNP* 44.8% *Endostatin* 18.8%	*NT-pro BNP* 74.9% *Endostatin* 90.8%

IS ischemic stroke, *HS* hemorrhagic stroke, *ICH* intracerebral hemorrhage, *AIS* acute ischemic stroke, *AHS* acute hemorrhagic stroke, *hsCRP* high sensitive C-reactive protein, *RBP4* retinol binding protein 4, *GFAP* glial fibrillary acidic protein, *ELISA* enzyme linked immunosorbent assay, *NSE* neuron-specific enolase, *APC-PCI* activated protein C-protein C inhibitor complex, *BNP* brain natriuretic peptide, *SRM* mass spectrometry-based selective reaction monitoring, *qRT-PCR* quantitative real-time polymerase chain reaction

onset. Only IL-6 (IS: 4.0 [0.8–12.3]; mimic: 1.2 [0.0–2.4]; $p < 0.001$), S100B (IS: 30.4 [0.0–115.2]; mimic: 2.3 [0.0–20.6]; $p < 0.001$), and MMP-9 (IS: 63.3 [29.7–122.8]; mimic: 33.8 [15.4–60.8]; $p < 0.001$) were found to be significantly elevated in 188 IS as compared to 90 stroke mimics.

A study published by Airas et al. in 2008 [31] observed the levels of vascular adhesion protein-1 (VAP-1) to be significantly elevated in 20 IS cases as compared to 20 mimics (IS: 652 ± 224 ng/ml, Mimics: 542 ± 104 ng/ml; $p < 0.05$) within 6 h of symptom onset.

Doehner et al. in 2012 [32] studied the plasma levels of neuropeptides proenkephalin A (PENK-A) and protachykinin (PTA) in 124 IS, 16 TIA, and 49 mimic cases. PENK-A concentration was significantly elevated in patients with stroke (median [IQR]): 123.8 pmol/l [93–160.5]) compared to patients with TIA (114.5 pmol/l [85.3–138.8]) and with mimics (102.8 pmol/l [76.4–137.6]; both groups vs. stroke $p < 0.05$). However, no significant difference was observed in PTA levels between these groups. Ahn et al. in 2011 [33] compared the usefulness of albumin-adjusted ischemia-modified albumin index (IMA index) to the ischemia modified albumin (IMA) in early detection of IS from mimics in a small sample of 28 IS and 24 mimics. IMA index was found to be more sensitive (sensitivity: 95.8%, specificity: 96.4, cutoff: 91.4 U/ml) than conventional IMA (sensitivity: 87.5%, specificity: 89.3%, cutoff: 98 U/ml) to diagnose IS from mimics within 6 h of symptom onset. Meng et al. in 2011 [34] assayed several plasma biomarkers including antithrombin III (AT III), thrombin–antithrombin III (TAT), fibrinogen, D-dimer, and high-sensitivity C-reactive protein (hsCRP) to differentiate 152 IS patients from 46 stroke mimics within 4.5 h of symptom onset. Only AT-III (sensitivity: 97.37, specificity: 93.62, cutoff: 210%) and fibrinogen (sensitivity: 96.05, specificity: 82.61, cutoff: 4 g/l) were able to differentiate IS from mimics with adequate sensitivities and specificities. Dambinova et al. in 2012 [35] tested the potential of NR2 peptide to differentiate IS from stroke mimics in a sample of 101 IS and 91 mimics within 72 h of symptom onset. NR2 peptide was found to have a sensitivity of 92% and specificity of 96% to differentiate IS from stroke mimics at a cutoff value of 1.0 μg/l. Dassan et al. in 2012 [36] found limited clinical utility of serum vascular endothelial growth factor (VEGF) in being able to differentiate IS from stroke mimics within 24 h of symptom onset. In a small sample of 29 IS and 15 mimics, VEGF was found to have a modest sensitivity of 69% and specificity of 73% to differentiate IS from mimics at a cutoff value of 1026 pg/ml. Wendt et al. in 2015 [37] did not find copeptin to be an appropriate biomarker to discriminate between stroke and mimics. No statistically significant difference was observed in the copeptin levels between stroke and mimics (11.5 (5.3–29.3) vs. 8.2 (3.3–32.8), $p = 0.15$).

2.2.2 Panel of Biomarkers

Laskowitz et al. in 2009 [38] published a multicenter study at 17 centers to test the diagnostic performance of a biomarker panel comprising of MMP-9, BNP, D-dimer and S100B to diagnose IS. In 293 IS and 361 stroke mimic cases, they observed a sensitivity of 85% and a specificity of 34% for the 25th percentile while a sensitivity of 36% and a specificity of 84% for the 75th percentile of the biomarker panel to diagnose IS from stroke mimics within 24 h of symptom onset. The performance of this panel was also assessed by a different group in a study published by Sibon et al. in 2009 [39]. This group compared the accuracy of the Triage stroke panel consisting of D-dimer, BNP, MMP-9 and S100B to the triaging nurse for diagnosis of stroke from mimics. In 126 stroke (85 IS, 33 TIA, 13 HS) and 65 mimic cases, the biomarker panel was found to have an equivalent sensitivity of 92.9% and a specificity of 23.8% to differentiate and diagnose total stroke from stroke mimics to that of the nurse. A few years later, Knauer et al. in 2012 [40] tested the potential of this panel of biomarkers and they did not recommend the use of BNP, D-dimer, MMP-9, and S100B as a panel to distinguish IS from mimics. Receiver operating characteristic (ROC) curve analysis observed low discriminating power of the panel of biomarkers with an area under the curve (AUC) of 0.59. The model had a high sensitivity of 92% but a low specificity of 14% while a low sensitivity of 14% and a high specificity of 86% to differentiate IS from mimics within 6 h of symptom onset at cutoff values of 1.3 and 5.9 ng/ml respectively.

Montaner et al. in 2011 [41] found a panel of 6 plasma biomarkers namely caspase-3, D-dimer, sRAGE, chimerin, secretagogin and MMP-9 which differentiated total stroke (IS = 776 + HS = 139) from 90 stroke mimics. Within 24 h of symptom onset, the model had a sensitivity of 17% and specificity of 98% in the first quartile (cutoff value of 0.87) and a sensitivity of 82% and specificity of 59% in the last quartile (cutoff value of 0.97) to differentiate total stroke from mimics. In the blood samples obtained within 3 h, the model had a sensitivity of 87% and specificity of 55% in the first quartile (cutoff value of 0.49) while a sensitivity of 28% and specificity of 99% in the last quartile (cutoff value of 0.87) for differentiating total stroke from mimics. When the levels of secretagogin and chimerin were low and the levels of caspase-3, D-dimer, sRAGE, and MMP-9 were high, the model had a 100% probability of predicting stroke.

Sharma et al. in 2014 [27] assayed a total of 262 serum biomarkers in 57 IS patients and 37 stroke mimics and identified a panel of five biomarkers including eotaxin, EGFR, S100A12, TIMP-4, and prolactin which differentiated IS patients from mimics with a discriminative capacity of $C = 0.92$ within 24 h of the onset of event.

Table 2 lists the studies to determine the blood-based biomarkers for differentiating ischemic stroke from stroke mimics.

Table 2

Studies determining blood-based biomarkers for the differentiation of ischemic stroke from stroke mimics

S. no	Study	Sample size	Protein profile	Methodology	Duration of stroke	Conclusion	Limitations of the study	Sensitivity	Specificity
1.	Sharma (2014) (USA) [27]	57—IS 37—Mimics	262 Serum Biomarkers	ELISA	Within 24 h	Five biomarkers (cotaxin, EGFR, S100A12, TIMP-4, and prolactin) differentiated stroke from stroke mimics	Sample size not large enough to perform external validation	90%	84%
2.	Glickman (2011) (USA) [28]	34—IS 29—Mimics	BNP, CRP, D-dimer, MMP-9, S100B (Serum)	Assay	5 h (3.5–8 h)	CRP has high discriminating capacity and MMP-9 and S100B have moderate discriminating capacity to distinguish IS from mimics	Less sample size, validation phase not done	–	–
3.	Montaner (2011) (Spain) [41]	776—IS 90—Mimics	CRP, D-dimer, sRAGE, MMP-9, S100B, BNP, caspase-3, neurotrophin-3, chimerin, secretagogin	Sandwich ELISA	Within 24 h (<6 h) (<3 h)	A combination of caspase-3, D-dimer, sRAGE, chimerin, secretagogin, MMP-9 differentiated stroke from mimics. Best association with a model of caspase-3 and D-dimer	Validation phase not done	*Caspase-3* 52% *D-dimer* 81% *sRAGE* 48% *Chimerin* 79% *Secretagogin* 67% *MMP-9* 65%	*Caspase-3* 73% *D-dimer* 38% *sRAGE* 65% *Chimerin* 32% *Secretagogin* 48% *MMP-9* 65%

(continued)

Table 2
(continued)

S. no	Study	Sample size	Protein profile	Methodology	Duration of stroke	Conclusion	Limitations of the study	Sensitivity	Specificity
4.	Dambinova (2012) (USA) [35]	101—IS 71—No stroke 20—Mimics 52—HC 48—Risk factor controls	NR-2 peptide	Rapid magnetic particle ELISA	Within 72 h	NR-2 peptide distinguishes stroke from mimics	Validation phase not done, blood sampling within a narrow time window is required	92%	96%
5.	Dassan (2012) (UK) [36]	29—IS 15—Mimics 15—HC	Serum VEGF	ELISA	24 h	VEGF has limited clinical utility for diagnosis of IS from mimics	Less sample size, validation phase not done	69%	73%
6.	Gonzalez-Garcia (2012) (Cuba) [29]	44—IS 11—Mimics 79—High-risk control	Serum NSE and S100B	Immunoassay	12–48 h	NSE and S100B cannot efficiently distinguish stroke from mimics	Small sample size, validation phase not done	*NSE* 53% *S100B* 55%	*NSE* 64% *S100B* 64%
7.	Knauer (2012) (Germany) [40]	100—IS 49—Mimics	BNP, D-dimer, MMP-9, S100B	Sandwich Fluorescence Immunoassay Triage stroke panel	<6 h	No single or combination of biomarker distinguished IS from mimics	Low specificity of the biomarker panel	*Cutoff:* 2.3 86% *Cutoff:* 2.5 84%	*Cutoff:* 2.3 33% *Cutoff:* 2.5 35%
8.	Wendt (2015) (Germany) [37]	287—IS 90—Mimics	Copeptin	Immune fluorescent assay	15 to more than 180 minutes	Copeptin does not discriminate between stroke and mimics	Sensitivity and specificity data not reported	–	–

#	Study	Sample	Biomarker	Method	Time	Finding	Limitation	First Quartile / third Quartile	First Quartile / third Quartile
9.	Laskowitz (2009) (USA) [38]	293—IS 361—Mimics Validated 87—IS 152—Mimics	MMP-9, D-dimer, S100B, BNP	Triage stroke panel (fluorescent immunoassay)	9.3 h (4.5–18.2 h)	A model of MMP-9, D-dimer, S100B, BNP can differentiate stroke from mimics	The biomarker levels may be elevated due to presence of other comorbidities	First Quartile 86% third Quartile 35%	First Quartile 37% third Quartile 86%
10.	Ahn (2011) (South Korea) [33]	28—IS 24—Non-stroke/mimics	Albumin adjusted IMA index, IMA	Albumin Cobalt binding test	Within 6 h	IMA index more sensitive than conventional IMA to diagnose stroke	Small sample size, no validation phase	IMA Index 95.8% IMA 87.5%	IMA Index 96.4% IMA 89.3%
11.	Airas (2008) (Finland) [31]	20—IS 20—Mimics	sVAP-1	Time resolved Immunofluorometric ELISA assay	Within 6 h	sVAP-1 levels higher in stroke patients as compared to age-sex matched control (mimic) group	Small sample size, no validation phase	–	–
12.	Doehner (2012) (Germany) [32]	124—IS 49—Non-stroke/mimics	PENK-A, PTA (plasma)	Sandwich ELISA	4.5 h (1.3–19.8 h)	Only PENK-A levels were higher in stroke patients as compared to non-stroke group	Small sample size, validation in other populations required	–	–
15.	Sibon (2009) (France) [39]	126—Stroke 63—Mimics	Triage stroke panel MMX (D-dimer, BNP, MMP-9, S100β)	Fluorescence immunoassay	–	MMX stroke panel of four biomarkers distinguished stroke from mimics	Very low specificity, no individual differentiation of IS or HS from mimics	92.9% (87–96.2%)	23.8% (14.9–35.6%)

(continued)

Table 2
(continued)

S. no	Study	Sample size	Protein profile	Methodology	Duration of stroke	Conclusion	Limitations of the study	Sensitivity	Specificity
16.	An (2013) (Korea) [30]	188—Stroke 90—Mimics	NSE, VSNL-1, hFABP, Ngb, S100B, GFAP, IL-6, TNFα, MMP-9, PAI-1	ELISA	6–24 h	A panel of IL-6, S100B, MMP-9 distinguished stroke from mimics but with low discrimination ability	Weak discrimination ability of true stroke from mimics	–	–
17.	Meng (2011) (China) [34]	152—IS 46—Non-stroke/mimics	AT-III, TAT, Fibrinogen, D-dimer, hsCRP (Plasma)	Immunoturbidimetry assay, Chromogenic assay, ELISA	Within 4.5 h	Plasma AT-III and fibrinogen distinguished stroke from non-stroke	No validation phase performed, non-stroke group not clearly defined	*AT-III* 97.37% *Fibrinogen* 96.05%	*AT-III* 93.62% *Fibrinogen* 82.61%

IS ischemic stroke, *HC* healthy control, *ELISA* enzyme linked immunosorbent assay, *qRT-PCR* quantitative real-time polymerase chain reaction, *CRP* C-reactive protein, *NSE* neuron-specific enolase, *MMP* matrix metalloproteinase, *BNP* B-type natriuretic peptide, *IMA* ischemia modified albumin, *sVAP-1* serum vascular adhesion protein-1, *PENK-A* precursor neuropeptides proenkephalin A, *PTA* protachykinin

2.3 Blood Biomarkers for Mixed Cases of Ischemic Stroke with Hemorrhagic Transformation

2.3.1 Individual Biomarkers

Montaner et al. performed an exploratory study in 2001 to determine the relationship between MMP-2 and MMP-9 and hemorrhagic transformation (HT) after cardioembolic (CE) subtype of IS [42]. They recruited 39 CE stroke patients within 12 h of symptom onset in which 16 patients suffered from HT. No significant difference was observed between the levels of MMP-2 and MMP-9 and presence or absence of HT. However, MMP-9 levels were found to be significantly higher in patients with late HT (5-to-7-day CT scan) as compared to early HT (occurring at <48 h) or no HT. In patients with late HT, the MMP-9 levels were found to be 240.4 ± 111.2 ng/ml, in early HT were 87.6 ± 65.4 ng/ml and in no HT were 157.6 ± 126 ng/ml (p value $= 0.05$). In 2003, the same authors conducted a study to determine the association between the pretreatment levels of MMP-2 and MMP-9 and risk of HT in stroke patients treated with tPA within 3 h of symptom onset [43]. The study included 41 CE stroke patients out of which 15 patients had HT. The mean levels of MMP-2 and MMP-9 did not differ in terms of presence or absence of HT. However, MMP-9 was found to have a sensitivity of 100% and specificity of 78% to detect the presence of parenchymal hemorrhage (PH) (a subtype of HT) with a positive and negative predictive value of 67% and 100%, respectively, at a cutoff point of 191.3 ng/ml.

Castellanos et al. in 2003 observed high levels of plasma MMP-9 within 24 h of symptom onset to be an independent predictor of HT (OR 12; 95% CI 3–51; $p < 0.001$) [44]. Out of 250 IS patients, HT was observed in 38 patients. MMP-9 level ≥ 140 ng/ml had a sensitivity and specificity of 87% and 90%, respectively, to predict HT. MMP-9 < 140 ng/ml had a high negative predictive value of 97%. In the consecutive year, the same author conducted a retrospective, hypothesis-generating study and determined the association of elevated plasma levels of cellular fibronectin (c-Fn) and plasma levels of MMP-9 with tPA-induced HT in IS patients within 6 h of symptom onset [45]. The study recruited 87 IS patients who received tPA out of which 26 patients had an HT. Plasma c-Fn ≥ 3.6 µg/ml had a sensitivity and specificity of 100% and 96% and a positive predictive value and a negative predictive value of 44% and 100% to predict hemorrhagic infarction type-2 (HI-2) and parenchymal hemorrhage (PH) after IS before tPA administration. After conducting the post hoc analysis, plasma MMP-9 level ≥ 140 ng/ml had a sensitivity and specificity of 81% and 88% and a positive predictive value and a negative predictive value of 41% and 98% to predict HI-2 and PH after IS before tPA administration. In 2007, the same group validated the predictive capacity of c-Fn and MMP-9 biomarkers using the same cutoff values in a new set of patients within 3 h of symptom onset [46]. A total of 134 IS patients were recruited in the study. Twenty patients had HT out of which 12 patients suffered from PH. c-Fn ≥ 3.6 µg/ml had a sensitivity of 95% and a specificity of 63% to predict HT. The sensitivity increased to 99% for the prediction of PH. MMP-9 $\geq$ 140 ng/ml had a sensitivity and specificity of 75% each to predict HT. However, the sensitivity increased to 92% for the prediction of PH. When c-Fn and MMP-9 biomarkers were used in combination, they had a sensitivity of 75% and a specificity of 89% to predict HT whereas a sensitivity of 92% and specificity of 87% to predict PH. The potential of MMP-9 as a biomarker to detect HT was also confirmed by Banhawy et al. in 2010. They recruited 35 IS patients within 24 h of symptom onset and 35 control subjects. Eight patients in the study had HT. MMP-9 $\geq$ 900 ng/ml had a sensitivity of 100% and a specificity of 81.48% to detect HT [47].

Trouillas et al. in 2004 studied the early coagulation parameters in 157 IS patients for predicting early HT after thrombolysis [48]. Forty-two patients in the studied had HT out of which 11 patients had early PH. Fibrinogen degradation products (FDP) > 100 ng/ml had a sensitivity and specificity of 36.4% and 84.2% to predict early PH at 2 h after tPA administration. Moreover, FDP > 200 ng/ml had a sensitivity and specificity of 27.3% and 93% to predict early PH at 2 h after tPA administration.

Cocho et al. in 2006 studied the pretreatment levels of hemostatic markers (fibrinogen, prothrombin fragments 1 + 2, Factor XIII, Factor VII, α_2 antiplasmin, plasminogen activator inhibitor-1 (PAI-1), and thrombin-activatable fibrinolysis inhibitor) to predict symptomatic intracranial hemorrhage (SICH) in IS patients treated with tPA within 3 h of symptom onset [49]. Eight out of 114 IS patients developed SICH. However, none of the hemostatic markers had the potential to detect SICH.

Foerch et al. in 2007 analyzed the role of serum S100B as a biomarker to detect HT in 275 tPA induced IS patients within 6 h of symptom onset [50]. HT occurred in 80 patients (45 = HI and 35 = PH). A pretreatment S100B value of >0.23 μg/l had a sensitivity and specificity of 46% and 82% to detect PH.

Mendioroz et al. in 2009, observed that activated protein C (APC) level > 176% had a sensitivity and specificity of 81.8% and 84.1% to detect PH (in 13 PH patients) at 2 h after tPA administration in 199 IS patients with a positive predictive value of 81.8% and a negative predictive value of 84.6% [51].

Choi et al. in 2012 analyzed the potential of serum ferritin level in a large sample of 752 IS patients enrolled within 24 h of symptom onset to detect the presence of HT [52]. In 90 patients with HT, ferritin level > 144.8 ng/ml had a sensitivity and specificity of 74% each to detect HT. While, ferritin level > 164.1 ng/ml had a sensitivity of 85% and specificity of 75% to detect PH in 32 patients with PH. Ferritin level > 171.8 ng/ml had a sensitivity of 84% and a specificity of 77% to detect symptomatic HT (sHT) in 37 patients with sHT.

Kazmierski et al. in the same year assayed the circulating tight-junction (TJ) proteins (occludin [OCLN], claudin 5 [CLDN5], zonula occludens 1 [ZO1]), standard markers of BBB breakdown (S100B, neuron-specific enolase [NSE]), and molecules involved in

BBB disintegration (MMP-9 and VEGF) to determine their role in detecting HT after IS within 178 h of symptom onset in 458 IS patients with 33 HT. They concluded that serum levels of TJ proteins and S100B and VEGF could be an effective marker for screening patients with HT. NSE > 24.05 μg/ml had the highest specificity of 94.8% and S100B > 11.89 pg/ml had the highest sensitivity of 92.9% to detect HT in IS patients [53].

Guo et al. in 2016 examined the potential of neutrophil-to-lymphocyte ratio (NLR) to detect HT after thrombolysis in 189 stroke patients. NLR at a cutoff value of 10.59 was found to have a sensitivity of 78.6% and specificity of 79.5% to detect PH in 28 patients with PH while a sensitivity of 76.5% and specificity of 75.6% to detect SICH in 17 patients with SICH at 12–18 h after tPA administration [54].

Hernandez-Guillamon et al. in 2010 observed high levels of vascular adhesion protein-1/semicarbazide-sensitive amine oxidase (VAP-1/SSAO) in 48 patients with HT as compared to 92 patients with no HT in a total of 141 tPA induced IS patients admitted within 3 h of symptom onset (PH: 3.41 ± 1.07; HI: 2.79 ± 0.75 and non-HT: 2.48 ± 0.92; $p = 0.001$) [55].

Rodriguez-Gonzalez et al. in 2013 analyzed the levels of platelet derived growth factor (PDGF)-AA, AB, BB, and CC in 129 IS patients to examine their potential to detect HT. HT was observed in 47 IS patients out of which 25 had PH. They observed that PDGF-CC levels ≥ 175 ng/ml predicted the development of PH with a sensitivity and specificity of 90% and 88%, respectively [56].

Chen et al. in 2016 determined the significance of plasma immunoproteasome subunits (low molecular mass peptide 2 [LMP2], multicatalytic endopeptidase complex-like 1 [MECL-1], LMP7, interleukin-1β [IL-1β], and high-sensitivity C-reactive protein [Hs-CRP]) to predict early HT in 316 IS patients within 72 h of symptom onset. Forty-two patients experienced HT. The optimal cutoff points as determined by the receiver operating characteristic analysis to predict HT were 988.3 pg/ml for LMP2 (sensitivity: 92.9%, specificity: 90.1%), 584.7 pg/ml for MECL-1 (sensitivity: 90.5%, specificity: 81.0%), and 509.0 pg/ml for LMP7 (sensitivity: 97.6%, specificity: 90.9%) [57].

2.3.2 Panel of Biomarkers

Ribo et al. in 2004 assessed the pretreatment levels of fibrinolysis inhibitors (plasminogen activator inhibitor [PAI]-1; lipoprotein a [Lp(a)]; thrombin-activated fibrinolysis inhibitor [TAFI]; and homocysteine) on the development of symptomatic HT in 77 IS patients treated with tPA within 3 h of symptom onset [58]. Out of 17 patients who presented with HT, 6 patients experienced SICH. A combination of PAI-1 < 21.4 ng/ml and TAFI >180% had a sensitivity of 75% and a specificity of 97.6% to predict SICH, with a positive and negative predictive value of 75% and 97.6% respectively. Table 3 enlists all the studies conducted till date to determine the blood-based biomarkers for ischemic stroke with mixed cases of hemorrhagic transformation.

Table 3
List of studies on blood biomarkers for determination of mixed cases of ischemic stroke with hemorrhagic transformation

S. no	Study	Sample size	Blood biomarkers	Duration of stroke	Conclusion	Limitations of the study	Sensitivity	Specificity
1.	Montaner (2001) (Spain) [42]	39—CE (16—HT)	MMP-2, MMP-9	Within 12 h	MMP-9 levels significantly higher in patients with late HT as compared to early HT or no HT	Only CE stroke patients recruited, small sample size with less cases of HT	–	–
2.	Montaner (2003) (Spain) [43]	41—CE (15—HT)	MMP-2, MMP-9	Within 3 h	MMP-9 was found to have a sensitivity of 100% and specificity of 78% to detect the presence of PH	Only CE stroke patients recruited, small sample size with less cases of PH	MMP-9 100%	MMP-9 78%
3.	Castellanos (2003) (Spain) [44]	250—IS (38—HT)	MMP-9	Within 24 h	MMP-9 level ≥140 ng/ml had a sensitivity and specificity of 87% and 90% to predict HT	Less patients included within 3 hrs., MMP-9 levels might be increased as a result of acute-phase reaction/prior systemic causes	MMP-9 87%	MMP-9 90%
4.	Castellanos (2004) (Spain) [45]	87—IS (26—HT)	MMP-9, c-Fn	Within 6 h	Plasma c-Fn ≥ 3.6 µg/ml had a sensitivity and specificity of 100% and 96% and plasma MMP-9 level ≥140 ng/ml had a sensitivity and specificity of 81% and 88% to predict HI-2 and PH before IS after tPA administration	Data were obtained from a post hoc analysis, small sample size, methodology adopted was analytically slow	c-Fn 100% MMP-9 81%	c-Fn 96% MMP-9 88%

#	Study	Sample	Biomarkers	Time	Description	Limitation	Sensitivity	Specificity
5.	Castellanos (2007) (Spain) [46]	134—IS (20—HT)	c-Fn, MMP-9	Within 3 h	c-Fn ≥ 3.6 µg/ml had a sensitivity of 95% and a specificity of 63% and MMP-9 ≥ 140 ng/ml had a sensitivity and specificity of 75% each to predict HT. In combination, they had a sensitivity of 75% and a specificity of 89% to predict HT	Small sample size with less cases of HT and PH, low PPVs of the biomarkers (ranged from 20% to 55%)	*c-Fn* 95% *MMP-9* 75% *c-Fn + MMP-9* 75%	*c-Fn* 63% *MMP-9* 75% *c-Fn + MMP-9* 89%
6.	Banhawy (2010) (Egypt) [47]	35—IS (8—HT)	MMP-9	Within 24 h	MMP-9 ≥ 900 ng/ml had a sensitivity of 100% and a specificity of 81.48% to detect HT	Small sample size with less cases of HT, cases within 3 h required	*MMP-9* 100%	*MMP-9* 81.48%
7.	Ribo (2004) (Spain) [58]	77—IS (17—HT)	PAI-1, Lpa, TAFI, homocysteine	Within 3 h	A combination of PAI-1 < 21.4 ng/ml and TAFI > 180% had a sensitivity of 75% and a specificity of 97.6% to predict SICH	Less number of patients with SICH (only 6)	*PAI-1 + TAFI* 75%	*PAI-1 + TAFI* 97.6%
8.	Trouillas (2004) (France) [48]	157—IS (42—HT)	FDP	Within 24 h	FDP > 100 ng/ml had a sensitivity and specificity of 36.4% and 84.2% while FDP > 200 ng/ml had a sensitivity and specificity of 27.3% and 93% to predict early PH at 2 h after tPA	Less sensitivity of the biomarker at both cutoffs, cases within 3 h are required	*FDP > 100* 36.4% *FDP > 200* 27.3%	*FDP > 100* 84.2% *FDP > 200* 93%

(continued)

Table 3
(continued)

S. no	Study	Sample size	Blood biomarkers	Duration of stroke	Conclusion	Limitations of the study	Sensitivity	Specificity
9.	Cocho (2006) (Spain) [49]	114—IS (8—HT)	Fibrinogen, prothrombin fragments 1+2, Factor XIII, Factor VII, α_2 antiplasmin, PAI-1, TAFI	Within 3 h	None of the hemostatic markers had the potential to detect SICH	Very small number of SICH cases	–	–
10.	Foerch (2007) (Germany) [50]	275—IS (80—HT)	S100B	Within 6 h	A pretreatment S100B value of >0.23 µg/l had a sensitivity and specificity of 46% and 82% to detect PH	Less sensitivity of the biomarker, cases within 3 h required	*S100B 46%*	*S100B 82%*
11.	Mendioroz (2009) (Spain) [51]	199—IS (13—HT)	APC	Within 3 h	APC level > 176% had a sensitivity and specificity of 81.8% and 84.1% to detect PH at 2 hours after tPA	Less number of patients with PH	*APC 81.8%*	*APC 84.1%*
12.	Choi (2012) (Korea) [52]	752—IS (90—HT)	Ferritin	Within 24 h	Ferritin > 144.8 ng/ml had a sensitivity and specificity of 74% each to detect HT while ferritin > 164.1 ng/ml had a sensitivity of 85% and specificity of 75% to detect PH (in 32 patients) Ferritin > 171.8 ng/ml had a sensitivity of 84% and a specificity of 77% to detect sHT (in 37 patients)	A retrospective study, confounding not managed, cases within 3 h required	*Ferritin > 144.8 74%* *Ferritin > 164.1 85%* *Ferritin > 171.8 84%*	*Ferritin > 144.8 74%* *Ferritin > 164.1 75%* *Ferritin > 171.8 77%*

No.	Study	N	Biomarker	Sampling time	Findings	Limitations	Sensitivity	Specificity
13.	Kazmierski (2012) (Poland) [53]	458—IS (33—HT)	TJ proteins (OCLN, CLDN5, ZO1), S100B, NSE, MMP-9, VEGF	Within 178 h	Serum levels of TJ proteins and S100B and VEGF could be an effective marker for screening patients with HT	Less patients with HT, cases within 3 h required	*NSE* 23.5% *S100B* 92.9% *MMP-9* 68.7% *VEGF* 53.3% *OCLN* 58.6% *CLDN5* 64.3% *ZO1* 56.7%	*NSE* 94.8% *S100B* 48.1% *MMP-9* 45.3% *VEGF* 82.4% *OCLN* 67.5% *CLDN5* 65.8% *ZO1* 56%
14.	Guo (2016) (China) [54]	189—IS (45—HT)	NLR	Within 4.5 h	NLR at a cutoff value of 10.59 had a sensitivity of 78.6% and specificity of 79.5% to detect PH in 28 patients while a sensitivity of 76.5% and specificity of 75.6% to detect SICH in 17 patients at 12–18 h after tPA	Small sample size, a control group not experiencing reperfusion therapy was not included	*NLR-10.59 (for PH)* 78.6% *NLR-10.59 (for SICH)* 76.5%	*NLR-10.59 (for PH)* 79.5% *NLR-10.59 (for SICH)* 75.6%
15.	Hernandez-Guillamon (2010) (Spain) [55]	141—IS (48—HT)	VAP-1/SSAO	Within 3 h	The mean levels of VAP-1/SSAO were significantly higher in patients with HT as compared with patients with no HT	Sensitivity and specificity of the biomarker not reported, small sample size	–	

(continued)

Table 3
(continued)

S. no	Study	Sample size	Blood biomarkers	Duration of stroke	Conclusion	Limitations of the study	Sensitivity	Specificity
16.	Chen (2016) (China) [57]	316—IS (42—HT)	LMP2, MECL-1, LMP7, IL1β, Hs-CRP	Within 72 h	High plasma levels of immunoproteasome subunits LMP2, MECL-1, and LMP7 proteins are independently associated with early HT in acute IS	Cases within 3 h required, presence of selection bias, small sample size with less HT cases	LMP2 92.9% MECL-1 90.5% LMP7 97.6%	LMP2 90.1% MECL-1 81% LMP7 90.9%

HT hemorrhagic transformation, *IS* ischemic stroke, *CE* cardio embolic stroke, *MMP* matrix metalloproteinase, *c-Fn* cellular fibronectin, *HI* hemorrhagic infarction, *PH* parenchymal hemorrhage, *SICH* symptomatic intracranial hemorrhage, *tPA* tissue plasminogen activator, *PPV* positive predictive value, *PAI-1* plasminogen activator inhibitor-1, *Lpa* lipoprotein a, *TAFI* thrombin-activated fibrinolysis inhibitor, *FDP* fibrinogen degradation products, *APC* activated protein C, *TJ* tight Junction proteins, *OCLN* occludin, *CLDN5* claudin 5, *ZO1* zonula occludens 1, *NSE* neuron-specific enolase, *VEGF* vascular endothelial growth factor, *NLR* neutrophil to lymphocyte ratio, *VAP-1/SSAO* vascular adhesion protein-1/ semicarbazide-sensitive amine oxidase, *LMP* low molecular mass peptide, *MECL-1* multicatalytic endopeptidase complex-like 1, *IL* interleukin, *Hs-CRP* high sensitive C-reactive protein

2.4 Blood Biomarkers for the Diagnosis of Transient Ischemic Attack

Diagnosis of transient ischemic attack (TIA) is primarily based on clinical history of the patient and thus can be challenging for a physician. Neuroimaging techniques are mainly used to diagnose or exclude an HS and not for the diagnosis of a TIA. Substantial interobserver disagreement has been observed in diagnosing a TIA among experienced neurologists, with Cohen's kappa statistics varying from 0.65 to 0.78 [59]. A blood biomarker approach would be extremely useful which can rapidly detect transient brain ischemia in early phase after symptom onset and therefore would have the potential to be used as a point-of-care test especially in the Emergency Departments. However, the current literature on the blood biomarkers for the diagnosis of TIA and differentiating it from IS and mimics is very scarce. The studies are very limited with a significant lack of validation phases to test the diagnostic performance of the discovered biomarkers.

Recently a study conducted by Penn et al. [60] observed a proteomic biomarker panel of 16 proteins which significantly differentiated TIA/minor IS from stroke mimics in 545 patients suspected of TIA within 36 h of symptom onset. Nine out of these 16 proteins (L-selectin; insulin-like growth factor-binding protein 3; coagulation factor X; serum paraoxonase/lactonase 3; thrombospondin-1; hyaluronan-binding protein 2; heparin cofactor 2; apolipoprotein B-100; and von Willebrand factor) were significant univariate predictors of TIA [60]. In a very small sample of 3 TIA patients and 57 control subjects, Ren et al. did not observe the median levels of GFAP (TIA: 0.004 (0.004–0.09; Controls: 0.004 (0.004–0,02)) and UCH-L1 (TIA: 0.02 (0.02–0.24); Controls: 0.05 (0.02–0.13) to be significantly different between the two groups within 4.5 h of symptom onset [18]. Recum et al. [61] determined the usefulness of copeptin biomarker to differentiate between IS, TIA and mimics in 36 patients (20-IS, 9-TIA, 7-mimics) within 4.5 h of symptom onset. They observed considerably higher levels of copeptin with IS [19.1 (11.2–48.5)] as compared to TIA [9.4 (5.4–13.8)]. However, the range of values was found to be extremely broad in case of mimics [33.3 (7.57–255.7)]. Copeptin had a diagnostic accuracy of 63% with a sensitivity of 80% [61]. George et al. [62] used a mass-spectrometry based proteomics approach to discover blood biomarkers for the diagnosis of TIA within 48 h of symptom onset. In a cohort of 20 TIA, 15 minor stroke and 12 mimics as control group, they observed ceruloplasmin (sensitivity: 80%, specificity: 83.3%), component C8 gamma (C8γ) (sensitivity: 74%, specificity: 83%), and platelet basic protein (PBP) (sensitivity: 71%, specificity: 91%) to be significantly different between the ischemic group (TIA + minor stroke) and control group. However, in the validation phase comprising 22 TIA, 20 minor stroke, and 12 controls, only PBP was found to be significantly different with a sensitivity of 91% and specificity of 57% to diagnose TIA [62]. Table 4 illustrates the studies conducted till now on blood-based biomarkers for the diagnosis of TIA.

3   Inference from Literature

The present scenario offers substantial promise for the development of a blood-based protein biomarker test for stroke diagnosis and differentiation in years to come. Proteomic studies in stroke are still in their infancy and developing a robust diagnostic biomarker for stroke still requires considerable effort.

So far, 21 studies have been conducted to determine blood-based biomarkers for differentiation of IS from HS. However, only one biomarker (i.e., GFAP) has been replicated in several studies and has retained its significance in diagnosing ICH and distinguishing it from IS across several different populations. A meta-analysis of five studies conducted by Zhang et al. also concluded that serum GFAP could be used as a potential biomarker to distinguish ICH from IS within 24 h of symptom onset with a sensitivity and specificity of 81.1% and 95.2% respectively [63]. However, its potential to differentiate ICH from IS within 6 h of symptom onset still needs to be validated for the biomarker to be applicable in clinical settings. Recently, a systematic review by Misra et al. did not recommend the use of any blood biomarker in clinical settings yet for differentiating IS from HS [64]. The review concluded that adequately powered studies with proper validation phases and narrow onset to blood collection time are warranted for the biomarkers to have clinical usefulness.

The diagnostic confidence in stroke is also limited in some patients who present with stroke like symptoms and are termed as "stroke mimics." These mimics are often misdiagnosed in clinical practice [65]. A blood biomarker approach can aid a physician to differentiate these mimics from real stroke cases to improve the certainty of diagnosis. However, in the 17 studies conducted on this research question, no promising biomarker has been discovered to date. The observed sensitivity and specificity have been extremely low for the biomarkers to have any clinical usefulness. More studies with large sample size are required within the appropriate time frame for tPA administration to be possible in real stroke cases.

Along with the misdiagnosis of stroke, the administration for tPA therapy is also hampered due to the risk of HT in IS patients. Sixteen studies on blood biomarkers have been published thus far to determine mixed cases of ischemic stroke with HT and the results have been promising. MMP-9 has turned out to be an

Table 4

Studies determining blood-based biomarkers for the diagnosis of transient ischemic attack

S. no	Study	Sample size	Blood biomarkers	Duration of TIA	Conclusion	Limitations of the study	Sensitivity	Specificity
1.	Penn (2018) (Canada) [60]	545—TIA	141 proteins (plasma)	36 h	A panel of 16 candidate protein biomarkers distinguished TIA from mimics	No validation phase performed	–	–
2.	Ren (2016) (China) [18]	3—TIA 57—Controls	GFAP and UHC-L1	4.5 h	GFAP and UHC-L1 did not differentiate TIA from control subjects	Very small sample size, no validation phase	–	–
3.	Recum (2015) (Germany) [61]	9—TIA 7—Mimics 20—IS	Copeptin	4.5 h	Copeptin levels higher in IS patients compared to TIA but the broad range of values in stroke-mimics limits diagnostic accuracy	Very small sample size, no validation phase, low specificity	80%	44%
4.	George (2015) [62]	20—TIA 15—Minor stroke 12—Mimics Validated 22—TIA 20—Minor stroke 12—Mimics	161 proteins (serum)	48 h	PBP was found to be a potential biomarker for TIA diagnosis	Less sample size in discovery and validation phases.	*Ceruloplasmin* 80% *C8γ* 74% *PBP* 71% *Validate PBP* 91%	*Ceruloplasmin* 83.3% *C8γ* 83% *PBP* 91% *Validate PBP* 57%

IS ischemic stroke, *TIA* transient ischemic attack, *GFAP* glial fibrillary acidic protein, *UCH-L1* ubiquitin C-terminal hydrolase, *PBP* platelet basic protein, *C8γ* component C8 gamma

independent predictor of HT after IS. Several studies have found the levels of MMP-9 to be significantly raised in IS patients with HT as compared to IS patients without HT. The role of MMP-9 in causing HT has been linked to its actions on microvascular integrity [66]. MMP activation plays a major role in basal lamina degradation thereby leading to secondary HT of the ischemic area [67–71]. Animal studies have also observed a significant increase in MMP-9 levels in HT cases of brain infarction [72]. Studies conducted on rat models have shown a significant decrease in the incidence and severity of tPA induced HT after administration of MMP inhibitor [73, 74]. However, further studies are warranted to improve the sensitivity and specificity of MMP-9 biomarker before implementing it in routine clinical practice.

4   Conclusion

The use of blood biomarkers could be an effective alternative in near future for diagnosing ischemic stroke in resource-limited settings, strengthening and confirming the clinical diagnosis and aiding management of stroke patients in appropriate care units. However, based on the studies conducted thus far, implementation of these biomarkers in current clinical practice is not recommended yet. Further well-designed studies are required to validate and test the diagnostic performance of potential biomarkers in different populations.

References

1. Feigin VL, Roth GA, Naghavi M, Parmar P, Krishnamurthi R, Chugh S, Mensah GA, Norrving B, Shiue I, Ng M, Estep K, Cercy K, Murray CJL, Forouzanfar MH, Global Burden of Diseases, Injuries and Risk Factors Study 2013 and Stroke Experts Writing Group (2016) Global burden of stroke and risk factors in 188 countries, during 1990–2013: a systematic analysis for the Global Burden of Disease Study 2013. Lancet Neurol 15:913–924. https://doi.org/10.1016/S1474-4422(16)30073-4

2. Lozano R, Naghavi M, Foreman K, Lim S, Shibuya K, Aboyans V, Abraham J, Adair T, Aggarwal R, Ahn SY, Alvarado M, Anderson HR, Anderson LM, Andrews KG, Atkinson C, Baddour LM, Barker-Collo S, Bartels DH, Bell ML, Benjamin EJ, Bennett D, Bhalla K, Bikbov B, Bin Abdulhak A, Birbeck G, Blyth F, Bolliger I, Boufous S, Bucello C, Burch M, Burney P, Carapetis J, Chen H, Chou D, Chugh SS, Coffeng LE, Colan SD, Colquhoun S, Colson KE, Condon J, Connor

MD, Cooper LT, Corriere M, Cortinovis M, de Vaccaro KC, Couser W, Cowie BC, Criqui MH, Cross M, Dabhadkar KC, Dahodwala N, De Leo D, Degenhardt L, Delossantos A, Denenberg J, Des Jarlais DC, Dharmaratne SD, Dorsey ER, Driscoll T, Duber H, Ebel B, Erwin PJ, Espindola P, Ezzati M, Feigin V, Flaxman AD, Forouzanfar MH, Fowkes FGR, Franklin R, Fransen M, Freeman MK, Gabriel SE, Gakidou E, Gaspari F, Gillum RF, Gonzalez-Medina D, Halasa YA, Haring D, Harrison JE, Havmoeller R, Hay RJ, Hoen B, Hotez PJ, Hoy D, Jacobsen KH, James SL, Jasrasaria R, Jayaraman S, Johns N, Karthikeyan G, Kassebaum N, Keren A, Khoo J-P, Knowlton LM, Kobusingye O, Koranteng A, Krishnamurthi R, Lipnick M, Lipshultz SE, Ohno SL, Mabweijano J, MacIntyre MF, Mallinger L, March L, Marks GB, Marks R, Matsumori A, Matzopoulos R, Mayosi BM, McAnulty JH, McDermott MM, McGrath J, Mensah GA, Merriman TR, Michaud C, Miller M, Miller TR, Mock C,

Mocumbi AO, Mokdad AA, Moran A, Mulholland K, Nair MN, Naldi L, Narayan KMV, Nasseri K, Norman P, O'Donnell M, Omer SB, Ortblad K, Osborne R, Ozgediz D, Pahari B, Pandian JD, Rivero AP, Padilla RP, Perez-Ruiz F, Perico N, Phillips D, Pierce K, Pope CA, Porrini E, Pourmalek F, Raju M, Ranganathan D, Rehm JT, Rein DB, Remuzzi G, Rivara FP, Roberts T, De León FR, Rosenfeld LC, Rushton L, Sacco RL, Salomon JA, Sampson U, Sanman E, Schwebel DC, Segui-Gomez M, Shepard DS, Singh D, Singleton J, Sliwa K, Smith E, Steer A, Taylor JA, Thomas B, Tleyjeh IM, Towbin JA, Truelsen T, Undurraga EA, Venketasubramanian N, Vijayakumar L, Vos T, Wagner GR, Wang M, Wang W, Watt K, Weinstock MA, Weintraub R, Wilkinson JD, Woolf AD, Wulf S, Yeh P-H, Yip P, Zabetian A, Zheng Z-J, Lopez AD, Murray CJL, AlMazroa MA, Memish ZA (2012) Global and regional mortality from 235 causes of death for 20 age groups in 1990 and 2010: a systematic analysis for the Global Burden of Disease Study 2010. Lancet Lond Engl 380:2095–2128. https://doi.org/10.1016/S0140-6736(12)61728-0

3. Nor AM, Davis J, Sen B, Shipsey D, Louw SJ, Dyker AG, Davis M, Ford GA (2005) The Recognition of Stroke in the Emergency Room (ROSIER) scale: development and validation of a stroke recognition instrument. Lancet Neurol 4:727–734. https://doi.org/10.1016/S1474-4422(05)70201-5

4. Hand PJ, Kwan J, Lindley RI, Dennis MS, Wardlaw JM (2006) Distinguishing between stroke and mimic at the bedside: the brain attack study. Stroke 37:769–775. https://doi.org/10.1161/01.STR.0000204041.13466.4c

5. Libman RB, Wirkowski E, Alvir J, Rao TH (1995) Conditions that mimic stroke in the emergency department. Implications for acute stroke trials. Arch Neurol 52:1119–1122

6. Grotta JC, Chiu D, Lu M, Patel S, Levine SR, Tilley BC, Brott TG, Haley EC, Lyden PD, Kothari R, Frankel M, Lewandowski CA, Libman R, Kwiatkowski T, Broderick JP, Marler JR, Corrigan J, Huff S, Mitsias P, Talati S, Tanne D (1999) Agreement and variability in the interpretation of early CT changes in stroke patients qualifying for intravenous rtPA therapy. Stroke J Cereb Circ 30:1528–1533

7. Allard L, Lescuyer P, Burgess J, Leung K-Y, Ward M, Walter N, Burkhard PR, Corthals G, Hochstrasser DF, Sanchez J-C (2004) ApoC-I and ApoC-III as potential plasmatic markers to distinguish between ischemic and hemorrhagic stroke. Proteomics 4:2242–2251. https://doi.org/10.1002/pmic.200300809

8. Lopez MF, Sarracino DA, Prakash A, Athanas M, Krastins B, Rezai T, Sutton JN, Peterman S, Gvozdyak O, Chou S, Lo E, Buonanno F, Ning M (2012) Discrimination of ischemic and hemorrhagic strokes using a multiplexed, mass spectrometry-based assay for serum apolipoproteins coupled to multimarker ROC algorithm. Proteomics Clin Appl 6:190–200. https://doi.org/10.1002/prca.201100041

9. Walsh KB, Hart K, Roll S, Sperling M, Unruh D, Davidson WS, Lindsell CJ, Adeoye O (2016) Apolipoprotein A-I and paraoxonase-1 are potential blood biomarkers for ischemic stroke diagnosis. J Stroke Cerebrovasc Dis 25:1360–1365. https://doi.org/10.1016/j.jstrokecerebrovasdis.2016.02.027

10. Kim MH, Kang SY, Kim MC, Lee WI (2010) Plasma biomarkers in the diagnosis of acute ischemic stroke. Ann Clin Lab Sci 40:336–341

11. Katsanos AH, Makris K, Stefani D, Koniari K, Gialouri E, Lelekis M, Chondrogianni M, Zompola C, Dardiotis E, Rizos I, Parissis J, Boutati E, Voumvourakis K, Tsivgoulis G (2017) Plasma glial fibrillary acidic protein in the differential diagnosis of intracerebral hemorrhage. Stroke 48:2586–2588. https://doi.org/10.1161/STROKEAHA.117.018409

12. Rozanski M, Waldschmidt C, Kunz A, Grittner U, Ebinger M, Wendt M, Winter B, Bollweg K, Villringer K, Fiebach JB, Audebert HJ (2017) Glial fibrillary acidic protein for prehospital diagnosis of intracerebral hemorrhage. Cerebrovasc Dis 43:76–81. https://doi.org/10.1159/000453460

13. Foerch C, Curdt I, Yan B, Dvorak F, Hermans M, Berkefeld J, Raabe A, Neumann-Haefelin T, Steinmetz H, Sitzer M (2006) Serum glial fibrillary acidic protein as a biomarker for intracerebral haemorrhage in patients with acute stroke. J Neurol Neurosurg Psychiatry 77:181–184. https://doi.org/10.1136/jnnp.2005.074823

14. Foerch C, Niessner M, Back T, Bauerle M, De Marchis GM, Ferbert A, Grehl H, Hamann GF, Jacobs A, Kastrup A, Klimpe S, Palm F, Thomalla G, Worthmann H, Sitzer M, BE FAST Study Group (2012) Diagnostic accuracy of plasma glial fibrillary acidic protein for differentiating intracerebral hemorrhage and cerebral ischemia in patients with symptoms of acute stroke. Clin Chem 58:237–245. https://doi.org/10.1373/clinchem.2011.172676

15. Dvorak F, Haberer I, Sitzer M, Foerch C (2009) Characterisation of the diagnostic

window of serum glial fibrillary acidic protein for the differentiation of intracerebral haemorrhage and ischaemic stroke. Cerebrovasc Dis 27:37–41. https://doi.org/10.1159/000172632

16. Undén J, Strandberg K, Malm J, Campbell E, Rosengren L, Stenflo J, Norrving B, Romner B, Lindgren A, Andsberg G (2009) Explorative investigation of biomarkers of brain damage and coagulation system activation in clinical stroke differentiation. J Neurol 256:72–77. https://doi.org/10.1007/s00415-009-0054-8

17. Xiong L, Yang Y, Zhang M, Xu W (2015) The use of serum glial fibrillary acidic protein test as a promising tool for intracerebral hemorrhage diagnosis in Chinese patients and prediction of the short-term functional outcomes. Neurol Sci 36:2081–2087. https://doi.org/10.1007/s10072-015-2317-8

18. Ren C, Kobeissy F, Alawieh A, Li N, Li N, Zibara K, Zoltewicz S, Guingab-Cagmat J, Larner SF, Ding Y, Hayes RL, Ji X, Mondello S (2016) Assessment of serum UCH-L1 and GFAP in acute stroke patients. Sci Rep 6:24588. https://doi.org/10.1038/srep24588

19. Llombart V, García-Berrocoso T, Bustamante A, Giralt D, Rodriguez-Luna D, Muchada M, Penalba A, Boada C, Hernández-Guillamon M, Montaner J (2016) Plasmatic retinol-binding protein 4 and glial fibrillary acidic protein as biomarkers to differentiate ischemic stroke and intracerebral hemorrhage. J Neurochem 136:416–424. https://doi.org/10.1111/jnc.13419

20. Luger S, Witsch J, Dietz A, Hamann GF, Minnerup J, Schneider H, Sitzer M, Wartenberg KE, Niessner M, Foerch C, BE FAST II and the IGNITE Study Groups (2017) Glial fibrillary acidic protein serum levels distinguish between intracerebral hemorrhage and cerebral ischemia in the early phase of stroke. Clin Chem 63:377–385. https://doi.org/10.1373/clinchem.2016.263335

21. Zhou S, Bao J, Wang Y, Pan S (2016) S100β as a biomarker for differential diagnosis of intracerebral hemorrhage and ischemic stroke. Neurol Res 38:327–332. https://doi.org/10.1080/01616412.2016.1152675

22. Roudbary SA, Saadat F, Forghanparast K, Sohrabnejad R (2011) Serum C-reactive protein level as a biomarker for differentiation of ischemic from hemorrhagic stroke. Acta Med Iran 49:149–152

23. Montaner J, Mendioroz M, Delgado P, García-Berrocoso T, Giralt D, Merino C, Ribó M, Rosell A, Penalba A, Fernández-Cadenas I,

Romero F, Molina C, Alvarez-Sabín J, Hernández-Guillamon M (2012) Differentiating ischemic from hemorrhagic stroke using plasma biomarkers: the S100B/RAGE pathway. J Proteome 75:4758–4765. https://doi.org/10.1016/j.jprot.2012.01.033

24. Kavalci C, Genchallac H, Durukan P, Cevik Y (2011) Value of biomarker-based diagnostic test in differential diagnosis of hemorrhagic-ischemic stroke. Bratisl Lek Listy 112:398–401

25. Bustamante A, López-Cancio E, Pich S, Penalba A, Giralt D, García-Berrocoso T, Ferrer-Costa C, Gasull T, Hernández-Pérez M, Millan M, Rubiera M, Cardona P, Cano L, Quesada H, Terceño M, Silva Y, Castellanos M, Garces M, Reverté S, Ustrell X, Marés R, Baiges JJ, Serena J, Rubio F, Salas E, Dávalos A, Montaner J (2017) Blood biomarkers for the early diagnosis of stroke: the Stroke-Chip Study. Stroke 48:2419–2425. https://doi.org/10.1161/strokeaha.117.017076

26. Rainer TH, Wong KS, Lam W, Lam NYL, Graham CA, Lo YMD (2007) Comparison of plasma beta-globin DNA and S-100 protein concentrations in acute stroke. Clin Chim Acta Int J Clin Chem 376:190–196. https://doi.org/10.1016/j.cca.2006.08.025

27. Sharma R, Macy S, Richardson K, Lokhnygina Y, Laskowitz DT (2014) A blood-based biomarker panel to detect acute stroke. J Stroke Cerebrovasc Dis 23:910–918. https://doi.org/10.1016/j.jstrokecerebrovasdis.2013.07.034

28. Glickman SW, Phillips S, Anstrom KJ, Laskowitz DT, Cairns CB (2011) Discriminative capacity of biomarkers for acute stroke in the emergency department. J Emerg Med 41:333–339. https://doi.org/10.1016/j.jemermed.2010.02.025

29. González-García S, González-Quevedo A, Peña-Sánchez M, Menéndez-Saínz C, Fernández-Carriera R, Arteche-Prior M, Pando-Cabrera A, Fernández-Concepción O (2012) Serum neuron-specific enolase and S100 calcium binding protein B biomarker levels do not improve diagnosis of acute stroke. J R Coll Physicians Edinb 42:199–204. https://doi.org/10.4997/JRCPE.2012.302

30. An S-A, Kim J, Kim O-J, Kim J-K, Kim N-K, Song J, Oh S-H (2013) Limited clinical value of multiple blood markers in the diagnosis of ischemic stroke. Clin Biochem 46:710–715. https://doi.org/10.1016/j.clinbiochem.2013.02.005

31. Airas L, Lindsberg PJ, Karjalainen-Lindsberg M-L, Mononen I, Kotisaari K, Smith DJ, Jalkanen S (2008) Vascular adhesion protein-1 in

human ischaemic stroke. Neuropathol Appl Neurobiol 34:394–402. https://doi.org/10.1111/j.1365-2990.2007.00911.x

32. Doehner W, von Haehling S, Suhr J, Ebner N, Schuster A, Nagel E, Melms A, Wurster T, Stellos K, Gawaz M, Bigalke B (2012) Elevated plasma levels of neuropeptide proenkephalin a predict mortality and functional outcome in ischemic stroke. J Am Coll Cardiol 60:346–354. https://doi.org/10.1016/j.jacc.2012.04.024

33. Ahn JH, Choi SC, Lee WG, Jung YS (2011) The usefulness of albumin-adjusted ischemia-modified albumin index as early detecting marker for ischemic stroke. Neurol Sci 32:133–138. https://doi.org/10.1007/s10072-010-0457-4

34. Meng R, Li Z-Y, Ji X, Ding Y, Meng S, Wang X (2011) Antithrombin III associated with fibrinogen predicts the risk of cerebral ischemic stroke. Clin Neurol Neurosurg 113:380–386. https://doi.org/10.1016/j.clineuro.2010.12.016

35. Dambinova SA, Bettermann K, Glynn T, Tews M, Olson D, Weissman JD, Sowell RL (2012) Diagnostic potential of the NMDA receptor peptide assay for acute ischemic stroke. PLoS One 7:e42362. https://doi.org/10.1371/journal.pone.0042362

36. Dassan P, Keir G, Jäger HR, Brown MM (2012) Value of measuring serum vascular endothelial growth factor levels in diagnosing acute ischemic stroke. Int J Stroke 7:454–459. https://doi.org/10.1111/j.1747-4949.2011.00677.x

37. Wendt M, Ebinger M, Kunz A, Rozanski M, Waldschmidt C, Weber JE, Winter B, Koch PM, Nolte CH, Hertel S, Ziera T, Audebert HJ, STEMO Consortium (2015) Copeptin levels in patients with acute ischemic stroke and stroke mimics. Stroke 46:2426–2431. https://doi.org/10.1161/STROKEAHA.115.009877

38. Laskowitz DT, Kasner SE, Saver J, Remmel KS, Jauch EC, BRAIN Study Group (2009) Clinical usefulness of a biomarker-based diagnostic test for acute stroke: the Biomarker Rapid Assessment in Ischemic Injury (BRAIN) Study. Stroke 40:77–85. https://doi.org/10.1161/STROKEAHA.108.516377

39. Sibon I, Rouanet F, Meissner W, Orgogozo JM (2009) Use of the Triage Stroke Panel in a neurologic emergency service. Am J Emerg Med 27:558–562. https://doi.org/10.1016/j.ajem.2008.05.001

40. Knauer C, Knauer K, Müller S, Ludolph AC, Bengel D, Müller HP, Huber R (2012) A biochemical marker panel in MRI-proven hyperacute ischemic stroke-a prospective study. BMC Neurol 12:14. https://doi.org/10.1186/1471-2377-12-14

41. Montaner J, Mendioroz M, Ribó M, Delgado P, Quintana M, Penalba A, Chacón P, Molina C, Fernández-Cadenas I, Rosell A, Alvarez-Sabín J (2011) A panel of biomarkers including caspase-3 and D-dimer may differentiate acute stroke from stroke-mimicking conditions in the emergency department. J Intern Med 270:166–174. https://doi.org/10.1111/j.1365-2796.2010.02329.x

42. Montaner J, Alvarez-Sabín J, Molina CA, Anglés A, Abilleira S, Arenillas J, Monasterio J (2001) Matrix metalloproteinase expression is related to hemorrhagic transformation after cardioembolic stroke. Stroke 32:2762–2767

43. Montaner J, Molina CA, Monasterio J, Abilleira S, Arenillas JF, Ribó M, Quintana M, Alvarez-Sabín J (2003) Matrix metalloproteinase-9 pretreatment level predicts intracranial hemorrhagic complications after thrombolysis in human stroke. Circulation 107:598–603

44. Castellanos M, Leira R, Serena J, Pumar JM, Lizasoain I, Castillo J, Dávalos A (2003) Plasma metalloproteinase-9 concentration predicts hemorrhagic transformation in acute ischemic stroke. Stroke 34:40–46

45. Castellanos M, Leira R, Serena J, Blanco M, Pedraza S, Castillo J, Dávalos A (2004) Plasma cellular-fibronectin concentration predicts hemorrhagic transformation after thrombolytic therapy in acute ischemic stroke. Stroke 35:1671–1676. https://doi.org/10.1161/01.STR.0000131656.47979.39

46. Castellanos M, Sobrino T, Millán M, García M, Arenillas J, Nombela F, Brea D, Perez de la Ossa N, Serena J, Vivancos J, Castillo J, Dávalos A (2007) Serum cellular fibronectin and matrix metalloproteinase-9 as screening biomarkers for the prediction of parenchymal hematoma after thrombolytic therapy in acute ischemic stroke: a multicenter confirmatory study. Stroke 38:1855–1859. https://doi.org/10.1161/STROKEAHA.106.481556

47. Banhawy EE, Amer H, Younes K, Nada MAF, Helmy H, Hassan MI (2014) Plasma matrix metalloproteinase-9 (MMP-9) and hemorrhagic transformation in acute ischemic stroke. Egypt J Neurol Psychiat Neurosurg 51:159–166

48. Trouillas P, Derex L, Philippeau F, Nighoghossian N, Honnorat J, Hanss M, Ffrench P, Adeleine P, Dechavanne M (2004) Early fibrinogen degradation coagulopathy is predictive of parenchymal hematomas in

cerebral rt-PA thrombolysis: a study of 157 cases. Stroke 35:1323–1328. https://doi. org/10.1161/01.STR.0000126040.99024.cf

49. Cocho D, Borrell M, Martí-Fàbregas J, Montaner J, Castellanos M, Bravo Y, Molina-Porcel L, Belvís R, Díaz-Manera J-A, Martínez-Domeño A, Martínez-Lage M, Millán M, Fontcuberta J, Martí-Vilalta J-L (2006) Pretreatment hemostatic markers of symptomatic intracerebral hemorrhage in patients treated with tissue plasminogen activator. Stroke 37:996–999. https://doi.org/10.1161/01. str.0000206461.71624.50

50. Foerch C, Wunderlich MT, Dvorak F, Humpich M, Kahles T, Goertler M, Alvarez-Sabín J, Wallesch CW, Molina CA, Steinmetz H, Sitzer M, Montaner J (2007) Elevated serum S100B levels indicate a higher risk of hemorrhagic transformation after thrombolytic therapy in acute stroke. Stroke 38:2491–2495. https://doi.org/10.1161/ STROKEAHA.106.480111

51. Mendioroz M, Fernández-Cadenas I, Alvarez-Sabín J, Rosell A, Quiroga D, Cuadrado E, Delgado P, Rubiera M, Ribó M, Molina C, Montaner J (2009) Endogenous activated protein C predicts hemorrhagic transformation and mortality after tissue plasminogen activator treatment in stroke patients. Cerebrovasc Dis 28:143–150. https://doi.org/10.1159/ 000225907

52. Choi K-H, Park M-S, Kim J-T, Nam T-S, Choi S-M, Kim B-C, Kim M-K, Cho K-H (2012) The serum ferritin level is an important predictor of hemorrhagic transformation in acute ischaemic stroke. Eur J Neurol 19:570–577. https://doi.org/10.1111/j.1468-1331.2011. 03564.x

53. Kazmierski R, Michalak S, Wencel-Warot A, Nowinski WL (2012) Serum tight-junction proteins predict hemorrhagic transformation in ischemic stroke patients. Neurology 79:1677–1685. https://doi.org/10.1212/ WNL.0b013e31826e9a83

54. Guo Z, Yu S, Xiao L, Chen X, Ye R, Zheng P, Dai Q, Sun W, Zhou C, Wang S, Zhu W, Liu X (2016) Dynamic change of neutrophil to lymphocyte ratio and hemorrhagic transformation after thrombolysis in stroke. J Neuroinflammation 13:199. https://doi.org/10.1186/ s12974-016-0680-x

55. Hernandez-Guillamon M, Garcia-Bonilla L, Solé M, Sosti V, Parés M, Campos M, Ortega-Aznar A, Domínguez C, Rubiera M, Ribó M, Quintana M, Molina CA, Alvarez-Sabín J, Rosell A, Unzeta M, Montaner J (2010) Plasma VAP-1/SSAO activity predicts intracranial hemorrhages and adverse neurological outcome after tissue plasminogen activator treatment in stroke. Stroke 41:1528–1535. https://doi.org/10.1161/STROKEAHA. 110.584623

56. Rodríguez-González R, Blanco M, Rodríguez-Yáñez M, Moldes O, Castillo J, Sobrino T (2013) Platelet derived growth factor-CC isoform is associated with hemorrhagic transformation in ischemic stroke patients treated with tissue plasminogen activator. Atherosclerosis 226:165–171. https://doi.org/10.1016/j.ath erosclerosis.2012.10.072

57. Chen X, Wang Y, Fu M, Lei H, Cheng Q, Zhang X (2017) Plasma immunoproteasome predicts early hemorrhagic transformation in acute ischemic stroke patients. J Stroke Cerebrovasc Dis 26:49–56. https://doi.org/10. 1016/j.jstrokecerebrovasdis.2016.08.027

58. Ribo M, Montaner J, Molina CA, Arenillas JF, Santamarina E, Quintana M, Alvarez-Sabín J (2004) Admission fibrinolytic profile is associated with symptomatic hemorrhagic transformation in stroke patients treated with tissue plasminogen activator. Stroke 35:2123–2127. https://doi.org/10.1161/01.STR. 0000137608.73660.4c

59. Fonseca AC, Canhão P (2011) Diagnostic difficulties in the classification of transient neurological attacks. Eur J Neurol 18:644–648. https://doi.org/10.1111/j.1468-1331.2010. 03241.x

60. Penn AM, Bibok MB, Saly VK, Coutts SB, Lesperance ML, Balshaw RF, Votova K, Croteau NS, Trivedi A, Jackson AM, Hegedus J, Klourfeld E, Yu AYX, Zerna C, Borchers CH, SpecTRA Study Group (2018) Verification of a proteomic biomarker panel to diagnose minor stroke and transient ischaemic attack: phase 1 of SpecTRA, a large scale translational study. Biomarkers 23:392–405. https://doi.org/10. 1080/1354750X.2018.1434681

61. von Recum J, Searle J, Slagman A, Vollert JO, Endres M, Möckel M, Ebinger M (2015) Copeptin: limited usefulness in early stroke differentiation? Stroke Res Treat 2015:768401. https://doi.org/10.1155/2015/768401

62. George PM, Mlynash M, Adams CM, Kuo CJ, Albers GW, Olivot J-M (2015) Novel TIA biomarkers identified by mass spectrometry-based proteomics. Int J Stroke 10:1204–1211. https://doi.org/10.1111/ijs.12603

63. Zhang J, Zhang C-H, Lin X-L, Zhang Q, Wang J, Shi S-L (2013) Serum glial fibrillary acidic protein as a biomarker for differentiating intracerebral hemorrhage and ischemic stroke in patients with symptoms of acute stroke: a systematic review and meta-analysis. Neurol

Sci 34:1887–1892. https://doi.org/10.1007/s10072-013-1541-3

64. Misra S, Kumar A, Kumar P, Yadav AK, Mohania D, Pandit AK, Prasad K, Vibha D (2017) Blood-based protein biomarkers for stroke differentiation: a systematic review. Proteomics Clin Appl 11. https://doi.org/10.1002/prca.201700007

65. Scott PA, Silbergleit R (2003) Misdiagnosis of stroke in tissue plasminogen activator-treated patients: characteristics and outcomes. Ann Emerg Med 42:611–618. https://doi.org/10.1016/S0196064403004438

66. Hamann GF, del Zoppo GJ, von Kummer R (1999) Hemorrhagic transformation of cerebral infarction—possible mechanisms. Thromb Haemost 82(Suppl 1):92–94

67. Anthony DC, Ferguson B, Matyzak MK, Miller KM, Esiri MM, Perry VH (1997) Differential matrix metalloproteinase expression in cases of multiple sclerosis and stroke. Neuropathol Appl Neurobiol 23:406–415

68. Clark AW, Krekoski CA, Bou SS, Chapman KR, Edwards DR (1997) Increased gelatinase A (MMP-2) and gelatinase B (MMP-9) activities in human brain after focal ischemia. Neurosci Lett 238:53–56

69. Montaner J, Alvarez-Sabín J, Molina C, Anglés A, Abilleira S, Arenillas J, González MA, Monasterio J (2001) Matrix metalloproteinase expression after human cardioembolic stroke: temporal profile and relation to neurological impairment. Stroke 32:1759–1766

70. Rosenberg GA, Navratil M, Barone F, Feuerstein G (1996) Proteolytic cascade enzymes increase in focal cerebral ischemia in rat. J Cereb Blood Flow Metab 16:360–366. https://doi.org/10.1097/00004647-199605000-00002

71. Romanic AM, White RF, Arleth AJ, Ohlstein EH, Barone FC (1998) Matrix metalloproteinase expression increases after cerebral focal ischemia in rats: inhibition of matrix metalloproteinase-9 reduces infarct size. Stroke 29:1020–1030

72. Heo JH, Lucero J, Abumiya T, Koziol JA, Copeland BR, del Zoppo GJ (1999) Matrix metalloproteinases increase very early during experimental focal cerebral ischemia. J Cereb Blood Flow Metab 19:624–633. https://doi.org/10.1097/00004647-199906000-00005

73. Lapchak PA, Chapman DF, Zivin JA (2000) Metalloproteinase inhibition reduces thrombolytic (tissue plasminogen activator)-induced hemorrhage after thromboembolic stroke. Stroke 31:3034–3040

74. Sumii T, Lo EH (2002) Involvement of matrix metalloproteinase in thrombolysis-associated hemorrhagic transformation after embolic focal ischemia in rats. Stroke 33:831–836

Chapter 7

Laser-Capture Microdissection for Measurement of Angiogenesis After Stroke

Mark Slevin, Xenia Sawkulycz, Laura Combes, Baoqiang Guo, Wen-Hui Fang, Yasmin Zeinolabediny, Donghui Liu, Glenn Ferris, and Anna Ludlaim

Abstract

Laser capture microdissection has been around for almost a decade now. It has shown great promise for identification and determination of differences between activity and health of cells or even a single cell compared with its neighboring, local, or adjacent tissue inhabitants. Here we will provide a background to its use in neurological and cardiovascular fields and indicate a detailed methodological approach for successful RNA capture and analysis by one of the many current profiling systems available for pattern recognition.

Key words Laser capture microdissection, Stroke, RNA/DNA, Angiogenesis

1 Introduction

Stroke is a leading cause of mortality worldwide [1–3], usually caused by an ischemic event—severe blockage of arteries and reduced blood flow to the brain, inflammation of tissue normally in the brain, or cardiovascular tissue [4]. While some individuals have a genetic predisposition to stroke, diagnosis of stroke is often poor, and finding an effective therapy is important [1]. Angiogenesis is the process in which blood vessels recover and regenerate after a loss of blood flow in the brain, branching off from preexisting vessels to form tube-like vascular structures and anastomosis [3–5] (Fig. 1).

The formation of new blood vessels involves a highly ordered cascade of events that is regulated all or in part by angiogenic factors as well as many mediators and molecules of the innate and

---

The original version of this chapter was revised. The correction to this chapter is available at https://doi.org/10.1007/978-1-4939-9682-7_21

Philip V. Peplow et al. (eds.), *Stroke Biomarkers*, Neuromethods, vol. 147, https://doi.org/10.1007/978-1-4939-9682-7_7,
© Springer Science+Business Media, LLC, part of Springer Nature 2020, corrected publication 2020

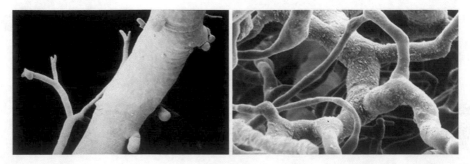

Fig. 1 Scanning electron microscopy of normal rat brains revealed a regular pattern of blood vessels. In the normal cortex, radially arranged penetrating arterioles and venules, were interconnected by an extensive microvascular network. The figure shows images 3 days after rat middle cerebral artery occlusion, the first vascular budding was visible at many sites. Angiogenesis was also observed in larger microvessels (50–150 μm). The smallest microvessels (4–10 μm) formed connections with the surrounding proliferating vessels

adaptive immune systems, in a response of leucocytes in chronic inflammation [2]. Hypoxia plays a role in angiogenesis through stimulus promotion with the growth factors (e.g., hypoxia-inducible transcription factor-1 (HIF-1) and vascular endothelial growth factor (VEGF)). The effect of angiogenesis is to restore the circulation reduced by the ischemic event [3]. Therapies to promote post-ischemic angiogenesis are critically important for the treatment of ischemic stroke [6].

According to Yin et al., several studies in patients with stroke demonstrated that angiogenesis occurs within 3–4 days after stroke and in post-mortem examinations. The brain tissues revealed that patients that suffered from stroke had increased levels of cerebral microvessels in the penumbral areas [6]. However, in some cases, angiogenesis can cause abnormal proliferation of structurally weak and leaky blood vessels leading to undesirable complications of the inflammatory processes [4]. Hypoxia also contributes to the abnormal process when immature neovessels are deprived of oxygen, and develop weak structures with smooth muscle cells [4]. Figure 1 shows Scanning electron micrographs of vascular casts show marked sprouting and anastomoses of two newly formed capillaries within days following blockage of blood vessel in the cerebral grey matter adjacent to ischemic infarction in a rat brain.

Understanding the differences within the extracellular matrix microenvironment that accounts for patent and nonpatent neovessel regeneration has important therapeutic implications.

2   Laser Capture Microdissection (LCM)

Laser capture microdissection (LCM) is a new, easy-to-use technique recently developed to harvest specific regions or cells of interest from complex, heterogeneous tissue samples that contain a mixture of cell populations at molecular level. Prior to the advent of this technology, comparison of sampling involved heterogeneous comparisons between core biopsies usually separated by scalpel which were unable to take into account absolute cellular content and type; hence, the ultimate detail gleaned from global or targeted gene analysis assays was flawed.

The LCM system is made up of an inverted microscope, an infrared laser and a laser control unit, a joy-stick controllable stage, a charge-coupled device (CCD) camera, and a monitor. It is based on the adherence of selected cells and focally melting the captured pathologic lesions by triggering of a quite low energy infrared laser pulse without destroying the unselected tissue area. This causes melting at a precise area of a thermoplastic membrane located between the laser and the sample. The melted membrane forms a composite with the selected tissue area, which can be removed easily by simply lifting the membrane.

The membrane rapidly resolidifies and forms a composite with the targeted cells. The adherence of these cells to the membrane allows for their subsequent removal from the rest of the tissue. LCM can also be used to extract DNA, RNA and proteins from the targeted tissue fragment. Most of the energy from the laser is absorbed by the membrane, so there is minimal damage to the tissue [7].

In circumstances of disease within a tissue, the abnormal cells may be surrounded by many other tissue elements, LCM can be applied to the cells of interest only at the precise level. Using LCM to isolate these cells allows for a sufficient number to be collected and to be molecularly analyzed without interruption of the other cell populations affecting results.

This technique can be utilized in determining cell-specific changes associated with angiogenesis, especially after stroke or in assessment of atherosclerotic plaque lesion stability as seen in Fig. 2 [4]. Understanding the perfect microenvironment associated with patent neovessel formation and function could help us to design more effective therapies to maximize reperfusion and tissue recovery or regeneration.

An example is shown in Fig. 2, where angiogenic microvessels in close proximity to inactive ones/areas without vessels are shown in a human heterogeneous carotid plaque (a and magnified in b). Also shown is a cluster of CD105-active-positive microvessels and a demonstration in one section of a region containing both active and quiescent vessels (c).

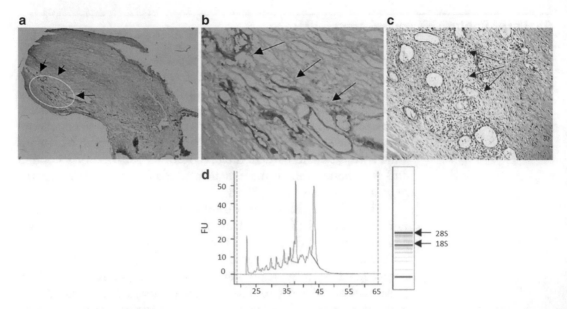

Fig. 2 Immunostaining, identification, and separation of angiogenic microvessels. (**a**) Low-power photomicrograph showing an area of neointima with a high concentration of CD105-positive blood vessels [taken from endarterectomy specimen 305 (age 53 and male with a complex, ulcerated plaque; online version: yellow insert); arrows; ×10]. (**b**) Higher power of the same sample showing a clearly labelled, compact group of blood vessels (arrows; ×100). (**c**) Example of CD105-positive immature and irregularly shaped vessels chosen as positive sample areas. (**d**) Bioanalysis of laser-cut areas was carried out using an Agilent Technologies 2100 analyzer and only samples without evidence of RNA degradation and RIN ≥ 6.0 were chosen for preamplification. Arrows and peaks show 18S and 28S ribosomal RNA. Figure taken from Slevin et al. (J Vasc Res. 2010;47(4):323–335) with agreement of Karger Publishers

LCM is significantly faster, more precise, and more versatile compared to previous microdissection methods [7]. Because LCM does not destroy the adjacent residual tissue, it allows multiple different tissue components to be taken from the same tissue lesion and compared using a battery of genetic, epigenetic, and even protein profiling systems. Our colleagues and ourselves have used LCM for examining gene expression patterns in heterogeneous plaque microenvironments in carotid and coronary arteries, for investigation of neuronal health and function following ischemic stroke, and also, comparing neurodegenerative (dementia) regions versus normal looking areas of the brain. All appropriately prepared histological cryo-tissue could theoretically be used in other disease processes for example in targeting the moment a cancer cell undergoes epithelial–mesenchymal transition and the capability to metastasize.

3 Detailed Methodology[1]

3.1 List of Consumables and Equipment

Tissue of interest
ddH$_2$O (DEPC treated)

Ethanol 70% and 100% (RNase Free)
RNase Zap decontamination solution (Fisher)
Cryostat mounting medium (OCT)
Polyethylene naphthalate (PEN) Membrane Frame Slides (fisher)
Superfrost Plus slides
Microcentrifuge tubes (RNase free)
RNase Zap decontamination solution (Fisher)
ABC elite kit (Vector)
DAB substrate kit (Vector)
CD105 Antibody
CD31 Antibody
Vector Hematoxylin QS (Vector)
Robot microbeam laser microscope
Total RNA Purification Kit (Norgen 48300)
β-Mercaptoethanol
Norgen Total RNA purification kit (reference)
Qubit Fluorometer (Life Technologies)
Qubit RNA HS Assay Kit (Fisher)
Bioanalyzer (Agilent)
Agilent RNA 6000 Pico Kit (Agilent)

3.2 Preparation of Tissues and Slides

- Clean cryostat and section area with 100% ethanol and RNase Zap.
- Cut fresh brain tissue or tissue of interest into small cubes of approximately 1 cm^3 (small pieces may be embedded into Optimal Cutting Temperature (OCT) medium and mold). These blocks should be flash-frozen in liquid Nitrogen.
- Serial sections are cut on a cryostat at experiment 6 μm.

---

[1] Note ∗ indicates a developmental issue or suggestive initiative.

- First section on Superfrost Plus slide for immunohistochemistry (IHC) (CD105—angiogenesis marker).

- Second section on polyethylene naphthalate (PEN) frame slide (laser dissection). Do not allow this section to dry out and store in the cryostat (approximately −20 °C) while cutting other section.

- Third section on Superfrost Plus Slide for IHC (CD31—mature blood vessel marker).

- Fourth section on Superfrost Plus Slide (spare slide).

- Allow sections on Superfrost slides to air dry for 5 min.

- Then move the slides to −20 °C either placing them in dry ice or keeping them in the cryostat.

- Store slides and sample at −80 °C, until ready to use.

   *Notes: *An automated system for equivalent orientation of serial sections is advised otherwise it is difficult to identify exact morphological regions (matching) within the two slides of interest-these can then be overlaid using a computer image analysis platform for best and accurate results.*

3.3 Identification of Region of Interest

Using slides 1 and 3 (Glass Superfrost Plus slides), set up an IHC experiment.

- Acclimatize slides to room temperature.

- Fix slides with 4% paraformaldehyde (PFA) for 5 min and rinse with PBS (Never allow slides to dry out always store in PBS).

- Block for endogenous peroxidases. Incubate slides in 0.5% v/v H_2O_2–methanol for 30 min at room temperature.

- Block for nonspecific antibody binding. Incubate slide in 2% serum PBS for 20 min (Vector ABC kit).

- Incubate with Primary Antibody (either CD105 [8] or CD31 on appropriate slide) for 30 min.

- Incubate with secondary antibody (1:50) for 30 min (Vector ABC kit).

- Incubate with avidin–biotin complex (ABC) (1:50) for 30 min (Vector ABC kit).

- Develop staining with 3′-diaminobenzidine (DAB) substrate solution for 3–10 min (Vector).

- Counterstain with hematoxylin, dehydrate, clear, and then mount with distyrene–tricresyl phosphate–xylene (DPX).

- Analyze sections pathologically to identify regions of interest for laser dissection (Fig. 2).

   *Notes: *A rapid immunostaining program can be used-overall time approximately 5 min and it is possible to directly obtain useful* *quality RNA from these but really this is hit and miss-the processing of the sections even from the simplest PBS wash all have a negative impact on the final RNA/DNA quality and quantity.*

3.4 Slide Preparation for LDM

- Once appropriate slides and regions of a slide are identified using IHC on serial slides 1 and 3, serial slide 2 (PEN frame slide) can then be used for analysis.

- Remove the slide from −80 °C freezer and fix sample in ice-cold 70% ethanol for 5 min.

- Stain the slide with Vector Hematoxylin QS for 30 s.

- Wash the slide by dipping in diethyl pyrocarbonate (DEPC) water, up and down (×10). Then repeat with fresh DEPC water.

- Dehydrate the slide by dipping in 70% ethanol, up and down (×10). Then repeat with 100% ethanol.

- Leave the slide 100% ethanol for at least 1 min (safe step if you need to stain other PEN slides).

- Air-dry the slides in a clean RNase-free environment (e.g., RNA prep area or hood).

- Store the slides in RNase-free slide box and immediately proceed to LDM.

3.5 Laser Capture Microdissection

Please follow manufacturer's instructions on how to set up the software for the microdissection.

- Turn the microscope on and allow the bulb and lasers to warm up.

- Open the laser capture imaging software.

- The polythyelene naphthalate membrane slides are placed on the microscope and visual inspection is performed.

- Using a robot microbeam laser microscope LCM is performed. The slide is placed on the microscope and the laser diameter is set to 15 μm [4].

- The laser will then capture the region of interest which has been outlined on the computer and catapults the tissue into the collection cap.

- Verification step (if needed you can verify that indeed the cells have been captured in the microcentrifuge cap lid by viewing under the optical eyepiece.

- Samples are then transferred into RNase-free microcentrifuge tube containing 600 μL of RLT buffer (RNeasy Lysis Buffer; Norgen, Total RNA purification kit + β-mercaptoethanol) [9].

- Serial sections of interest can be pooled (up 25 mg) to increase RNA yield by adding to RLT buffer.

- Homogenize samples in RL buffer.
- Store samples in RL buffer at -80C for a short time then proceed to RNA isolation.

*Notes: *The laser cutting is not always as straightforward as it should be, thicker tissue sections or "hard" tissues may not cut at the first try—you may need to repeat the cutting cycle or increase the laser intensity; however,this also may "burn" the section, leading to RNA damage and so on.*

3.6   RNA Isolation

For efficient RNA isolation of total RNA and miRNA we use Norgen—Total RNA Purification Kit (Cat. 48300) [9]. Other suitable kits are available from other manufacturers.

- Allow sample in RL Buffer containing β-mercaptoethanol to acclimatize to room temperature before beginning RNA extraction.
- Follow RNA isolation protocol provided with Norgen RNA Isolation kit, starting at point 1B [9].
- The purified RNA samples should be aliquoted and stored at −70 °C.

3.7   RNA Quality Control

The RNA can be quantified using a Qubit Fluorometer and quality of the RNA can be characterized using an Agilent Bioanalyzer.

*Note: *The bioanalysis stage is also sensitive-it is recommended to have an experienced technical officer in charge of the equipment who will supervise the analysis of samples due to the value and lack of replicability of the material.*

3.7.1   Qubit Fluorometer

Using Qubit RNA HS Assay Kit [10], we can accurately measure the concentration of RNA sample concentrations between 250 pg/μL and 100 ng/μL.

3.7.2   Bioanalyzer

Using Agilent RNA 6000 Pico assay kit [11] we can characterize as little as 50 pg of RNA to produce a RIN score (RNA Integrity Number). A RIN value greater than 6 is ideal [4].

- Take an aliquot of RNA from −80 °C freezer on ice and gently thaw on a finger.
- Follow the Qubit RNA HS Assay Kit protocol [10] and measure sample concentration.
- Follow the Agilent RNA 6000 Pico assay kit protocol [11] to characterize RNA quality.

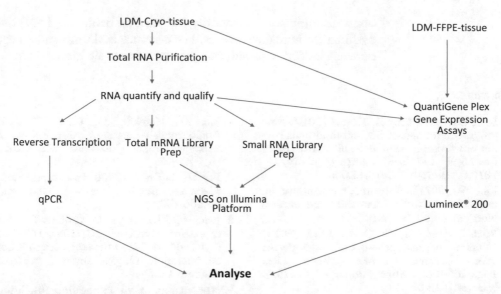

Fig. 3 Flow diagram showing the main processes of RNA extraction, processing, amplification, and possible routes of global analysis

3.8 Downstream Applications

The RNA can be used for analysis using several analysis streams including the following:

RT-qPCR.

RNAseq.

miRNAseq.

QuantiGene Plex Gene Expression Assays.

A flow diagram is provided as Fig. 3 showing the main preparation and analysis steps.

∗Note: *Preamplification of RNA is usually required and care must be taken to ensure that ALL RNA/DNA samples amplify at the same rate and testing before and after PCR is essential including housekeeping genes.*

4   Conclusion

With the need for replicable production of high enough-quality RNA and the method's demand for sufficient quantities of RNA and subsequent amplification strategies and choice of the most appropriate housekeeping genes being tissue and cell specific, obtaining reliable, reproducible, and important data using LCM as a tool is difficult and time-consuming; that is the reality. However, no other methodology can allow for precise unentanglement of critical microenvironmental pathological network pathways that often discriminate and identify the reasons behind cell and tissue dysfunction in major diseases. Perseverance and time-gained experience is indeed required, but the rewards can be high.

References

1. Li Y, Zhong W, Jiang Z, Tang X (2019) New progress in the approaches for blood-brain barrier protection in acute ischemic stroke. Brain Res Bull 144:46–57. https://doi.org/10.1016/j.brainresbull.2018.11.006

2. Libby P (2007) Inflammatory mechanisms: the molecular basis of inflammation and disease. Nutr Rev 65(12 Pt 2):S140–S146

3. Ruan L, Wang B, ZhuGe Q, Jin K (2015) Coupling of neurogenesis and angiogenesis after ischemic stroke. Brain Res 1623:166–173. https://doi.org/10.1016/j.brainres.2015.02.042

4. Slevin M, Turu MM, Rovira N, Luque A, Baldellou M, Krupinski J, Badimon L (2010) Identification of a 'snapshot' of co-expressed angiogenic markers in laser-dissected vessels from unstable carotid plaques with targeted arrays. J Vasc Res 47(4):323–335. https://doi.org/10.1159/000265566

5. Moulton KS (2006) Angiogenesis in atherosclerosis: gathering evidence beyond speculation. Curr Opin Lipidol 17(5):548–555. https://doi.org/10.1097/01.mol.0000245261.71129.f0

6. Yin KJ, Hamblin M, Chen YE (2015) Angiogenesis-regulating microRNAs and ischemic stroke. Curr Vasc Pharmacol 13(3):352–365

7. Fend F, Raffeld M (2000) Laser capture microdissection in pathology. J Clin Pathol 53:666–672

8. Piao M, Tokunaga O (2006) Significant expression of endoglin (CD105), TGF beta-1 and TGFbeta R-2 in the atherosclerotic aorta: an immunohistological study. J Atheroscler Thromb 13:82–89

9. Norgen total RNA Purification Kit. https://norgenbiotek.com/product/total-rna-purification-plus-kit. Accessed 2 Dec 2019

10. Qubit RNA HS Assay Kit. https://www.thermofisher.com/order/catalog/product/Q32852. Accessed 2 Dec 2019

11. Agilent RNA 6000 Pico Kit. https://www.agilent.com/en/product/bioanalyzer-automated-electrophoresis/bioanalyzer-rna-kits-reagents/bioanalyzer-high-sensitivity-rna-analysis-228255#productdetails. Accessed 2 Dec 2019

Part III

Translational Innovative Methods

Blood-Borne Biomarkers of Hypertension Predicting Hemorrhagic and Ischemic Stroke

Alina González-Quevedo, Marisol Peña Sánchez, Sergio González García, María Caridad Menéndez Saínz, and Marianela Arteche Prior

Abstract

Arterial hypertension is the most prevalent noncommunicable disease worldwide and has long been recognized as a major risk factor for stroke, cognitive decline, and dementia. The main goal for prevention is to identify high-risk patients, targeting the modifiable risk factors, and among these, hypertension is the most powerful one. Screening for subclinical brain deterioration is rarely performed because it requires very expensive and not widely available neuroimaging techniques. The main challenge is to detect asymptomatic brain lesions with noninvasive and cost-effective techniques that can be easily performed and interpreted for widespread screening in the community. In this chapter we present an update on the status of blood-based biomarkers explored as alternatives for early detection of brain damage in neurologically asymptomatic subjects (community studies, arterial hypertension, diabetes mellitus, and elderly individuals) and longitudinal studies to explore their value as long-term predictors of incident acute cerebrovascular events. This would undoubtedly have a very positive effect for primary stroke prevention. Numerous blood biomarkers have been investigated with very controversial results. Nevertheless, blood-based brain-specific biomarkers are beginning to stand out in this field and will probably be able to offer much more in the future for the detection of asymptomatic CSVD and in the long-term prediction of acute cerebrovascular events, due to the fact that they can more directly represent what is occurring in the brain.

Key words Arterial hypertension, Cerebral small vessel disease, Stroke, Blood biomarkers, White matter hyperintensities, Lacunar infarct, S100B protein, Neuron-specific enolase, NMDA receptor, Inflammatory biomarkers

1 Introduction

According to the Global Burden of Disease (GBD) 2013 Study on worldwide stroke burden [1] in 2013, stroke was the second most common cause of death (11.8%), preceded only by ischemic heart disease (14.8%). It causes twice as more deaths as AIDS, malaria, and tuberculosis together and is a major cause of acquired disability in adults. Although a tendency towards a decline in stroke incidence, prevalence, mortality, and disability-adjusted life-years rates (DALY) was reported from 1990 to 2013, the overall stroke

Philip V. Peplow et al. (eds.), *Stroke Biomarkers*, Neuromethods, vol. 147, https://doi.org/10.1007/978-1-4939-9682-7_8,
© Springer Science+Business Media, LLC, part of Springer Nature 2020

burden (absolute number of people affected or disabled) has increased in men and women of all ages. In 1–12% of patients the first episode of stroke occurs below the age of 45 years. Stroke incidence from 15–45 years of age is approximately 6–20 cases per 100,000 inhabitants per year. Nevertheless, a very preoccupying fact is that although a 20.1% decline in the number of total stroke deaths has been reported among younger adults in developed countries, this is totally opposite to what has occurred in developing countries where a 36.7% increase has been observed [1].

In Cuba, cerebrovascular diseases are the third cause of death. In 2017, there were 9399 deaths by stroke (crude rate: 88.1 and age adjusted rate: 40.6 per 100,000 inhabitants). Although crude rates were similar in men and women, men exhibited a higher age-adjusted rate than women (46 and 35.7 respectively) [2].

Additionally, stroke leads to a high frequency of sequelae (five million worldwide). It constitutes the main cause of disability-adjusted life-years rate (DALYS): seventh worldwide and fifth in Cuba and the second cause of dementia. Cognitive decline appears in 30–50% of patients, and 1/3 of them develop post-stroke depression. Thus, stroke prevention must be a primary goal for healthcare systems throughout the world. It has been reported that prevention is approximately ten-fold more cost-effective than treating stroke [3].

The main goal for prevention is to identify high-risk patients, targeting the modifiable risk factors, and among these, hypertension is the most powerful one, only surpassed by age over which there is no control. Approximately 54% of strokes worldwide can be attributed to hypertension and individuals with hypertension are 3–4 times more likely to suffer a stroke than those without hypertension [4].

Additionally, arterial hypertension is one of the most prevalent diseases, affecting one billion individuals worldwide and causing considerable morbidity. The overall prevalence of hypertension in US adults was reported as 29.0%, increasing to 64.9% in those over 50 years of age [5]. In Cuba the estimated prevalence for 2017 was 22.5% [2], but this value is considered to be underestimated, with a suspected prevalence over 30% [6].

In patients with essential hypertension (HT) clinical evaluation includes the identification of target organ damage, which comprises basic and optional laboratory tests to assess mainly cardiorenal damage. The 2017 American Guideline for the Prevention, Detection, Evaluation, and Management of High Blood Pressure in Adults [7], specifically addresses this issue, but brain assessment is not included, due to the high costs brain MRI scanning imposes. Nevertheless, the brain is one of the major organs that suffer from the harmful effects of high blood pressure [8, 9], as much as or even more than the heart and kidneys [10–12].

Several studies have demonstrated that subclinical brain damage, mainly associated with cerebral small vessel disease (CSVD), is present with increasing frequency in aging [13, 14] and in asymptomatic individuals suffering from arterial hypertension [15–19] and diabetes mellitus [20, 21]. The toll of CSVD expresses itself in the future development of cognitive decline, dementia and stroke [22, 23]. As brain MRI scanning—the gold standard for CSVD diagnosis—is very costly and not readily available, the search for alternative methods demonstrating asymptomatic neurological damage by HT is imperative in order to identify individuals at higher risks in the population. Several methods have been evaluated that are less expensive and more available than MRI: ambulatory blood pressure monitoring [18, 24, 25], quantitative retinal microvascular assessment [26, 27], quantitative electroencephalography [28, 29], carotid ultrasonography [10, 30, 31], and neurocognitive assessment [9, 18, 32, 33]. Blood-based biomarkers have also been evaluated in this context, ranging from routine laboratory tests, through inflammatory, hemostatic, endothelial function, neurohormonal, and oxidative stress biomarkers to brain-specific proteins [34–37]. To date, none of these techniques has passed into clinical practice for different reasons (not widely available, require high expertise for interpretation, and studies with insufficient sample numbers, among others), but there is no doubt that finding one or more biomarkers would facilitate decision making for the clinicians in charge of these patients. The idea of employing blood-based biomarkers for predicting CSVD in hypertensive subjects is very attractive, because their clinical application would be cost-effective and minimally invasive, offering the opportunity to help in identifying patients who require referrals for neuroimaging studies or follow-up examination.

In this chapter we focus on blood biomarkers related to inflammation, endothelial dysfunction, oxidative stress, hemostasis, and brain-specific proteins, among others, and their possible predictive value for CSVD and for stroke in asymptomatic individuals.

2 Cerebral Small Vessel Disease (CSVD)

CSVD refers to a syndrome that comprises clinical and imaging findings that are thought to result from pathologies in perforating cerebral arterioles, capillaries, and venules [38]. The hallmark of the disease is the presence of white matter lesions (WMHs), enlarged perivascular spaces (EPVSs), lacunar infarcts and cerebral microbleeds (CMBs), which are visualized on a brain magnetic resonance imaging (MRI) study [22, 39].

CSVD is a disease which progresses throughout life, tripling the risk of stroke, doubling the risk of dementia, and in general increasing the risk of disability and death [40]. Furthermore, it is closely related to age, hypertension and other vascular risk factors. A recent meta-analysis demonstrated that the combination of two imaging findings was more strongly associated with the risk of stroke occurrence than only one finding [41]. In fact, individual and combinations of CSVD findings are strongly associated with future stroke, both ischemic and hemorrhagic, all-cause dementia, depression, and all-cause mortality [41].

Subtle and more spatially extensive blood–brain barrier (BBB) leakage has been reported in patients with CSVD compared with age- and sex-matched controls without clinically overt cerebrovascular diseases, supporting the generalized nature of CSVD and the conceptual idea of spatially diffuse microvascular endothelial failure [42].

2.1 White Matter Lesions

White matter hyperintensities (WMHs) are a wide spectrum of confluent hyperintensities without tissue cavitation, generally located surrounding ventricles, which varies from mild perivascular tissue damage to more severe ischemic damage, the later characterized by extensive myelin and axonal loss due to endothelial dysfunction, disturbance of the microcirculation, and BBB breakdown [43]. They can be visualized on fluid-attenuated inversion recovery (FLAIR) and proton density T2-weighted images, without prominent hypointensity on T1-weighted images [44]. WMHs are the most common and frequently the earliest tissue lesions that can be observed, with a prevalence close to 80% or greater in individuals 60 years of age or older [45]. Population-based studies have reported a prevalence of WMHs ranging from 51% among people aged 44–48 years up to 95% in people aged 60–90 years [45, 46].

The occurrence and severity of WMHs is strongly associated with age and hypertension [47, 48], however other risk factors including diabetes mellitus (DM), atherosclerosis, hyperlipidemia, smoking, and hyperhomocysteinemia have also been associated [49, 50]. Long-term HT results in lipohyalinosis of the media and thickening of the vessel walls, with further narrowing of the lumen of the arterioles and small perforating arteries that are derived from cortical and leptomeningeal arteries and nourish the deep WMHs [51, 52]. In addition, hypertension increases blood vessel fibrosis, altering the distribution of extracellular matrix proteins and resulting in stiffening of the vessel walls and a reduction in cerebral blood flow, especially at times of increased need [53]. An important risk factor, DM promotes changes similar as those produced by hypertension in cerebral small vessels, and elevated fasting glucose and DM diagnosis have been associated with the detection of WMHs [54, 55]. A longitudinal study has shown an expedited WMH progression in patients with diabetes [56].

2.2 Enlarged Perivascular Spaces

Perivascular or Virchow–Robin spaces are spaces that surround perforating arterioles and venules as they travel from the subarachnoid space through the brain parenchyma and serve as an important drainage system for interstitial fluids in the brain. Their increased size or enlargement denotes early CSVD in subjects with vascular overload. In T2-weighted MRI study, they are visualized as punctate or linear signal intensities, similar to the image of cerebrospinal fluid [57, 58].

Long-lasting HT during decades damages the blood vessels and initiates the expression of hypoxia-sensitive genes and molecular cascades during its hypoxic phase. Inflammation is ultimately induced by the release of cytokines, matrix metalloproteinases (MMP), and cyclooxygenase-2, and these, in turn increase the permeability of the BBB resulting in the induction of the expression of adhesion molecules in endothelial cells and thereby contributing to leukocyte and platelet activation and adhesion with additional microvascular occlusion [59, 60]. The disarrangement of the BBB leads to leakage of plasma components through the BBB into the vessel wall and perivascular space, promoting their enlargement [61, 62].

2.3 Lacunar Infarcts

Silent lacunar infarcts are defined as brain lesions of 3–15 mm in diameter caused by the occlusion of a perforating artery, with several parenchymal locations, such as internal capsule, basal ganglia, corona radiata, thalamus, and brainstem [63]. Its prevalence varies from 8% to 28% [20]; however, community studies have reported up to 68% of lacunar infarcts [64] Different risk factors, promote the occurrence of lacunar infarcts in multiple sites: HT facilitates the occurrence of new lacunar infarcts in the deep white matter, hyperhomocysteinemia with lacunar infarcts in basal ganglia, and hyperlipidemia always leads to isolated lacunar infarcts in the deep gray nuclei/internal capsule [65, 66].

Silent brain lacunar infarcts are closely related to the occurrence and progression of cognitive decline, and dementia [67, 68], specifically small, deep brain infarcts cause strategic infarct dementia, and the lesion has also been associated with dementia risk [69], cognitive decline, dementia, gait disturbance, urinary incontinence, and disability [20, 64–70].

2.4 Cerebral Microbleeds

Cerebral microbleeds (CMBs) are homogeneous, small (diameter < 10 mm), round, or ovoid hypointensities evident on susceptibility-weighted imaging or T2$*$ Gradient-Recall Echo MRI sequences [71]. These signs correspond to areas of hemosiderin deposits that are caused by the prior leakage of blood from small arteries, arterioles, and/or capillaries [72]. CMBs can be classified according to their location to be either lobar (cortical and subcortical hemispheres) or deep (thalamus, basal ganglia, internal capsule) or infratentorial.

CMBs occur more frequently with increasing age and vascular risk factors [73] such as HT, DM, and smoking [74, 75]. Previous data show that HT increases the expression of the cytokine tumor necrosis factor-α (TNF-α) [76], which is a pivotal regulatory cytokine that is secreted primarily by macrophages and microglia. In addition, higher levels of TNF receptor 2 promotes the pathogenesis of CMB [77], and finally APOE ε4 has been found to be associated with lobar microbleeds [74, 75, 78]. Multiple population-based studies have investigated the epidemiology of CMBs, and overall, the prevalence ranges between 3.1% and 23.5% [74].

3   Mechanisms Involved in the Pathogenesis of Hypertension-Induced Stroke

Hypertension affects more than 60% of individuals aged 65 years or older and more than 80% of people older than 85 years; it is the main risk factor for CSVD and stroke [1, 79]. There is a delay between the onset of hypertension and a hypertensive complication. During this period, a series of changes occur in the cardiovascular system including cerebral circulation. Hypertension is related to multiple pathophysiological processes, such as inflammation, vascular remodeling and oxidative stress, which contribute to the pathogenesis of stroke in hypertension [4].

Most of the lesions found in CSVD can be considered as hypertensive vascular lesions, as it promotes the development of atherosclerotic plaques in cerebral arteries and arterioles, which may lead to arterial occlusions and ischemic injury. Once initiated, CSVD typically progresses slowly over a period of many years, associated with age, HT, and other vascular risk factors [4].

Risk factors for stroke are grouped as non-modifiable, modifiable, and potentially modifiable, and the main goal in prevention is to identify high-risk patients who could develop subclinical CSVD and to target the modifiable risk factors, such as HT, DM, hyperlipidemia, obesity, atrial fibrillation, and smoking, in order to avoid stroke [80, 81]. HT is the most powerful modifiable factor; it is estimated that approximately 54% of strokes worldwide can be attributed to hypertension alone [82], and hypertensive subjects are 3–4 times more likely to suffer a stroke than those without hypertension [83].

HT influences the cerebral circulation with a reduction in cerebral blood flow, causing important adaptive vascular changes in cerebral blood vessels: vascular stiffening, increased pulse pressure, lipohyalinosis of penetrating arteries and arterioles, and favoring the development of atherosclerotic plaques in cerebral arteries [84, 85]. Specifically, HT induces hypertrophy, hyperplasia, and rearrangement and remodeling of smooth muscle cells in cerebral arteries, with further narrowing of the vessel lumen and increased pulse pressure [84]. Biochemical pathways to promote vascular remodeling include growth factors, oxidative stress, reactive oxygen and nitrogen species (ROS and NOS), activation of matrix metalloproteinases, and nitric oxide synthesis [84, 86].

HT alters endothelium dependent relaxation of cerebral blood vessels by decreasing the release of vasodilators. In addition, an increased vascular tone secondary to endothelial dysfunction enhances cerebral blood flow reduction. Endothelial dysfunction also results in overproduction of nitric oxide, which may increase the permeability of cerebral vessels and attenuate myogenic auto-regulation, causing an increase in perfusion pressures in order to maintain the adequate cerebral blood flow [87]. Finally, remodeling and hypertrophy in hypertensive subjects contribute to the shift in autoregulation by reducing the vascular lumen and increasing cerebrovascular resistance [87, 88].

Several other processes, in addition to changes in cerebral blood flow and endothelial dysfunction, are related to the pathogenesis of stroke induced by HT. Some of the most important are oxidative stress and inflammation. Changes in the antioxidant capacity of the cell, or excess generation of ROS, can result in oxidative stress. Oxidative stress in the cerebral blood vessels plays a critical role in the pathogenesis of hypertension [89, 90], through angiotensin II, via activation of NADPH oxidase in the brain vasculature [91] and through the control of specific neurohumoral changes [92]. In addition, HT itself can result in oxidative stress on the basis that ROS production in cerebral small vessels is increased in angiotensin II-induced hypertension [93, 94].

There is increasing evidence supporting the role of vascular inflammation in the pathogenesis of hypertension. Sustained hypertension stimulates inflammatory reactions in cerebral blood vessels because of the production of chemokines (CXCL-8, CCL2, CCL3), cytokines (IL-1, IL-6, TNF-α), and adhesion molecules (ICAM, VCAM, selectins); proliferation of lymphocytes; and recruitment of neutrophils and other immune cells. These inflammatory reactions lead to intracellular activation signaling cascades (transcription factors NFkβ, MAPk) promoting the transcription of inflammatory genes, and perpetuating the inflammatory response [94, 95]. Conversely, inflammation causes oxidative stress, since activated immune cells have been shown to produce ROS and express Angiotensin II, resulting in oxidative stress and hypertension [96].

In addition to hypertension, other risk factors such as dyslipidemia, diabetes, smoking, and obesity accelerate endothelial dysfunction and the occurrence of atherosclerotic processes, through the accumulation and modifications of low-density lipoproteins, microvascular complications, abnormal proliferation of endothelial cells, and increased vascular permeability [97]. In fact, endothelial dysfunction, oxidative stress, and inflammation are some of the most important pathological processes linked to arterial hypertension in a self-perpetuating vicious cycle [4] with devastating consequences on brain parenchyma, leading to CSVD and stroke.

4 Blood-Borne Biomarkers as Predictors of CSVD in Asymptomatic Individuals

Blood biomarkers have been employed frequently to evaluate the pathophysiology of arterial hypertension, its progression and complications, as well as therapeutic efficacy and adverse effects of drugs [98, 99]. While their use for the evaluation of heart and kidney damage is well known and has been established in the different Hypertension Guidelines during the last two decades, this is not the case for the evaluation of brain damage. The BBB imposes an important obstacle for the passage of substances from and to the brain. Although BBB leakage is present in many neurological conditions, what the damaged brain tissue adds to the concentrations of most substances in the blood—comprised of the input from all peripheral organs—is very sparse in general (except for brain-specific molecules). Thus, blood biomarkers have not been introduced in neurological clinical practice as it has in diseases of the heart, kidneys, liver, etc. Nevertheless, numerous studies have been conducted employing blood biomarkers for diagnosis and follow-up, and most specifically as biomarkers of prognosis and risk evaluations in different neurological settings. The possibility of detecting subtle changes in blood concentrations of different substances or of quantifying them at concentrations at micro or nano levels, that is, as in brain-specific proteins, is undoubtedly increasing with the greater availability and sensitivity of current methods.

4.1 Clinical Laboratory Markers

An inverse relationship has been reported between serum bilirubin levels and the prevalence of cardiovascular diseases, as well as with the risk of other atherosclerotic diseases [100]. For the first time, a graded, inverse relationship between serum levels of total bilirubin and increased prevalence of silent cerebral infarctions was shown in middle-aged Chinese individuals (2865 subjects undergoing medical checkup) [101]. In another study including 1121 apparently healthy Japanese adults (62 ± 10 years) low total bilirubin levels were associated with a high prevalence of severe deep white matter lesions [102]. The previous findings suggest that total bilirubin could be a novel biochemical indicator for CSVD regardless of classical cardiovascular risk factors and point towards a possible vasoprotective effect of mildly elevated total bilirubin levels. Nevertheless, it should be noted that in large artery atherosclerotic strokes hyperbilirubinemia was reported and found to be indicative of poor prognosis [103]. Bilirubin is known for its antioxidant properties, but high levels of bilirubin are neurotoxic. The mechanisms through which moderately increased bilirubin could exert vasoprotective effects are currently not understood, but inhibition of lipid oxidation, of immune reactions and inflammatory processes, of cell migration and proliferation, of apoptosis, and its relation to enhanced heme oxygenase-1 activity have been suggested. Additional large and long-term observational studies in different ethnic groups are required to establish the role of serum total bilirubin as a causal risk factor for coronary heart disease and ischemic stroke [100].

Plasma creatine kinase (CK) activity was introduced 50 years ago to identify female carriers of Duchenne muscular dystrophy, but presently this has been substituted with genetic testing. It is still assayed in some neurological settings in the initial steps of evaluation and for follow-up of many neuromuscular diseases. Currently there has been a shift of attention towards its role as a cardiovascular risk factor [104].

CK is an enzyme amply expressed in the mitochondrion and cytosol that catalyzes the transfer of a phosphoryl group from creatine phosphate to adenosine diphosphate, thereby forming creatine and adenosine triphosphate (ATP). It transports creatine phosphate to subcellular locations of high-energy demands, thus connecting sites of ATP production (glycolysis and mitochondrial oxidative phosphorylation) with subcellular sites of ATP consumption (myosin ATPase and myosin light chain kinase at the contractile proteins, and Ca^{2+}-ATPase and Na^+/K^+-ATPase at cellular membranes [105].

High resting plasma and intracellular CK activities have been reported in healthy men, in obesity and in individuals of West African ancestry [106]. Additionally, plasma CK was found to be a main predictor of blood pressure in the general population, independent of age, sex, body mass index, or ancestry [106–108]. High plasma CK activity is thought to enhance bleeding risk through reduction of circulating adenosine diphosphate [109]. Furthermore, it has been proposed that CK enhances ATP buffer capacity for cardiovascular contractility and renal sodium retention [106, 110].

Taking these findings into account, the CK system has been considered a target for pharmacologic intervention in cardiovascular disease. Horjus et al. [105] reviewed existing therapeutic strategies and concluded that the available trials showed an improvement in abnormal heart rhythm and shortness of breath in heart disease, but the effect on mortality, progression of myocardial infarction, and ejection fraction were unclear.

Despite the abundant evidence associating plasma CK activity with cardiovascular diseases and specifically with arterial hypertension, to our knowledge there are no studies evaluating its activity in target organ damage to the brain. Nevertheless, one study including 5026 initially healthy Japanese (mean age: 54.5 years) without a history of myocardial infarction or stroke found that after a follow-up of 11.8 years, serum CK levels in initially healthy subjects predicted first-ever myocardial infarction in the future, whereas no relationship was observed between s-CK levels and the risk for stroke [111].

In line with this, we examined the results obtained in our laboratory with 54 asymptomatic hypertensive patients (59.6 ± 9.2 years) and 28 controls (54.2 ± 12.1 years). Significantly higher total CK (CK-NAC) and its cardiac isoform (CK-MB) activities were observed in patients with respect to controls (CK-NAC: 104.1 ± 73.2 vs. 72.5 ± 19.2 IU/L; $t = 2.033$, $p = 0.045$ and CK-MB: 20.0 ± 11.7 vs. 12.4 ± 5.7 IU/L; $t = 3.219$, $p = 0.002$). In 27 hypertensive patients, 3 T brain MRI scanning was available, and a highly significant association of classical brain MRI lesions (white matter hyperintensities: Fazekas 2–3 and/or silent lacunar infarcts and/or microbleeds) with higher levels of CK-MB activity was evidenced (Fig. 1). No association with brain MRI variables was observed for total CK activity.

These findings, although limited due to the small sample size, suggest that serum CK-MB could be a biomarker of asymptomatic CSVD if future studies in larger populations replicate these results. Hence, this mostly abandoned test for striated muscle damage seems to be gaining momentum in the evaluation of hypertension, bleeding, and cardiovascular disease, and it might also prove to be useful in predicting subclinical brain damage in asymptomatic individuals with vascular risk factors.

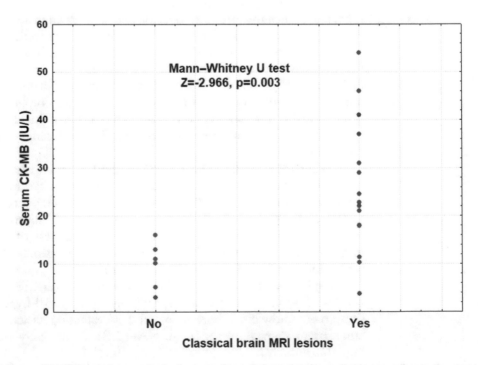

Fig. 1 Serum CK-MB levels in neurologically asymptomatic hypertensive patients according to the presence or absence of classical brain MRI lesions

Gamma-glutamyl transferase (GGT) is a glycoprotein, widely distributed in plasma membranes of various cells and organ tissues. Its most widespread use is as a biomarker of liver injury or excessive alcohol intake, and is associated with cardiovascular disease and its risk factors. It has been reported that baseline serum GGT level is associated with an increased risk of hypertension in the general population, consistent with a linear dose–response relationship [112].

Some studies, including a meta-analysis, also reported the association of GGT with stroke; however, most of these had important limitations and currently, GGT is not widely used to evaluate risk of stroke in clinical practice [113]. Therefore, very recently Yang et al. conducted a national survey in South Korea, with 456,100 eligible participants, of whom 1.64% developed stroke. GGT was independently correlated with increased risk of stroke after adjustment for alcohol consumption and stroke risk factors. The authors concluded that in the Korean population GGT was a novel biomarker predicting stroke risk, independently from alcohol consumption and other risk factors [113].

4.2 Inflammatory Markers

Low-grade systemic inflammation underlies the pathophysiology of HT. C-reactive protein (CRP) and interleukin-6 (IL-6) are probably the most extensively studied inflammatory biomarkers in relation to CSVD, although results are not consistent among all populations [114–116].

CRP is an unspecific biomarker reflecting a systemic inflammatory state, and it has been associated with the risk and severity of ischemic stroke in healthy and in prehypertensive populations [117–120].

Cross-sectional studies have established significant, independent associations between blood inflammatory biomarkers and brain MRI findings related to CSVD [77, 114, 121]. Nevertheless, evidence relating inflammatory markers and brain damage is controversial. Table 1 summarizes results reported for the association of several blood biomarkers with the presence of brain MRI lesions in neurologically asymptomatic individuals. The most frequently explored biomarkers were CRP and IL-6. Blood CRP levels in relation to brain MRI lesions were evaluated in nine studies; half of them reported elevated levels with more severe WMH, while the other half found no association (WMH levels were not tested in one study). Only three studies tested silent lacunar infarcts (two of them found elevated CRP—67%, no changes—33%) and microbleeds (one of them reported elevated CRP—33%, no changes—67%). Blood IL-6 levels were measured in five studies, half of them reported elevated levels with more severe WMH, while the other half found no association (WMH levels were not tested in one study). Silent lacunar infarcts and microbleeds were tested in only one study, where higher levels were associated with the presence of

Table 1
Association of several blood-based biomarkers with the presence of brain MRI lesions in neurologically asymptomatic subjects

| Biomarkers | Author (year) [reference] | Participants | Sample characteristics | Brain MRI | | |
				WML	Lacunar infarcts	Microbleeds
C-reactive protein (CRP)	Hoshi et al. (2005) [122]	194	Neurologically asymptomatic patients	↔	↑	n/t
	Wright et al. (2009) [123]	527	Stroke-free community-based study	↔	n/t	n/t
	Kim et al. (2011) [124]	1586	Asymptomatic subjects	↑	n/t	n/t
	Miwa et al. (2011) [121]	131	Asymptomatic subjects	n/t	n/t	↑
	Satizabal et al. (2012) [118]	1841	Community-based elderly	↑	n/t	n/t
	Miralbell et al. (2012) [125]	86	Dementia-free subjects—no history of vascular disease (aged 50–65 years)	↑	n/t	n/t
	Abe et al. (2014) [115]	228	Healthy volunteers	↑[a]	n/t	n/t
	Shoamanesh et al. (2015) [77]	1763	Community-based study	↔	↔	↔
	Mitaki et al. (2016) [116]	519	Neurologically normal subjects	↔	↑	↔
Interleukin-6 (IL-6)	Hoshi et al. (2005) [122]	194	Neurologically asymptomatic patients	↔	↑	n/t
	Miwa et al. (2011) [121]	131	Asymptomatic subjects	n/t	n/t	↑
	Satizabal et al. (2012) [118]	1841	Community-based elderly	↑	n/t	n/t
	Abe et al. (2014) [115]	228	Healthy volunteers	↑[a]	n/t	n/t

(continued)

Table 1
(continued)

Biomarkers	Author (year) [reference]	Participants	Sample characteristics	Brain MRI		
				WML	Lacunar infarcts	Microbleeds
Interleukin-18 (IL-18)	Miwa et al. (2011) [121]	131	Asymptomatic subjects	n/t	n/t	↑
Lipoprotein-associated phospholipase A2 ([Lp-PLA2)	Wright et al. (2009) [123]	527	Stroke-free community-based	↑	n/t	n/t
	Shoamanesh et al. (2015) [77]	1763	Community-based	↑	↑	n/t
Myeloperoxidase	Wright et al. (2009) [123]	527	Stroke-free community-based	↑	↔	n/t
	Shoamanesh et al. (2015) [77]	1763	Community-based	↓	↓	↑
Intercellular adhesion molecule 1 (ICAM-1)	Shoamanesh et al. (2015) [77]	1763	Community-based	↑	↑	↔
	Han et al. (2009) [126]	175	Asymptomatic subjects (age (60 years)	↑	n/t	n/t
Plasminogen activator inhibitor (PAI)	Miralbell et al. (2012) [125]	86	Dementia-free subjects—no history of vascular disease (aged 50–65 years)	↑	n/t	n/t
sRAGE	Hudson (2011) [127]	1102	Stroke-free, community-based, >55 years	↓	↓	n/t
N-terminal brain natriuretic peptide (NT-proBNP)	Reinhard et al. (2012) [128]	1501	CVD–stroke-free diabetic patients	↑	↑	n/t
	Vilar et al. (2016) [129]	278	Asymptomatic hypertensive patients	↑	↑	n/t

(continued)

Table 1
(continued)

Biomarkers	Author (year) [reference]	Participants	Sample characteristics	Brain MRI		
				WML	Lacunar infarcts	Microbleeds
Cardiac troponin T	Reinhard et al., 2012 [128]	1502	CVD–stroke-free diabetic patients	↑	↑	n/t
Tumor necrosis factor receptor 2	Shoamanesh et al.	1763	Community-based	↔	↔	↑
Osteoprotegerin	(2015)			↑	↑	↔
Tumor necrosis factor α (TNFα)	[77]			↔	↔	↔
Asymmetrical dimethylarginine (ADMA)	Pikula et al. (2009) [130]	2013	2013 stroke-free Framingham offspring	↑	n/t	n/t
Leukocyte count	Kim et al. (2011) [124]	1586	Asymptomatic subjects (mean age: 53.6 years)	↑	n/t	n/t

CVD cardiovascular disease (coronary heart disease or heart failure), ↑ increased biomarker concentration, ↓ decreased biomarker concentration, ↔ no change in biomarker concentration; *n/t* not tested
[a]Only for periventricular hyperintensity, not for deep-and-subcortical white matter hyperintensity

these lesions. Miwa et al. [121] reported significantly higher levels of hsCRP, IL-6, and IL-18 in patients with microbleeds after adjusting for age, sex, cardiovascular risk factors, presence of silent lacunar strokes, and WMHs.

These results show that the interrelation between these two biomarkers (CRP and IL-6) and brain MRI lesion load is very controversial through the different investigations. WMH were the most frequently evaluated lesions; nevertheless, the results are not clear. Many factors could be responsible: study design, methods employed for biomarker determinations, sample size, age, and different ethnic composition, among others.

Elevations of the inflammatory markers lipoprotein-associated phospholipase A2 and myeloperoxidase, but not CRP, were found to be associated with a greater burden of WMH [131]. On the other hand, Shoamanesh's study in a large middle-aged, community-based sample is perhaps one of the most extensive concerning the possible role of inflammation in the pathogenesis of CSVD [77]. Assessing the association of a panel of 15 systemic inflammatory markers with MRI findings they reported increased levels of tumor necrosis factor (TNF) receptor 2 and myeloperoxidase in relation with cerebral microbleeds, but not with WMH or lacunar infarcts. Higher levels of osteoprotegerin, intercellular adhesion molecule-1 (ICAM-1) and lipoprotein-associated phospholi-pase A2, as well as lower myeloperoxidase, were observed in participants with greater WMH volumes and silent cerebral infarcts (as opposed to the results obtained by the previously mentioned work), while no associations were observed between cerebral microbleeds and osteoprotegerin, ICAM-1, and lipoprotein-associated phospholipase A2. Additionally, no changes in CRP or TNF-levels accompanied brain MRI lesions (Table 1).

Leukocyte count was employed as a marker of systemic inflammation in a study including 1586 asymptomatic individuals. Elevated blood leukocyte count was associated with moderate to severe WMH, independently of age, hypertension, and DM [124].

Lower serum levels of soluble receptors for advanced glycation end-products (sRAGE) were associated with higher prevalence of silent brain infarcts and WMH in stroke-free Hispanic and African-American subjects in the USA, suggesting that sRAGE may be predictive of asymptomatic CSVD, particularly in ethnic/racial groups with increased risk for cerebrovascular disease [127].

ICAM-1 participates in inflammatory endothelial activation and is also involved in the pathogenesis of cerebral small and large vessel diseases. An association of high levels of ICAM-1 with more severe deep subcortical and periventricular WMH was reported by Han et al. in 175 elderly subjects without neurological impairment [126]. Six years later, a community-based study of 1763 subjects supported these results and also added an association with the presence of lacunar infarcts [77].

Hyperhomocysteinemia is considered an independent risk factor of cardiovascular disease, including myocardial infarction and atherosclerosis, as well as arterial and venous thrombosis. Homocysteine can lead to endothelial damage, thus contributing to vascular pathology, and it is known to be associated with MRI lesions and cognitive status in symptomatic CSVD (transient ischemic attack, subcortical stroke) [132]. Plasma homocysteine levels measured in a middle-aged cohort from the Framingham Offspring study, showed a strong inverse association with total cerebral brain volume, while higher homocysteine related with silent lacunar infarcts but not with WMH volumes [133]. The Rotterdam Scan study obtained similar results, evidencing a continuous and graded association with silent lacunar infarcts, although homocysteine levels were also higher in those with more WMHs [134]. Homocysteine was found to be a stronger predictor for diffuse CSVD than for isolated lacunar infarction [135], while an association with progression of WMHs was reported in a cohort with symptomatic atherosclerotic disease [65].

4.3 Oxidative Stress Markers

Endothelial dysfunction appears as a result of persistent ischemia–reperfusion and inflammation generating oxidative stress from unbalanced free radical formation that leads to peroxynitrite formation, lipid peroxidation, protein modification, matrix metal-loproteinase (MMP) activation, and DNA damage [114].

It has been suggested that oxidative stress is also related with CSVD and stroke risk. Nevertheless, this association still remains unclear, especially due to the overlap between systemic inflammation, endothelial dysfunction, and oxidative stress. To our knowledge, direct biomarkers of oxidative stress (oxidative damage products and antioxidant enzymes) have not been investigated in relation to asymptomatic CSVD.

A study including 131 "stroke-free" individuals who were followed during an average of 13.8 years measured 8-iso-prostaglandin F2-α, 4-hydroxinonenal conjugated with mercapturic acid, 8-hydroguanosine, and 8-nitroguanine, and found that only higher 8-iso-prostaglandin F2-α levels were associated with the risk of stroke and increased risk stratification for stroke [136].

4.4 Hemostatic Biomarkers

Due to the close relation between CSVD, endothelial dysfunction, and platelet activation, during the last years several studies have evaluated markers of coagulation and fibrinolysis in relation to CSVD. The most studied markers are those related to changes in coagulation/fibrinolysis ratio, promoting a prothrombotic status, and some of these mediators are fibrinogen, plasminogen activator inhibitor (PAI), tissue plasminogen activator (t-PA), D-dimer, thrombomodulin, and von Willebrand factor.

Fibrinogen is an indicator of increased blood coagulation, plasma viscosity, and platelet activation, and raised levels have been associated with WMH and lacunar infarct [137], with leukoaraiosis and periventricular hyperintensity in patients with stroke and atrial fibrillation [138, 139], and with WMHs in nondiabetic patients with lacunar noncardiogenic ischemic stroke [140]. In the Rotterdam Study the authors found that fibrinogen levels are associated with WMHs in subjects over 65 years of age in the general population [141]. Plasminogen activator inhibitor (PAI) was independently associated with the presence of WMH and/or lacunar infarcts in some studies [125, 142]. The authors hypothesized that PAI-1 may play a role in maintaining the BBB through positive actions on tight junctions in cerebral endothelial cells [143]. t-PA is locally expressed within the brain and release of t-PA in the neurovascular unit increases permeability of the BBB [144, 145]. Van Overbeek et al. found an association between plasma t-PA-activity and progression of periventricular WMH [142]. Higher t-PA serum levels have been reported in lacunar stroke patients with WMH compared to patients without WMH [146], and in patients with lacunar stroke compared to non-stroke individuals [147]. D-dimer levels were independently associated with subclinical lacunar infarcts in the Atherosclerosis Risk in Communities cohort [148] and with total cerebral volume, but not with WMHs or silent brain infarct in the Framingham study

[34]. Other studies employing small cohorts demonstrated increased D-dimer levels in CSVD, for example in lacunar infarcts and clinically manifest Binswanger disease, but the relation was not confirmed after multivariate analyses [149, 150]. Serum thrombomodulin has been associated with the presence and number of lacunar infarctions and with WMHs, especially with the coexistence of microalbuminuria in multiple studies, indicating an interesting association with renal impairment [151, 152]. High levels of von Willebrand factor (vWF)—an endothelial activation marker and an important performer in thrombosis—have been related with the presence and number of silent deep lacunar infarcts, with periventricular but not deep WMHs, and with WMH occurrence and progression in several population studies [148, 149, 153].

Other hemostatic markers have been less studied in CSVD, with non-concluding results. Results with tissue factor pathway inhibitor (TFPI) are controversial: higher [151] and lower [154] levels have been reported in patients with lacunar stroke compared with controls. In one study, the levels of TFPI were independently associated with the extent of WMH after multivariate analysis [151]. The soluble complex thrombin–antithrombin (TAT) was independently associated with the presence of WMH only in one small-sized study [149] and was found to be higher in a group of patients with subcortical vascular dementia and severe leukoaraiosis [150] Factor VII has been found to be increased in CSVD, whereas antithrombin III, a plasma protein that inactivates thrombin, was found to be reduced [147, 155].

Differences in the methodological conception of studies could influence the heterogeneity observed in the collected data. Different groups of subjects have been employed in studies: population-based and hospital cohorts, which differ in median age of enrollment, clinical status, and vascular comorbidities. Specifically, in hospital-based cohorts, the clinical status varies from asymptomatic and hypertensive subjects to previous symptomatic lacunar stroke patients [156]. In addition, the detection of MRI findings could be visual or volume-rated, where the expertise of the technician or the machine software could make major differences. Finally, different technical approaches employed in multiple laboratories for blood marker quantification, ranging from manual assays to automatic platforms and from turbidimetric/nephelometric methods to enzyme-linked immunosorbent methods, limit the reproducibility of the results.

4.5 Specific Brain Damage Markers

Blood-based brain-specific biomarkers have been employed to assess degree of brain injury and as outcome predictors in different clinical conditions (stroke, traumatic brain injury, and cardiac arrest, among others) [157, 158], and also as end points for evaluating the effect of new therapies [159].Among these biomarkers, brain-specific proteins (neuronal or glial in origin) have received growing attention in clinical neurological research [157–159], as well as autoantibodies to brain-specific proteins [160, 161]. The brain, chronically affected by CSVD, could be leaking brain-specific molecules into the blood stream at a rate much lower than in acute brain injuries (stroke, brain trauma, and global ischemia due to cardiac arrest). Studies employing autoantibodies to n-methyl-D-aspartate (NMDA) and α-amino-3-hydroxy-5-methyl-4-isoxazole-propionic acid (AMPA) receptors have shown promise as biomarkers of brain damage for detection of asymptomatic CSVD [34, 160, 162–164].

In CSVD the permeability of the BBB is compromised as a result of the slight and maintained injury to brain parenchyma caused by sustained HT, DM, and other vascular comorbidities, facilitating the slow release of cytosolic contents of brain cells to the bloodstream. Several studies have measured brain-derived proteins in cerebrospinal fluid as markers of CSVD (published elsewhere); however, fewer studies have evaluated the same proteins in blood. Brain natriuretic peptide (BNP), neurofilament light chain (NfL), neuron-specific enolase (NSE), S100B protein, and β-amyloid are the most studied brain-derived markers in subjects with asymptomatic CSVD.

N-terminal pro-brain natriuretic peptide (NT-proBNP) is released together with BNP from cardiomyocytes in response to myocardial wall stress and is considered a marker of subclinical cardiac injury. BNP and NT-proBNP have been reported to be associated with both WMHs and silent lacunar infarcts [165]. In a study including 278 hypertensive patients free of stroke or dementia, NT-proBNP and cardiac troponin-T—another cardiovascular biomarker—were found to be independently associated with silent cerebrovascular lesions [129]. Reinhard et al. had previously reported very similar results with proBNP in a large sample of stroke-free diabetic patients (Table 1), thus supporting the notion that NT-proBNP could be a surrogate marker of vascular brain damage in hypertension [128].

A recent study demonstrated a twofold increased level of serum-NfL in CSVD subjects compared with healthy controls, and an association was observed with both imaging and clinical features of CSVD, suggesting a possible utility for assessing CSVD burden in asymptomatic subjects [166]. Higher NfL serum levels have also been associated with the occurrence of small subcortical infarcts and with new-CSVD-related MRI lesions, thus also presenting it as a putative marker of active CSVD [167].

A cross-sectional study conducted by our group included measuring serum levels of NSE and S-100B in 101 hypertensive patients and seeking association with brain MRI lesions indicative of CSVD. Higher levels of NSE but not of S100B were associated with more severe WMH [35, 168]. Furthermore, antibodies against brain-specific proteins have also been measured in blood as indicators of CSVD. Serum autoantibodies against the subunit NR2 of the NMDA receptor (NR2Ab) were explored in hypertensive subjects, and higher serum NR2Ab levels were related with more severe brain MRI lesions, particularly WMH denoting CSVD [164]. Serum NSE concentration could be a filter in identifying asymptomatic hypertensive patients with putative subclinical brain damage for subsequent brain MRI scanning and a prognostic marker for the occurrence of acute vascular events in the central nervous system.

S100B, a family of Ca^{2+} binding proteins, abundantly expressed in brain glia, have been independently associated with CSVD; Gao et al. suggest that increased S100B does not simply reflect the degree of brain injury but also plays an important role in the pathogenesis of CSVD [169]. Moreover, higher levels of S100B were independently associated with the presence and number of CMBs in patients with first-ever acute lacunar stroke [170]. A recent report demonstrated that microRNAs (miRNAs) could be used to differentiate patients with CSVD, specifically miR-409-3p, miR-502-3p, miR-486-5p, and miR-45 [171]. miRNAs are a class of small noncoding RNA molecules that are highly stable in biofluids because they are protected by membranes in exosomes and other microparticles and by binding to specific proteins [172].

A small cross-sectional study associated high plasma levels of Aβ1–40 with increased WMH and lacunes in cerebral amyloid angiopathy [173] and high plasma Aβ1–40 and Aβ1–42 levels with increased WMH and lacunes [173] but only in APOE e4 carriers [174].

A recent report concluded that higher serum levels of cystatin C, a cysteine proteinase inhibitor, are independently associated with the long-term progression rate of the cerebral WMH volume, independently of kidney function [175]. Previous studies have proposed that serum cystatin C levels might reflect the functional status of cerebral penetrating arterioles and the activity of neuronal degeneration processes, as cystatin C is mainly secreted from neurons, astrocytes, and microglia, it is very concentrated in the brain [176, 177] and is deposited in brain parenchyma and the walls of microvessels, inducing further neuronal and vascular degeneration [178, 179].

The use of brain-derived markers could be a promising option, due to their higher accuracy as a neurochemical expression of brain damage considering that they are secreted and/or released from specifically damaged brain tissue. As blood-based biomarkers, they represent a more accessible and less expensive tool, enabling physicians to take action earlier against the increased burden of stroke. However additional issues should be considered in order to obtain a precise result, such as the presence of the anatomical barrier, the reactive response of the brain to injury and the existence of extracerebral sources of the brain-derived markers.

5   Predicting Risk of Stroke in Asymptomatic Individuals

Potential blood biomarkers for predicting the risk of future strokes in patients with asymptomatic disease have been studied, as their introduction in clinical practice would be helpful in guiding management decisions, especially for individuals at highest risks.

During a 12-year follow-up of 1462 subjects in the Framingham sub-study of healthy elderly subjects, CRP in the top quartile predicted an increased risk of stroke [180]. Similar results were also reported in the Rotterdam Scan Study [181], while the risk for silent lacunar infarct was found to be 1.85 and 2 times higher with increased levels of high sensitivity CRP (hsCRP) and IL-6 respectively [122]. Particularly in this sense, hsCRP seems to be a very useful biomarker. An investigation conducted in a cohort of 10,456 healthy "cardiovascular disease-free" men (65 ± 8.9 years of age), that was followed during 15 years, individuals with hsCRP >3 mg/ L showed a 40% higher stroke risk than those with hsCRP > 1 mg/ L, after adjusting for blood pressure and cardiovascular risk factors [117].

A very large prospective study was conducted in postmenopausal women ($n = 972$; incident first ever stroke cases), where plasma CRP, IL-6, TNF-α, neopterin, E-selectin, VCAM, Factor VII, prothrombin fragment 1 + 2, tPA, plasminogen activator inhibitor-1 antigen (PAI-1), fibrinogen, and homocysteine, as well as fasting plasma glucose and lipids were measured at inclusion [182]. Of all the individual biomarkers examined, CRP was the only independent single predictor of ischemic stroke after adjustment for other biomarkers and standard stroke risk factors. The Biomarker Risk Score identified a gradient of increasing stroke risk with a greater number of elevated inflammatory/hemostasis biomarkers. No evidence for an association between stroke and levels of E-selectin, fibrinogen, tumor necrosis factor-alpha, vascular cell adhesion molecule-1, prothrombin fragment 1 + 2, Factor VIIC, or plasminogen activator inhibitor-1 antigen was encountered. This study supports the need for further evaluation of multiple-biomarker panels to develop approaches for stratifying an individual's risk of stroke [182]. Very recently in a 2-year follow-up study of 123 elderly persons (age: 72.2 ± 8 years) Staszewski et al. reported that IL-6, platelet factor-4 (PF-4), CD40 ligand (sCD40 L), and homocysteine were associated with progressive brain MRI lesions of CSVD, suggesting that endothelial dysfunction modulates the radiological progression of CSVD through different inflammatory mechanisms [183].

Midlife systemic inflammation was found to be associated with late life brain MRI lesions indicative of CSVD in a bi-ethnic prospective cohort study (1485 participants, baseline age, 56 years— 28% African ancestry). A strong association was demonstrated between CRP and periventricular white matter microstructural integrity, but CRP was not associated with the presence of cerebral infarcts or microbleeds. The authors concluded that midlife systemic inflammation may promote the development of chronic microangiopathic structural white matter abnormalities in the elderly [184].

Asymmetric dimethylarginine (ADMA), an inhibitor of endothelial nitric oxide synthase, is a marker of endothelial dysfunction. Pikula et al. [130] included 2013 stroke-free Framingham offspring (age, 58 ± 9.5 years) in whom they measured plasma ADMA levels, followed during approximately 5 years and performed subsequent brain magnetic resonance imaging measures of subclinical brain injury. They found that higher concentrations of ADMA were associated with a greater risk of silent lacunar infarcts after adjustment for traditional stroke risk factors but not of WMHs and proposed that ADMA may be a potentially useful new biomarker of subclinical vascular brain injury [130]. A few years later this group evaluated an eight biomarker panel comprising inflammatory (CRP), hemostatic (D-dimer and plasminogen activator inhibitor-1), and neurohormonal activity (aldosterone/renin, natriuretic peptide, and N-terminal pro-atrial natriuretic peptide), and endothelial function (homocysteine and urinary albumin–creatinine) in 3127 "stroke-free" subjects, who were followed for an average period of 9.2 years. The biomarker panel was associated with the incidence of stroke/TIA and with total brain volume but not with extensive WMH [34].

von Willebrand coagulation factor has also been associated with the risk of stroke. The Rotterdam Study demonstrated that the risk of stroke increased with the elevation of this biomarker in 6250 "stroke-free" individuals ≥55 years of age, who were followed during 5 years [185].

Serum enzyme activities of CK and GGT have also been evaluated in relation to stroke risk. In an urban Japanese population-based cohort study including 5026 initially healthy subjects (mean age 54.5 years), that were followed for 11.8 years on average, CK levels were useful to predict first-ever myocardial infarction in the future, but no relationship was observed with the risk for stroke [111]. GGT was evaluated in 456,100 eligible participants and was found to be independently correlated with increased risk of stroke after adjustment for alcohol consumption and stroke risk [113].

A longitudinal study of the cohort including 101 hypertensive patients, where our group had reported that higher blood levels of NSE, but not S100B were associated with more severe WMHs, provided additional information suggesting that increased NSE could also be a useful predictor for the occurrence of acute central nervous system vascular events in these previously asymptomatic hypertensive patients [35].

In the Three-City Dijon Study, with over 1600 participants, lower β-amyloid (Aβ) serum levels were associated with the progression of WMH volume. Lower levels of Aβ1–40 and Aβ1–42 were related with increased rate of progression of total and periventricular WMHs in CSVD [186]. The association of low plasma Aβ protein levels with increased WMH observed in this study could indicate an increased deposition of this protein in cerebral vessel walls, resulting in impaired cerebral blood flow [187].

6   Future Perspectives

Presently in clinical practice blood biomarkers have not gained use for stroke diagnosis and prognosis, given the more extensive availability of neuroimaging technology in the clinical settings that receive these patients and the many factors that confuse interpretation of most serum biomarker levels [188]. Nevertheless, very promising results have been obtained in the differentiation between TIA and stroke measuring blood levels of autoantibodies to NR2 peptides of the NMDA receptors [160, 161].

The potential application of blood biomarkers for the detection of subclinical brain damage and for predicting long-term stroke risk in asymptomatic individuals with vascular risk factors opens a new perspective that has attracted many investigators. The main challenge is to detect asymptomatic brain lesions with noninvasive and cost-effective techniques that can be easily performed and interpreted for widespread screening in the community. This would undoubtedly have a very positive effect for primary stroke prevention.

In this line of thought, numerous blood biomarkers have been investigated with very controversial results. In spite of this, there seem to be some unspecific blood biomarkers which have accumulated evidence as predictors of asymptomatic CSVD in cross-sectional studies or of incident TIA or stroke in longitudinal studies. According to Vilar et al. [114], considering the following selection criteria (two or more replications within independent studies and more than 1000 patients tested), to date, the best performing blood biomarkers in association with asymptomatic CSVD are CRP with WMH and IL-6, homocysteine, urine albumin–creatinine ratio, and amyloid β (1–42) peptide (all associated with silent brain infarcts and WMH).

Notwithstanding, blood-based brain-specific biomarkers will probably be able to offer much more in the detection of asymptomatic CSVD and in the long-term prediction of acute cerebrovascular events than measures of degenerative or inflammatory changes. The latter mainly correspond to unspecific peripheral processes, many of which are undoubtedly related to vascular changes in the central nervous system but do not directly represent what is occurring in the brain. This imposes a great challenge for researchers working in the field of brain-specific biomarkers and subclinical brain damage, who have to make a transition from current exploratory studies to large-scale, well-designed, cross-sectional, and prospective research in order to integrate them into clinical practice.

References

1. Feigin VL, Roth GA, Naghavi M et al (2016) Global burden of stroke and risk factors in 188 countries, during 1990–2013: a systematic analysis for the Global Burden of Disease Study 2013. Lancet Neurol 15(9):913–924

2. Ministerio de Salud Pública (2018) Anuario Estadístico de Salud de la República de Cuba. Ministerio de Salud Pública, Habana. http://bvscuba.sld.cu/anuario-estadistico-de-cuba/

3. Spence JD (2018) Stroke prevention: editorial to accompany June issue of SVN. Stroke Vasc Neurol 3:e000171

4. Yu J-G, Zhou R-R, Cai G-J (2011) From hypertension to stroke: mechanisms and potential prevention strategies. CNS Neurosci Ther 17(5):577–584. https://doi.org/10.1111/j.1755-5949.2011.00264.x46

5. Yoon SS, Fryar CD, Carroll MD (2015) Hypertension prevalence and control among adults: United States, 2011–2014. NCHS Data Brief no. 220. National Center for Health Statistics, Hyattsville, MD

6. Bonet Gorbea M, Varona Pérez P (2015) Encuesta Nacional de factores de riesgo y actividades preventivas de enfermedades no transmisibles. Cuba 2010–2011. Editorial Ciencias Médicas, Havana, pp 140–165. http://www.bvs.sld.cu/libros/encuesta_nacional_riesgo/hipertension.pdf

7. Whelton PK, Carey RM, Aronow WS et al (2018) 2017 ACC/AHA/AAPA/ABC/ACPM/AGS/APhA/ASH/ASPC/NMA/PCNA guideline for the prevention, detection, evaluation, and management of high blood pressure in adults. J Am Coll Cardiol 71(6):1269–1324. https://doi.org/10.1016/j.jacc.2017.11.006

8. Scuteri A (2012) Brain injury as end-organ damage in hypertension. Lancet Neurol 11 (12):1015–1017

9. Sierra C, López-Soto A, Coca A (2011) Connecting cerebral white matter lesions and hypertensive target organ damage. J Aging Res 2011:438978. https://doi.org/10.4061/2011/438978

10. González-García S, Hernández-Díaz Z, Quevedo-Sotolongo L, Peña-Sánchez M, Pino Peña Y, Fernández-Carriera R et al (2014) Resistive cerebral blood flow as a potential marker of subclinical brain damage in essential hypertension. World J Cardiovasc Dis 4(4):169–178

11. Henskens LH, van Oostenbrugge RJ, Kroon AA, Hofman PA, Lodder J, de Leeuw PW (2009) Detection of silent cerebrovascular disease refines risk stratification of hypertensive patients. J Hypertens 27(4):846–853

12. Mancia G, Fagard R, Narkiewicz K, Redón J, Zanchetti A, Böhm M et al (2013) 2013 ESH/ESC guidelines for the management of arterial HT: the task force for the management of arterial HT of the European Society of Hypertension (ESH) and of the European Society of Cardiology (ESC). J Hypertens 31 (7):1281–1357

13. Longstreth WT Jr, Bernick C, Manolio TA, Bryan N, Jungreis CA, Price TR (1998) Lacunar infarcts defined by magnetic resonance imaging of 3660 elderly people: the cardiovascular Health Study. Arch Neurol 55 (9):1217–1225

14. Longstreth WT, Manolio TA, Arnold A, Burke GL, Bryan N, Jungreis CA et al (1996) Clinical correlates of white matter findings on cranial magnetic resonance imaging of 3301 elderly people. The Cardiovascular Health Study. Stroke 27(8):1274–1282

15. Sierra C (2014) Essential hypertension, cerebral white matter pathology and ischemic stroke. Curr Med Chem 21(19):2156–2164

16. Liao D, Cooper L, Cai J, Toole JF, Bryan NR, Hutchinson RG et al (1996) Presence and severity of cerebral white matter lesions and hypertension, its treatment, and its control. The ARIC study. Stroke 27(12):2262–2270

17. de Leeuw FE, De Groot JC, Oudkerk M, Witteman JC, Hofman A, van Gijn J, Breteler MM (2002) Hypertension and cerebral WML in a prospective cohort study. Brain 125:765–772

18. van Boxtel M, Henskens LH, Kroon AA, Hofman PA, Gronenschild EH, Jolles J et al (2006) Ambulatory blood pressure, asymptomatic cerebrovascular damage and cognitive function in essential hypertension. J Hum Hypertens 20(1):5–13

19. Gottesman RF, Coresh J, Catellier DJ, Sharrett AR, Rose KM, Coker LH et al (2010) Blood pressure and white-matter disease progression in a biethnic cohort: Atherosclerosis Risk in Communities (ARIC) study. Stroke 41 (1):3–8

20. Vermeer SE, Longstreth WT Jr, Koudstaal PJ (2007) Silent brain infarcts: a systematic review. Lancet Neurol 6(7):611–619

21. Sanahuja J, Alonso A, Diez J, Ortega E, Rubinat E, Traveset A et al (2016) Increased burden of cerebral small vessel disease in patients with type 2 diabetes and retinopathy. Diabetes Care 39(9):1614–1620. https://doi.org/10.2337/dc15-2671

22. Rincon F, Wright CB (2014) Current pathophysiological concepts in cerebral small vessel disease. Front Aging Neurosci 6:24

23. Shi Y, Wardlaw JM (2016) Update on cerebral small vessel disease: a dynamic whole-brain disease. Stroke Vasc Neurol 1(3):83–92. https://doi.org/10.1136/svn-2016-000035

24. Prabhakaran S, Wright CB, Yoshita M, Delapaz R, Brown T, De Carli C et al (2008) Prevalence and determinants of subclinical brain infarction: the Northern Manhattan Study. Neurology 70(6):425–430

25. Sierra C (2011) Associations between ambulatory blood pressure parameters and cerebral white matter lesions. Int J Hypertens 2011:478710. https://doi.org/10.4061/2011/478710

26. Cheung CY, Tay WT, Mitchell P, Wang JJ, Hsu W, Lee ML et al (2011) Quantitative and qualitative retinal microvascular characteristics and blood pressure. J Hypertens 29 (7):1380–1391

27. Grassi G, Schmieder RE (2011) The renaissance of the retinal microvascular network assessment in hypertension: new challenges. J Hypertens 29(7):1289–1291

28. Brown Martínez M, Valdés-González Y, GonzálezOrtiz E, Hernández-González G, Valdés Sosa P, Galán García L et al (2014) Use of electroencephalography to identify asymptomatic cerebrovascular lesions among hypertensives. Rev Cubana Invest Bioméd 33:231–240

29. Hernández-González G, Bringas-Vega ML, Galán-García L et al (2011) Multimodal quantitative neuroimaging databases and methods: the Cuban Brain Mapping Project. Clin EEG Neurosci 42(3):149–159

30. Heliopoulos I, Artemis D, Vadikolias K, Tripsianis G, Piperidou C, Tsivgoulis G (2012) Association of ultrasonographic parameters with subclinical white-matter hyperintensities in hypertensive patients. Cardiovasc Psychiatry Neurol 2012:616572

31. Kurata M, Okura T, Watanabe S, Higaki J (2005) Association between carotid hemodynamics and asymptomatic white and gray matter lesions in patients with essential hypertension. Hypertens Res 28 (10):797–803

32. Appel J, Potter E, Bhatia N, Shen Q, Zhao W, Greig MT (2009) Association of white matter hyperintensity measurements on brain MR imaging with cognitive status, medial temporal atrophy, and cardiovascular risk factors. Am J Neuroradiol 30:1870–1876

33. Van Dijk EJ, Prins ND, Vrooman HA et al (2008) Progression of cerebral small vessel disease in relation to risk factors and cognitive consequences: Rotterdam Scan study. Stroke 39(10):2712–2719

34. Pikula A, Beiser AS, DeCarli C et al (2012) Multiple biomarkers and risk of clinical and subclinical vascular brain injury: the Framingham Offspring Study. Circulation 125 (17):2100–2107

35. González-Quevedo A, González-García S, Hernández-Díaz Z et al (2016) Serum neuron specific enolase could predict subclinical brain damage and the subsequent occurrence of brain related vascular events during follow up in essential hypertension. J Neurol Sci 363:158–163

36. Shoamanesh A, Preis SR, Beiser AS et al (2016) Circulating biomarkers and incident ischemic stroke in the Framingham Offspring Study. Neurology 87:1206–1211

37. Shoamanesh A, Preis SR, Beiser AS et al (2015) Inflammatory biomarkers, cerebral microbleeds and small vessel disease. Framingham Heart Study. Neurology 84 (8):825–832

38. Moody DM, Brown WR, Challa VR, Anderson RL (1995) Periventricular venous collagenosis: association with leukoaraiosis. Radiology 194(2):469–476

39. Wardlaw JM, Smith EE, Biessels GJ et al (2013) Neuroimaging standards for research into small vessel disease and its contribution to ageing and neurodegeneration. Lancet Neurol 12(8):822–838

40. Pantoni L (2010) Cerebral small vessel disease: from pathogenesis and clinical

characteristics to therapeutic challenges. Lancet Neurol 9(7):689–701

41. Rensma SP, van Sloten TT, Launer LJ, Stehouwer CDA (2018) Cerebral small vessel disease and risk of incident stroke, dementia and depression, and all-cause mortality: a systematic review and meta-analysis. Neurosci Biobehav Rev 90:164–173

42. Zhang CE, Wong SM, van de Haar FJ et al (2017) Blood–brain barrier leakage is more widespread in patients with cerebral small vessel disease. Neurology 88:1–7

43. Teng Z, Dong Y, Zhang D et al (2017) Cerebral small vessel disease and post-stroke cognitive impairment. Int J Neurosci 127 (9):824–830

44. Kern KC, Wright CB, Bergfield KL et al (2017) Blood pressure control in aging predicts cerebral atrophy related to small-vessel white matter lesions. Front Aging Neurosci 9:132

45. de Leeuw FE, de Groot JC, Achten E et al (2001) Prevalence of cerebral white matter lesions in elderly people: a population based magnetic resonance imaging study. The Rotterdam Scan Study. J Neurol Neurosurg Psychiatry 70(1):9–14

46. Lin Q, Huang W-Q, Ma Q-L et al (2017) Incidence and risk factors of leukoaraiosis from 4683 hospitalized patients: a cross-sectional study. Med Baltim 96(39):e7682

47. Birns J, Jarosz J, Markus HS, Kalra L (2009) Cerebrovascular reactivity and dynamic autoregulation in ischaemic subcortical white matter disease. J Neurol Neurosurg Psychiatry 80 (10):1093–1098

48. de Groot JC, de Leeuw FE, Oudkerk M et al (2001) Cerebral white matter lesions and subjective cognitive dysfunction: the Rotterdam Scan Study. Neurology 56(11):1539–1545. http://www.ncbi.nlm.nih.gov/pubmed/11402112

49. Jung S, Mono ML, Findling O et al (2012) White matter lesions and intra-arterial thrombolysis. J Neurol 259(7):1331–1336. https://doi.org/10.1007/s00415-011-6352-y

50. Vernooij MW, Ikram MA, Tanghe HL et al (2007) Incidental findings on brain MRI in the general population. N Engl J Med 357 (18):1821–1828. https://doi.org/10.1056/NEJMoa070972

51. Pantoni L, Garcia JH (1995) The significance of cerebral white matter abnormalities 100 years after Binswanger's report. A review. Stroke 26(7):1293–1301

52. Mok V, Kim JS (2015) Prevention and management of cerebral small vessel disease. J Stroke 17(2):111–122. https://doi.org/10.5853/jos.2015.17.2.111

53. Faraco G, Iadecola C (2013) Hypertension: a harbinger of stroke and dementia. Hypertens (Dallas, TX 1979) 62(5):810–817. https://doi.org/10.1161/HYPERTENSIONAHA.113.01063

54. Manschot SM, Brands AMA, van der Grond J et al (2006) Brain magnetic resonance imaging correlates of impaired cognition in patients with type 2 diabetes. Diabetes 55 (4):1106–1113

55. van Harten B, Oosterman JM, Potter van Loon B-J, Scheltens P, Weinstein HC (2007) Brain lesions on MRI in elderly patients with type 2 diabetes mellitus. Eur Neurol 57 (2):70–74. https://doi.org/10.1159/000098054

56. Taylor WD, MacFall JR, Provenzale JM et al (2003) Serial MR imaging of volumes of hyperintense white matter lesions in elderly patients: correlation with vascular risk factors. AJR Am J Roentgenol 181(2):571–576. https://doi.org/10.2214/ajr.181.2.1810571

57. Potter GM, Doubal FN, Jackson CA et al (2015) Enlarged perivascular spaces and cerebral small vessel disease. Int J Stroke 10 (3):376–381. https://doi.org/10.1111/ijs.12054

58. Yang S, Qin W, Yang L et al (2017) The relationship between ambulatory blood pressure variability and enlarged perivascular spaces: a cross-sectional study. BMJ Open 7 (8):e015719. https://doi.org/10.1136/bmjopen-2016-015719

59. Iadecola C (2013) The pathobiology of vascular dementia. Neuron 80(4):844–866. https://doi.org/10.1016/j.neuron.2013.10.008

60. Wallin A, Ohrfelt A, Bjerke M (2012) Characteristic clinical presentation and CSF biomarker pattern in cerebral small vessel disease. J Neurol Sci 322(1–2):192–196. https://doi.org/10.1016/j.jns.2012.07.068

61. Wardlaw JM (2010) Blood-brain barrier and cerebral small vessel disease. J Neurol Sci 299 (1–2):66–71. https://doi.org/10.1016/j.jns.2010.08.042

62. Doubal FN, MacLullich AMJ, Ferguson KJ et al (2010) Enlarged perivascular spaces on MRI are a feature of cerebral small vessel disease. Stroke 41(3):450–454. https://doi.org/10.1161/STROKEAHA.109.564914

63. Li Y, Liu N, Huang Y et al (2016) Risk factors for silent lacunar infarction in patients with transient ischemic attack. Med Sci Monit 22:447–453. http://www.ncbi.nlm.nih.gov/pubmed/26864634

64. Launer LJ, Hughes TM, White LR (2011) Microinfarcts, brain atrophy, and cognitive function: the Honolulu Asia Aging Study Autopsy Study. Ann Neurol 70(5):774–780

65. Kloppenborg RP, Nederkoorn PJ, Grool AM et al (2017) Do lacunar infarcts have different aetiologies? Risk factor profiles of lacunar infarcts in deep white matter and basal ganglia: the second manifestations of arterial disease-magnetic resonance study. Cerebrovasc Dis 43(3–4):161–168

66. Rutten-Jacobs LCA, Markus HS, Young UK (2017) Lacunar Stroke DNA Study. Vascular risk factor profiles differ between magnetic resonance imaging-defined subtypes of younger-onset lacunar stroke. Stroke 48 (9):2405–2411

67. Debette S, Markus HS (2010) The clinical importance of white matter hyperintensities on brain magnetic resonance imaging: systematic review and meta-analysis. BMJ 341: c3666

68. Debette S, Beiser A, DeCarli C et al (2010) Association of MRI markers of vascular brain injury with incident stroke, mild cognitive impairment, dementia, and mortality: the Framingham Offspring Study. Stroke 41 (4):600–606

69. Koga H, Takashima Y, Murakawa R et al (2009) Cognitive consequences of multiple lacunes and leukoaraiosis as vascular cognitive impairment in community-dwelling elderly individuals. J Stroke Cerebrovasc Dis 18 (1):32–37

70. Baezner H, Blahak C, Poggesi A et al (2008) Association of gait and balance disorders with age-related white matter changes: the LADIS Study. Neurology 70(12):935–942

71. Kwon H-M, Lim J-S, Kim YS et al (2014) Cerebral microbleeds are associated with nocturnal reverse dipping in hypertensive patients with ischemic stroke. BMC Neurol 14(1):8. https://doi.org/10.1186/1471-2377-14-8

72. Yates PA, Villemagne VL, Ellis KA, Desmond PM, Masters CL, Rowe CC (2014) Cerebral microbleeds: a review of clinical, genetic, and neuroimaging associations. Front Neurol 4:205. https://doi.org/10.3389/fneur.2013.00205

73. Fladt J, Kronlage C, De Marchis GM (2018) Cerebral white matter disease and microbleeds in acute ischemic stroke: impact on recanalization therapies. A review of the literature. Neurosci Lett. https://doi.org/10.1016/j.neulet.2018.09.003

74. Romero JR, Preis SR, Beiser A et al (2014) Risk factors, stroke prevention treatments, and prevalence of cerebral microbleeds in the Framingham Heart Study. Stroke 45 (5):1492–1494. https://doi.org/10.1161/STROKEAHA.114.004130

75. Cordonnier C, Al-Shahi Salman R, Wardlaw J (2007) Spontaneous brain microbleeds: systematic review, subgroup analyses and standards for study design and reporting. Brain 130(Pt 8):1988–2003. https://doi.org/10.1093/brain/awl387

76. Granger JP (2006) An emerging role for inflammatory cytokines in hypertension. Am J Physiol Heart Circ Physiol 290(3): H923–H924. https://doi.org/10.1152/ajpheart.01278.2005

77. Shoamanesh A, Preis SR, Beiser AS, Vasan RS, Benjamin EJ, Kase CS et al (2015) Inflammatory biomarkers, cerebral microbleeds, and small vessel disease. Framingham Heart Study. Neurology 84(8):825–832

78. Poels MMF, Vernooij MW, Ikram MA et al (2010) Prevalence and risk factors of cerebral microbleeds: an update of the Rotterdam Scan Study. Stroke 41(10, Supplement 1): S103–S106. https://doi.org/10.1161/STROKEAHA.110.595181

79. Writing Group Members D, Mozaffarian D, Benjamin EJ et al (2016) Heart disease and stroke statistics – 2016 update: a report from the American Heart Association. Circulation 133(4):e38–e360

80. Pandian JD, William AG, Kate MP et al (2017) Strategies to improve stroke care services in low- and middle-income countries: a systematic review. Neuroepidemiology 49 (1–2):45–61. https://doi.org/10.1159/000479518

81. Marzona I, Avanzini F, Lucisano G et al (2017) Are all people with diabetes and cardiovascular risk factors or microvascular complications at very high risk? Findings from the Risk and Prevention Study. Acta Diabetol 54 (2):123–131. https://doi.org/10.1007/s00592-016-0899-0

82. Lawes CM, Vander HS, Rodgers A, International Society of Hypertension (2008) Global burden of blood-pressure-related disease, 2001. Lancet 371(9623):1513–1518. https://doi.org/10.1016/S0140-6736(08)60655-8

83. Gorelick PB (2002) New horizons for stroke prevention: PROGRESS and HOPE. Lancet Neurol 1(3):149–156

84. Benjo A, Thompson RE, Fine D et al (2007) Pulse pressure is an age-independent predictor of stroke development after cardiac surgery. Hypertens (Dallas, TX 1979) 50 (4):630–635. https://doi.org/10.1161/HYPERTENSIONAHA.107.095513

85. O'Callaghan CJ, Williams B (2000) Mechanical strain-induced extracellular matrix production by human vascular smooth muscle cells: role of TGF-beta(1). Hypertens (Dallas, TX 1979) 36(3):319–324

86. Harrison DG, Widder J, Grumbach I et al (2006) Endothelial mechanotransduction, nitric oxide and vascular inflammation. J Intern Med 259(4):351–363. https://doi.org/10.1111/j.1365-2796.2006.01621.x

87. Cipolla MJ, Liebeskind DS, Chan S-L (2018) The importance of comorbidities in ischemic stroke: impact of hypertension on the cerebral circulation. J Cereb Blood Flow Metab. https://doi.org/10.1177/0271678X18800589

88. Chrissobolis S, Sobey CG (2006) Recent evidence for an involvement of rho-kinase in cerebral vascular disease. Stroke 37 (8):2174–2180. https://doi.org/10.1161/01.STR.0000231647.41578.df

89. Vaziri ND, Rodríguez-Iturbe B (2006) Mechanisms of disease: oxidative stress and inflammation in the pathogenesis of hypertension. Nat Clin Pract Nephrol 2(10):582–593. https://doi.org/10.1038/ncpneph0283

90. Vaziri ND (2004) Roles of oxidative stress and antioxidant therapy in chronic kidney disease and hypertension. Curr Opin Nephrol Hypertens 13(1):93–99

91. Sindhu RK, Roberts CK, Ehdaie A, Zhan C-D, Vaziri ND (2005) Effects of aortic coarctation on aortic antioxidant enzymes and NADPH oxidase protein expression. Life Sci 76(8):945–953. https://doi.org/10.1016/j.lfs.2004.10.014

92. Hirooka Y (2008) Role of reactive oxygen species in brainstem in neural mechanisms of hypertension. Auton Neurosci 142 (1–2):20–24. https://doi.org/10.1016/j.autneu.2008.06.001

93. Kazama K, Anrather J, Zhou P et al (2004) Angiotensin II impairs neurovascular coupling in neocortex through NADPH oxidase-derived radicals. Circ Res 95 (10):1019–1026. https://doi.org/10.1161/01.RES.0000148637.85595.c5

94. Girouard H, Park L, Anrather J, Zhou P, Iadecola C (2007) Cerebrovascular nitrosative stress mediates neurovascular and endothelial dysfunction induced by angiotensin II. Arterioscler Thromb Vasc Biol 27 (2):303–309. https://doi.org/10.1161/01.ATV.0000253885.41509.25

95. Malone K, Amu S, Moore AC, Waeber C (2018) The immune system and stroke: from current targets to future therapy. Immunol Cell Biol. https://doi.org/10.1111/imcb.12191

96. Nava M, Quiroz Y, Vaziri N, Rodriguez-Iturbe B (2003) Melatonin reduces renal interstitial inflammation and improves hypertension in spontaneously hypertensive rats. Am J Physiol Renal Physiol 284(3): F447–F454. https://doi.org/10.1152/ajprenal.00264.2002

97. Pooja Naik LC, Sajja RK, Naik P, Cucullo L (2014) Diabetes mellitus and blood-brain barrier dysfunction: an overview. J Pharmacovigil 02(02):125. https://doi.org/10.4172/2329-6887.1000125

98. Shere A, Eletta O, Goyal H (2017) Circulating blood biomarkers in essential hypertension: a literature review. J Lab Precis Med 2:99. https://doi.org/10.21037/jlpm.2017.12.06

99. Wu O, Leng JH, Yang FF et al (2017) A comparative research on obesity hypertension by the comparisons and associations between waist circumference, body mass index with systolic and diastolic blood pressure, and the clinical laboratory data between four special Chinese adult groups. Clin Exp Hypertens 40 (1):16–21. https://doi.org/10.1080/10641963.2017.1281940

100. Targher G (2014) Risk of ischemic stroke and decreased serum bilirubin levels. Is there a causal link? Arterioscler Thromb Vasc Biol 34:702–704

101. Li R-Y, Cao Z-G, Zhang J-R et al (2014) Decreased serum bilirubin is associated with silent cerebral infarction. Arterioscler Thromb Vasc Biol 34:946–951

102. Higuchi S, Kabeya Y, Uchida J et al (2018) Low Bilirubin levels indicate a high risk of cerebral deep white matter lesions in apparently healthy subjects. Sci Rep 8:6473. https://doi.org/10.1038/s41598-018-24917-8

103. Wang Y, Xu S, Pan S et al (2018) Association of serum neuron-specific enolase and bilirubin levels with cerebral dysfunction and prognosis in large-artery atherosclerotic strokes. J Cell Biochem. https://doi.org/10.1002/jcb.27281

104. Cardon MW (2017) 50 years ago in the Journal of Pediatrics: an assessment of the creatine kinase test in the detection of carriers of Duchenne Muscular Dystrophy. J Pediatr 186:63

105. Horjus DL, Oudman I, van Montfrans GA, Brewster LM (2011) Creatine and creatine analogues in hypertension and cardiovascular disease. Cochrane Database of Syst Rev 11: CD005184.   https://doi.org/10.1002/14651858.CD005184.pub2

106. Brewster LM, Mairuhu G, Bindraban NR, Koopmans RP, Clark JF, van Montfrans GA (2006) Creatine kinase activity is associated with blood pressure. Circulation 114:2034–2039

107. Brewster LM, Coronel CMD, Sluiter W et al (2012) Ethnic differences in tissue creatine kinase activity: an observational study. PLoS One 7(3):e32471

108. Brewster LM, Seedat YK (2013) Why do hypertensive patients of African ancestry respond better to calcium blockers and diuretics than to ACE inhibitors and β-adrenergic blockers? A systematic review. BMC Med 11:141

109. Horjus DL, Nieuwland R, Boateng KB et al (2014) Creatine kinase inhibits ADP-induced platelet aggregation. Sci Rep 9:6551

110. Karamat FA, Horjus DL, Haan YC et al (2015) The acute effect of beta-guanidinopropionic acid versus creatine or placebo in healthy men (ABC Trial): a randomized controlled first-in human trial. Br J Pharmacol 16:56

111. Watanabe M, Okamura T, Kokubo Y et al (2009) Elevated serum creatine kinase predicts first-ever myocardial infarction: a 12-year population-based cohort study in Japan, the Suita study. Int J Epidemiol 38(6):1571–1579. https://doi.org/10.1093/ije/dyp212

112. Kunutsor SK, Apekey TA, Cheung BMY (2015) Gamma-glutamyltransferase and risk of hypertension: a systematic review and dose-response meta-analysis of prospective evidence. J Hypertens 33:2373–2381

113. Yang W, Kim CK, Kim DY et al (2018) Gamma-glutamyl transferase predicts future stroke: a Korean Nationwide Study. Ann Neurol 83(2):375–386. https://doi.org/10.1002/ana.25158

114. Vilar-Bergua A, Riba-Llena I, Nafría C et al (2015) Blood and CSF biomarkers in brain subcortical ischemic vascular disease: involved pathways and clinical applicability. J Cereb Blood Flow Metab 36(1):55–71. https://doi.org/10.1038/jcbfm.2015.68

115. Abe A, Nishiyama Y, Harada-Abe M et al (2014) Relative risk values of age, acrolein, IL-6 and CRP as markers of periventricular hyperintensities: a cross-sectional study. BMJ Open 4(8):e005598

116. Mitaki S, Nagai A, Oguro H, Yamaguchi S (2016) C-reactive protein levels are associated with cerebral small vessel-related lesions. Acta Neurol Scand 133(1):68–74

117. Jiménez MC, Rexrode KM, Glynn RJ et al (2015) Association between high-sensitivity C-reactive protein and total stroke by hypertensive status among men. J Am Heart Assoc 4:e002073

118. Satizabal CL, Zhu YC, Mazoyer B et al (2012) Circulating IL-6 and CRP are associated with MRI findings in the elderly: the 3C-Dijon Study. Neurology 78(10):720–727

119. Kaptoge S, Di Angelantonio E, Lowe G et al (2010) C-reactive protein concentration and risk of coronary heart disease, stroke, and mortality: an individual participant meta-analysis. Lancet 375:132–140

120. Tanaka F, Makita S, Onoda T et al (2010) Prehypertension subtype with elevated C-reactive protein: risk of ischemic stroke in a general Japanese population. Am J Hypertens 23:1108–1113

121. Miwa K, Tanaka M, Okazaki S et al (2011) Relations of blood inflammatory marker levels with cerebral microbleeds. Stroke 42:3202–3206

122. Hoshi T, Kitagawa K, Yamagami H et al (2005) Relations of serum high sensitivity C-reactive protein and interleukin-6 levels with silent brain infarction. Stroke 36:768–772

123. Wright CB, Moon Y, Paik MC, Brown TR, Rabbani L, Yoshita M et al (2009) Inflammatory biomarkers of vascular risk as correlates of leukoariosis. Stroke 40(11):3466–3471

124. Kim CK, Lee SH, Kim BJ et al (2011) Elevated leukocyte count in asymptomatic subjects is associated with a higher risk for cerebral white matter lesions. Clin Neurol Neurosurg 113(3):177–180

125. Miralbell J, Soriano JJ, Spulber G et al (2012) Structural brain changes and cognition in relation to markers of vascular dysfunction. Neurobiol Aging 33:e9–e17

126. Han JH, Wong KS, Wang YY et al (2009) Plasma level of sICAM-1 is associated with the extent of white matter lesion among asymptomatic elderly subjects. Clin Neurol Neurosurg 111:847–851

127. Hudson BI, Moon YP, Kalea AZ et al (2011) Association of serum soluble receptor for advanced glycation endproducts with subclinical cerebrovascular disease: the Northern Manhattan Study (NOMAS). Atherosclerosis 216(1):192–198

128. Reinhard H, Garde E, Skimminge A et al (2012) Plasma NT-proBNP and white matter hyperintensities in type 2 diabetic patients. Cardiovasc Diabetol 11:119. https://doi.org/10.1186/1475-2840-11-119

129. Vilar-Bergua A, Riba-Llena I, Penalba A et al (2016) N-terminal pro-brain natriuretic peptide and subclinical brain small vessel disease. Neurology 87:1–7

130. Pikula A, Boger RH, Beiser AS, Maas R, De Carli C, Schwedhelm E et al (2009) Association of plasma ADMA levels with MRI markers of vascular brain injury: Framingham offspring study. Stroke 40:2959–2964

131. Wright CB, Shah NH, Mendez AJ et al (2016) Fibroblast growth factor 23 is associated with subclinical cerebrovascular damage. The Northern Manhattan Study. Stroke 47:923–928

132. Pavlovic AM, Pekmezovic T, Obrenovic R et al (2011) Increased total homocysteine level is associated with clinical status and severity of white matter changes in symptomatic patients with subcortical small vessel disease. Clin Neurol Neurosurg 113:711–715

133. Seshadri S, Wolf PA, Beiser AS et al (2008) Association of plasma total homocysteine levels with subclinical brain injury: cerebral volumes, white matter hyperintensity, and silent brain infarcts at volumetric magnetic resonance imaging in the Framingham Offspring Study. Arch Neurol 65:642–649

134. Rosenberg GA, Bjerke M, Wallin A (2014) Multimodal markers of inflammation in the subcortical ischemic vascular disease type of vascular cognitive impairment. Stroke 45:1531–1538

135. Hassan A, Hunt BJ, O'Sullivan M et al (2004) Homocysteine is a risk factor for cerebral small vessel disease, acting via endothelial dysfunction. Brain 127:212–219

136. Lin HJ, Chen ST, Wu HY et al (2015) Urinary biomarkers of oxidative and nitrosative stress and the risk for incident stroke: a nested case-control study from a community-based cohort. Int J Cardiol 183:214–220. https://doi.org/10.1016/j.ijcard.2015.01.043

137. Aono Y, Ohkubo T, Kikuya M et al (2007) Plasma fibrinogen, ambulatory blood pressure, and silent cerebrovascular lesions: the Ohasama Study. Arterioscler Thromb Vasc Biol 27(4):963–968. https://doi.org/10.1161/01.ATV.0000258947.17570.38

138. Martí-Fàbregas J, Valencia C, Pujol J, García-Sánchez C, Martí-Vilalta J-L (2002) Fibrinogen and the amount of leukoaraiosis in patients with symptomatic small-vessel disease. Eur Neurol 48(4):185–190. https://doi.org/10.1159/000066161

139. Wei C-C, Zhang S-T, Liu J-F et al (2017) Association between fibrinogen and leukoaraiosis in patients with ischemic stroke and atrial fibrillation. J Stroke Cerebrovasc Dis 26(11):2630–2637. https://doi.org/10.1016/j.jstrokecerebrovasdis.2017.06.027

140. You C-J, Liu D, Liu L-L et al (2018) Correlation between fibrinogen and white matter hyperintensities among nondiabetic individuals with noncardiogenic ischemic stroke. J Stroke Cerebrovasc Dis 27(9):2360–2366. https://doi.org/10.1016/j.jstrokecerebrovasdis.2018.04.025

141. Breteler MM, van Swieten JC, Bots ML et al (1994) Cerebral white matter lesions, vascular risk factors, and cognitive function in a population-based study: the Rotterdam Study. Neurology 44(7):1246–1252

142. van Overbeek EC, Staals J, Knottnerus ILH et al (2016) Plasma tPA-activity and progression of cerebral white matter hyperintensities in lacunar stroke patients. PLoS One 11(3):e0150740. https://doi.org/10.1371/journal.pone.0150740

143. Dohgu S, Takata F, Matsumoto J et al (2011) Autocrine and paracrine up-regulation of blood–brain barrier function by plasminogen activator inhibitor-1. Microvasc Res 81(1):103–107. https://doi.org/10.1016/j.mvr.2010.10.004

144. Markus HS, Hunt B, Palmer K, Enzinger C, Schmidt H, Schmidt R (2005) Markers of endothelial and hemostatic activation and progression of cerebral white matter hyperintensities: longitudinal results of the Austrian Stroke Prevention Study. Stroke 36(7):1410–1414. https://doi.org/10.1161/01.STR.0000169924.60783.d4

145. Yepes M (2015) Tissue-type plasminogen activator is a neuroprotectant in the central nervous system. Front Cell Neurosci 9:304. https://doi.org/10.3389/fncel.2015.00304

146. Knottnerus ILH, Winckers K, Ten Cate H et al (2012) Levels of heparin-releasable TFPI are increased in first-ever lacunar stroke patients. Neurology 78(7):493–498. https://doi.org/10.1212/WNL.0b013e318246d6b7

147. Wiseman S, Marlborough F, Doubal F et al (2014) Blood markers of coagulation, fibrinolysis, endothelial dysfunction and inflammation in lacunar stroke versus non-lacunar stroke and non-stroke: systematic review and meta-analysis. Cerebrovasc Dis 37(1):64–75. https://doi.org/10.1159/000356789

148. Gottesman RF, Cummiskey C, Chambless L et al (2009) Hemostatic factors and subclinical brain infarction in a community-based sample: the ARIC study. Cerebrovasc Dis 28 (6):589–594

149. Kario K, Matsuo T, Kobayashi H et al (2001) Hyperinsulinemia and hemostatic abnormalities are associated with silent lacunar cerebral infarcts in elderly hypertensive subjects. J Am Coll Cardiol 37(3):871–877

150. Tomimoto H, Akiguchi I, Ohtani R et al (2001) The coagulation-fibrinolysis system in patients with leukoaraiosis and Binswanger disease. Arch Neurol 58(10):1620–1625

151. Hassan A, Hunt BJ, O'Sullivan M et al (2003) Markers of endothelial dysfunction in lacunar infarction and ischaemic leukoaraiosis. Brain 126(Pt 2):424–432

152. Wada M, Nagasawa H, Kurita K et al (2007) Microalbuminuria is a risk factor for cerebral small vessel disease in community-based elderly subjects. J Neurol Sci 255 (1–2):27–34. https://doi.org/10.1016/j.jns.2007.01.066

153. Nagai M, Hoshide S, Kario K (2012) Association of prothrombotic status with markers of cerebral small vessel disease in elderly hypertensive patients. Am J Hypertens 25 (10):1088–1094

154. Knottnerus ILH, Govers-Riemslag JWP, Hamulyak K et al (2010) Endothelial activation in lacunar stroke subtypes. Stroke 41 (8):1617–1622. https://doi.org/10.1161/STROKEAHA.109.576223

155. Isenegger J, Meier N, Lämmle B et al (2010) D-dimers predict stroke subtype when assessed early. Cerebrovasc Dis 29(1):82–86. https://doi.org/10.1159/000256652

156. Poggesi A, Pasi M, Pescini F, Pantoni L, Inzitari D (2016) Circulating biologic markers of endothelial dysfunction in cerebral small vessel disease: A review. J Cereb Blood Flow Metab 36(1):72–94. https://doi.org/10.1038/jcbfm.2015.116

157. Strathmann FG, Schulte S, Goerl K, Petron DJ (2014) Blood-based biomarkers for traumatic brain injury: evaluation of research approaches, available methods and potential utility from the clinician and clinical laboratory perspectives. Clin Biochem 47 (10–11):876–888

158. Gazzolo D, Li Volti G, Gavilanes AW, Scapagnini G (2015) Biomarkers of brain function and injury: biological and clinical significance. Biomed Res Int 2015:389023. https://doi.org/10.1155/2015/389023

159. Ehrenreich H, Hasselblatt M, Dembowski C et al (2002) Erythropoietin therapy for acute stroke is both safe and beneficial. Mol Med 8 (8):495–505

160. Dambinova SA, Khounteev GA, Izykenova GA et al (2003) Blood test detecting autoantibodies to N-methyl-D-aspartate neuroreceptors for evaluation of patients with transient ischemic attack and stroke. Clin Chem 49(10):1752–1762

161. Weissman JD, Khunteev GA, Heath R, Dambinova SA (2011) NR2 antibodies: risk assessment of transient ischemic attack (TIA)/stroke in patients with history of isolated and multiple cerebrovascular events. J Neurol Sci 300:97–102

162. Zerche M, Weissenborn K, Ott C et al (2015) Preexisting serum autoantibodies against the NMDAR subunit NR1 modulate evolution of lesion size in acute ischemic stroke. Stroke 46 (5):1180–1186

163. Dambinova SA, Maroon JC, Sufrinko AM et al (2016) Functional, structural, and neurotoxicity biomarkers in integrative assessment of concussions. Front Neurol 7:PMC5050199. https://doi.org/10.3389/fneur.2016.00172

164. González-García S, González-Quevedo A, Hernandez-Diaz Z et al (2017) Circulating autoantibodies against the NR2 peptide of the NMDA receptor are associated with subclinical brain damage in hypertensive patients with other pre-existing conditions for vascular risk. J Neurol Sci 375:324–330. https://doi.org/10.1016/j.jns.2017.02.028

165. Dadu RT, Fornage M, Virani SS et al (2013) Cardiovascular biomarkers and subclinical brain disease in the atherosclerosis risk in communities study. Stroke 44 (7):1803–1808. https://doi.org/10.1161/STROKEAHA.113.001128

166. Duering M, Konieczny MJ, Tiedt S et al (2018) Serum neurofilament light chain levels are related to small vessel disease burden. J Stroke 20(2):228–238. https://doi.org/10.5853/jos.2017.02565

167. Gattringer T, Pinter D, Enzinger C et al (2017) Serum neurofilament light is sensitive to active cerebral small vessel disease.

Neurology 89(20):2108–2114. https://doi.org/10.1212/WNL.0000000000004645

168. González-Quevedo A, García SG, OF C et al (2011) Increased serum S-100B and neuron specific enolase – potential markers of early nervous system involvement in essential hypertension. Clin Biochem 44 (2–3):154–159. https://doi.org/10.1016/j.clinbiochem.2010.11.006

169. Gao Q, Fan Y, Mu L-Y et al (2015) S100B and ADMA in cerebral small vessel disease and cognitive dysfunction. J Neurol Sci 354 (1–2):27–32. https://doi.org/10.1016/j.jns.2015.04.031

170. Xiao L, Sun W, Lan W et al (2014) Correlation between cerebral microbleeds and S100B/RAGE in acute lacunar stroke patients. J Neurol Sci 340(1–2):208–212. https://doi.org/10.1016/j.jns.2014.03.006

171. Prabhakar P, Chandra SR, Christopher R (2017) Circulating microRNAs as potential biomarkers for the identification of vascular dementia due to cerebral small vessel disease. Age Ageing 46(5):861–864. https://doi.org/10.1093/ageing/afx090

172. Karp X, Ambros V (2005) Developmental biology. Encountering microRNAs in cell fate signaling. Science 310 (5752):1288–1289. https://doi.org/10.1126/science.1121566

173. Gurol ME, Irizarry MC, Smith EE et al (2006) Plasma beta-amyloid and white matter lesions in AD, MCI, and cerebral amyloid angiopathy. Neurology 66(1):23–29. https://doi.org/10.1212/01.wnl.0000191403.95453.6a

174. Van Dijk EJ, Prins ND, Vermeer SE et al (2004) Plasma amyloid, apolipoprotein E, lacunar infarcts, and white matter lesions. Ann Neurol 55(4):570–575. https://doi.org/10.1002/ana.20050

175. Lee W-J, Jung K-H, Ryu YJ et al (2017) Cystatin C, a potential marker for cerebral microvascular compliance, is associated with white-matter hyperintensities progression. PLoS One 12(9):e0184999. https://doi.org/10.1371/journal.pone.0184999

176. Palsdottir A, Snorradottir AO, Thorsteinsson L (2006) Hereditary cystatin C amyloid angiopathy: genetic, clinical, and pathological aspects. Brain Pathol 16(1):55–59

177. Levy E, Sastre M, Kumar A et al (2001) Codeposition of cystatin C with amyloid-beta protein in the brain of Alzheimer disease patients. J Neuropathol Exp Neurol 60(1):94–104

178. Yang S, Cai J, Lu R et al (2017) Association between serum cystatin C level and total magnetic resonance imaging burden of cerebral small vessel disease in patients with acute lacunar stroke. J Stroke Cerebrovasc Dis 26 (1):186–191. https://doi.org/10.1016/j.jstrokecerebrovasdis.2016.09.007

179. Weller RO, Djuanda E, Yow H-Y, Carare RO (2009) Lymphatic drainage of the brain and the pathophysiology of neurological disease. Acta Neuropathol 117(1):1–14. https://doi.org/10.1007/s00401-008-0457-0

180. Rost NS, Wolf PA, Kase CS et al (2001) Plasma concentration of C-reactive protein and risk of ischemic stroke and transient ischemic attack: the Framingham study. Stroke 32 (11):2575–2579

181. van Dijk EJ, Prins ND, Vermeer SE et al (2005) C-reactive protein and cerebral small-vessel disease: the Rotterdam Scan Study. Circulation 112(6):900–905

182. Kaplan RC, McGinn AP, Baird AE et al (2008) Inflammation and hemostasis biomarkers for predicting stroke in postmenopausal women: the Women's Health Initiative Observational Study. J Stroke Cerebrovasc Dis 17(6):344–355

183. Staszewski J, Piusińska-Macoch R, Brodacki B et al (2018) Il-6, PF-4, sCD40 l, and homocysteine are associated with the radiological progression of cerebral small-vessel disease: a 2-year follow-up study. Clin Interv Aging 13:1135–1141

184. Walker KA, Power MC, Hoogeveen RC et al (2017) Midlife systemic inflammation, late-life white matter integrity, and cerebral small vessel disease. The Atherosclerosis Risk in Communities Study. Stroke 48:3196–3202

185. Wieberdink RG, van Schie MC, Koudstaal PJ et al (2010) High von Willebrand factor levels increase the risk of stroke: the Rotterdam study. Stroke 41:2151–2156

186. Kaffashian S, Tzourio C, Soumare A et al (2014) Plasma ß-amyloid and MRI markers of cerebral small vessel disease: Three-City Dijon Study. Neurology 83(22):2038–2045. https://doi.org/10.1212/WNL.0000000000001038

187. Johnson NA, Jahng G-H, Weiner MW et al (2005) Pattern of cerebral hypoperfusion in Alzheimer disease and mild cognitive impairment measured with arterial spin-labeling MR imaging: initial experience. Radiology 234(3):851–859. https://doi.org/10.1148/radiol.2343040197

188. Maas MB, Furie KL (2009) Molecular biomarkers in stroke diagnosis and prognosis. Biomark Med 3(4):363–383

Chapter 9

RNA Gene Expression to Identify the Etiology of Acute Ischemic Stroke: The Biomarkers of Acute Stroke Etiology (BASE) Study

Edward C. Jauch, W. Frank Peacock IV, Judy Morgan, Jeff June, and James Ireland

Abstract

Acute ischemic stroke affects over 800,000 US adults annually, with hundreds of thousands more experiencing a transient ischemic attack. Emergent identification, evaluation, and expeditious reperfusion therapies have made significant impact on reducing stroke morbidity and mortality. Once efforts to restore reperfusion have been addressed, identifying the stroke etiology and initiating secondary prevention strategies are essential for decreasing further morbidity and mortality of cerebrovascular disease. Despite advances in imaging and diagnostic testing, identifying stroke etiology remains challenging even in the most capable centers, with roughly 30% of patients discharged from the hospital without a clear cause. Without a defined stroke etiology, additional and perhaps unnecessary testing is performed, and initiating optimal prevention strategies for the specific stroke etiology are delayed. The Biomarkers of Acute Stroke Etiology (BASE) study is a multicenter observational study to identify serum markers defining the etiology of acute ischemic stroke. Blood samples are collected at arrival, 24, and 48 h later, and RNA gene expression is utilized to identify stroke etiology marker candidates. The BASE study began in January 2014. At the time of writing, there are 22 recruiting sites. Enrollment is ongoing, recruiting over 1670 patients as of January 2019. The BASE study could potentially aid in focusing the initial diagnostic evaluation to determine stroke etiology, with more rapidly initiated targeted evaluations and secondary prevention strategies.

Key words Ischemic stroke, Stroke etiology, RNA expression, Biomarkers

1 Introduction

Acute ischemic stroke (AIS) remains a leading cause of mortality and morbidity in the USA, affecting over 800,000 adults annually and leaving many with permanent disability [1]. Furthermore, hundreds of thousands of Americans experience a transient ischemic attack which often precedes a major stroke and serves as a warning for future ischemic events [2]. Despite resolved symptoms, experiencing a TIA (transient ischemic attack) is associated with a stroke risk of up to 20% within 90 days. Collectively, previous stroke

Philip V. Peplow et al. (eds.), *Stroke Biomarkers*, Neuromethods, vol. 147, https://doi.org/10.1007/978-1-4939-9682-7_9,
© Springer Science+Business Media, LLC, part of Springer Nature 2020

Table 1
Subtype of acute ischemic strokes, TOAST classification, and suggested clinical management

Stroke subtype	Typical clinical management
Large artery atherosclerosis	Antiplatelet therapy, major risk factor modification (hypertension, diabetes, hyperlipidemia, smoking cessation)
Cardioembolic	Anticoagulation
Small vessel occlusion (lacunar)	Antiplatelet therapy, major risk factor modification (hypertension, diabetes, hyperlipidemia, smoking cessation)
Stroke of other determined etiology	Interventions specific to cause
Stroke of undetermined etiology (two or more causes identified; negative evaluation; incomplete evaluation)	Antiplatelet therapy (aspirin), further diagnostic testing

TOAST Trial of Org 10,172 in Acute Stroke Treatment [12]

and TIA confer an annual recurrent stroke risk of 3–4%. Equally concerning, the 2-year mortality rate for cryptogenic stroke has been reported to be as high as 39%. Emergent evaluation, prompt acute treatment, and identification of stroke or TIA etiology for specific secondary prevention are critical for decreasing further morbidity and mortality of cerebrovascular disease.

Key to secondary prevention is stroke etiology identification; stroke etiology determines the most effective interventions to prevent future strokes (Table 1) [2]. For all patients experiencing cerebrovascular disease, general modifiable risk factor (smoking, hypertension, dyslipidemia, glucose disorders, obesity, etc.) modification should be addressed. More specific interventions target specific mechanisms of the stroke. Anticoagulation therapy is indicated for cardioembolic stroke. This is in contradistinction to atherogenic strokes where antiplatelet agents are recommended. Currently, the diagnosis of ischemic stroke etiology is determined from a combination of patient history, clinical assessment, cerebrovascular imaging, and cardiovascular evaluation (Table 2). Many challenges exist in determining stroke etiology. Not all patients are cared for in centers with access to stroke expertise. Since much of the diagnosis remains physician judgment, lack of access to stroke expertise may limit identifying stroke etiology. Furthermore, diagnostic modalities may not be available at all facilities. Magnetic resonance imaging available is not universal; transthoracic echocardiography and transesophageal echocardiography for detailed cardiac evaluation are also limited and potentially not cost-effective for noncardioembolic strokes. However, even with extensive testing over the course of several days, identifying the cause of an acute stroke is challenging. Strokes of unclear etiology, or cryptogenic

Table 2
Suggested diagnostic evaluation of ischemic stroke to determine etiology[a]

Diagnostic evaluation	Suggested approach
Brain imaging	• Brain MRI in patients with cryptogenic stroke • Brain CT when stroke mechanisms known
Cardiac imaging	• TTE on all patients with AIS • TEE with bubble study on patients <50 years old if TTE nondiagnostic
Cardiac monitoring	• Thirty day noninvasive cardiac monitoring for patients with cryptogenic stroke and ≥40 years old • Implantable cardiac monitor if 30 day monitor does not reveal atrial fibrillation or flutter
Hypercoagulable testing	• Serum hypercoagulable workup in patients with no or minimal risk factors
Malignancy screening	• Age appropriate screening • CT of chest/abdomen/pelvis when systemic symptoms suggestive of cancer
Vascular imaging	• Intracranial and extracranial vascular imaging in all patients with AIS • MRA with fat-suppressed images if cervical artery dissection suspected

AIS acute ischemic stroke, *CT* computed tomography, *MRA* magnetic resonance angiography, *MRI* magnetic resonance imaging, *TEE* transesophageal echocardiography, *TTE* transthoracic echocardiography
[a]Adapted from Yaghi et al. [13]

strokes, represent a significant risk as optimal prevention measures cannot be identified. Therefore, there is a great need to identify the pathogenesis of acute ischemic stroke in order to implement targeted and effective preventative measures.

The pursuit of blood-borne stroke diagnostic tests for diagnosing acute ischemic stroke and determining stroke etiology has spanned decades. Many have investigated individual or combinations of serum proteins and yet no study has demonstrated sufficient diagnostic robustness to be useful in routine clinical practice [3]. More recent studies have suggested whole blood RNA expression may help differentiate ischemic stroke mechanisms [4–8]. The majority of RNA in whole blood is from circulating leukocytes, monocytes, and neutrophils. In very early samples obtained close to stroke ictus, these circulating immune cells likely reflect prestroke biologic activity, including inflammatory states, immune modulation, states of coagulation balance, and overall signaling. Samples obtained later in the course of acute ischemic stroke likely reflect more common pathways associated with response to the infarction. Thus, understanding the time course of gene expression is critical to be able to use the signatures for a specific clinical question.

Jickling and colleagues were amongst the first to perform detailed whole blood gene expression studies in patients with acute ischemic stroke [5]. In an early study, a 40-gene profile differentiated large-vessel stroke from cardioembolic stroke with reported 95% sensitivity and specificity. In the same study, a different 37-gene expression signature further classified patients with cardioembolic stroke due to atrial fibrillation vs. other cardioembolic sources with >90% sensitivity and specificity. More recently, Jickling and colleagues looked at the differential gene expression with the additional information of physical infarct location to identify the likely etiology of stroke previously classified as cryptogenic [9]. Using Affymetrix® U133 Plus 2.0 microarrays (Affymetrix, Thermo Fisher Scientific, Waltham, MA) and the sample profiles from the prior study, combining stroke location with the gene expression signatures identified 58% of patients previously classified with cryptogenic stroke as being cardioembolic, 18% to be large vessel stroke, 12% lacunar, and 12% unclassified. These findings encouraged further study.

Based on these promising studies, the Biomarkers of Acute Stroke Etiology (BASE) study was initiated by Ischemia Care (ISCDX, Oxford, OH, USA) and partnered with leading academic medical centers for study sites. BASE (NCT02014896) is a multi-center observational study utilizing RNA gene expression from the Ischemia Care diagnostic platform to identify the etiology of acute ischemic stroke. As noted before when stroke or TIA occur, the immune system changes gene expression in multiple cell types, thus activating innate and adaptive immune responses. Previous studies suggest that differential gene expression profiles are a function of stroke subtype, with each subtype producing a unique gene expression "signature" [4–8]. The Ischemia Care diagnostic platform consists of whole blood biomarker tests to determine the etiology of ischemic stroke (ISCDX, Oxford, OH, USA) by measuring acute ischemic stroke gene expression changes. For example, the ISCDX test based on previous gene signatures distinguishes between cardioembolic and large artery, as well as lacunar, atherosclerotic stroke using a signature of 40 unique genes. A patient's pattern of gene regulation can determine if the stroke etiology is that of a cardioembolic or large artery atherosclerotic source. Further, a separate 37 gene signature can differentiate cardioembolic strokes caused by atrial fibrillation (AF) or other cardioembolic sources. Ultimately, for most patients, the diagnostic expression pattern clearly identifies stroke etiology.

The primary objective of the BASE study is to confirm the diagnostic accuracy of the ISCDX test to identify stroke subtypes in patients with acute ischemic stroke. This manuscript describes the methodology employed in the BASE study to identify stroke etiology in patients presenting with acute stroke.

2   Methods

BASE is an ongoing prospective multicenter convenience sample study, registered as NCT02014896 and approved by each participating Institutional Review Board [10]. Patients with acute ischemic stroke who meet the inclusion and exclusion criteria (Table 3) are enrolled in the Emergency Department (ED) and blood samples are drawn. Control samples consist of 100 non-stroke ED patients matched on clinical risk factors of age, race, gender, smoking history, diabetes, hypertension, atrial fibrillation, and hyperlipidemia.

BASE initially enrolled acute stroke patients within 8 h of symptom onset or the time of last known normal. However, after patient enrollment reached 650, evaluation time was lengthened to 24 ($\pm$ 6) h. This was from a planned interim data analysis determining the 24 ($\pm$ 6) h window from symptom onset was most predictive for identifying stroke cause using blood biomarkers, was most consistent with the time a stroke patient would present, and represented the window for which a blood test for stroke would be used clinically.

Typically, prior to enrollment, patients are evaluated by the local stroke team or ED physicians, have undergone baseline laboratory testing and cerebrovascular imaging, and may receive intravenous thrombolysis and/or endovascular therapies. Approximately 2.5 mL of blood is drawn into two PreAnalytiX® PAXgene® blood RNA tubes (Qiagen, Venlo, Netherlands) within 24 ($\pm$ 6) h of stroke onset. Additional draws occur at 24 ($\pm$ 6), and 48 ($\pm$ 6) h, or at ED/hospital discharge, whichever comes first. Longer collection periods were considered but were challenged by the amount of RNA response making it difficult to identify

Table 3
Inclusion and exclusion criteria for BASE

Inclusion criteria
• Suspected acute ischemic stroke within 24 ($\pm$ 6) h of last known normal or symptom onset
• Baseline CT normal, without hemorrhage or alternate explanation for symptoms
• >18 years old
• Informed consent obtained

Exclusion criteria
• Central nervous system infection within 30 days
• Serious head trauma within 30 days
• Any ischemic or hemorrhagic stroke within 30 days
• Active cancer (not in remission)
• Autoimmune disease (e.g., lupus)
• Acute systemic infection
• Major surgery within 90 days

diagnostic patterns consistent with the primary objective of this study. Eligible control subjects are patients presenting without a potentially neurologic complaint and have blood drawn within 6 h of ED presentation.

PAXgene® tubes can be kept at room temperature for up to 24 h, and then are frozen at −20 °C, until shipped on dry ice to the Ischemia Care CLIA laboratory (Middletown, OH) where the ISCDX testing is performed. The entire sample from one tube will be used to perform the ISCDX test. The second tube is stored at −80 °C for future testing.

Analysis for RNA expression is performed by Affymetrix® human gene ST array plates. These provide whole-genome coverage, including protein coding and long intergenic non-coding RNA (lincRNA) transcripts. Whole genome arrays thus have the ability to provide a complete profile of mRNA expression.

The microarray procedure is performed as follows:

1. RNA extraction: Total RNA is extracted from blood collected in PAXgene® RNA tubes which are used to specifically preserve the integrity of the RNA.

2. cDNA synthesis and labeling: Multiple complementary DNA (cDNA) copies are made of each RNA. cDNAs are fragmented to sizes for optimal hybridization to the probes on the microarrays and labeled so they can be stained and detected after hybridization to the microarray.

3. Microarray hybridization: Amplified, fragmented and labeled cDNAs are hybridized overnight to an Affymetrix® U133 plus 2.0 microarray. Each microarray contains probes for the majority of expressed RNAs from the human genome.

4. Microarray staining and scanning: After hybridization, each microarray is washed, stained with fluorescence that binds to the labels previously attached to each cDNA, then rinsed.

5. Microarray data is normalized using the "Signal Space Transformation with probe Guanine Cytosine Count Correction" algorithm (Thermo Fisher, Waltham, MA) to control for normal experimental variability. Array quality control metrics are checked against acceptable ranges and rejected when not within range. As an additional quality check, sex-specific expressed genes are checked against the clinically recorded sex to identify potential sample mix-ups.

All study patients undergo standard clinical assessments as determined by the local treating physician, including baseline biochemistry and neuroimaging. Vascular imaging is not required before enrollment. Control patients do not receive specified imaging or biochemistry assays as part of the study protocol. The extent of the inpatient diagnostic evaluation is per standard of care and is determined by the physician team caring for the patient [11]. In the United States, the standard of care generally consists of a 12-lead ECG, bedside ECG monitoring, standard transthoracic echocardiogram (or transesophageal echocardiogram as indicated), and outpatient cardiac event monitoring (Table 3). Specific study data collected include patient demographics, past medical history, social history, medications, ED evaluation information (including the National Institutes of Health stroke scale [NIHSS]), neurologic symptom duration and onset time, baseline cerebrovascular imaging, laboratory tests, electrocardiogram, and cardiac monitoring. Cardiac evaluations during admission are collected, as are data from other studies performed to treat or determine stroke etiology.

Ischemic stroke etiology is determined locally using all sources of clinical information, according to the Trial of Org 10,172 in Acute Stroke Treatment (TOAST) classification (Table 2). TOAST is a well validated classification system with five subtypes of ischemic stroke. A cardioembolic stroke diagnosis requires at least one source of cardiac emboli and the exclusion of large or small vessel causes of stroke. Cardioembolic sources include AF, acute myocardial infarction, prosthetic valves, and/or cardiomyopathy. Patients with AF are identified using electrocardiogram, echocardiogram, and cardiac monitoring. Cardiac monitoring is performed as standard of care (either during hospitalization or using outpatient event monitoring in unconfirmed suspected AF).

The diagnosis of large-vessel stroke requires >50% stenosis of ipsilateral extracranial or major intracranial artery (middle cerebral artery, posterior cerebral artery, basilar artery) presumed due to atherosclerosis determined by ultrasound, computed tomography angiography, magnetic resonance angiography, or digital subtraction angiography, and is further supported by the absence of acute infarction in other vascular territories. Because of a lack of standardization for defining stroke in the setting of ulcerated plaques of less than <50% narrowing, sites were not asked to categorize these, rather they are placed in the cryptogenic category.

The diagnosis of small-vessel stroke requires symptoms corresponding to a subcortical infarction <15 mm in longest diameter on brain imaging, typically identified on MRI, and the exclusion of other stroke mechanisms. Often, patients will present with classic lacunar syndromes (pure motor hemiparesis). Stroke caused by other uncommon etiologies refers to atypical but specific causes of ischemic stroke (e.g., nonatherosclerotic arteriopathies, vascular dissections, or hypercoagulable states).

Finally, strokes with an extensive work up and remaining of unknown origin are referred to as cryptogenic, patients with multiple stroke etiologies identified are placed in their own category, and those with insufficient information are categorized as such.

The final gold standard diagnosis is determined by an adjudication committee consisting of two vascular neurologists independently reviewing all available data and blinded to ICDX testing results. All diagnostic impression data is used for the TOAST criteria. In cases of unresolved diagnostic disagreement, a third vascular neurologist serves as a tie breaker. The BASE study is non-interventional, with treating physicians blinded to genomic test results, thus presenting minimal to no patient risk. Categorical data will be analyzed by chi-square testing and linear continuous data by Student's t-test. Univariate analysis will identify significant outcome predictors for multivariable modeling. RNA expression data will be presented as heat maps with multiple comparison corrections.

Statistical performance of cut points, determined by multivariable modeling and by heat map identification, will be presented using sensitivity, specificity, and C statistic, as well as positive and negative predictive values and likelihood ratios. Net reclassification improvement and integrated discrimination improvement will evaluate the change in clinical diagnosis using RNA expression.

The robust BASE trial methodology is predicted to allow for the generation of a number of RNA expression signatures that will be of clinical significance. First it will differentiate between strokes resulting from embolic causes, as compared to stroke caused by large artery thrombosis. This important determination dichotomizes treatment strategies into the divergent categories of either anticoagulants or antiplatelet medication.

Secondly, RNA expression signatures are likely to separate true stroke presentations from that of stroke mimics. Patients are enrolled in BASE before inpatient admission and thus much of the inpatient diagnostic evaluation will not have been performed at the time of enrollment. It is anticipated that a significant number of patients will be enrolled with stroke mimics, other medical conditions, such as seizures, hypoglycemia, and migraine headache, that present with signs and symptoms concerning for acute ischemic stroke. If such an expression signature can be found, as a point of care test, this distinction would have massive health care economic benefits as the current standard results in patients suffering from stroke mimics receive expensive, complicated, and unnecessary evaluations that are ultimately negative.

A third potential outcome from the BASE trial is identifying patients with less common causes of stroke, such as vasculopathies, malignancies, and hypercoagulable states. These others causes of ischemic stroke have greatly different treatment approaches and earlier diagnosis of stroke etiology may allow for early primary treatment and secondary prevention.

As is the case with all investigations, the potential for methodologic limitations may exist. Because the medical care in patients enrolled in BASE is determined by the physician caring for the patient, and not by a defined protocol, variations in stroke etiology evaluation may occur (e.g., not all patients may receive echocardiographic bubble studies), and how this may impact outcomes will be unmeasured. Furthermore, the timing of RNA expression measurement was arbitrarily chosen at 24 h as this is a time for which marker analysis is obtainable and is still within a clinically relevant window.

Arguments for alternative timing could be effected and may provide an impetus for future investigations. Finally, the decision to define a vascular stenosis <50% to be more likely associated with embolic events represents a compromise for consistent diagnosis but may ultimately not be definitive in its accuracy. BASE is funded by Ischemic Care, LLC, whose involvement includes providing funding for blood and data collection, assay performance, laboratory analysis, gold standard diagnostic adjudication, and statistical analysis.

The BASE study began in January 2014. At the time of writing, there are 22 recruiting sites. Enrollment is ongoing, recruiting over 1670 patients as of January 2019. To date, 1670 subjects have been enrolled with over three discrete time point samples drawn over a 0–48 h window, when available.

3   BASE Pilot Study

As part of the larger BASE study, a pilot study was designed to investigate gene expression in differentiating cardioembolic vs large artery atherosclerotic stroke. The pilot study was performed using the methodology described above. Samples utilized for the pilot study came from the larger adjudicated BASE cohort.

3.1   Methods

Subject enrollment, sample acquisition, RNA processing and analysis were performed as described for the overall BASE study. For the pilot study, the Affymetrix® GeneChip Human Transcriptome Array (HTA 2.0), with >6 million probes was used. This allowed for the analysis of extensive protein coding content (44,699 genes, 245,349 transcripts, 560,472 exons, 296,058 exon clusters, and 339,146 splice junction probesets) as well as non-protein-coding content (22,829 genes), 40,914 transcripts, 109,930 exons, and 82,444 exon clusters). For the pilot study, significantly differentially expressed genes were identified by Bayes moderated t-statistic contrasting large artery vs. cardioembolic stroke. Statistical analyses were performed using R (R Foundation for Statistical Computing, Vienna, Austria). Differentially expressed genes were utilized in a multilayer perceptron neural network.

3.2   Results

In the pilot study 32 stroke subjects were analyzed. Baseline median NIHSS was 9 (Interquartile range 4, 14), median age was 68.6 years, and with 50% of subjects male. The median time from symptoms onset to sample draw was 320 min. Patients in the pilot

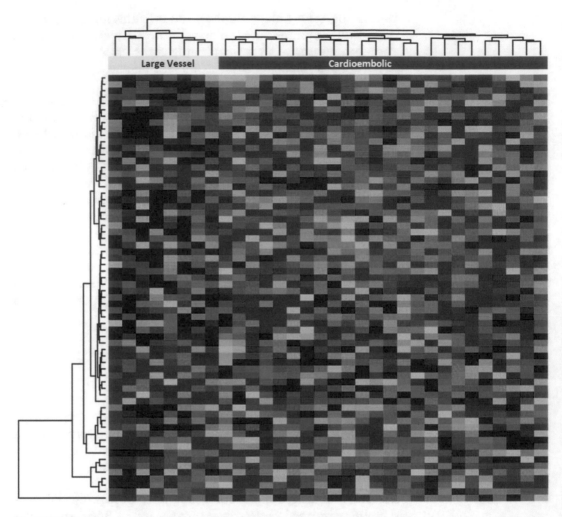

Fig. 1 Baseline samples with 66 differentially expressed genes. Hierarchical cluster plot of genes significantly different between cardioembolic stroke and large-vessel stroke. Subjects are shown on the x-axis and genes are shown on the y-axis. Red indicates a high level of gene expression and blue indicates a low level of gene expression. Subjects are cohorted by diagnosis

study had significant comorbidities, 41% with atrial fibrillation, 22% with congestive heart failure, 22% with prior stroke, and 25% with coronary artery disease. In this cohort, stroke etiology was 25% large artery stroke and 75% cardioembolic stroke.

In the preliminary pathway analysis, 66 genes, comprised of 17 protein and 49 non-protein-coding (long non-coding RNA and microRNA) genes, were differentially expressed (Fig. 1). The National Center for Biotechnology Information reference sequence database (RefSeq) (http://www.ncbi.nlm.nih.gov/RefSeq/) identified nine protein genes with known association with stroke, brain function, blood or cardiovascular disease, and three putative genes with limited information.

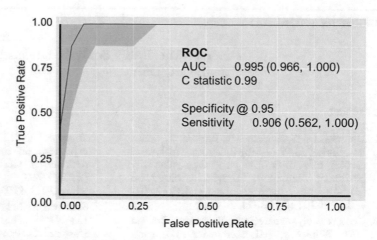

Fig. 2 Eight-hour draw samples receiver operating characteristic. *AUC* area under curve, *ROC* receiver operating curve

Based on the 66 differentially expressed genes, the receiver operating characteristic (ROC) demonstrated the diagnostic ability of the panel to classify stroke subtype with a 95% specificity and 90% sensitivity (Fig. 2). The area under the ROC curve was 0.995 and the C statistic was 0.99. While these results were very encouraging, it is important to note that this was a well adjudicated and defined cohort, and small vessel and cryptogenic strokes, and stroke mimics were not included.

4 Conclusions

The BASE pilot study demonstrated that RNA expression can accurately differentiate large artery stroke from cardioembolic stroke in the acute setting. With this information, early stroke subtype determination may help guide diagnostic selection and the early initiation of targeted secondary stroke prevention strategies.

The BASE study began in January 2014. At the time of writing there are 22 recruiting sites across the USA and over 1670 patients have been enrolled to date. It is anticipated that the analysis of the differential gene expression in the full BASE cohort will result in the identification of a series of unique stroke biomarkers. It is anticipated that the identification of stroke etiology will be possible, thus differentiating large artery atherogenic events from stroke caused by cardiogenic emboli, as well as further subclassification of cardioembolic stroke into AF and non-AF etiologies. Because determination of cryptogenic stroke etiology remains challenging, it is estimated that the 250,000 annual US stroke patients whose stroke etiology is unknown may benefit from the findings of BASE.

Acknowledgment

Funding: The BASE Study is funded by Ischemia Care, LLC, Oxford, OH, USA.

References

1. Dhamoon MS, Sciacca RR, Rundek T, Sacco RL, Elkind MSV (2006) Recurrent stroke and cardiac risks after first ischemic stroke: the Northern Manhattan Study. Neurology 66 (5):641–646. https://doi.org/10.1212/01. wnl.0000201253.93811.f6

2. Kernan WN, Ovbiagele B, Black HR, Bravata DM, Chimowitz MI, Ezekowitz MD, Fang MC, Fisher M, Furie KL, Heck DV, Johnston SC, Kasner SE, Kittner SJ, Mitchell PH, Rich MW, Richardson D, Schwamm LH, Wilson JA (2014) Guidelines for the prevention of stroke in patients with stroke and transient ischemic attack: a guideline for healthcare professionals from the American Heart Association/American Stroke Association. Stroke 45 (7):2160–2236. https://doi.org/10.1161/ STR.0000000000000024

3. Laskowitz DT, Kasner SE, Saver J, Remmel KS, Jauch EC, Group BS (2009) Clinical usefulness of a biomarker-based diagnostic test for acute stroke: the Biomarker Rapid Assessment in Ischemic Injury (BRAIN) Study. Stroke 40 (1):77–85. https://doi.org/10.1161/ STROKEAHA.108.516377

4. Stamova B, Xu H, Jickling G, Bushnell C, Tian Y, Ander BP, Zhan X, Liu D, Turner R, Adamczyk P, Khoury JC, Pancioli A, Jauch E, Broderick JP, Sharp FR (2010) Gene expression profiling of blood for the prediction of ischemic stroke. Stroke 41(10):2171–2177. https://doi.org/10.1161/STROKEAHA. 110.588335

5. Jickling GC, Xu H, Stamova B, Ander BP, Zhan X, Tian Y, Liu D, Turner RJ, Mesias M, Verro P, Khoury J, Jauch EC, Pancioli A, Broderick JP, Sharp FR (2010) Signatures of cardioembolic and large-vessel ischemic stroke. Ann Neurol 68(5):681–692. https://doi.org/ 10.1002/ana.22187

6. Tian Y, Stamova B, Jickling GC, Liu D, Ander BP, Bushnell C, Zhan X, Davis RR, Verro P, Pevec WC, Hedayati N, Dawson DL, Khoury J, Jauch EC, Pancioli A, Broderick JP,

Sharp FR (2012) Effects of gender on gene expression in the blood of ischemic stroke patients. J Cereb Blood Flow Metab 32 (5):780–791. https://doi.org/10.1038/ jcbfm.2011.179

7. Stamova B, Jickling GC, Ander BP, Zhan X, Liu D, Turner R, Ho C, Khoury JC, Bushnell C, Pancioli A, Jauch EC, Broderick JP, Sharp FR (2014) Gene expression in peripheral immune cells following cardioembolic stroke is sexually dimorphic. PLoS One 9(7):e102550. https://doi.org/10.1371/jour nal.pone.0102550

8. Jickling GC, Stamova B, Ander BP, Zhan X, Liu D, Sison SM, Verro P, Sharp FRI (2012) Prediction of cardioembolic, arterial, and lacunar causes of cryptogenic stroke by gene expression and infarct location. Stroke 43 (8):2036–2041. https://doi.org/10.1161/ STROKEAHA.111.648725

9. Jickling GC, Ander BP, Stamova B, Zhan X, Liu D, Rothstein L, Verro P, Khoury J, Jauch EC, Pancioli AM, Broderick JP, Sharp FR (2013) RNA in blood is altered prior to hemorrhagic transformation in ischemic stroke. Ann Neurol. https://doi.org/10.1002/ana. 23883

10. Jauch EC, Barreto AD, Broderick JP, Char DM, Cucchiara BL, Devlin TG, Haddock AJ, Hicks WJ, Hiestand BC, Jickling GC, June J, Liebeskind DS, Lowenkopf TJ, Miller JB, O'Neill J, Schoonover TL, Sharp FR, Peacock WF (2017) Biomarkers of Acute Stroke Etiology (BASE) study methodology. Transl Stroke Res. https://doi.org/10.1007/s12975-017-0537-3

11. Jauch EC, Saver JL, Adams HP Jr, Bruno A, Connors JJ, Demaerschalk BM, Khatri P, McMullan PW Jr, Qureshi AI, Rosenfield K, Scott PA, Summers DR, Wang DZ, Wintermark M, Yonas H. Guidelines for the early management of patients with acute ischemic stroke: a guideline for healthcare professionals from the American Heart Association/

American Stroke Association. Stroke 2013. doi:https://doi.org/10.1161/STR. 0b013e318284056a

12. Adams HP Jr, Bendixen BH, Kappelle LJ, Biller J, Love BB, Gordon DL, Mearsh EE III (1993) Classification of subtype of acute ischemic stroke. Definitions for use in a multicenter clinical trial. TOAST. Trial of Org 10172 in acute stroke treatment. Stroke 24(1):35–41. https://doi.org/10.1161/STR. 0000000000000024

13. Yaghi S, Elkind MSV (2014) Cryptogenic stroke: a diagnostic challenge. Neurol Clin Pract 4(5):386–393. https://doi.org/10. 1212/CPJ.0000000000000086

Chapter 10

Neurotoxicity Biomarker Assay Development

Galina A. Izykenova, German A. Khunteev, Ivan I. Krasnjuk, Vladimir L. Beloborodov, and Svetlana A. Dambinova

Abstract

Neurotoxicity biomarker concept implying NR2 peptide and antibodies is presented in experimental models of transient cerebral ischemia and intracerebral hematoma. Various platforms of NR2 peptide and antibodies assays are developed to improve diagnostic certainty of cerebral ischemia. The assay bench testing results and methods are described. There are automatic analyzer, ELISA, and point-of-care tests discussed for laboratory and bedside applications.

Key words Neurotoxicity, Biomarkers, NR2 peptide, Antibodies, Immunoassay

1 Introduction

Research implying animal models of cerebral ischemia have shown overexpression of N-methyl-aspartate (NMDA) receptors that generally has been perceived as the major pathway for the lethal influx of Ca^{2+} [1]. NMDA receptors are thought to be a major contributor to the post-ischemic elevation of Ca^{2+} permeability. This is supported by the observation that NMDA-triggered toxicity in neuronal cultures is dependent on the presence of extracellular Ca^{2+} because prior application of neuroprotective NMDA receptor antagonists blocks NMDA triggered increase in Ca^{2+} permeability [2]. Therefore, these receptors have become principal biomarkers of neurotoxicity cascade underlying cerebral ischemia.

Acute cerebral ischemia results in excitatory neurotransmitter (glutamate) release and activation of ionotropic NMDA receptors localized on synapses and brain microvessel surfaces [3]. Activated NMDA receptors contribute to disruption of the blood-brain barrier (BBB) with influx of calcium into neurons [4]. Under these conditions, NMDA receptors undergo an activation-dependent post-translational modification by extracellular and intracellular proteases [5] leading to the production of fragment peptides that can be detected in the bloodstream [6]. The release of the abundant

Philip V. Peplow et al. (eds.), *Stroke Biomarkers*, Neuromethods, vol. 147, https://doi.org/10.1007/978-1-4939-9682-7_10,
© Springer Science+Business Media, LLC, part of Springer Nature 2020

amount of NR2 peptide from brain into the blood might be the objective evidence of cerebral ischemia.

The development of an immunoassay detecting NR2 peptide in plasma is remarkably complex, calling for a demanding multidisciplinary approach, knowledge, and skills. The principal objectives of immunoassay are (1) generating rare reagents, including key antigen and corresponding polyclonal or monoclonal antibodies; (2) selecting a testing format and technology; (3) optimizing an assay conditions; and (4) assay validation and feasibility studies. To meet these objectives, assay designers need to balance many factors, including those imposed by quality, cost, and time.

This chapter is devoted to the proof of the neurotoxicity concept in experimental models of stroke, biomarker assay development, and following bench testing to improve diagnostic certainty of cerebrovascular onsets.

2   Neurotoxicity Biomarkers in Experimental Stroke

The NR2 peptide was found to be a degradation product of original presynaptic and extrasynaptic NMDA receptors located on microvessels and neurons [7]. Under cerebral ischemia, both presynaptic and extrasynaptic receptors are involved in stimulation of NR2A and NR2B mRNA expression [6]. N-terminal extracellular peptide fragments of NMDA receptor normally cleaved by serine proteases are peripherally located in the bloodstream and are bound by natural antibodies [8] present in small amounts in healthy organisms. However, most C-terminal NMDA receptors metabolized by calpain or caspases are digested inside neurons [9].

2.1   NMDA Receptors in Transient Cerebral Ischemia

Early neurotoxicity in the cortex of rat with middle cerebral artery occlusion (MCAo) was detected with major involvement of synaptic (NR2A) and extrasynaptic (NR2B) subunits of NMDA receptors [6]. Indeed, immunohistochemical staining of cortical slices by hen antibodies against NR2 peptide clearly demonstrated a diminishing peptide density (Fig. 1a) which cannot be compensated by upregulated NR2 mRNA expression in the cortex and hippocampus (Fig. 1b, c). Furthermore, the upregulation of NR2 mRNA expression executes programmed cell death that was clearly demonstrated histochemically when neurons changed shape and shrunk after the reversible ischemia (Fig. 1d). At the same time, there was no NR2 mRNA expression detected in white blood cells, indicating that NR2 subunits belong exclusively to brain and are not produced in the bloodstream. The monitoring of BBB integrity revealed reduced NR2 immunoreactivity in brain, a significant NR2 peptide trafficking in to blood within the first 2 h, reaching maximum at 24 h after MCAo, and indicating compromised BBB (Fig. 2a).

The NR2 peptide release might be a tissue-based evidence of early neurotoxicity underlying transient cerebral ischemia. The

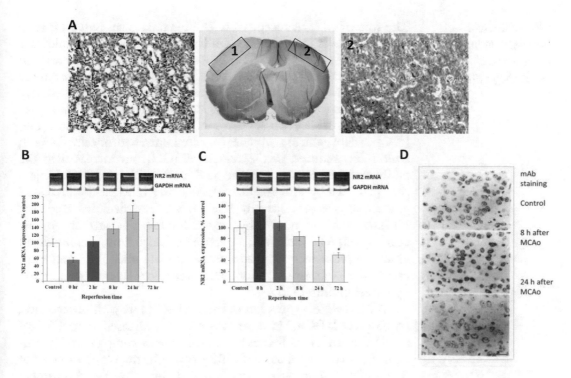

Fig. 1 (**a**) NR2 antibodies (**b**, ×400) staining of 5 mm coronal slices from rat with MCAo (1 h of MCAo and 24 h of reperfusion): 1—infarcted area; and 2—contralateral area, NR2 mRNA expression in (**b**) cortex of MCAo rats, (**c**) hippocampus of MCAo rats; and (**d**) neuronal damage in induced cerebral ischemia

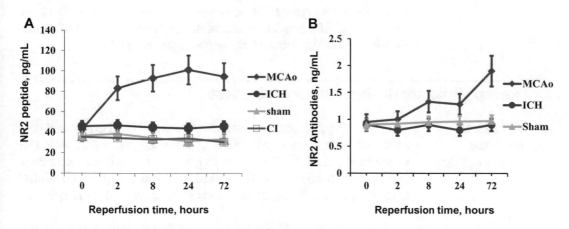

Fig. 2 Accumulation of NR2 peptide (**a**) and NR2 antibodies (**b**) in the blood of experimental animals

circulating NR2 antibodies measured in the blood support the brain origin of NR2 peptide as a "foreign" antigen that is capable of activating the immune system to generate antibodies (Fig. 2b). Significant amount of NR2 antibodies appeared with a delay of 72 h in MCAo rats and persisted for weeks to months [6].

2.2 Glutamate Receptors in Induced Intracerebral Hemorrhage (ICH)

The role of NMDA receptors in ICH is unknown. Based on scattered reports of increases of glutamate in brain and CSF following ICH, Sharp et al. explored the possibility that local increases in blood flow might be related to glutamate release [10]. To examine the mechanisms of changes in glucose utilization, Sprague Dawley rats subjected to lysed blood injections into striatum were pretreated with NMDA antagonist MK-801. The results showed that [^{14}C]-2-deoxyglucose uptake (injected intraperitoneally 1–72 h after ICH) reduced in the region of ICH, but increased in the perihematomal region, peaking at 3 h after the lysed blood injection [10]. The pretreatment with MK-801 blocked the increased glucose uptake produced by ICH. It was also demonstrated that NMDA and α-amino-3-hydroxy-5-methyl-4-isoxazolepropionic acid (AMPA) but not glutamate injections upregulated glucose uptake in perihematoma brain early after ICH and indicated the ionotropic glutamate receptor-dependent glucose hypermetabolism [11].

NR2 mRNA expression in the cortex of rats with intracortical rat blood cells (RBC) infusion was steadily decreased up to 47–56% at all time points studied in hematoma cortex compared with controls. Low concentrations of NR2 peptide in the blood samples of RBC-infused rats were accompanied by a reduced immune response (Fig. 2a). NR2 antibodies monitored in the blood sera of rats with intracerebral hematoma did not vary significantly compared to sham-operated rats (Fig. 2b).

The combined data suggest that NMDA receptor function is down-regulated after ICH compared to cerebral ischemia and the detection of NR2 peptide or antibodies in blood could be of benefit for discriminating between these two types of stroke.

3   NR2 Peptide Assays Development for Acute Stroke

3.1 NR2 Peptide Enzyme-Linked Immunoassay (ELISA)

The NR2 Peptide test is a magnetic particle-based enzyme-linked immunosorbent assay (MP-ELISA) intended for the quantitative determination of NR2 peptide fragment of NMDA receptors in plasma. The test is intended to be used in conjunction with clinical evaluation and radiological methods for diagnosis of acute ischemic stroke.

Concentrations of NR2 peptide are determined immunochemically in a blood assay.

--NR2 Antibodies ←Plasma NR2 peptide ← NR2 Antibodies-HRP

In the first incubation step, NR2 peptides in the sample react with immobilized NR2 antibodies. In the second incubation step, horseradish peroxidase-labeled antibodies against NR2 peptides are added to the reagent mixture. Formed immunocomplex is determined using 3,3′,5,5′-tetramethylbenzidine (TMB) detection reaction. An acidic stopping solution is then added. The color converts from blue to yellow. The intensity of the color is directly proportional to the concentration of NR2 peptide in the sample. A dose response curve of the absorbance measured at 450 nm or using dual wave measurement at 450 nm and 630 nm vs. concentration is generated. NR2 peptide concentrations in the plasma samples are determined directly from this calibration curve. Results of a typical standard calibration curve with optic density (OD) readings at 450 vs. 630 nm are shown on the x-axis against NR2 peptide concentrations (ng/mL) shown on the y-axis (Fig. 3a).

3.2 NR2 Peptide Automatic Analyzer (AA) Assay

The assay transfer to automated analyzer platform employed paramagnetic particles and chemiluminescent technology. A modification is made to accommodate unique reagents for "on-board" detection using ready-to-use reagent packs. Preformed with streptavidin supermagnetic particles (0.8 μm) are bound to carrier antibodies (Ab) for the quantitative detection of NR2 peptide (Fig. 3b). The analyte in plasma competes with

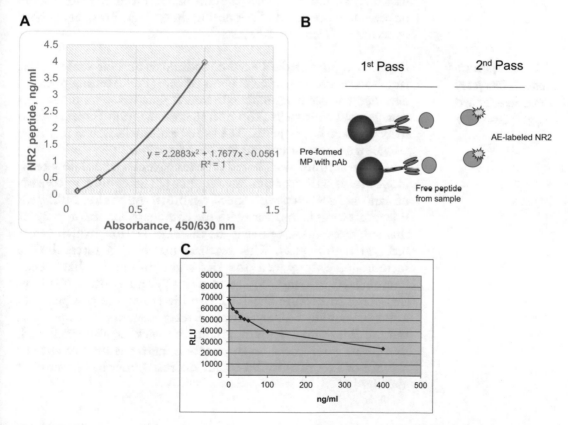

Fig. 3 (a) MP-ELISA calibration curve of NR2 peptide assay, (b) scheme of automatic analyzer (AA) format for NR2 peptide detection, (c) NR2 peptide AA assay calibration curve where RLU—relative light units

chemiluminescent labeled synthetic peptide for preformed with microparticles (MP)-polyclonal Ab (competitive assay). Translation from flat surface of microplate to microparticles increases reaction surface with simultaneous reduction in reagents consumption. That allows to speed up the turnaround time of laboratory testing from to 30 min.

Programmed AA separates, aspirates, and washes three times the cuvettes using buffer (PBS/Tween 20). Then it dispenses Base reagent to initiate the chemiluminescent reaction that automatically measured and reported in relative light units (RLU) according to software to the selected option, as described in the system operating instructions. Results of a standard calibration curve with RLU readings are shown on the y-axis against NR2 peptide concentrations (ng/mL) shown on the x-axis (Fig. 3c).

The anti-NR2 peptide assay kit uses a ready-to-use pack consisting solid phase compartment filled with 100 μg/mL latex paramagnetic MP preformed with biotinylated pAb. The second compartment of ready-to-use pack contains 10 ng/mL NR2—acridinium ester (AE) Lite reagent. The separate ancillary pack holds PBS-T buffer is necessary for reagent dilutions and pack for base reagents used to initiate the chemiluminescent reaction should be presented on-board of AA during the testing. Reagents in both packs are sufficient for 100 tests.

3.3 Nanoparticles Lateral Flow (nLF) Procedure for NR2 Peptide Detection

Recognizing the medical need of bedside testing for acute stroke, largely for neuro-critical care, point-of-care (POC) testing platform development has been initiated. Lateral flow based immunostrips engage gold nanoshells (150 nm) that dramatically increase sensitivity of detection reagents. The use of nanoshells results in a large visible signal even at low analyte concentrations.

The test principle is based on nanoparticles lateral flow technique and is a direct immunoassay [12]. The conjugate consists of nanoshells labeled with specific capture antibodies. NR2 peptide in the sample interacts with the conjugate to form a complex that migrates along the strip to the next section, which is the reaction matrix [12]. This reaction matrix is a nitrocellulose membrane, onto which the detection antibodies have been immobilized in bands to create test (T) and control (C) lines that serve to capture the NR2 peptide-conjugate complex and unbound conjugate, respectively. Excess reagents move past the capture lines and are entrapped in the wick or absorbent pad. Results are interpreted on the reaction matrix as the presence or absence of lines of captured conjugate, read either by eye or using a reader [12].

4   NR2 Antibody (Ab) Tests to Assess Chronic Ischemic Events

4.1   NR2 Ab ELISA Assay

The detection of the NR2 Ab on the microplate (MTP) loaded with synthetic NR2 peptide (antigen) is performed with horseradish peroxidase (HRP)-labeled Protein A (PrA) that reacts with the heavy chain of the human IgG from sample that is captured by antigen (Fig. 4a).

The detection of the formed immunocomplex—fixed on the MTP—is then determined by the HRP/tetramethylbenzidine (TMB)/H_2O_2-enzymatic reaction which is stopped by diluted acid. A color change from blue to yellow occurs after the addition of acid and the obtained color is measured photometrically at 450 nm referenced against 630 nm. Results of a typical standard calibration curve with optical density (OD) readings at 450 nm vs. 630 nm are shown on the x-axis against NR2 Ab concentrations (ng/mL) shown on the y-axis (Fig. 4b).

4.2   Automatic Analyzer Platform for NR2 Ab Assay

The automated analyzer platform for NR2 Ab also employs para-magnetic particles and chemiluminescent technology. Free antibodies from serum sample react with biotinylated NR2 peptide from ancillary pack then on-board coupling allowed before solid-phase streptavidin supermagnetic particles (0.8 μm) is added (Fig. 5a). Then anti-human antibodies or PrA labeled with acridinium ester bind in the first pass complex (direct sandwich assay). Programmed analyzer separates, aspirates, and washes three times the cuvettes using buffer. Then it dispenses base reagent to initiate the chemiluminescent reaction that automatically measured and reported in RLU according to software to the selected option. Results of a

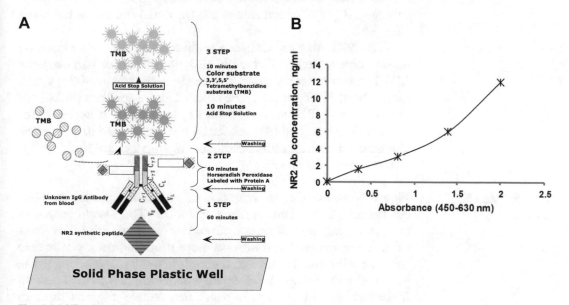

Fig. 4 (**a**) NR2 Ab ELISA scheme, (**b**) NR2 Ab ELISA calibration curve

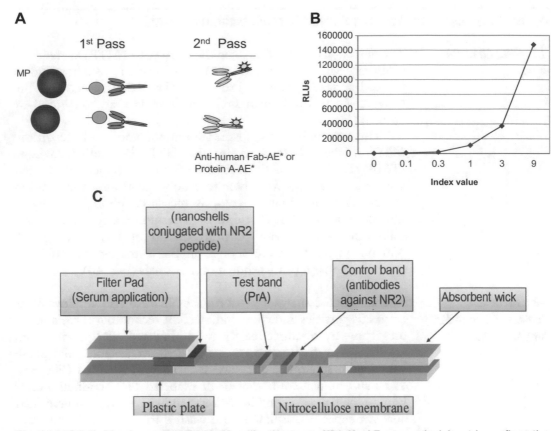

Fig. 5 (**a**) NR2 Ab AA scheme, (**b**) NR2 Ab AA calibration curve, NR2 Ab nLF assay principle: strip configuration

standard calibration curve with RLU readings are shown on the y-axis against NR2 Ab concentrations (ng/mL) shown on the x-axis (Fig. 5b).

The NR2 Ab assay kit uses a ready-to-use pack consisting solid phase compartment filled with 100 μg/mL latex paramagnetic MP. The second compartment of ready pack contains 0.5 μg/mL biotinylated NR2 peptide. The separate three ancillary packs hold (1) 10 ng/mL Protein A reagent, (2) buffer that is necessary for reagent dilutions, and (3) pack for base reagents used to initiate the chemiluminescent reaction. Reagents in all packs are sufficient for 100 on-board tests.

4.3   NR2 Ab LF Design

Figure 5c shows the configuration of nLF immunoassay for NR2 Ab detection in serum and whole blood. The assay consists of several overlapping zones, constituted by segments of various filters of different material and nitrocellulose mounted on a plastic card using an adhesive. To run a test, sample (20–30 μL) is added to the proximal end of strip, the filter pad where sample is treated making it compatible with the rest of test. Then sample migrates through this region to the conjugate pad containing gold nanospheres

(80 nm) with attached modified NR2 peptide. The sample remo-
bilizes the dried conjugate, and the NR2 Ab in the serum interacts
with the conjugate as both migrate into the next section of the
nitrocellulose strip or the reaction matrix. This reaction matrix
contains one band of Protein A (test) and the second one of specific
polyclonal or monoclonal antibodies against NR2 peptide (control
band) where they serve to capture the analyte and the conjugate as
they migrate by the capture lines. Excess reagents move past the
capture lines and are entrapped in the wick or absorbent pad.
Results are interpreted on the reaction matrix as the presence or
absence of lines of captured conjugate, read either by eye or using a
reader.

5   NR2 Peptide and Antibody Assays Analytical Data

5.1   Original NR2 Peptide and Antibody Stability in Blood Samples

The recovery study of neurotoxicity biomarkers in blood samples
(plasma and serum) from individuals is performed at approximate
storage times spent by: (1) phlebotomist as a daily routine of blood
draws in in-patient wards with storage on cold pack (+4 °C, 4 h),
(2) reference laboratory representative for regular samples pick up
at medical office's outdoor storage containers. There is an adjustment of time of storage (24 h) made at room temperature (RT) accounting summer temperatures on South East of USA (35–40 °C)
and 12-h shift. (3) The point of storage at −20 °C for 2 months is
chosen as an average time of temporary refrigerated storage for
research purposes.

The recovery data for plasma samples are in range of 12–14%
($r = 0.992$) compared to baseline of fresh plasma at +4 °C with
mean recovery of 86% (Table 1). After storage at RT for 4 h, the
mean recovery for NR2 peptide was about 79% in plasma samples
($r = 0.984$, Table 1) revealing lower recovery for samples containing high amounts of NR2 peptide (65–70%). A better recovery is
obtained from plasma samples stored frozen (Table 1).

On the basis of these results for NR2 peptide biomarker, it is
recommended specimen collection is made in tubes containing
EDTA (not heparinized) with immediate storage on cold packs.
Samples should be centrifuged (3500 × g, 4 °C for 5 min) within
1 h of blood drawing. For storage up to a week, processed samples
are stored in an aliquot tube at or below −20 °C. For longer
storage, freeze samples at −80 °C.

Processed plasma samples may be frozen and thawed only once
without significantly affecting the amount of NR2 peptide. To
transport samples, shipment boxes in bigger Styrofoam box filled
with dry ice (−78.5 °C, pellets or slabs not less than 3 lb per box)
should be prepared.

NR2 antibodies in samples are stable for up to 3 h at room
temperature. Serum samples may be tested within 3 days of

Table 1
Summary of plasma stability

Parameters	+4 °C, 24 h	RT, 4 h	−20 °C, 2 months
N	38	41	15
Mean recovery, %	85.5	78.6	90.4
Standard deviation (SD)	7.4	8.1	6.1
Correlation coefficient (r)	0.989	0.992	0.987

collection when stored at 2–8 °C. It is recommended collecting blood samples in serum collection tube (e.g., gel separation). Samples should be centrifuged and separated at 2500 × g for 5 min within 3 h of venipuncture. Store all serum samples refrigerated (2–8 °C) immediately upon collection; however, for storage up to 3 months, processed aliquoted serum samples should be frozen at or below −20 °C and for longer storage at –80 °C.

Processed serum samples may be frozen and thawed up to two times without significantly affecting the amount of NR2 antibodies. To transport samples domestically, samples could be shipped at 2–8 °C on ice pack. For international delivery, frozen samples should be shipped on dry ice.

5.2 NR2 Peptide/ Antibodies Assays Cut Offs, Turnaround Time, and Stability

5.2.1 Assays Cut Off Values and Detection Limits

NR2 peptide assays. Cut off values for NR2 peptide test varies from 0.5 to 1.5 ng/mL depending on age (children or adults), acuteness of cerebrovascular event (hours since onset), and consequently from clinical setting (emergency room, acute care, surgery, physician's office). The minimum detection limit is 0.1 ng/mL, as calculated by interpolation of the mean plus 2 standard deviations of 20 replicates of the 0 ng/mL NR2 peptide ELISA and AA assay as well.

NR2 antibody assays. The reference interval calculated from the samples (92%) was found to be 0.87–2.0 ng/mL according to guidance for establishing reference intervals NCCLS Standard C28-A2 (http://www.zxyjhjy.com/upload/attached/file/20170406/20170406120112_8797.pdf).

Recent studies [13, 14] have demonstrated an increased risk of cerebrovascular disease (associated with NR2 antibody values >2.0 ng/mL (upper 90% of populations with stroke-like symptoms) vs. <2.0 ng/mL (lower 90% of the studied populations with non-stroke and non-transient ischemic attack (TIA). Therefore, a more conservative approach for identifying individuals with a significantly increased risk for TIA and ischemic stroke attributable to NR2 antibodies may be the 90th percentile value of the population. The cut off for NR2 antibody test is 2.0 ng/mL.

Table 2
Summary of turnaround times for various assay formats

Test	Sample	Format	Turnaround information	
			Number of tests	Duration
NR2 peptide	Plasma	MP-ELISA	89	1 h
		AA	1200	1 h
		LF	1	<2 min
NR2 Ab	Serum	ELISA	89	2 h
		AA	1200	1 h
		LF	1	<2 min

The minimum detection limit is <0.5 ng/mL, as calculated by interpolation of the mean plus 2 standard deviations of 20 replicates of the 0 ng/mL NR2 Antibody.

5.2.2 Turnaround Time

There are three test formats for each of the neurotoxicity biomarkers and turnaround time varies from minutes to hours (Table 2). Testing time for ELISA and AA testing includes respectively manual and automatic loading of samples and reagents, two incubations, two wash procedures, and color development.

5.2.3 Unique Reagents Stability

ELISA format. NR2 peptide unopened test kits should be stored at 2–8 °C up to 6 months upon receipt while NR2 Ab test kits may be stored refrigerated for 12 months. Opened test kits will remain stable until the indicated expiration date provided they are stored as indicated.

AA format. Both NR2 peptide and antibody reagent packs shelf life at 2–8 °C up to 6 months upon receipt. On "board" stability (inside working analyzer) for the same reagent packs is 8 h while NR2 antigen/antibody calibrator packs might be stored more than 14 days.

6 Bench Testing Studies

6.1 NR2 Peptide MP-ELISA and AA Performance Characteristics

Bench testing has been conducted to evaluate the performance of the NR2 peptide and antibody assays before use for clinical studies. The results presented below showed that devices satisfy the diagnostic test performance requirements. Sensitivity, assay precision, linearity, interfering substances, interfering antibodies, cross-reactivity, spiking recovery, dilution recovery, hook effect, and reference interval were assessed according to corresponding NCCLS guidelines EP5-A (http://www.zxyjhjy.com/upload/attached/file/20170406/20170406154608_8336.pdf).

6.1.1 Assay Precision

For *NR2 peptide ELISA assay* each calibrator ($n = 3$) and control ($n = 2$) sample is tested in duplicate, twice a day for at least 20 consecutive days. Runs performed on daily basis are separated by 2 h. Within-run, between-run, between-day, and total precision were determined by testing calibrators containing NR2 peptide, negative and positive controls, negative serum samples, and positive serum samples. Calibrator performance data showed good standard deviations (<0.1), and coefficient variations (<10%) in all assay characteristics (Table 3).

NR2 peptide automatic analyzer repeatability as expected is better within and between runs (CV < 2.5 and 3.4% correspondently) and generally below 7.2% for reagent runs between days (company's proprietary report).

For *NR2 Ab ELISA* assay total and within-run precision was determined by testing the matrix control, containing monoclonal antibodies against NR2 peptide (12 ng/mL), and two human serum controls with normal and high NR2 antibody concentrations, according to NCCLS Guidelines EP5-A. The samples were assayed in quadruplicate, using a single lot of reagents for 20 days. A three-point calibration curve was run in duplicate (Table 4).

There is within run and between run range <9% obtained for *NR2 Ab AA* forward sandwich 2-pass assay format.

6.1.2 Linearity

A normal human plasma (NHP) and human plasma with high titer of NR2 peptide were mixed to obtain six serial dilutions from low to high range of NR2 peptide determined by ELISA. Four replicates were analyzed per dilution. Regression analysis yielded a good linearity in the interval from a negative sample to a highest, positive sample ($r^2 = 0.997$).

A normal human plasma (NHP) and human plasma with high titer of NR2 peptide were mixed to obtain four serial dilutions from low to high range of NR2 peptide to assess linearity for AA format. Two replicates were analyzed per dilution Regression analysis yielded a good linearity with $r^2 = 0.989$ (company's proprietary report).

For NR2 Ab ELISA assay, pool of normal serum samples is spiked with NR2 antibodies to reach a concentration at maximum detection (200 ng/mL). Six serially dilutions of serum samples were prepared by diluting prepared serum sample with normal serum. Samples were analyzed in duplicates. The regression analyses showed $r^2 = 0.935$.

6.1.3 Interfering Substances

The interfering substances were tested in pooled normal human plasma (pNHP). Commercially available substances are diluted in distilled water and spiked into pNHP. Interference tests are performed using spiked levels of albumin 5 g/dL, bicarbonate 130 mg/dL, bilirubin 15 mg/dL, cholesterol 250 mg/dL, creatinine 5 ng/mL, and hemoglobin 20 g/dL. No interference was observed in spiked samples containing triglycerides, cholesterol,

Table 3
NR2 peptide ELISA assay precision results

Precision	Calibrator 1			Calibrator 2			Calibrator 3			Negative control			Positive control		
	SD	Variance	CV	SD	Variance	CV	SD	Variance	CV	SD	Variance	CV	SD	Variance	CV
Repeatability (within-run)	0.045	0.002	3.80	0.074	0.005	3.37	0.098	0.010	2.70	0.019	0.002	4.14	0.104	0.011	3.17
Between-run, within day	0.018	0.000	1.55	0.072	0.005	3.31	0.106	0.011	2.92	0.008	0.001	1.73	0.051	0.003	1.55
Between-day	0.100	0.010	8.45	0.016	0.000	0.72	0.040	0.002	1.11	0.009	0.000	2.00	0.047	0.002	1.43
Within device	0.111	0.012	9.39	0.105	0.011	4.78	0.150	0.022	4.13	0.022	0.001	4.91	0.125	0.016	3.81

Table 4
NR2 Ab assay precision results

Sample	ng/mL	OD	Total % CV	Within-run % CV
Sample, 12 ng/mL	N/A	2.17 ± 0.091	6.9	1.5
Serum control low anti-NR2 activity	1.64 ± 0.03	N/A	8.9	5.0
Serum control high anti-NR2 activity	20.28 ± 0.41	N/A	9.2	5.6

bilirubin, and albumin. Optical densities of all samples, including phosphate buffered solution are low (between 0.02 and 0.044 OD). However, a significant interference is detected with hemoglobin: ODs are between 3.063 and 3.389. It was no significant interference for used substances in ELISA and AA assay formats assessing NR2 antibodies in serum samples.

6.1.4 *Cross-Reactivity*

Eleven serum (plasma) samples known to have elevated levels of Rheumatoid factor (RF) antibodies are analyzed for interference in the NR2 antibody (NR2 peptide) test. The mean recovery of NR2 antibodies from samples containing heterophilic antibodies is 105%, with individual recoveries ranging from 89% to 112%. The mean recovery of NR2 peptide from samples containing heterophilic antibodies is 99%, with individual recoveries ranging from 95% to 101%. Therefore, no evidence of significant interference is observed due to the heterophilic antibodies tested in the NR2 antibody and NR2 peptide tests.

6.1.5 *Dilution Recovery*

Six serum (plasma) samples with known high NR2 antibody (high NR2 peptide) levels are intermixed with six serum (plasma) samples and known low NR2 antibody (Calibrator #2) for NR2 antibody or NR2 peptide percent dilution recovery and is determined as the measured value divided by the expected value, multiplied by 100. Mean dilution recovery for NR2 peptide trials ranged from 92% to 100% and for NR antibodies 98–105%.

6.1.6 *Hook Effect*

Serum (plasma) samples containing excessively high levels of NR2 antibodies (NR2 peptide, >200 ng/mL) are studied to determine if a hook effect occurred. The NR2 antibodies (NR2 peptide) are spiked in normal human plasma or serum at a concentration approximately 20 times greater than the detection level. No hook effects were observed for either test kit.

7 Conclusion: Coordinated Utilization of NR2 Peptide and Antibody Testing

The transfer to personalized medicine is requiring specific diagnostics developed and proved from experimental research to clinical applications for stroke. This is a broad field that implies devices or kits for use (1) of brain-specific biomarkers, (2) an immunoassay capable of testing for a biomarker; (3) a method of diagnosis based on the comparison of patient data (medical history, neuroimaging, laboratory results) with normal levels; (4) the power to follow up after treatment(s) until biomarker counts show it has been effective; (5) the testing of an individual for a preexisting conditions and a patient's suitability for a particular treatment.

Rapid diagnosis of both TIA and stroke is essential for optimal patient treatment and outcome. Clinical use of blood tests that can reliably detect or exclude a cerebrovascular event, predict consequences, or forecast recovery and outcome, might improve diagnostic certainty of TIA/stroke. Early identification of TIA based on a blood assay that can detect neurotoxicity biomarkers, such as the NR2 peptide and NR2 antibody, has the potential to become a key component of a successful treatment strategy and outcome monitoring.

Advances in analytical assay technologies have made it possible to develop a rapid, cost-effective brain panel that can be used to predict a cerebrovascular event in a target population (e.g., those with preexisting conditions) and to select a high-risk group for the immediate attention of a stroke specialist. Timely assessment of TIA/stroke includes educating non-stroke specialists (primary care physicians, cardiologists, vascular surgeons, anesthesiologists, and diabetologists) in new approaches to early diagnosis with the use of appropriate blood tests.

Rapid results of blood tests have the potential to shorten the time patients suspected of having a TIA or stroke are seen by neurologists. This can benefit the healthcare system, not only by improving stroke management but by saving lives and reducing costs resulting from disabilities worldwide.

8 Methods

8.1 Assessments of Neurotoxicity Biomarkers

8.1.1 MCAo Model

Adult male Sprague Dawley rats (250–300 g) are anesthetized with chloral hydrate (400 mg/kg, ip). Transient ligation of the right middle cerebral artery (MCA) and bilateral common carotids arteries (CCAs) was performed for 1 h [15]. After recovery from anesthesia, animals were sacrificed at 0, 2, 8, 24, and 72 h allowing varied reperfusion periods after ischemia.

8.1.2 Induced Experimental Hematoma

Sprague-Dawley rats are subjected to lysed rat blood cell (RBC, 50 μL) infusion into the deep cerebral cortex 2.5 mm below the surface of the skull of the right MCA territory through a 26-G needle over 5 min. Controls for needle insertion received 20 μL saline infusions. The injection of rat blood cells into the brain usually do not cause edema within 48 h.

8.1.3 Determination of Vascular Permeability

Sodium fluorescein (6 mg/mL, 200 μL) is injected through the rat tail vein at 0, 2, 8, 24, 72 h after reperfusion. Thirty minutes later, rats were reanesthetized (except for 0 h), perfused with PBS (20 mL), and the brains removed. Hemispheres were homogenized in 0.5 M borate buffer (pH, 10) and centrifuged (2500 × g) for 15 min at 4 °C. Ethanol (1.2 mL) was added to the supernatant and then centrifuged (10,000 × g) for 20 min at 4 °C. The fluorescence of the supernatant was measured at 485/330 nm using fluoroscopic techniques (Tecan, Raleigh, NC).

8.1.4 Infarct Determination Using Triphenyl Tetrazolium Chloride (TTC) Staining

Anesthetized sham and MCAo rats are decapitated immediately after MCAo and brains removed, sliced into 2.0-mm-thick sections, and incubated in 2% TTC and saline for 15 min at 25 °C. Then sections should be transferred to a 5% formaldehyde solution. The area of infarction in each slice are digitized as described [15].

8.1.5 Tissue Preparation for Immunohistochemistry

Digitized TTC stained brain slices (2.3 mm from bregma) should be post-fixed in 10% formaldehyde. Vibratome brain sections of 5 μm thick are stained with Nissl and examined by light microscopy. Each second section corresponding to the Nissl-stained section is used for immunocytochemical staining. The fixed tissue sections are pretreated with 0.3% H_2O_2 in 10% methanol, rinsed in buffer, and preincubated with 2% normal goat serum. Tissue sections with primary antibodies against NMDA receptor subunits are incubated at 4 °C overnight. After washing in buffer, sections should be incubated with secondary biotin-labeled goat antibodies against corresponding monoclonal or polyclonal IgG, rinsed in buffer, and incubated in avidin–biotin complex. After washing, the color is developed by treating with 3,3'-diaminobenzidine (DAB). Negative controls sections are treated with non-immune goat serum.

8.1.6 RNA Isolation from Peripheral Mononuclear Cells and Brain

Experimental animals are anesthetized with chloral hydrate (400 mg/kg, ip). Approximately 5 mL of whole (heparinized) blood is withdrawn from the left ventricle of the heart. Mononuclear cells are isolated from whole blood using QIAamp RNA Blood mini kit (Qiagen, Valencia, CA). Brains are removed, frozen, and sectioned in cryostat producing coronal 700 μm serial sections. Brain regions (about 20 mg) of interest are taken from the area of ischemic damage in cortex and adjacent CA1 area of hippocampus. Templates of the outlined areas are drawn on the contralateral non-ischemic side; this procedure allowed comparison between equivalent brain areas.

8.1.7 Reverse Transcription (RT)

Total RNA is isolated from rat hippocampus or cortex by RNA-gents® kit (Promega, Southampton, UK). RNA samples are quantified by spectrophotometry. The RT controls for isolated RNAs are performed to rule out contaminations of RT samples by use of Reverse Transcription System kit (Promega, Southampton, UK). The total RNA (10 μg) is reverse-transcribed with a random hexamer.

8.1.8 Polymerase Chain Reaction (PCR)

PCR reactions are performed with 40 pmol of each NR2 specific primers [7], and 100 ng of total RNA using master mix (Qiagen) in eighteen cycles (94 °C—2 min, 55 °C—2 min, 72 °C—2 min), and then 4 °C using a programmable Gene Cycler (Bio-Rad Lab., Hempstead, UK). Amplified products (5 μL, 600 bp) were separated by electrophoresis in 7.5% agarose gel and visualized by staining with ethidium bromide. All PCR amplifications were assessed relatively to a glyceraldehyde-3-phosphate dehydrogenase (GAPDH) standard (600 bp), which demonstrated similar expressions in ischemic vs. control tissue.

8.1.9 NR2 Peptide Detection by Western-Blot

Protein samples (10 μg), purified from brain tissue (during RNA isolation) or plasma samples are separated using 10% SDS-PAGE. Following electrophoresis, proteins are transferred to nitrocellulose. Following a brief rinse with buffer, the blots are blocked, and polyclonal antibodies NR2 peptide are applied for 1.5 h. After rinsing in the buffer membranes are incubated with horseradish-conjugated secondary antibodies. Western blots are then developed using the enhanced chemiluminescence (ECL) method (Amersham, Arlington Heights, IL).

8.1.10 Detection of NR2 Peptide and Antibodies in Blood Samples

Blood samples (150 μL), collected from the tail veins of anesthetized rats are centrifuged and the resulting plasma and serum used for peptide and antibody assessments (CIS Biotech, Inc., Atlanta, GA) as described below.

8.2 MP-ELISA Materials and Procedure

The reagents in one kit are sufficient for 96 wells (CIS Biotech, Inc., Atlanta, GA, USA) and each test kit contains reagents (Table 5).

There are additional materials required: distilled or deionized water, magnetic separator for microtiter plate, vacuum aspirator, precision pipettes calibrated to deliver 10–1000 μL, a multichannel dispenser or repeating dispenser, vortex-mixer, incubating plate shaker (37 °C), and microtiter plate reader.

8.2.1 Specimen Collection and Preparation

Collect blood samples in lavender (purple) head collection tubes (contains K_2EDTA) immediately place on ice. Prepare all samples by separating plasma using standard separation procedures using centrifuge (preferably at 2–8 °C). Store all plasma samples refrigerated (2–8 °C) immediately upon collection. Test plasma within 1 day of collection when stored at 2–8 °C. For longer storage of up to a week, freeze processed samples stored in an aliquot tube at or below −20 °C. For longer storage freeze samples at −80 °C. Processed plasma samples may be frozen and thawed only once

Table 5
Reagents supplied in NR2 antibody test kit

Product number	Kit components	Quantity
P-C07N(Ab)	1. Antibody preloaded magnetic particles (Ab-NR2-MP), concentrated	250 μL
MP-C06N	2. Anti-NR2 peptide antibody conjugated with horseradish peroxidase (concentrated)	300 μL
MP-C01N, MP-C02N, MP-C03N	3. Calibrators C1, C2, C3, ready to use, each	300 μL
MP-C04N	4. Control negative (CN), ready-to-use	300 μL
MP-C05N	5. Control positive (CP), ready-to-use	300 μL
UP-003	6. PBS-T, to prepare 500 mL	1 tabl.
MP-C11	7. 96 well no-binding Microtiter plate	1
UP-001	8. 3,3′,5,5′-Tetramethylbenzidine (TMB) Liquid Substrate System for ELISA, ready to use solution	15 mL
UP-002	9. Stop reagent (powder)	0.6 g
	10. Package Insert	1

without significantly affecting the amount of NR2 peptide. When transporting samples, ship prepared boxes in bigger Styrofoam box filled with dry ice (−78.5 °C, pellets or slabs not less than 3 lb per box).

Assay Procedure includes preparatory steps where working concentration of reagents should be prepared and assay method.

Preparatory Steps

1. Buffer: Dissolve PBS-T buffer tablet in 500 mL of distilled or deionized (DI) water. Mix thoroughly on magnetic stirrer to reach complete solubility. Once prepared, the PBS-T buffer may be stored refrigerated for up to 30 days.

2. Plasma samples: Thaw the plasma samples at 2–8 °C. Do not thaw the samples at room temperature or in warm water bath. It takes about 4 h to thaw 1 mL of plasma sample from −80 °C to +4 °C. *Vortex* samples well before analysis.

3. Microtiter plate: Bring the MTP to room temperature. It is recommended that each calibrator, control, and plasma sample to be run in duplicate.

4. Stop reagent: Add 15 mL deionized or distilled water to the bottle labeled Stop Solution. Mix well; at least 5 min. Once dissolved, the Stop Reagent may be refrigerated for up to 30 days.

Procedure (for 1 MTP-96 wells)

Prepare *working mixture 1*: Add 3.9 mL of PBS-T buffer to the fitted tube. Resuspend magnetic particles (MP) by vortexing the tube for 1–2 min using gentle vortex. Add 100 µL of resuspended magnetic particles to previously prepared 3.9 mL of PBS-T buffer. *Note*: MP solution must be well suspended before adding. Mix well, avoid foaming.

Prepare *working mixture 2*: Add 3.8 mL of PBS-T buffer to the fitted tube. Add 200 µL of anti-NR2 peptide HRP-conjugated antibodies. Mix well, avoiding foaming. Preparation of calibrators, controls, and samples. Vortex vials with calibrators and controls gently for 1–3 min. Make sure that the solutions are well suspended before adding into wells. Vortex tubes with plasma samples for 1 min.

1. Pipet 20 µL of each calibrator, negative (N) and positive (P) controls, and prepared plasma samples into each well.

2. Pipet 40 µL of previously prepared *working mixture 1* into each well (it is recommended but not necessary to use any standard mechanical 8 or 12 multichannel pipette). Make sure that *working mixture 1* is mixed well. Gently shake the plate to ensure mixing. Incubate the plate at 37 °C for 5 min on a plate shaker and do not discard content of the wells.

3. Add 40 µL of previously prepared *working mixture 2* into each well. Make sure that *working mixture 2* is mixed well. Gently shake the plate to ensure mixing. Seal the MTP with the included sealing film. Incubate the plate at 37 °C for 30 min on a plate shaker.

4. Washing procedure: After 30 min of MTP incubation, place the MTP on a magnetic separator. Wait for 2 min, until all particles are concentrated to the bottom corner of wells closest to the magnet and the solution appears clear. Aspirate the liquid carefully pointing an aspiration needle to the wall opposite from concentrated magnetic particles. For aspiration procedure it is possible to use a reverse way of dispenser. Remove MTP from the magnetic separator. Add 200 mL of PBS-T buffer in each well pointing the liquid to the magnetic particles pellet. Make sure that no pellet is remained and magnetic particles are homogeneously resuspended. Shake MTP on a shaker at 37 °C for 2 min. Place MTP back on the magnetic separator. Wait for about 2 min. Aspirate the PBS-T buffer as described above. Repeat the washing procedure with PBS-T buffer two more times (three times total). Repeat the washing procedure using distilled water instead of PBS-T buffer and do not discard content of the wells.

5. Color reaction: Remove MTP from the magnetic separator. Add 100 µL of TMB substrate into each well pointing the

liquid to the magnetic particles pellet. Gently shake the plate to ensure mixing and incubate at room temperature for 10 min in the dark for color development. Do not use the shaker and do not discard content of the wells.

6. Stop Solution: Shake MTP on the shaker for 5–10 s. Immediately place MTP on the magnetic separator for 1 min. Do not remove the MTP from magnetic separator until the plate is ready to read in a Microplate Reader.

7. Add 100 µL of stop solution per each well pointing the liquid to the MTP wall opposite to magnetic particle pellet. This stops the reaction and turns the blue color into yellow. It is important to make sure that the blue color completely changes into yellow; if not, gently shake MTP along with magnetic separator for 20–30 s to ensure mixing.

8. Reading Optical Density: Wipe moisture from the bottom of the plate using a paper towel. Read the optical density (O.D.) within 10 min at 450 nm vs. 630 nm dual wave using a microplate reader.

Procedural Notes: All test reagents should be stored at 2–8 °C. Prior to their use allow the reagents to reach room temperature (about 30–40 min). For accurate results, the dispensing of samples, calibrators, and controls must be precise using calibrated pipettes. Always have the next step reagent ready 2–3 min ahead.

8.2.2   Calculation of Results Using Microsoft® Office Excel

Plot Optical Densities on the Horizontal Axis NR2 peptide concentrations marked for each calibrator on the Vertical Axis. Find Trendline which best fits the data and corresponding equation. Use r^2 value as close as possible to 1.0 while looking for the best fit. Using the Trendline equation, calculate the peptide concentrations in your samples: in the equation replace sign(s) "X" with optical density of your sample. The result of calculation is the peptide concentration in the sample.

8.3   NR2 Ab ELISA Method

8.3.1   Reagent Preparation and Storage

Upon receipt, store unopened test kits at 2–8 °C. Once opened, keep the microwell strips sealed in the foil pouch with desiccant to minimize exposure to moisture. Once opened, test kits will remain stable until the indicated expiration date, provided that they are stored as requested.

Dissolve the PBS-T buffer tablet in 500 mL of distilled or deionized water. Mix thoroughly on magnetic stirrer to reach complete solubility. Once prepared, the PBS-T buffer may be stored refrigerated for up to 30 days.

8.3.2   Assay Procedure   *Preparatory Steps*

1. Dissolve the PBS-T tablet in 500 mL of distilled or deionized water. Mix thoroughly on magnetic stirrer to reach complete solubility. Once prepared, the PBS-T buffer may be stored refrigerated for up to 30 days.

2. Thaw the serum samples at 2–8 °C. Do not thaw the samples at room temperature or in a warm water bath.

   Note: it takes about 4 hours to thaw 1 mL of serum sample from −80 °C to +4 °C.

   Dilute serum samples 1:50 in the PBS-T buffer. Pipet 980 μL PBS-T buffer into 1.5–3 mL volume tubes and add exactly 20 μL of each serum sample. Mix thoroughly. Avoid foaming.

3. Bring the MTP to room temperature. Calculate the number of microplate (MTP) strips required. It is recommended that each calibrator, control, and serum sample be run in duplicate. Remove the MTP frame with the strips from the foil pouch. Leave the required number of microwell strips in the frame and place the remaining strips back in the pouch. Ensure that the foil pouch is completely resealed with the desiccant. Store at 2–8 °C.

4. Stop reagent. Add 15 mL deionized or distilled water to the white bottle. Mix well for at least 10 min. Once dissolved, the stop reagent may be refrigerated for up to 30 days.

5. Ancillary buffer. Dilute the entire amount (1.5 mL) of 10× ancillary buffer in 13.5 mL of PBS-T buffer for a final concentration of 1×.

6. The minimal volume per one strip is 800 μL of each of the Protein A-HRP, TMB, and stop solution reagents plus at least 200 μL of dead volume.

   NOTE: It is recommended but not essential to use any standard mechanical 8- or 12-channel pipettor.

Procedure

1. Wash the MTP with 200 μL of PBS-T buffer once for 5 min at 37 °C on a shaker. Discard content and tap the inverted plate against a paper towel placed on a flat surface.

2. Calibrators, controls, and sample incubation. Vortex calibrators and controls vials gently for 2 s. Pipet 100 μL negative (NC), positive (PC) controls, and prepared serum samples (S_1–S_{42}) into each well according to the following scheme:

	1	2	3	4	5	6	7	8	9	10	11	12
A	PBS-T	PBS-T	S_3	S_3	S_{11}	S_{11}	S_{19}	S_{19}	S_{27}	S_{27}	S_{35}	S_{35}
B	NC	NC	S_4	S_4	S_{12}	S_{12}	S_{20}	S_{20}	S_{28}	S_{28}	S_{36}	S_{36}
C	PC	PC	S_5	S_5	S_{13}	S_{13}	S_{21}	S_{21}	S_{29}	S_{29}	S_{37}	S_{37}
D	C1	C1	S_6	S_6	S_{14}	S_{14}	S_{22}	S_{22}	S_{30}	S_{30}	S_{38}	S_{38}
E	C2	C2	S_7	S_7	S_{15}	S_{15}	S_{23}	S_{23}	S_{31}	S_{31}	S_{39}	S_{39}
F	C3	C3	S_8	S_8	S_{16}	S_{16}	S_{24}	S_{24}	S_{32}	S_{32}	S_{40}	S_{40}
G	S_1	S_1	S_9	S_9	S_{17}	S_{17}	S_{25}	S_{25}	S_{33}	S_{33}	S_{41}	S_{41}
H	S_2	S_2	S_{10}	S_{10}	S_{18}	S_{18}	S_{26}	S_{26}	S_{34}	S_{34}	S_{42}	S_{42}

If using only few samples, use this scheme (where PBS-T buffer):

	1	2
A	PBS-T	PBS-T
B	NC	NC
C	PC	PC
D	C1	C1
E	C2	C2
F	C3	C3
G	S_1	S_1
H	S_2	S_2

3. Incubate MTP at 37 °C for 30 min on a shaker.

4. During the MTP incubation, prepare 1:1000 (v:v) dilution of Protein A-HRP in 1× ancillary buffer. Vortex Protein A-HRP gently for 20 s before dilution. For example, to prepare a sufficient amount of Protein A-HRP working solution for 12 strips, mix 15 μL Protein A-HRP with 15 mL of 1× ancillary buffer. Mix thoroughly using a vortex-mixer. The Protein A-HRP working solution is stable for only 1 h. The Protein A-HRP working solution cannot be stored.

5. After 30 min of MTP incubation at 37 °C, discard the content and tap the plate. Wash with 200 μL of PBS-T buffer three times for 3 min on a shaker at 37 °C. Discard the content and tap inverted plate against the paper towel placed on a flat surface.

6. Add 100 μL Protein A-HRP working solution (from **step 4**) into each well.

7. Incubate the plate at 37 °C for 30 min on a plate shaker. Discard the content and tap the plate as mentioned above in **step 5**.

8. Wash three times with 200 µL PBS-T buffer at 37 °C for 3 min on a shaker. Discard content and tap the plate each time as in **step 5**.

9. Wash one time with 200 µL deionized or distilled water at 37 °C for 3 min on a shaker. Discard content and tap the plate. DO NOT USE TAP WATER.

10. Add 100 µL TMB substrate into each well. Gently shake the plate to ensure mixing and incubate at room temperature for 10 min in darkness to achieve color development. Do not use the shaker. DO NOT DISCARD THE CONTENTS!

11. Add 100 µL Stop Reagent per each well to stop the reaction. The blue color should change to yellow. Gently shake the plate on a flat surface for 20–30 s to ensure mixing.

    NOTE: Make sure that the blue color completely changes to yellow.

12. Wipe moisture from the bottom of the plate using a paper towel.

13. Read the optical density (OD) within 10 min at 450 nm/ 630 nm dual wave using a microplate reader.

Procedural Notes: Store all test reagents at 2 °C–8 °C. Allow the reagents to equilibrate to room temperature prior to use (about 40 min). Bring the MTP to room temperature. Store the unused strips in the resealable foil pouch with desiccant to minimize exposure to moisture. *Optional*: Use MTP lid (not supplied) to reduce evaporation.

– The volume of PBS-T buffer required during the washing procedure is 200 µL per well.

– The standard volume of all other solutions added to each well is 100 µL.

– Discard the contents from the MTP. Then, with the MTP inverted, tap it three times on a paper towel placed on a flat surface.

– DO NOT ALLOW THE MICROWELL STRIP TO DRY DURING THE PROCEDURE. This affects the accuracy of the results.

– For accurate measurement, dispensing of samples, calibrators, and controls must be precise. Pipet carefully using only calibrated equipment.

Always have the next reagent ready 2–3 min prior each next step.

8.3.3 Calculation of Results

Plot the absorbance obtained for each calibrator on the x-axis vs. the NR2 antibody concentration in ng/mL on the y-axis. Use a point-to-point curve fit with appropriate computer software to construct the standard calibration curve. Using the absorbance value for each sample, determine concentration of NR2 antibodies (ng/mL) from calibration curve.

References

1. Tang N, Wu J, Zhu H, Yan H, Guo Y, Cai Y, Yan H, Shi Y, Shu S, Pei L, Lu Y (2017) Genetic mutation of GluN2B protects brain cells against stroke damages. Mol Neurobiol 55(4):2979–2990. https://doi.org/10.1007/s12035-017-0562-y

2. Tremblay R, Chakravarthy B, Hewitt K, Tauskela J, Morley P, Atkinson T, Durkin JP (2000) Transient NMDA receptor inactivation provides long-term protection to cultured cortical neurons from a variety of death signals. J Neurosci 20(19):7183–7192. PMID:11007874

3. Sharp CD, Fowler M, Jackson TH 4th, Houghton J, Warren A, Nanda A, Chandler I, Cappell B, Long A, Minagar A, Alexander JS (2003) Human neuroepithelial cells express NMDA receptors. BMC Neurosci 4:28. PMID:14614784

4. Choi DW (2005) Neurodegeneration: cellular defenses destroyed. Nature 433 (7027):696–698. https://doi.org/10.1038/433696a

5. Yuan H, Vance KM, Junge CE, Geballe MT, Snyder JP, Hepler JR, Yepes M, Low CM, Traynelis SF (2009) The serine protease plasmin cleaves the amino-terminal domain of the NR2A subunit to relieve zinc inhibition of the N-methyl-D-aspartate receptors. J Biol Chem 284(19):12862–12873. https://doi.org/10.1074/jbc.M805123200

6. Gappoeva MU, Izykenova GA, Granstrem OK, Dambinova SA (2003) Expression of NMDA neuroreceptors in experimental ischemia. Biochemistry (Mosc) 68(6):696–702. PMID:12943515

7. Dambinova SA (2002) Rapid multiple panel of biomarkers in laboratory blood tests for TIA/stroke PCT Int. Appl. no. WO 2002012892, WO02/12892 A2

8. Gastaldi M, Thouin A, Vincent A (2016) Antibody-mediated autoimmune encephalopathies and immunotherapies. Neurotherapeutics 13(1):147–162. https://doi.org/10.1007/s13311-015-0410-6

9. Bi X, Chang V, Molnar E, McIlhinney RA, Baudry M (1996) The C-terminal domain of glutamate receptor subunit 1 is a target for calpain-mediated proteolysis. Neuroscience 73 (4):903–906. PMID:8809808

10. Sharp F, Liu DZ, Zhan X, Ander BP (2008) Intracerebral hemorrhage injury mechanisms: glutamate neurotoxicity, thrombin, and Src. Acta Neurochir 105:43–46. PMID:19066080

11. Ardizzone TD, Lu A, Wagner KR, Tang Y, Ran R, Sharp FR et al (2004) Glutamate receptor blockage attenuates glucose hypermetabolism in perihematomal brain after experimental intracerebral hemorrhage in rat. Stroke 35 (11):2587–2591. https://doi.org/10.1161/01.STR.0000143451.14228ff

12. Izykenova GA, Baldwin R, Oldenburg SJ (2018) Development of novel test platforms for the assessment of brain injury. In: Peplow PV, Dambinova SA, Gennarelli TA, Martinez B (eds) Acute brain impairment: scientific discoveries and translational research. Royal Society of Chemistry, London, pp 315–326

13. Stanca DM, Mărginean IC, Soriţău O, Dragoş C, Mărginean M, Mureşanu DF, Vester JC, Rafila A (2015) GFAP and antibodies against NMDA receptor subunit NR2 as biomarkers for acute cerebrovascular diseases. J Cell Mol Med 19(9):2253–2261. https://doi.org/10.1111/jcmm.12614

14. Kidher E, Patel VM, Nihoyannopoulos P, Anderson JR, Chukwuemeka A, Francis DP, Ashrafian H, Athanasiou T (2014) Aortic stiffness is related to the ischemic brain injury biomarker N-Methyl-D-aspartate receptor antibody levels in aortic valve replacement. Neurol Res Int 2014:970793. https://doi.org/10.1155/2014/970793

15. Wang Y, Chang CF, Morales M, Chiang YH, Harvey BK, Su TP, Tsao LI, Chen S, Thiemermann C (2003) Diadenosine teraphosphate protects against injuries induced by ischemia and 6-hydroxydopamine in rat brain. J Neurosci 23(21):7958–7965. PMID:12944527

Chapter 11

Glutamate Receptor Peptides as Potential Neurovascular Biomarkers of Acute Stroke

Svetlana A. Dambinova, J. D. Mullins, J. D. Weissman, and A. A. Potapov

Abstract

In this chapter different scenarios of biomarkers evaluating ischemic stroke caused by small vessel occlusions and leading to cerebral infarction are considered. Results of assays detecting glutamate receptor (GluR) peptides alone or combined into biomarkers panel to assess microvessel and small vessel strokes are explored in case report studies. A clinical protocol of neurovascular biomarkers implying GluR peptides is suggested for translational research assessing the severity of acute ischemic events based on structural location of cerebral infarction. It is proposed that the combination of clinical, biochemical, and radiological data might increase the diagnostic certainty of suspected acute ischemic stroke due to small occlusions in selected patients for timely and personalized therapy.

Key words Neurovascular biomarkers, Glutamate receptors, Acute ischemic stroke, Small vessel occlusion, Cerebral infarction, Lacunar stroke

1 Introduction

Acute cerebral vessel occlusion has a critical effect on normal brain function. A stroke is the result of alterations in the structure and function of cerebral blood vessels with associated reduction in cerebral blood flow (CBF) and oxygen delivery [1, 2]. There are few validated and available brain vascular biomarkers that might be measured in peripheral fluids and reflect an early pathophysiology of acute ischemic stroke (IS) due to small vessel occlusion (SVO) causing cerebral infarction. Computed tomography (CT) and CT angiography is used to confirm blockage of a large vessel occlusion and cerebral hemorrhage at emergency rooms. The diagnosis of IS as caused by SVO requires additional magnetic resonance (MR) angiography creating 1- to 2-h delays before treatment. It is well known that 6-h treatment window is imperative for favorable outcome after stroke ("time is brain").

The detection of IS-SVO and cerebral infarction present a challenge in biomarker design because acute IS may occur in the

Philip V. Peplow et al. (eds.), *Stroke Biomarkers*, Neuromethods, vol. 147, https://doi.org/10.1007/978-1-4939-9682-7_11,

setting of chronic or subacute stroke. Current clinical strategies rely on a "last known well" time point noted by the patient or family. This information is frequently not available with the stroke being noticed only at the time of "waking up." Such cases are not, unfortunately, amenable to modern therapies even with diffusion MRI and CT perfusion methods. A blood biomarker of acute-only cerebral infarction or IS caused by SVO might expand treatment options in patients with "wake-up" strokes, particularly if it is codeveloped with an interventional treatment modality [3]. The marker should be detectable (i.e., be sensitive to) and associated with an area of salvageable ischemic penumbra. From other side, the level of marker should not be affected (i.e., be specific to) by prior stroke or areas of core infarction.

Future translational acute stroke biomarker research should be developed conjointly with therapy so that key therapeutic decisions can be facilitated [4]. This might include (1) confirmation of diagnostic accuracy using reference standards, (2) confirmation of the ischemic penumbra volume, (3) identification of the effects of preexisting and comorbid conditions, (4) confirmation of the time of onset or "last known intact," and (5) ruling out stroke "mimics."

This chapter outlines a translational research strategy for assessment of acute stroke severity using neurovascular biomarkers. It is anticipated that variables such as stroke severity, stroke origin (embolic, thrombotic, hemorrhagic, venous) throb, brain vascular distribution and comorbid conditions will require stratification of study protocols as therapeutic decisions will likely be affected by these factors. The cost and statistical power of research will be accordingly affected.

2   Current Vascular Biomarkers Associated with Large and Small Vessel Occlusions

Ischemic stroke caused by large intracranial vessels occlusion (basilar artery, carotid terminus, and middle cerebral artery) trigger loss of blood flow to significant portions of the brain. Mechanical thrombectomy involves stenting or clot extraction. Current interventions may include intravenous as well as selective arterial delivery of recombinant tissue plasminogen activator (rTPA) for acute IS caused by small vessel occlusion. Both approaches require identification of the type of thrombosis within a maximum time period of 6 h, preferably much earlier.

Distinct levels of cerebral arterial/venous network are closely related to neurovascular units (NVU) comprising cerebral microvasculature and nervous cells interacting with blood–brain barrier (BBB) and extracellular matrix [5]. The acute IS takes place when mechanical occlusion of cerebral blood vessels causes metabolic alterations in NVU and is associated with cerebral tissue viability and salvageable area. The penumbra that surrounds an ischemic core [6] could be assessed using neurovascular biomarker(s).

Degradation of NVU function within evolving penumbra is reflected by the insufficiency of glucose supply as an energy source and is presented as a transient symptom. Following an ischemic onset, energy deficit could be transient or reversible and implies ion balance disturbances, shift pH to lower values, and accumulation of extracellular glutamate due to disruption of glutamate–glutamine cycle [7]. A prolonged state of energy deficit might lead to formation of small-sized lesions that, depending on severity, could result in acute cortical infarction [8].

The search for vascular biomarkers of acute stroke has naturally involved investigation of molecular aspects of blood–brain barrier (BBB) disruption (protein S-100), glial scarring and gliosis (glial fibrillary acidic protein, GFAP), and shifts in neuronal–glial ratio (neuron-specific enolase, NSE). The utility of a biomarker cannot be predicted a priori and must be validated by clinical trials.

Most clinical trials testing the specificity and sensitivity of blood-borne biomarkers for acute ischemic stroke due to small vessel occlusion have failed or have been inconclusive due to diverse design of study protocols, absence of confirmatory neuroimaging, and lack of biomarker specificity for acute as compared to subacute or chronic stroke or non-stroke conditions [9]. Some of these failures are not surprising when it is considered that the physical and biochemical aspects of the stroke process would not be expected to be involved and detectable in the early stages with many of these biomarkers. Future translational studies should be focused on core–penumbra ratios with prompt estimation to protect brain function.

3 Glutamate Receptors Subtypes in Cerebral Vascular Regulation

Brain vessels are regulated by major excitatory neurotransmitters (glutamate, aspartate) that escape into the extracellular fluid from synapses [10]. Normally glutamate transporters take up glutamate and neutralize it into glutamine by astrocytic glutamine synthase [11]. Cerebral ischemia distorts this glutamate–glutamine cycle with extracellular glutamate overload that triggers neurotoxicity.

Most cortical neuronal tracts from NVU project into deeper layers of gray matter (Fig. 1a). Subcortical pathways are ascending to the cortical layers and hypothetically could be saturated by glutamatergic structured fibers (Fig. 2b). There are three major ionotropic GluR subtypes: N-methyl-D-aspartate (NMDA), α-amino-3-hydroxy-5-methyl-4-isoxazolepropionic acid (AMPA), and kainate receptors, categorized by structural similarity, electrophysiological functioning, and selective pharmacological properties [12]. It was shown that neuronal and glial subtypes of glutamate

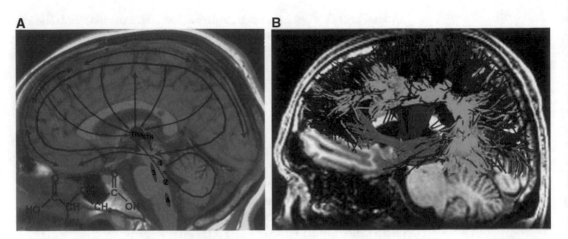

Fig. 1 Glutamate neuroreceptor localization and distribution in Central Nervous System. (**a**) Glutamatergic pathways, (**b**) hypothetical tracking of glutamatergic neuronal fibers where red-colored fibers represents NMDA receptors, green—AMPA receptors, and blue—kainite receptors

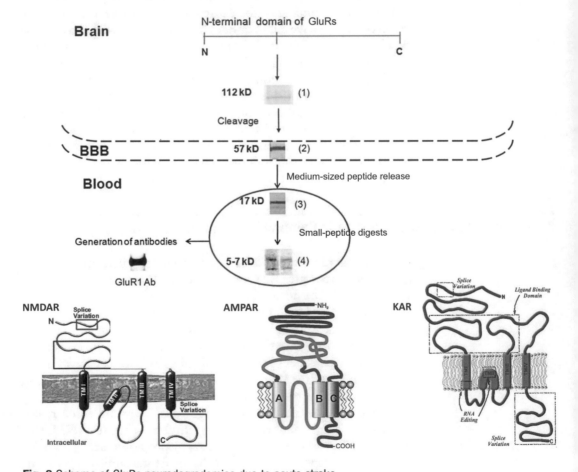

Fig. 2 Scheme of GluRs neurodegradomics due to acute stroke

receptors are located on the extracellular endothelial surface of cerebral vessels [13–15] of cortical-subcortical areas (Fig. 2c, d) [16–18].

Cerebral blood flow (CBF) autoregulation is controlled by excitatory amino acid receptors including glutamate (GluR) and its subtypes of NMDAR, AMPAR, and kainic acid (KAR) receptors [12, 19]. The dominant role of GluR subtypes in neurotoxicity due to stroke has been demonstrated in acute ischemic stroke [20–22]. Considering the structural distribution of these GluR subtypes in specific levels of the NVU, future translational study design should find a correlation between a panel of related biomarkers, clinical information, and neuroimaging findings including diffusion- and perfusion-weighted imaging.

Molecular diversity of GluR genes encoding various receptor subunits is responsible for the pharmacological and functional heterogeneity associated with predominant subtype locations in the cerebral vascular territory [23]. In addition, the extracellular N-terminal domains of the ionotropic GluRs control diverse receptor functions (i.e., subunit assembly, receptor trafficking, channel gating, agonist potency, and allosteric modulation) [24–26]. Several divergent features of the GluRs also derive from natural tendency of certain subunits to assemble in heteromeric complexes: NR1/NR2A/NR2B [27], GluR1–4 [28], GluK1/GluK5, and GluK2/GluK5 [29, 30].

Generally, acute cerebral ischemia affects GluRs that is abundant and specific to a certain brain vascular territory. Proteases then cleave the N-terminal domains of original receptor complexes [31]. Multiple layers of digested endogenous products might be revealed in the brain tissue, CSF, and blood that could fuse into small- to large-sized (2–57 kDa) receptor complexes like NR2A/NR2B (Fig. 3) [12]. Smaller peptide fragments may penetrate the BBB compromised by cerebral ischemia and enter the bloodstream. It seems that mostly active neuropeptides or "neurodegradomes" containing immune epitopes (antigens) of 2–7 kDa could be positively associated with ischemic stroke, subcortical hemorrhagic transformations, and spinal cord ischemia [21, 32–34].

4 NR2 Peptide as a Biomarker of Cortical Microvessel Occlusion

NR2 subtypes of NMDA receptors are located on the surface of neuroendothelial cells [13, 35]. These subtypes are mostly dependent on glucose metabolism as a source of energy and involved in the regulation of NVU functions predominantly in grey matter [21]. Metabolic alterations of glucose and glutamate amounts in NVU are often associated with risk factors of cortical lesions caused by cardiovascular diseases [36] and diabetes [37].

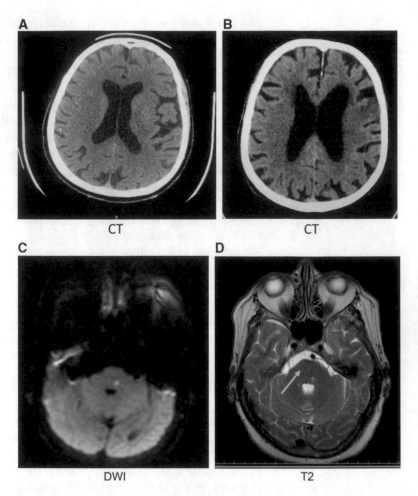

Fig. 3 Radiological images of case reports with NR2 peptide detections. (**a**) CT of acute cardioembolic stroke with diabetes mellitus as preexisting condition, (**b**) CT of acute atherothrombotic stroke, (**c**) DWI of acute stroke of unknown origin in vertebral area, and (**d**) T2-weighted image for the same patient

Preliminary studies of the NR2 peptide as a neurovascular biomarker demonstrated its utility in detecting acute ischemic stroke vs. cerebral hemorrhage and stroke mimics [21, 22]. A significant correlation of NR2 peptide values and new ischemic cortical lesion with volumes below 100 mL was found and associated mostly with cardioembolic stroke [21, 38]. The NR2 peptide also increases during carotid endarterectomy and stenting due to microemboli with following postoperative neurological deficits [39]. Some cases of unknown stroke origin or cryptogenic stroke have extremely elevated NR2 peptide concentrations [40]. Subjects with vascular risk factors such as diabetes and atherosclerosis showed elevated NR2 peptide levels that may indicate cerebral microvessel dysfunction in the areas of NMDAR pathways. Possible brain lesion formation in cortical and subcortical areas might be seen (*see* Appendix 1).

NR2 peptide assay indicated the capability of recognizing acute ischemic stroke with a sensitivity of 92–98% and specificity of 72–97% at 0.5 and 1.0 ng/mL cutoffs respectively [21]. It was also demonstrated that a statistically significant increase of NR2 peptide occurred in acute IS compared to non-stroke ($P < 0.0001$). Additionally the NR2 peptide assay notably discriminated stroke mimics from TIA and ischemic stroke. The best cutoff value for TIA diagnosis was 1.0 ng/mL (sensitivity, 83%) at which a positive predictive value of 73% was achieved [22].

Further, the high diagnostic power for the NR2-peptide assay in assessing neurological adverse events (NAE) with a sensitivity of 99% was calculated [22]. It was shown that NR2-peptide concentrations for NAE after carotid endarterectomy and carotid artery stenting ($n = 7$) were significantly increased preoperatively, intraoperatively, and postoperatively ($P < 0.05$) compared to those without NAE.

It would appear that measurement of the NR2 peptide for patients with suspected stroke symptoms presented to emergency room (ER) within 24 h of the onset may be beneficial for management decision. Further translational study is needed as to the best use of this data. This neurovascular biomarker may have a potential clinical use for assessment of acute cortical strokes of embolic and thrombotic subtypes in patients suffering cardiovascular disorders or diabetes mellitus as preexisting conditions.

5 AMPAR Peptide as an Indicator of Subcortical Lesions Caused by Small Vessel Occlusion

Biomarkers of subacute stroke, completed stroke, and prior stroke (possibly derived from previously failed acute stroke biomarkers) may find an application in identifying acute stroke cases with large core infarcts.

These cases with large core to penumbra ratios might be at increased risk of hemorrhagic transformation or poor outcome if thrombolysis or revascularization procedures are employed. Alternatively, the more chronic biomarkers may find use in the optimization of secondary stroke prevention treatment strategies or in situations where multiple subclinical small vessel infarctions over time are not amenable to the more modern acute stroke therapies. Optimal acute cerebral vascular biomarkers will correlate with imaging and enhance accuracy of diagnosis, improve prognosis, and help to monitor therapy [4].

It is known that AMPAR subtypes (GluR1, GluR2/3, and GluR4) are primarily distributed in the forebrain and subcortical pathways (hippocampus, amygdala, thalamus, hypothalamus, and brainstem) [41]. GluR1/2 has been shown to be expressed by microglia [42] in close proximity to small blood vessels [43]. It was also demonstrated that the GluR1 subtype of AMPAR was mainly located in dendrites and axons of brain white matter, while GluR2/3 and GluR4 belong to spinal cord white matter [44]. The oxygen-dependent ionotropic AMPA receptor subtypes are expressed at the astrocyte–small vessel interface [14] or white matter microglia [15, 17] vulnerable to excitotoxic events following acute stroke.

Due to co-localization of NMDAR and AMPAR in certain subcortical areas it would be plausible to detect not only increased NR2 but also AMPAR peptides. Considering prior investigations [45, 46], the detection of AMPAR peptide would be the most effective for assessment of lacunar lesions and hemorrhagic transformations particularly in deep subcortical structures of hindbrain.

Hypertension and hypoxia may affect arterial microvessel velocity and contractility as well as causing significant dysfunction—vasospasm in small arteries of subcortical structures [47, 48]. Increased microvascular tone is a compensatory mechanism against hypertension enhancing autoregulation and redistribution of blood flow and pressure from major arteries at the base of the brain to small resistant vessels [23]. These small vessels might be directly damaged by high pressure that can induce lacunar lesions or microbleedings in subcortical areas [49]. Additionally, hypoxia or multiple subtle brain injuries may be associated with subcortical microbleeding (Appendix 2), and the subsequent acute inflammation causing focal white matter hyperintensities [32].

Abnormally elevated AMPAR peptide values were observed in patients with uncontrolled chronic hypertension ($P < 0.01$) and prior multiple concussions. White matter fiber tracts were often absent in frontal and parietal areas of brain of concussed subjects [32]. These individuals showed hypoperfusion abnormalities in subcortical and basal ganglia regions on MRI accompanied by microscopic hemorrhage and lacunar infarctions (unpublished data).

It is possible that AMPAR peptide detection in suspected acute stroke would aid in assessment of white matter lesions (lacunar stroke) caused by small vessel occlusion as well as hemorrhagic transformations. Simultaneous measurements of AMPAR and NR2 peptides might help in assessment of possible location of subcortical lesions prior to CT/MRI and could reflect the severity of stroke.

6   Panel of GluR Peptide Biomarkers for Assessment of Acute Stroke Severity

The diagnosis of acute stroke is currently largely based on clinical assessment and non-contrast head CT [50]. However, CT is insensitive to early ischemic changes [51, 52]. Although advanced neuroimaging, such as diffusion-weighted MRI (DW-MRI) or perfusion head CT, are more sensitive to detect early ischemia, they are not readily available in most hospitals [52–54]. In addition, MRI is expensive and may be contraindicated in a significant number of patients due to metal implants, medical instability, obesity, and/or severe claustrophobia [53]. Similarly, patients with renal impairment or contrast allergies cannot undergo perfusion CT. In these cases, diagnosis and treatment rest on clinical symptoms that introduce the risk of inappropriate use of thrombolytic therapy, adverse outcome, and increased treatment cost.

Preexisting conditions like atherosclerosis and cardioembolic conditions might produce large infarcts in gray and white matter [8], while lacunar lesions of different etiology are found in subcortical areas and deep white matter [55]. Diabetes is another risk factor that could produce cortical lesions due to metabolic alterations in neurovascular unit. Ischemic lesions in cryptogenic strokes are often found in deep subcortical and brainstem vascular territories [56]. To understand the disease origin, it would be plausible to add specific blood assay(s) to clinical and radiological examinations.

A rapid panel of GluR peptide assays [57] might help to assess the severity of cortical/subcortical infarctions particularly at risk of hemorrhagic transformations [58]. The progressing acute ischemic stroke is difficult to predict at ER presentation. It was estimated about 30% patients would recover, up to 6% will have unfavorable outcome, while approximately 46–50% might have unpredictable outcome [59]. One of the scenarios could result in symptoms worsening because of microlesions might expand to deeper subcortical structures and/or merge forming larger volume lesions.

It was reported that up to 20% of patients presenting with stroke-like symptoms have no cerebral ischemia but present with the so-called "stroke mimics," including complicated migraine, postictal paresis, psychological disturbance and other etiologies [60]. These patients should not receive thrombolytics. If a biomarker assay(s) could accurately differentiate stroke mimics from acute IS and TIA, potentially dangerous treatment with thrombolytics could be avoided. Furthermore, costs due to unnecessary hospital admissions, extensive neuroimaging and diagnostic studies could theoretically be reduced.

The selection of potential neurovascular biomarkers based on structural distribution of glutamate receptors family (NMDAR and AMPAR) in certain neurovascular territories is explored (Appendixes 1 and 2). Results of GluR peptide assays in acute IS support the premise that preexisting conditions affect the acute cerebral infarction development. Indeed, diabetic and cardioembolic conditions showed a tendency to worsen an acute cortical infarction that accompanied drastically increased NR2 peptide values. Hypertension and prior concussions/mild TBI could sequentially trigger subcortical (lacunar) strokes or hemorrhagic transformations that correlated with elevated AMPAR peptide levels in peripheral fluids. Multiple preexisting conditions might affect the severity of acute ischemic stroke that could be more accurately assessed by a panel of GluR peptide biomarkers [61].

The combined use of neuroimaging and a rapid panel of GluR peptide assays might improve diagnostic certainty of cortical/subcortical infarctions at ER. Additionally, the rapid assay panel could speed up patients' stratification that could significantly improve stroke outcomes and increase efficiency of ambulance services and stroke center management.

7   Methods

7.1   Translational Study Protocol

A translational study protocol to assess the utility of GluR peptide assays for acute stroke is considered. The protocol could be used as a template for various design scenarios of clinical studies when the project involves any physical contact or medical interventions with participants. It would be the best approach if the investigation is planned to be a multicenter, prospective, and blinded clinical study.

The initial requirements including purpose, objectives, hypothesis, specific aims, and background information for selected neurovascular biomarkers should be addressed in the research protocol.

7.1.1   Objectives

To investigate diagnostic values of GluR (NR2, AMPAR, KAR) peptides for assessment of acute ischemic events in ER within 12 h after the onset. This study intended to analyze diagnostic accuracy of each biomarker depending on the level of arterial/venous network involved (capillary, small vessels, arterial, or venous circulation) in the formation of cerebral lesions defined by MRI. The secondary objective is to examine the possibility to differentiate the severity of acute ischemic stroke using the biomarkers related to preexisting conditions.

7.1.2   Aims

1. Examine GluR peptides levels in a healthy population and subjects with preexisting conditions.

2. Examine changes in GluR peptide levels that occur in patients presented to ER with symptoms suggestive of acute stroke (stroke vs. non-stroke).

3. Examine GluR peptide values and diagnostic utility in patients with acute ischemic stroke relative to standard stroke protocol (USA).

4. Examine GluR peptide levels and their relationship to MRI and neuroimaging modalities.

Based on proposed Aims the following hypotheses should be tested.

7.1.3 Hypotheses

1. GluR peptides values will not be elevated in healthy volunteers.

2. GluR peptides levels will show greater increase in acute ischemic stroke than acute cerebral hemorrhage and stroke mimics.

3. GluR peptides amounts will correlate and demonstrate at least similar diagnostic utility compared to standard stroke workup.

4. GluR peptides levels will triage acute IS patients for MRI.

The relevant prior experience and gaps in current knowledge should be addressed showing how the biomedical study will add to current information. The background evaluation should delineate relevant preliminary data and references for each biomarker.

7.2 Inclusion and Exclusion Criteria

As a general approach the correct study cohort should selected based on (1) disorder subtype, (2) preexisting, and (3) "mimic" conditions or concomitant diseases, (4) race, (5) age, and (6) gender. Then control group should include not only age-, race-, and gender-matched healthy volunteers but contain additional group of persons with preexisting conditions (atherosclerosis, diabetes, and hypertension) and stroke mimics (migraine, palsies, and postictal paresis).

It is known the stroke causes lesion(s) in certain brain vascular territory(ies) where the definite subtype of GluR would be preferably located. Then the study groups should be considered from lesion location points (cortical or subcortical and vertebrae), stroke subtype (cardioembolic, thrombotic, cryptogenic). It is required the data from MRI diffusion and diffusion tensor modalities will be acquired to assess hyperintensities in cortical and subcortical areas due to acute stroke (please *see* Chapter 13).

Additionally, it is necessary indicate (1) the time since onset when sample drawn, (2) prior medications used (cross-reaction or factor diminishing biomarker levels), and (3) type of sample (plasma or serum) to assess the peptide or protein.

The information about "critical values" of the test is needed for the blood test to speed up the diagnosis of acute ischemic stroke prior to CT imaging—for example, at ER (*see* Appendix 3). Critical values are test results that fall significantly outside the normal range and may represent life-threatening values even if obtained from routine tests (e.g., "panic values" or "red-line values") and must be conveyed immediately to the physician or other health care professional so that therapeutic measures can be instituted rapidly.

The following eligibility criteria are designed to select subjects for whom participation in the study is considered appropriate. All relevant medical and nonmedical conditions should be taken into consideration when deciding whether a subject is suitable for the study. Subject eligibility should be reviewed and documented by an appropriate member of the investigator's study team before being included in the study. The enrollment starts after informed consent is obtained and all inclusion/exclusion criteria met.

7.2.1 Study Population   The study will consist of patients with acute stroke-like symptoms (including cerebral hemorrhage and stroke mimics) and/or suspected cerebral infarction (not necessary first) according to Cincinnati prehospital stroke scale (Cincinnati code) admitted through emergency medical services (EMS) [62].

Control group will include age- and gender-matched healthy volunteers and subjects with preexisting conditions.

7.2.2 Inclusion Criteria   To be included in the study the following criteria must be met:

- Patient or legal representative has read, understood, signed, and dated the Informed Consent prior to initiation of the study procedures.

- Male or female patients 21 years of age or older. There is no upper maximum age.

- Patient or legal representative must present or report with ongoing and/or a history of neurologic symptoms suggestive of stroke within 12 h of symptom onset with and without history of preexisting conditions (atherosclerosis, diabetes, and cardiovascular disease).

- A blood sample must be collected within 12 h of symptom onset.

7.2.3 Exclusion Criteria   If excluding vulnerable populations (pregnant women, illiterate or non-English speaking individuals) a scientific rationale for the exclusion should be provided. Inconvenience or cost is not an acceptable rationale.

The subject will be excluded from the study if the following criteria are met:

- Presentation outside the 12-h time window of symptom onset or if the onset time cannot be identified as likely within this time frame.

- Patients with traumatic brain injury.

- Patients who have a contraindication to having a brain MRI (claustrophobia, pacemaker, defibrillator, neurostimulator, a metal implant of any kind or any metallic foreign bodies such as bullets, shrapnel, metal slivers).

- Pregnant women will be excluded from testing which involves the use of MRI and CT scans specific to this protocol.

- Patients with a history of severe psychiatric illness.

- Patients with a history of alcohol or substance abuse.

- Patients who may be unable or unlikely to complete the study.

- Currently enrolled in another investigational study or partici-
  pated in an investigational study in the 30 days prior to the
  present study start.

- Patient has received rt-PA treatment after current admission to
  the hospital and before blood drawn for the peptide assays.

*7.2.4 Number of
Research Participants*

The study will recruit 800 participants and proceed in two phases.
The intent of the first phase is to estimate optimal cutoffs for each
time window from which the blood sample for the GluRs peptides
levels quantification is drawn (i.e., 0–6, 6–12 h) for discriminating
ischemic vs. stroke mimics and cerebral hemorrhage. The intent of
the second phase is to validate each peptide as a diagnostic test
using the cutoff values identified in the first phase. According to
sample size calculation there should be $n = 100$ patients per group
and time window (Appendix 4) to be enrolled in at least 3–4
Centers of Excellence for Stroke ERs to complete the study within
2 years.

Practical experience shows that written informed consent usually could not be obtained from about 30% to 50% enrolled patients
due to ER conditions. Therefore, the above cited number of study
participant should be increased by 50% from initial enrollment.
Consider the use of electronic database where the recruitment at
each participating side should be reflected in numbers per day.
Additionally, Cincinnati code sensitivity ranges from 44% to 95%
and specificity showed a wide bracket (23–96%) due to the scale
reproducibility variations between EMS teams in the same region of
country [62]. This code is designed to assess large vessel occlusions
and does not detect up to 38% of posterior cerebral circulation
strokes. The use of Cincinnati scale provides about 57.6% correctly
recognized cases of acute stroke [63] with about 25% of patients
arriving at a hospital within 3.5 h, and less than 65% arriving within
8 h of symptom onset [64].

**7.3 Recruitment
Methods**

Patient enrollment starts when all training of study personnel and
certification requirements are met. Staff are trained on how to
effectively administer informed consent and provide adequate
information for each potential participant as well as the procedures
for the study. A standardized consent form should contain a clear
and concise study description in language at an acceptable reading
level and will be customized with center-specific information. All
key personnel should be required to provide proof of Human
Subjects Education prior to beginning of screening. A properly
constituted, valid Institutional Review Board (IRB) must review
and approve the protocol, the investigator's informed consent
document, any subsequent changes to the protocol, and related
patient information and recruitment materials before the start of
the study. It is the responsibility of the principal investigators at

each site to ensure that written informed consent is obtained from the patient or the legal representative before any activity or procedure is undertaken that is not part of routine care and that each participant receives a copy of the signed consent.

Enrollment starts after informed consent is obtained and ends after the site diagnosis has been entered into the electronic standardized Case Report Form (CRF) following the diagnostic workup. This process typically occurs within 24–72 h of study enrollment but in single cases may take a longer time. Group enrollment will be closed as soon as the maximum number of patients has been accrued for each group and time window. There is no randomization. Different diagnostic categories (IS, intracerebral hemorrhage, stroke mimic, or control) and time points from symptom onset will be accrued by group quota.

7.4 Study Design, Procedures and Timeline

Study design and related research procedures to be performed should be described. Study safety or minimizing risk monitoring should be included. The major procedures are intended to perform: neurological, radiological assessment of patients and blood sample withdrawal on admission to ER.

7.4.1 Neurological Assessments

Each patient arriving at the ED with stroke-like symptoms and/or suspected cerebral infarction will be identified and screened for the study. The onset time of stroke symptoms will be obtained from the patient and, whenever possible, verified by a third-party history, and in every case by the documentation by EMS personnel and the ER documentation forms. If the time of onset cannot be determined, for instance, if the patient awoke with stroke symptoms from sleep, the time last seen normal or being normal will be used as onset time. Upon enrollment evaluations will be performed on each symptomatic patient (Table 1).

Patients' diagnoses will be determined by the attending physician based on data collected. This will include medical and neurological history, NIHSS scores at two time points, TOAST (Trial of ORG 10172 in Acute Stroke Treatment) classification, and all imaging data performed at the hospital for that episode. Neuroimaging analysis of each imaging study will be performed by independent board-certified neuroradiologists and the results will be provided in the database. The final diagnosis of ischemic or hemorrhagic stroke and stroke mimics will be determined by an expert neurologist who will be blinded to both GluR peptides results, and the diagnosis made at each clinical site. The expert will review clinical assessments and imaging reports to determine the final diagnosis that will be recorded for each case. If the site and final diagnoses coincide, it will be accepted as the final diagnosis for each patient. In cases that remain questionable a consensus opinion of a panel of stroke experts (clinical PIs) will determine the final diagnosis.

Table 1
Schedule of events for symptomatic patients

Procedure	ED presentation within 12 h of the onset
Informed consent	×
Inclusion/exclusion criteria	×
Demographics and medical history	×
Current medication usage	×
Clinical laboratory tests[a]	×
NIH Stroke Scale	×
TOAST classification	×
Stroke symptom questionnaire[b]	×
Performed imaging[c]	×
Blood draws: GluR peptide assays (plasma)	×
Patient's final diagnosis[d]	×
Study related adverse events (AE)	×

[a]Clinical Laboratory Tests include a complete blood count, coagulation studies, basic metabolic panel including serum glucose, renal function, and electrolytes
[b]The questionnaire has been validated to be sensitive and specific survey tool
[c]Imaging should be performed preferably within 24 h, but accepted if performed within 72 h. All enrolled patients should have an MRI
[d]The final diagnosis will be performed centrally by an expert review of the results of all tests performed. Reviewer will be blinded from all peptide data and the hospital discharge diagnosis

7.4.2 Neuroimaging

Brain MRI with DW sequences can identify even subtle ischemic brain injury soon after the onset of ischemia. Ischemic lesions can be visualized in the acute phase as an area of high signal on DWI with a corresponding area of restricted diffusion on the apparent diffusion coefficient (ADC) map. It is anticipated that about 60–70% of all study participants at each site will have a brain MRI within 48–72 h during their routine stroke workup.

MRI data will be collected at each study site for each enrolled patient and the interpretation obtained at each study site will be transmitted as de-identified DICOM images to the electronic study databank. A stroke analysis will be performed by a board-certified radiologist and imaging studies will be analyzed using Analyze software (http://www.mayo.edu/bir/Software/Analyze/Analyze. html). Acute strokes will be quantified based on the diffusion weighted images using a semi-automated region-of-interest tracing approach and their locations will be recorded. A board-certified neuroradiologist will correlate any identified DWI lesions with corresponding MR information (ADC map, structural T1, T2, and FLAIR) to verify lesion volume and designation as acute stroke. Additional information capturing the presence of hemorrhage, prior infarcts, and any clinically significant findings will be evaluated.

7.4.3 Biomarker Evaluations

All study participants will have a blood draw at the time of their initial presentation to measure GluR peptides levels in plasma. After the initial draws subsequent blood samples will be collected in participants who have arrived within 3 h of the onset to further characterize the time course of GluR peptides. This study will emphasize enrollment of stroke patients presenting within 12 h of symptom onset when acute intervention is most effective. De-identified blood samples will be sent to the Laboratory for blinded analyses by laboratory personnel. The results of assays will be uploaded to the database for further data analysis by biostatistician(s). Randomly, for about every 100 participants enrolled, duplicate blood samples will be sent to the Laboratory for quality assurance. Each peptide concentrations will be measured by a magnetic particle-based (MP) enzyme-linked immunosorbent assay (ELISA) intended for rapid quantitative determination of each peptide fragment of GluR in blood with the turnaround time of 30 min (*see* Chapter 10). After measurements have been performed, blood samples will be stored at -70 °C.

7.5 Data Analysis Plan

Initial analyses will include basic frequency of each peptide values in control groups of healthy volunteers, subjects with preexisting conditions and stroke mimics. The examination of the influence of demographic factors on baseline GluR peptides (gender, race, age, etc.; likely through t test, ANOVA, or Pearson correlation depending on the nature of the demographic factors) will be examined.

There are expected some nonparametric analyses (likely chi-square) examining basic frequencies of GluR peptides (elevated/not-elevated) for subjects with preexisting conditions or stroke mimics vs. healthy volunteers, frequencies of GluR peptides levels (elevated/not-elevated) in all control groups vs. hemorrhagic stroke, frequencies of GluR peptides levels (elevated/not-elevated) in all control groups vs. ischemic stroke, and frequencies of GluR peptides amounts (elevated/not-elevated) in stroke (hemorrhagic and ischemic) and MRI findings (abnormal/normal).

Then each peptide (NR2, AMPAR, and KAR) critical cutoffs at 0–6 and 6–12 h after the onset to diagnose ischemic cerebral events will be established. In the first phase, a receiver operating characteristics (ROC) curve (sensitivity vs. specificity) for corresponding time window by varying the cutoff value for distinguishing IS from non-ischemic events will be constructed. The final diagnosis of acute ischemic stroke is based on MRI data as a gold standard.

The partial area under the ROC curve for the region with *specificity* between 0.75 and 0.95, denoted by area under the curve ($pAUC_{0.75}^{0.95}$) as a global measure of the diagnostic effectiveness of each peptide will be used. It is considered that the partial area rather than the commonly used total AUC is of clinical interest since only the region with high specificity. The standardized partial area $spAUC_{0.75}^{0.95} = \frac{1}{0.20} pAUC_{0.75}^{0.95}$ can be interpreted as the *average sensitivity* level when the specificity level lies within the range [0.75, 0.95]. To test the global null hypothesis that the GluR peptide does not have adequate accuracy at any of the time windows, we will evaluate whether the average $pAUC_{0.75}^{0.95}$ across the time windows is at least 0.8.

At the end of the first phase, the empirical estimate for the ROC curve is constructed and its standard errors should be estimated [65]. Moreover, we will use the perturbation technique to construct the confidence band for the ROC curve *see* [66]. If the confidence intervals or band for the estimated ROC curves are not tight enough for $n = 100$ per time window, then it will need additional participants enrolled in the first phase to improve the precision. Since the adjustment is only made considering the sampling variability and is not for time window comparisons, no statistical adjustment for type I error is necessary even if the sample size is increased.

After approximately half of the target participants are enrolled, a futility analysis by calculating the conditional power for testing the null hypothesis that $spAUC_{0.75}^{0.95}$ across the 2-time windows is less than or equal to 0.8 versus the alternative hypothesis that it is at least 0.8 should be conducted. This conditional power is the probability of rejecting the null hypothesis at the end of this first phase, given the data at the halfway point.

As soon as the first part of study has set up the optimal cutoffs for each time window, the next point should be determining whether the GluR peptides assays can adequately differentiate brain ischemia from stroke mimics, non-ischemic stroke, and controls.

Based on preliminary data, a working hypothesis is that the GluR peptides are found at significantly higher levels in patients with cerebral ischemia than in individuals with stroke mimics or in controls at a specificity of >85% at an optimal cutoff points calculated in the first part of the study for 0–6 and 6–12 h points (could be the same numbers). The analysis will investigate the diagnostic performance of GluR peptides in general. However, the analysis will consist of investigating the diagnostic performance of peptides levels for various subgroups:

- Time from symptom onset (0–6, 6–12 h).
- Lesion size (<2 mm, 2–5 mm, >5 mm).
- Lesion location.
- Stroke severity based on NIHSS (mild, moderate, and severe).

Obviously, the sample size for these subgroups will be much smaller than the sample sizes for the general analyses. Nevertheless, it is critical to investigate the diagnostic performance for these subgroups in order to determine if subgroups exist for which the GluR peptides levels have outstanding, or not so outstanding performance.

7.5.1 Potential Problems and Alternative Considerations

Specificity and sensitivity for IS diagnosis need to be acceptable for use of the neurovascular biomarkers in clinical practice. The aim is for a minimum diagnostic sensitivity of >80%, and specificity of >85% in the first phase of the study to have a neurovascular biomarker that has sufficient diagnostic accuracy for diagnosis of cerebral ischemic events (cortical, subcortical, or deep subcortical areas). In comparison, brain diffusion weighted imaging (DWI) which is the diagnostic standard for acute IS (mostly cortical) within 12 h of symptom onset, has a diagnostic sensitivity of 90% (range: 88–100%) and a specificity of 97% (range: 86–100%) [67, 68]. Diffusion tensor imaging (DTI), more often utilized image of deep white matter lesions due to acute stroke care, has a sensitivity and specificity of 85% [69]. The GluR peptides assays performed at ER for acute IS assessment could navigate neuroimaging to region of interest and, possibly, triage patients for endovascular treatment. This is especially important when MRI is not readily available or contraindicated which is a rather frequently encountered limitation in clinical practice affecting about 30–40% of all acute stroke patients [67].

7.6 Confidentiality of Patient's Data, Specimens, and Banking

The present part is usually required to maintain the confidentiality of the data concerning each participant. It is usually required by institutional (federal) guidelines and policies to de-identify patient data and describe the manner of electronic data storage (electronic CRFs). It is necessary to describe as well how data should be protected in computer databases and transferred between the clinical sites, the imaging reading center(s) and the data coordinating center (passwords, data encryption and security, types of accesses to hospital electronic records). The results of participation should be confidential and should not be released in any individually identifiable form without the prior consent of the participant unless required by law.

Confidentiality, storage issues, and banking of specimens should be reflected in the protocol. It is essential to report how samples shipped (if necessary), processed, and then eliminated or be stored frozen as de-identifiable samples (facility, temperatures, and duration).

The protocol should address risks, benefits, cost, and compensations (research related injury or any incentives) to research participants. Enrolled in the study subject's privacy interest protection, possibility to withdraw from research, and alternative participation should be outlined.

The data monitoring plan for completeness, accuracy, and adherence to the protocol should be described for each participating clinical site (community-based, multisite, or international).

8 Conclusion

The emergency pathway for suspected stroke is initiated by ambulance personnel, but early recognition is challenging due to heterogeneous clinical presentations and an absence of portable biomarker assay technology. The prehospital decision to rapidly redirect to ER relies solely upon a basic clinical symptom like Cincinnati Code which has wide range sensitivity (44–95%) and specificity (23–96%). With a score of two, it predicted patients receiving thrombolytic therapy with a high sensitivity of 96% and moderate specificity of 65% with favorable outcome [62].

The major concern at ER is to triage patients for emergent endovascular (IS due to large vessel occlusion) or thrombolytic therapy (IS caused by small vessel occlusion) that could not be defined based on clinical symptoms only [63]. As most ED in low income countries have limited or no access to urgent advanced imaging to confirm IS (e.g., MRI) it creates clinical diagnostic uncertainty and delays in urgent endovascular or thrombolytic treatment leading to serious neurological complications. A simple rapid blood assay detecting neurovascular biomarkers that might distinguish acute IS from non-stroke and select patients with large or small vessel occlusion would be needed.

Point of care GluR peptide assays panel is a promising approach that might be readily applied at ambulance setting and ER to triage patients according to severity of stroke. To become an objective tool aiding to practice, each potential neurovascular biomarker assay and panel should be further tested on a large multiracial population to prove the accuracy and efficacy in assessment of acute stroke vs. non-stroke and ischemic onset vs. hemorrhagic stroke. The use of the rapid blood testing could increase efficiency of ambulance, ER, and stroke center services.

Appendix 1: Case-Report Studies of NR2 Peptide Contents in Patients with Preexisting Conditions

Cardioembolic stroke. A 67-year-old female with prior aortic valve replacement surgery and diabetes mellitus (DM, 227 mg/dL glucose) was admitted to ER within 2 h of stroke onset (NIH stroke score 10) with ischemic lesions in right middle cerebral artery area (cardioembolic subtype) seen on CT (Fig. 3a). The measurement of NR2 peptide assay yielded in 4.71 ng/mL (preliminary cutoff of 0.5 ng/mL) depicted the acute cerebral ischemia.

Atherothrombotic stroke. A 82-year-old male presented with a history of stenosis of right (40%) and occlusion in left carotid artery, motor aphasia, dysphagia, and right hemiparesis (NIH stroke scale 10). He presented within 10 h post-symptom onset and had head CT that showed acute ischemic changes in territory of left middle cerebral artery (thrombotic subtype, Fig. 3b). As a part of an ongoing clinical study a blood sample was taken while he had the neurological deficit and showed elevated NR2 peptide of 2.34 ng/mL.

Stroke of unknown origin in vertebral region. A 52-year-old female patient presented after series of TIA within 2 weeks with last resolved with symptoms of dysarthria and hemihypesthesia within 12 h since onset. The CT was normal, and MRI showed acute lacunar infarct in the posterior-left sections of the Varoliev Bridge of hindbrain with hyperintensities on DWI and T2 (Fig. 3c, d). The major function of Varoliev Bridge is to pass nerve impulses to the cortex and to the spinal cord. NR2 peptide was 1.68 ng/mL and about three times higher than the normal levels in healthy controls.

Stroke due to neurovasculitis. Neurovasculitis and renal disorders may show elevated NR2 biomarker in certain cases due to NMDA receptors response to microvessel inflammation [43]. A 21-year-old woman "crystal" methamphetamine user presented with headaches and visual complaints and a T2-weighted MRI scan showed multiple bilateral subcortical areas of white matter abnormality. This was followed by a worsening in condition that resulted in admission to the hospital with left hemiparesis and delirium. MRI scans after admission showed acute right hemispheric infarction and diffuse vasospasm of cerebral vessels without the beading associated with typical vasculitis on magnetic resonance angiography (Fig. 4a). Initial ischemic lesion in left hemisphere revealed by routine CT (Fig. 4b) progressed further (Fig. 4c). Time course of NR2 peptide obtained in this case showed significant and prolonged elevation within a week suggesting an ongoing ischemic process (Fig. 4d).

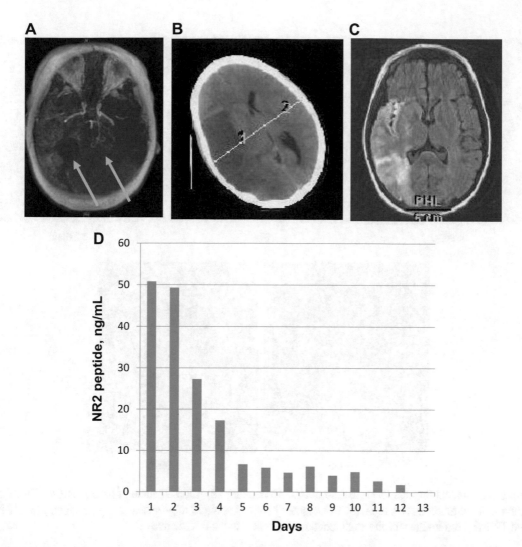

Fig. 4 Acute stroke due to neurovasculitis. (**a**) MRA brain on admission depicts vasospasm that creates large areas of the brain with no visible vessels (arrows), (**b**) routine CT scan on admission, (**c**) FLAIR shows region of stroke, (**d**) NR2 peptide monitoring (cutoff of 0.5 ng/mL)

Appendix 2: Case-Report Studies of AMPAR Peptide Contents in Patients with Subcortical Strokes Including Lacunar Lesions

Lacunar strokes. A 43-year-old hypertensive Afro-American woman presented with sudden symptoms of left-sided weakness, incoordination, unsteadiness, cerebellar ataxic dysarthria, and dysphonia. The apparent diffusion coefficient (ADC) sequence shows an acute bilateral pontine infarct (Fig. 5a). Drastically increased NR2 (9.32 ng/mL) and AMPAR (7.03 ng/mL) peptides were detected in the patient's plasma on admission.

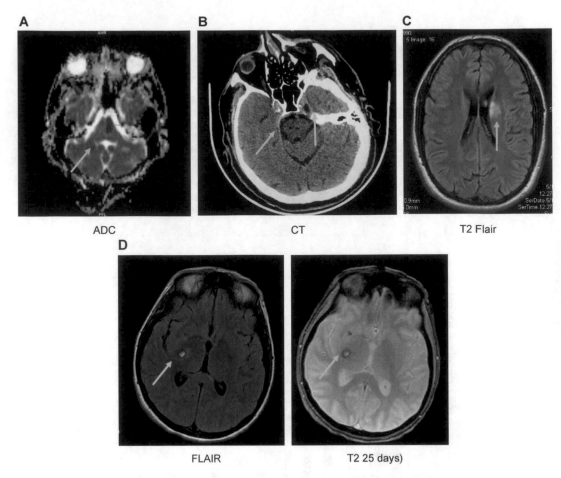

Fig. 5 Radiological findings for case reports (AMPAR peptide). (**a**) ADC sequence of lacunar stroke, (**b**) CT of acute IS in vertebral basilar area, (**c**) T2-weighted FLAIR of acute IS after snowboarding accident, (**d**) FLAIR and T2-weighted images for the same patient with hemorrhagic transformation

Stroke in vertebral basilar area. A 59-year-old Caucasian male patient presented with a history of hypertension and symptoms of left hemiparesis, imbalance during walking, facial asymmetry, and double vision within 15 h after the onset. The CT showed acute infarct in vertebral basilar area with multiple arachnoid cysts in basal nuclei in deep frontal lobe and brainstem (Fig. 5b). The elevated NR2 (1.8 ng/mL) and AMPAR (3.1 ng/mL) peptides were measured compared to that for controls (preliminary cutoffs are 0.5 and 0.4 ng/mL, respectively).

Ischemic stroke after snowboarding accident. The case of multiple concussions resolved in stroke after snowboarding accident with white matter lesions depicted on Fig. 5c. A 22-year-old Caucasian male had a seizure on the day of the accident with following admission to hospital. The transformation from initial normal CT image to subcortical lesions in left hemisphere was observed within the next 48 h and accompanied increased AMPAR peptide (1.6 ng/ mL at preliminary cutoff 0.4 ng/mL) and control level of NR2 peptide (0.5 ng/mL) that could be associated with cytotoxic edema formation in subcortical structures with presumably AMPAR localization (Figs. 1b and 5c).

Hemorrhagic transformation. The contusion acquired by a 31-year-old Caucasian woman occurred after a car accident and an impact with a tree. She had acute symptoms of incoordination, low alertness and was hyporeflexic. The subcortical areas of hemorrhage are seen on axial Flair and T2-weighed scans registered on seventh day after the injury (Fig. 5d). AMPAR peptide concentration of 7.2 ng/mL and NR2 peptide of 2.0 ng/mL are measured compared to non-injured controls (0.5–1.0 ng/mL range).

Appendix 3

Preliminary Critical Values of NR2 Peptide

Samples from apparently healthy males/females (52 M/102 F) and persons (42 M/35 F) with preexisting conditions (Table 2) in the clinically relevant age range of 30–70 years were evaluated using the NR2 peptide assay.

AMPAR Peptides Preliminary Reference Values

Total of 128 apparently healthy males/females (75 M/52 F) including 53 persons (28 M/25 F) with preexisting conditions (Table 4) in the clinically relevant age range of 30–70 years included to evaluate reference interval for AMPAR peptide assays.

Table 2
The distribution of persons with preexisting conditions for NR2 peptide assay

Preexisting condition	All subjects ($n = 231$)	
	N	% Absolute
Diabetes	24	10.4
Hypertension	16	6.9
Atherosclerosis	12	5.2
Cardiac diseases	20	8.7
Sleep apnea	5	2.2
Total	77	33.4

The reference population was represented by the following ethnic backgrounds: African-American ($n = 84$), Caucasian ($n = 142$), and Asian ($n = 5$). The distributions of NR2 peptide values across the entire population presented in the Table 3. The reference interval calculated from the samples (central 90th percentile) is revealed to be 0.1–1.0 ng/mL for both genders

Table 3
The distribution of NR2 peptide across enrolled population

NR2 peptide ng/mL	All subjects (*n* = 231)		
	N	% Absolute	% Population
<0.1	82	35.5	35.5
0.1–0.5	97	42.0	77.5
0.6–1.0	31	13.4	90.9
>1.0	21	9.1	100

Table 4
The distribution of persons with preexisting conditions for AMPAR peptide assay

Preexisting condition	Subjects (*n* = 53)	
	N	% Absolute
Diabetes	13	10.2
Hypertension	15	11.7
Atherosclerosis	14	10.9
Cardiac diseases	11	8.6
Total	53	41.4

The reference population was represented by the following ethnic backgrounds: African-American (*n* = 57), Caucasian (*n* = 68), and Asian (*n* = 3). The distributions of NR2 peptide values across the entire population presented in the Table 5

Table 5
The distribution of AMPAR peptides across population investigated

AMPAR peptides, ng/mL	All subjects (*n* = 128)		
	N	% Absolute	% Population
<0.1	54	42.2	42.2
0.1–0.5	49	38.3	80.5
0.6–1.0	18	14.0	94.5
>1.0	7	5.5	100

The reference interval calculated from the samples (central 90th percentile) is found to be 0.1–0.6 ng/mL for both genders

Table 6
Power of study calculations

	$n = 100$ per time window	$n = 150$ per time window	$n = 200$ per time window
$d_0 = 1.0$	0.066 (0.033)	0.052 (0.026)	0.047 (0.023)
$d_0 = 1.5$	0.049 (0.024)	0.038 (0.019)	0.033 (0.016)
$d_0 = 2.0$	0.046 (0.022)	0.035 (0.017)	0.031 (0.015)

Assuming that the true sensitivity and specificity are 0.85 and 0.95 for the Phase 2 and that the prevalence of ischemic event is 80%, then $n = 322$. If the expected lower bound of two-sided 97.5% confidence intervals for sensitivity and specificity are 0.80 and 0.82 respectively (i.e., greater or equal to 0.80), it indicates that $n = 322$ for Phase 2 is adequate

---

Appendix 4: Sample Size Calculations

The sample size calculation for the first phase is based on the following assumptions:

1. The prevalence rate of ischemic events in the enrolled cohort is 80%.

2. A transformation of each peptide measure is normally distributed for trial participants with and without ischemic events.

It is assumed that the standard deviation for the non-ischemic participant is one and the counterpart for the ischemic participants is d_0 without loss of generality. With n participants per time window in the first phase, the following table provides the standard error for the $pAUC_{0.75}^{0.95}$ estimate for each time window and the standard error of the average $pAUC_{0.75}^{0.95}$ (displayed in the parenthesis) when the true $spAUC_{0.75}^{0.95}$ is 0.90.

Results summarized in the Table 6 suggest that a sample size $n = 100$ per time window is adequate for the first phase of the trial with a power of no less than 86% for the above settings. With $n = 100$ per time window, the 95% confidence interval for the average $spAUC_{0.75}^{0.95}$ across the 3-time windows is expected to be (0.84, 0.96), which is sufficiently tight clinically.

References

1. Spence JD (1999) New approaches to atherosclerosis based on endothelial function. In: Bogousslavsky J, Fisher M (eds) Cerebrovascular disease, 4th edn. Current Medicine Inc., Philadelphia, PA, pp 1–14

2. Roach ES, Bettermann K, Jose Biller J (2010) Toole's cerebrovascular disorders, 6th edn. Cambridge University Press, Cambridge, p 422. https://doi.org/10.1017/CBO9781139644235

3. The personalized medicine report 2017: opportunity, challenges, and future. http://www.personalizedmedicinecoalition.org/Userfiles/PMC-Corporate/file/The_PM_Report.pdf_. Accessed on 19 Sept 2019

4. Ng GJL, The Ei Z, Ng MY, Quek AML, Seet RCS (2018) Resolving difficult case scenarios by incorporating stroke biomarkers in clinical decision-making. In: Peplow PV, Dambinova SA, Gennarelli TA, Martinez B (eds) Acute

brain impairment: scientific discoveries and translational research. Royal Society of Chemistry, London, pp 289–314. https://doi.org/10.1039/9781788012539-00289

5. Zhang JH, Badaut J, Tang J, Obenaus A, Hartman R, Pearce WJ (2012) The vascular neural network – a new paradigm in stroke pathophysiology. Nat Rev Neurol 8(12):711–716. https://doi.org/10.1038/nrneurol.2012.210

6. Barber PA (2013) Magnetic resonance imaging of ischemia viability thresholds and the neurovascular unit. Sensors (Basel) 13(6):6981–7003. https://doi.org/10.3390/s130606981

7. Terasaki Y, Liu Y, Hayakawa K, Pham LD, Lo EH, Ji X, Arai K (2014) Mechanisms of neurovascular dysfunction in acute ischemic brain. Curr Med Chem 21(18):2035–2042. PMID: 24372202

8. Sommer CJ (2017) Ischemic stroke: experimental models and reality. Acta Neuropathol 133(2):245–261. https://doi.org/10.1007/s00401-017-1667-0

9. Katan M, Elkind MSV (2018) The potential role of blood biomarkers in patients with ischemic stroke: an expert opinion. Clin Transl Neurosci. https://doi.org/10.1177/2514183X18768050

10. Iadecola C (2017) The neurovascular unit coming of age: a journey through neurovascular coupling in health and disease. Neuron 96(1):17–42. https://doi.org/10.1016/j.neuron.2017.07.030

11. Hayashi MK (2018) Structure-function relationship of transporters in the glutamate-glutamine cycle of the central nervous system. Int J Mol Sci 19(4):pii: E1177. https://doi.org/10.3390/ijms19041177

12. Dambinova SA (2012) Neurodegradomics: the source of biomarkers for mild traumatic brain injury. In: Dambinova SA, Hayes RL, Wang KKW (eds) Biomarkers for TBI. Royal Society of Chemistry, London, pp 66–86. https://doi.org/10.1039/9781849734745

13. Sharp CD, Fowler M, Jackson TH 4th, Houghton J, Warren A, Nanda A, Chandler I, Cappell B, Long A, Minagar A, Alexander JS (2003) Human neuroepithelial cells express NMDA receptors. BMC Neurosci 4:28. PMID: 14614784

14. Brand-Schieber E, Lowery SL, Werner P (2004) Select ionotropic glutamate AMPA/kainate receptors are expressed at the astrocyte-vessel interface. Brain Res 1007(1–2):178–182. PMID: 15064149

15. Liu H, Leak RK, Hu X (2016) Neurotransmitter receptors on microglia. Stroke Vasc Neurol 1(2):52–58. https://doi.org/10.1136/svn-2016-000012

16. Sanz-Clemente A, Nicoll RA, Roche KW (2013) Diversity in NMDA receptor composition: many regulators, many consequences. Neuroscientist 19(1):62–75. https://doi.org/10.1177/1073858411435129

17. Christensen PC, Samadi-Bahrami Z, Pavlov V, Stys PK, Moore GRW (2016) Ionotropic glutamate receptor expression in human white matter. Neurosci Lett 630:1–8. https://doi.org/10.1016/j.neulet.2016.07.030

18. Jin XT, Smith Y (2011) Localization and functions of kainate receptors in the basal ganglia. Adv Exp Med Biol 717:27–37. https://doi.org/10.1007/978-1-4419-9557-5_3

19. McConnell HL, Kersch CN, Woltjer RL, Neuwelt EA (2017) The translational significance of the neurovascular unit. J Biol Chem 292(3):762–770. https://doi.org/10.1074/jbc.R116.760215

20. Dambinova SA, Khounteev GA, Skoromets AA (2002) Multiple panel of biomarkers for TIA/stroke evaluation. Stroke 33(5):1181–1182. PMID: 11988587

21. Dambinova S, Bettermann K, Glynn T, Tews M, Olson D, Weissman JD, Sowell RL (2012) Diagnostic potential of the NMDA receptor peptide assay to distinguish acute ischemic stroke and stroke mimics. PLoS One 7:e42362. https://doi.org/10.1371/journal.pone.0042362

22. Dambinova SA (2008) Biomarkers for transient ischemic attack (TIA) and ischemic stroke. Clin Lab Int 32(7):7–10. https://www.clinlabint.com/fileadmin/pdf/digital_issues_archives/CLI_Nov08.pdf

23. Blanco PJ, Müller LO, Spence JD (2017) Blood pressure gradients in cerebral arteries: a clue to pathogenesis of cerebral small vessel disease. Stroke Vasc Neurol 2(3):108–117. https://doi.org/10.1136/svn-2017-000087

24. Yuan H, Hansen KB, Vance KM, Ogden KK, Traynelis SF (2009) Control of NMDA receptor function by the NR2 subunit amino-terminal domain. J Neurosci 29(39):12045–12058. https://doi.org/10.1523/JNEUROSCI.1365-09.2009

25. Möykkynen T, Coleman SK, Semenov A, Keinänen K (2014) The N-terminal domain modulates α-amino-3-hydroxy-5-methyl-4-isoxazolepropionic acid (AMPA) receptor desensitization. J Biol Chem 289(19):13197–13205. https://doi.org/10.1074/jbc.M113.526301

26. Sheng N, Shi YS, Nicoll RA (2017) Amino-terminal domains of kainate receptors determine the differential dependence on Neto auxiliary subunits for trafficking. Proc Natl Acad Sci U S A 114(5):1159–1164. https://doi.org/10.1073/pnas.1619253114

27. Hansen KB, Ogden KK, Yuan H, Traynelis SF (2014) Distinct functional and pharmacological properties of triheteromeric GluN1/GluN2A/GluN2B NMDA receptors. Neuron 81(5):1084–1096. https://doi.org/10.1016/j.neuron.2014.01.035

28. Greger IH, Watson JF, Cull-Candy SG (2017) Structural and functional architecture of AMPA-type glutamate receptors and their auxiliary proteins. Neuron 94(4):713–730. https://doi.org/10.1016/j.neuron.2017.04.009

29. Fisher MT, Fisher JL (2014) Contributions of different kainate receptor subunits to the properties of recombinant homomeric and heteromeric receptors. Neuroscience 278:70–80. https://doi.org/10.1016/j.neuroscience.2014.08.009

30. Paramo T, Brown PMGE, Musgaard M, Bowie D, Biggin PC (2017) Functional validation of heteromeric kainate receptor models. Biophys J 113(10):2173–2177. https://doi.org/10.1016/j.bpj.2017.08.047

31. Yuan H, Vance KM, Junge CE, Geballe MT, Snyder JP, Hepler JR, Yepes M, Low CM, Traynelis SF (2009) The serine protease plasmin cleaves the amino-terminal domain of the NR2A subunit to relieve zinc inhibition of the N-methyl-D-aspartate receptors. J Biol Chem 284(19):12862–12873. https://doi.org/10.1074/jbc.M805123200

32. Gennarelli T, Dambinova SA, Weissman JD (2018) Advances in diagnostics and treatment of neurotoxicity after sport-related injuries. In: Peplow PV, Dambinova SA, Gennarelli TA, Martinez B (eds) Acute brain impairment: scientific discoveries and translational research. Royal Society of Chemistry, London, pp 141–161. https://doi.org/10.1039/9781788012539-00141

33. Dambinova SA, Khounteev GA, Izykenova GA, Zavolokov IG, Ilyukhina AY, Skoromets AA (2003) Blood test detecting autoantibodies to N-methyl-D-aspartate neuroreceptors for evaluation of patients with transient ischemic attack and stroke. Clin Chem 49(10):1752–1762. PMID: 14500616

34. Ponomarev GV, Dambinova SA, Skoromets AA (2018) Neurotoxicity in spinal cord impairments. In: Peplow PV, Dambinova SA, Gennarelli TA, Martinez B (eds) Acute brain impairment: scientific discoveries and translational research. Royal Society of Chemistry, London, pp 198–213. https://doi.org/10.1039/9781788012539-00198

35. Sharp CD, Houghton J, Elrod JW, Warren A, Jackson TH 4th, Jawahar A, Nanda A, Minagar A, Alexander JS (2005) N-methyl-D-aspartate receptor activation in human cerebral endothelium promotes intracellular oxidant stress. Am J Physiol Heart Circ Physiol 288(4):H1893–H1899. https://doi.org/10.1152/ajpheart.01110.2003

36. Abraham HM, Wolfson L, Moscufo N, Guttmann CR, Kaplan RF, White WB (2016) Cardiovascular risk factors and small vessel disease of the brain: blood pressure, white matter lesions, and functional decline in older persons. J Cereb Blood Flow Metab 36(1):132–142. https://doi.org/10.1038/jcbfm.2015.121

37. Bettermann K, Slocomb J, Shivkumar V, Quillen D, Gardner TW, Lott ME (2017) Impaired retinal vasoreactivity: an early marker of stroke risk in diabetes. J Neuroimaging 27(1):78–84. https://doi.org/10.1111/jon.12412

38. Dambinova SA, Aliev KT, Bondarenko EV, Ponomarev GV, Skoromets AA, Skoromets AP, Skoromets TA, Smolko DG, Shumilina MV (2017) The biomarkers of cerebral ischemia as a new method for the validation of the efficacy of cytoprotective therapy. Zh Nevrol Psikhiatr Im S S Korsakova 117(5):62–67. https://doi.org/10.17116/jnevro20171175162-67

39. Brightwell RE (2007) Plasma biomarkers for the early diagnosis of stroke. 29th charing cross intern symposium, CX Innovations Showcase, London

40. Weissman JD, Ponomarev G, Heath R, Boiser J, Dambinova S (2018) Rapid point-of-care NR2 peptide test for acute stroke detection. abstractonline.com, Session P9, Abstract WP212, Jan 24

41. Hirai S, Hotta K, Kubo Y, Nishino A, Okabe S, Okamura Y, Okado H (2017) AMPA glutamate receptors are required for sensory-organ formation and morphogenesis in the basal chordate. Proc Natl Acad Sci U S A 114(15):3939–3944. https://doi.org/10.1073/pnas.1612943114

42. Beppu K, Kosai Y, Kido MA, Akimoto N, Mori Y, Kojima Y, Fujita K, Okuno Y, Yamakawa Y, Ifuku M, Shinagawa R, Nabekura J, Sprengel R, Noda M (2013) Expression, subunit composition, and function of AMPA-type glutamate receptors are changed in activated microglia; possible contribution of GluA2 (GluR-B)-deficiency under

pathological conditions. Glia 61(6):881–891. https://doi.org/10.1002/glia.22481

43. Torres-Platas SG, Comeau S, Rachalski A, Bo GD, Cruceanu C, Turecki G, Giros B, Mechawar N (2014) Morphometric characterization of microglial phenotypes in human cerebral cortex. J Neuroinflammation 11:12. https://doi.org/10.1186/1742-2094-11-12

44. Li S, Stys PK (2000) Mechanisms of ionotropic glutamate receptor-mediated excitotoxicity in isolated spinal cord white matter. J Neurosci 20 (3):1190–1198

45. Dambinova SA, Gill S, St. Onge L, Sowell R (2012) Biomarkers for subtle brain dysfunction. In: Dambinova SA, Hayes RL, Wang KKW (eds) Biomarkers for TBI. Royal Society of Chemistry, London, pp 134–147. https://doi.org/10.1039/9781849734745-00134

46. Dambinova SA, Maroon JC, Sufrinko AM, Mullins JD, Alexandrova EV, Potapov AA (2016) Functional, structural, and neurotoxicity biomarkers in integrative assessment of concussions. Front Neurol 7:172. https://doi.org/10.3389/fneur.2016.00172

47. Pires PW, Dams Ramos CM, Matin N, Dorrance AM (2013) The effects of hypertension on the cerebral circulation. Am J Physiol Heart Circ Physiol 304(12):H1598–H1614. https://doi.org/10.1152/ajpheart.00490.2012

48. Hoiland RL, Bain AR, Rieger MG, Bailey DM, Ainslie PN (2016) Hypoxemia, oxygen content, and the regulation of cerebral blood flow. Am J Physiol Regul Integr Comp Physiol 310(5):R398–R413. https://doi.org/10.1152/ajpregu.00270.2015

49. Zwank MD, Dummer BW, Danielson LT, Haake BC (2014) Lacunar stroke in a teenager after minor head trauma: case report and literature review. J Child Neurol 29(9):NP65–NP68. https://doi.org/10.1177/0883073813500850

50. Chalela JA, Kidwell CS, Nentwich LM et al (2007) Magnetic resonance imaging and computed tomography in the emergency assessment of patient's with suspected acute stroke: a prospective comparison. Lancet 369:293–298. https://doi.org/10.1016/S0140-6736(07)60151-2

51. van der Worp HB, van Gijn J (2007) Acute ischemic stroke. N Engl J Med 357:572–912. https://doi.org/10.1056/NEJMcp072057

52. Hand PJ, Kwan J, Lindley RI et al (2006) Distinguishing between stroke and mimic at the bedside: the brain attack study. Stroke 37:769–775. https://doi.org/10.1161/01.STR.0000204041.13466.4c

53. Hand PJ, Wardlaw JM, Rowat AM, Haisma JA, Lindley RI, Dennis MS (2005) Magnetic resonance brain imaging in patients with acute stroke: feasibility and patient related difficulties. J Neurol Neurosurg Psychiatry 76 (11):1525–1527. https://doi.org/10.1136/jnnp.2005.062539

54. von Kummer R, Dzialowski I (2007) MRI versus CT in acute stroke. Lancet 369 (9570):1341–1342. https://doi.org/10.1016/S0140-6736(07)60621-7

55. Kloppenborg RP, Nederkoorn PJ, Grool AM, De Cocker LJL, Mali WPTM, van der Graaf Y, Geerlings MI for the SMART Study Group (2017) Do lacunar infarcts have different etiologies? Risk factor profiles of lacunar infarcts in deep white matter and basal Ganglia: the second manifestations of arterial disease-magnetic resonance study. Cerebrovasc Dis 43:161–168. https://doi.org/10.1159/000454782

56. Radu RA, Terecoasă EO, Băjenaru OA, Tiu C (2017) Etiologic classification of ischemic stroke: where do we stand? Clin Neurol Neurosurg 159:93–106. https://doi.org/10.1016/j.clineuro.2017.05.019

57. Izykenova GA, Balswin R, Oldenburg SJ (2018) Development of novel test platforms for the assessment of brain injury. In: Peplow PV, Dambinova SA, Gennarelli TA, Martinez B (eds) Acute brain impairment: scientific discoveries and translational research. Royal Society of Chemistry, London, pp 315–326. https://doi.org/10.1039/9781788012539-00315

58. Hernandez-Diaz Z, Barroso-Garcia E, González-García S, González-Quevedo A, Reyes-Berazain A, Arteche-Prior M (2017) Confounding imaging findings in subacute-chronic cerebral infarction. Austin J Cerebrovasc Dis Stroke 4(3):1063–1068. ISSN:2381-9103

59. George MG, Tong X, McGruder H, Yoon P, Rosamond W, Winquist A, Hinchey J, Wall HK, Pandey DK (2009) Paul Coverdell National Acute Stroke Registry Surveillance - four states, 2005–2007. Centers for Disease Control and Prevention (CDC). MMWR Surveill Summ 58(7):1–23. PMID: 19893482

60. Gibson LM, Whiteley W (2013) The differential diagnosis of suspected stroke: a systematic review. J R Coll Physicians Edinb 43 (2):114–118. https://doi.org/10.4997/JRCPE.2013.205

61. González-García S, González-Quevedo A, Hernandez-Diaz Z et al (2017) Circulating autoantibodies against the NR2 peptide of the NMDA receptor are associated with subclinical brain damage in hypertensive patients with other pre-existing conditions for vascular risk.

J Neurol Sci 375:324–330. https://doi.org/10.1016/j.jns.2017.02.028

62. Glober NK, Sporer KA, Guluma KZ, Serra JP, Barger JA, Brown JF, Gilbert iGH, Koenig KL, Rudnick EM, Salvucci AA (2016) Acute stroke: current evidence-based recommendations for prehospital care. West J Emerg Med 17(2):104–128. https://doi.org/10.5811/westjem.2015.12.28995

63. Abboud ME, Band R, Jia J, Pajerowski W, David G, Guo M, Mechem CC, Messé SR, Carr BG, Mullen MT (2016) Recognition of stroke by EMS is associated with improvement in emergency department quality measures. Prehosp Emerg Care 20(6):729–736. https://doi.org/10.1080/10903127.2016.118260

64. Tong D, Reeves MJ, Hernandez AF, Zhao X, Olson DM, Fonarow GC, Schwamm LH, Smith EE (2012) Times from symptom onset to hospital arrival in the get with the guidelines—stroke program 2002 to 2009: temporal trends and implications. Stroke 43(7):1912–1917. https://doi.org/10.1161/STROKEAHA.111.644963

65. Hsieh F, Turnbull B (1996) Nonparametric and semiparametric estimation of the receiver operating characteristic curve. Ann Stat 24(1):25–40

66. Lin D, Wei LJ, Ying Z (1993) Checking the Cox model with cumulative sums of martingale-based residuals. Biometrika 80(3):557–572

67. Brunser AM, Hoppe A, Illanes S, Díaz V, Muñoz P, Cárcamo D, Olavarria V, Valenzuela M, Lavados P (2013) Accuracy of diffusion-weighted imaging in the diagnosis of stroke in patients with suspected cerebral infarct. Stroke 44(4):1169–1171. https://doi.org/10.1161/STROKEAHA.111.000527

68. Edlow BL, Hurwitz S, Edlow JA (2017) Diagnosis of DWI-negative acute ischemic stroke: a meta-analysis. Neurology 89(3):256–262. https://doi.org/10.1212/WNL.0000000000004120

69. Alcántara JP (2014) Diffusion tensor imaging in acute ischemic stroke: the role of anisotropy in determining the time of onset and predicting long-term motor outcome. Ph.D. thesis, University of Girona, Girona, Catalonia. http://hdl.handle.net/10803/132xxx

Chapter 12

Antibodies to NMDA Receptors in Cerebral and Spinal Cord Infarctions

G. V. Ponomarev, E. V. Alexandrova, Svetlana A. Dambinova, D. S. Asyutin, N. A. Konovalov, and A. A. Skoromets

Abstract

The present chapter provides a minireview of potential markers for assessment of cerebral and spinal cord infarctions. It has been suggested that cerebral and spinal cord infarction might be assessed using antibodies to immune active fragments of glutamate receptor biomarkers according to gradual presentation of symptoms and structural alterations defined by neuroimaging in certain brain and spinal cord structures. Antibodies to NR2 subtype of N-methyl-D-aspartate receptors (NMDAR) are detected in patients with acute and chronic conditions. The biomarker revealed ischemic events in recurrent cerebral and chronic spinal cord infarctions and correlated with lesions presence. NR2 antibodies might improve diagnostic certainty of infarctions in cervical and thoracic region of the spinal cord.

Key words Spinal cord, Ischemia, Injury, Biomarkers, Glutamate, NMDA, NR2, Antibodies

1 Introduction

Neurovascular dysfunction and insufficient cerebral blood flow (CBF) are prevalent causes in acute and chronic infarctions. There are a number of subtle asymptomatic events in the brain and spinal cord (SC) that due to preexisting conditions remain frequently undiagnosed and underestimated. Cortical infarctions or silent brain infarction ranges from 10% to 20% [1], while SC infarction is rare and constitutes less than 1% of all central nervous system ischemic events [2]. Both cerebral and SC nervous fibers consist specific neuroreceptors regulating fast signal transductions involved in vasoconstriction and vasodilatation of cerebral blood vessels providing nutrition to nervous tissue [3, 4]. Cerebral and SC infarctions affect multiple microvascular and capillary networks causing edema formation. The latter evokes immunochemical reactions that in more severe cases are followed by structural alterations with lesion(s) formation.

Philip V. Peplow et al. (eds.), *Stroke Biomarkers*, Neuromethods, vol. 147, https://doi.org/10.1007/978-1-4939-9682-7_12,
© Springer Science+Business Media, LLC, part of Springer Nature 2020

The spectrum of clinical symptoms of cerebral and SC infarction is broad with various degrees of sensorimotor and autonomic disturbances. The main causes of SCi are aortic interventions and pathologies that constitute around 30–33% of the cases [5], while nonsurgical etiologies are dominated by atherosclerosis, followed by arterial or cardiac embolism, global hypoxia, infection, hypercoagulable state, and vasculitis [6, 7]. In 20–30% of patients, the etiology remains unclear and less explored.

In contrast to cerebral ischemic stroke, SC infarction patients tend to be younger and more often female, suffering pain and sensory disturbances commonly located at the level of the spinal cord lesion. Further, symptoms often develop slower (within several days) than in cerebral ischemic stroke [8].

In current clinical practice, cerebral and spinal cord magnetic resonance imaging (MRI) is the gold standard imaging modality in acute myelopathies allowing successfully visualize macrostructural abnormalities like edema, lesions, and hemorrhage (*see* Chapter 13) [9–11]. However, despite clinical evidence of myelopathy, a number of subacute and subtle spinal pathologies are more difficult to visualize due to delay between symptom manifestation and ischemic lesions displaying on MRI [9, 12, 13].

In recent years, the role of ionotropic glutamate neuroreceptor subtypes in CBF regulation in distinct brain territories has become clear (*see* Chapter 11). It has been suggested that cerebral and spinal cord infarction might be assessed using antibodies to immune active fragments of glutamate receptor biomarkers according to gradual presentation of symptoms and structural alterations defined by neuroimaging in certain brain and spinal cord structures [14, 15].

The present chapter provides minireview of potential markers and results in detecting neurovascular biomarkers—antibodies to NR2 subtype of N-methyl-D-aspartate receptors (NMDAR) for comparable assessments of chronic cerebral and SC infarction.

2   Current Biomarkers of Acute and Chronic Cerebral and Spinal Cord Infarctions

Recent progress in molecular immunology shows a tendency for immune active biomarkers being used for diagnosis and prediction of cerebral infarctions [16–18]. It particularly concerns perioperative risks of ischemic stroke [19].

Although the exact pathophysiologic mechanisms that underlie spinal cord ischemia-reperfusion injury remain to be defined, inflammatory response is known to play an important role after ischemia and contributes to spinal cord injury [20, 21].

Another source of markers for acute cerebral and SC infarction might be magnetic resonance spectroscopy (MRS), a noninvasive tool for detection of in vivo metabolic reactions from nervous tissue that is sometimes referred as "virtual biopsy" [22].

2.1 Hematological Indicators of Acute Stroke

Short periods of ischemia and reperfusion (1–3 h) are characterized by mild edema and increased vascular permeability without micro-vascular injury [23]. Additionally, early ischemia activates an innate immune system [17] that further affects metabolic alterations in the central nervous system (CNS) [24].

Fast immune reactions in pre- and post-surgical periods are correlated with significantly higher levels of complement C3 and matrix metalloproteinase-9, and lower levels of interferon-γ and macrophage inflammatory protein-1β levels in cerebral infarction cases compared to controls [18]. In a study of the cerebrospinal fluid of stroke patients, increased levels of TNF, IL-1 and IL-6 were strongly associated with development of ischemic brain damage [25] and are considered as putative markers of stroke severity and neurologic outcome.

Acute ischemic stroke prompts a significant elevation of lactate (Lac) detected by proton MRS. The latter combined with a broad lipid resonance signal is associated with cell death [26, 27]. A weak Lac signal is detected in ischemic penumbra and hypoxic conditions depicting the shift to anaerobic glycolysis [24]. Contrary to Lac, N-acetyl aspartate (NAA) signal shows a tendency to a gradual reduction over several hours after the onset that might be due to activation of enzymatic reactions digesting NAA or to alterations of glutamate (Glu) and gamma-aminobutyric acid (GABA) balance [28]. It is also demonstrated that total choline (Cho) resonance in acute and chronic ischemia can be increasing or decreasing as a consequence of ischemic damage to myelin, edema, or cell loss. Initial reduction in creatine/phosphocreatine occurs immediately after the onset and is maintained for 10 days.

The genetical diagnosis of cerebrovascular infarctions is based on whole blood detection of sequence variants in the *NOTCH3* gene in patients with CADASIL (cerebral autosomal dominant arteriopathy with subcortical infarcts and leukoencephalopathy). In CARASIL (cerebral autosomal recessive arteriopathy with sub-cortical infarcts and leukoencephalopathy) patients that are characterized by gait and balance disturbance, lacunar stroke, mood changes, dementia, and alopecia, the HTRA1 sequencing test assesses sequence variants in the *HTRA1* gene (www.athenadiagnostics.com).

Plasma cardiac N-terminal prohormone of brain natriuretic peptide (NT-pro-BNP) level appears to be a useful biological marker for predicting in-hospital mortality rate for acute ischemic stroke [29]. The optimal cutoff levels for the NT-pro-BNP of 1583.50 pg/mL and a National Institute of Health Stroke Scale (NIHSS) score of 12.5 were established [30].

The biochemical marker D-dimer, a breakdown product of a cross-linked fibrin blood clot that indicates occurrence of plasmin-mediated lysis of cross-linked fibrin, has been evaluated for acute ischemic stroke [31]. The levels of D-dimer and NT-pro-BNP were positively correlated with NIHSS scores in acute stroke ($p < 0.05$). A regression analysis showed that the positive predictive values of D-dimer and NT-pro BNP were 3.65 and 6.96, respectively [31].

The quantitative determination of lipoprotein-associated phospholipase A_2 ($Lp-PLA_2$), which predicts risk of ischemic stroke associated with atherosclerosis, established area under the curve (AUC) assay performance of 0.732. The later combined with measurement of C-reactive protein (hs-CRP) slightly improved the AUC to 0.743 [32].

The other thrombotic marker homocysteine, the sulfinic analog of aspartate, has been associated with increased risk of first stroke among subjects with methylenetetrahydrofolate reductase C677T polymorphism with hazard ratio of 3.1 (1.1–9.2) [33]. The homocysteine abnormalities have been found in 20–40% of persons presenting with premature peripheral vascular diseases or stroke [34, 35].

The simultaneous assessment of four blood markers (BNP, D-dimer, matrix metallopeptidase 9 (MMP-9), and S100B protein) in a cohort of 585 patients with stroke vs. 361 mimics yielded 86% sensitivity and 37% specificity [36]. Samples were acquired from plasma within 24 h of stroke onset and evaluated by ELISA.

The blood ubiquitin C-terminal hydrolase-L1 (UCH-L1) and glial fibrillary acidic protein (GFAP) are increased early after stroke and distinct biomarker-specific release profiles are associated with stroke characteristics and type. The potential of GFAP as a tool for early rule-in of ICH was confirmed, while UCH-L1 was not clinically useful [37].

NR2 antibodies (Ab) levels proved to be beneficial for patients who underwent surgical revascularization with cardiopulmonary bypass and known to be at increased risk of stroke preoperatively and postoperatively [38, 39]. Additionally, increased NR2Ab levels in blood (> 1.8 ng/mL) were independently associated with aortic stiffness causing microvascular damage (odds ratio of 7.23) to cerebral cells in patients with severe aortic stenosis assigned for aortic valve replacement [40].

2.2 Proposed Biomarkers of Acute Spinal Cord Infarction

Unlike cerebral MRS, the application of spectroscopy in the spinal cord faces several challenges regarding signal quality and resolution [41]. These challenges are related to: (1) anatomic heterogeneity of surrounding tissues (meninges, cerebrospinal fluid, bone, and muscles) that induces strong susceptibility changes and distortion of spectral shape, (2) small diameter of the spinal cord (about 1 cm in the cervical area) restricts voxel size, (3) elongated shape in the craniocaudal direction limits attainable signal-to-noise ratio from the available tissue, (4) distant application of coils due to spinal cord being located in a canal within bony vertebrae. Moreover, dynamic changes of static field due to cardiac, respiratory and subject motion yields a less reproducible spectrum. Furthermore, the quality and the metabolite profile depend on the level of acquisition—the peak size of NAA, creatine (Cr), and Cho, as well as the amount of lipids and macromolecules contributing to the spectrum change dramatically in different areas including the pons, the medulla oblongata, the cervical and lumbar region of the spinal cord [42].

The total Cho/Cr and Cho/NAA ratios are increased in patients with cervical spondylosis and T2-hyperintensity compared with healthy controls (ANOVA, $p < 0.01$). At the same time, in patients with stenosis without T2 hyperintensity slightly increased levels of glutamate–glutamine complex and myoinositol glial cell marker are found. A linear correlation between Cho-NAA ratio and modified Japanese Orthopaedic Association (mJOA) score was also observed ($p < 0.01$). Additionally, the spinal canal space was significantly different between patients and controls ($p < 0.0001$).

A significantly elevated Cho/Cr was shown in patients with compressive myelopathy and T2 signal abnormalities. In patients with missing T2-hyperintensity, slightly elevated Cho/Cr was detected. All patients showed increased Cho/NAA indicating destructive processes represented by ongoing axonal loss, metabolic dysfunction, and increased membrane turnover [43]. In cervical spondylosis with cervical myelopathy (CSM), reduced NAA/Cr but no differences in Cho/Cr have been observed [44, 45]. In another study, the combination of diffusion MRI and MRS of the spinal cord has been suggested as a promising approach to predict neurologic impairment in patients with CSM [46]. Further studies are required to assemble a body of data to better understand the role of ischemia in the pathogenesis of myelopathies.

Immune system cascade in acute SC infarction often related to cytokines such as interleukin (IL)-1β, tumor necrosis factor (TNF)-α, and IL-10 are important mediators of the inflammatory reactions in spinal cord ischemia [47]. CSF analysis in a case report study of anterior spinal artery (ASA) infarct showed elevated antineutrophil cytoplasmic antibodies (cANCA) titers (28.2 U/l, normal range less than 15 U/l) suggesting the possibility of immune-mediated disease [48].

The advancement of MRI methods implying spinal cord tractography capturing longitudinal changes combined with magnetic resonance spectroscopy detecting metabolic alterations could provide prospective biomarkers of SCi progression [49].

In a study, six cases of cervical spine trauma with cord damage and pronounced neurological deficit showed significantly increased concentrations of both GFAP and neurofilament triplet protein (NFL) in the CSF [50]. Patients with tetrapareses showed higher values than those with incomplete injuries. Three of the 17 cases had increased levels of NFL and normal GFAP indicating neural damage within cases with neurological deficit [50].

Biochemical markers in cerebrospinal fluid (CSF) were investigated to identify patients with SC ischemia during thoracoabdominal aortic repair and/or a vulnerable spinal cord during the postoperative period [51]. During the post-operative period patients with spinal cord symptoms had significant increases of CSF biomarkers GFAP (571-fold), NFL (14-fold) and S100B (18-fold) compared to asymptomatic patients. GFAP increased before or in parallel with the onset of symptoms in the patients with delayed paraplegia. These biomarkers may indicate transient or permanent neurological deficit due to systemic injuries.

It was reported that combined detection of NR2 Ab and GFAP can differentiate ischemic vs hemorrhagic stroke at 12 h after onset (sensitivity of 94%, specificity of 91%) [52].

The level of GFAP autoantibodies (aAb) measured at 16 ± 7 days post-SCi was found to be significantly higher in patients that subsequently developed neuropathic pain (within 6 months post-SCi) than patients who did not ($T = 219$; $p = 0.02$) [53]. Collapsin response mediator protein-2 (CRMP2) autoantibody target has been identified in 23% of acute SCi patients as well. The presence of GFAP aAb and/or CRMP2 aAb increased the risk of subsequently developing neuropathic pain within 6 months of injury by 9.5 times ($p = 0.006$). This study data suggest that a link could be established between investigated antibodies and the development of neuropathic pain.

It is also proposed that microRNAs (miRs) are emerging candidates as diagnostic biomarkers for SCi. Indeed, miR-21 could be important link in hypertrophy-to-hyperplasia alterations of astrogliosis [54].

3   Prospective Biomarkers for Chronic Cerebral and Spinal Cord Infarctions

At present few studies are focused on specific biomarkers for acute conditions of SC infarction. Most of them are related to neurological adverse events after thoracic surgeries. The prediction of neurological adverse events is another area that continues to evolve. The translational research of biomarkers should be concentrated on near term prediction of conceivable ischemic or hemorrhagic outcomes. Biomarkers would have useful application in triaging patients that are at greater risk for near term infarction. It is an unmet medical need in search of markers that are capable of improving diagnostic certainty of cerebral and SC infarction for personalized treatment to avoid severe consequences.

3.1   Risk Assessment of Cerebral Infarctions

Prolonged ischemic condition causes inflammation and lesions exhibiting severely damaged vasculature and hemorrhagic transformations [23]. The severity of inflammation is related to humoral (adaptive) immune reactions when continuous overload of antigen (s) due to preexisting conditions [55, 56] might evoke generating specific antibodies [14, 38, 39, 57, 58]. The normal function of these antibodies is binding and removing excessive amounts of antigens in non-active immune complexes that are extracted from the organism [18]. Humoral or adaptive immune reactions should not be mistaken for autoimmune responses when the prolonged circulation of immune complexes in biological fluids combined with compromised blood brain barrier could affect immunocompetent cells and cause autoimmune responses or immune-mediated diseases [59, 60]. The latter could be associated with delayed consequences leading to autism spectrum disorders [61], cognitive deficit [60, 62, 63], vascular dementia, Alzheimer's disease [64], and epilepsy [65].

A number of antigens enter the biological fluids coinciding with cellular immune response after ischemic stroke (IS) [18]. However, there are few immune active antigens found to produce antibodies in humoral immune response to ischemic stroke [14, 66].

Over the past two decades, the pathogenicity of other thrombotic markers, such as antiphospholipid (aPL) protein antibodies, have been investigated [67]. The role of aPL has been recognized as a potentially important marker and/or cause of increased vascular risk associated with ischemic stroke. As part of the structural components of excitatory membranes containing glutamate receptors, aPL may also be involved in the neurotoxicity process.

It is necessary to note that arterial hypertension (HT) and other vascular preexisting conditions (retinopathy, overweight/obesity, diabetes mellitus, and dyslipidemia) might sensitize adaptive immune system to generate antibodies along with asymptomatic brain abnormalities increasing the occurrence of stroke and cognitive decline [56].

NR2 antibodies might be a predictive factor for transient ischemic attack (TIA) with sensitivity of 98% at a cutoff value of 2.0 ng/mL with risk ratio of 33 [14, 68]. If a patient with prior IS has a steadily increasing blood NR2 Ab that is maintained beyond the cutoff value of 2.0 ng/mL, then this patient has a risk of recurrent onset and should be directed for neuroprotection [14]. However, concentrations of NR2 Ab below 1.0 ng/mL within 5-day serial blood samples drawn from a patient who suffered acute IS with lesion volume more than 200 mL, it could lead to fatal outcome [69] possibly due to overload of adaptive immune system.

3.2 Comparative Study of NR2 Antibodies in Recurrent Ischemic Stroke and Spinal Cord Infarction

Earlier it was demonstrated that NR2 antibodies were present in patients with prior and multiple recurrent ischemic stroke and persisted for a substantial period [14]. It may be that SC infarction could be assessed using the same neurovascular biomarker of grey matter [15, 70]. This prospective study is focused on comparative assessment of humoral immune responses to NR2 subunit of NMDA receptors in patients with IS, SC infarctions, and post-traumatic myelopathy (PTM).

Thirty patients at the age of 56.6 ± 1.5 years with multiple recent ischemic stroke with mean NIHSS scores of 9–10. The vast majority of patients with cerebral infarction ($n = 29$) were examined within 3 weeks after the onset. Only one patient of this study group was examined in the subacute stage of the IS. T2-weighed MRI images from some patients of this group are shown in Fig. 1.

A total of 119 subjects were included in the study with demographic data summarized in Table 1. Patients with SC infarction ($n = 27$, age of 56.8 ± 3.3) of non-traumatic origin comprised subgroup Ia. Patients with post-traumatic myelopathy (PTM) composed subgroup Ib ($n = 25$, age of 46.7 ± 2.1). The average duration of myelopathy within 15–26 months is estimated.

Comparison of the SC subgroups revealed no significant difference in neurological manifestations between subgroups Ia and Ib ($p > 0.05$). Most patients in both groups (90%) are characterized by central lower paraparesis or paraplegia, with a gradient of motor

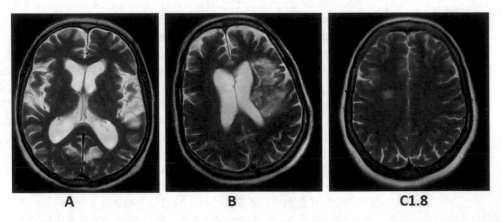

Fig. 1 T2-weighed MRI for: (**a**) a 83-year-old female, multiple lesions in white matter seen on MRI, (**b**) a 76-year-old male with recurrent ischemic stroke and lesions with hemorrhagic transformation in left frontal parietal area, (**c**) a 55-year-old female with chronic ischemic stroke in right middle cerebral artery area

Table 1
Enrolled subject demographic characteristics

Group	Age, mean $\pm$ SD	Male N	Male %	Female N	Female %
SC infarction	56.8 ± 3.3	17	63.0	10	37.0
Posttraumatic myelopathy (PTM)	46.7 ± 2.1	15	60.0	10	40.0
Radiculopathy (Rad)	57.0 ± 1.7	18	56.3	14	43.7
Ischemic stroke (IS)	56.6 ± 1.5	17	56.7	13	43.3
Healthy subjects	50.2 ± 3.6	8	53.3	7	46.7

and sensory loss involving the lower extremities more than the upper extremities. The neurological status of patients was significantly more severe in patients with PTM than in patients with SC infarction ($p < 0.05$). The course of myelopathy in the main subgroups is shown in Fig. 2.

Among the main causes of SC infarction (Ia subgroup), herniated intervertebral discs in 12 (44.4%) cases, arteriovenous malformations in 6 (22.2%) patients, and thrombotic coagulopathies in 3 (11.1%) cases were diagnosed. In 6 (22.2%) patients, the cause of SC infarction could not be established at the time of the study. The isolated lesion of cervical segments in 10 (37%), thoracic segments in 6 (22.2%), and lumbosacral in 2 (7.4%) were defined by MRI. There were observed lesions spread within cervical/thoracic areas ($n = 3$, 11.1%) and thoracolumbar region ($n = 6$, 22.2%).

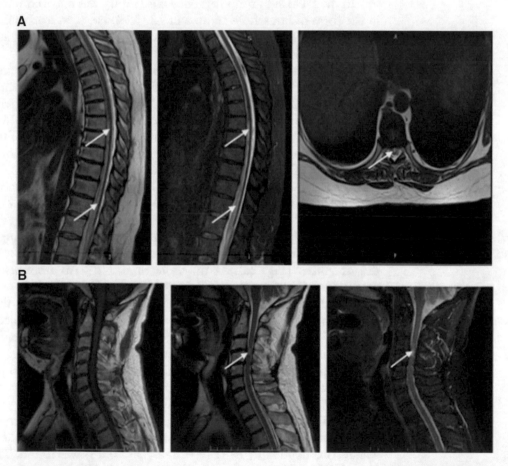

Fig. 2 Spinal cord MRI sequences (left-to-right): (**a**) thoracic spine area: T2 (1, 3) and STIR (2) of a 47-year-old female with spinal stroke in the ASA territory and Factor V Leiden thrombophilia (arrows indicate prolong hyperintensities on level of T7–L1). (**b**) cervical spine area T1 (1), T2 (2), and STIR (3) of a 56-year-old male patient with ischemic myelopathy due to protrusion on the C3–C4 (arrow indicates hyperintensity of lesion and edema on level C4)

Table 2
Lesion volumes in patients with SC infarction and PTM defined by MRI

Lesion volume, mL	SC infarction ($n = 27$)		PTM ($n = 20^a$)	
	N	%	N	%
<1.0	5	18.5	2	10
1–3	6	22.2	4	20
3–5	4	14.8	6	30
6–8	7	25.9	6	30
>8	5	18.5	2	10

[a]Number of patients with brain and spinal MRI

In the PTM subgroup Ib, spinal cord injury was more common in the form of household trauma in 13 (52%) cases, motor vehicle accidents in 7 (28%) cases, acts of violence in 3 (12%) patients, and sport-related injuries in 2 (8%) cases.

Spinal cord injuries by compression (66%), compression/fractures of vertebral bodies (13%), and dislocations and fracture-dislocations (21%) were characterized. In 8 (32%) patients with PTM, cervical and spinal cord injuries were diagnosed, 7 (28%) had thoracic and 5 (20%) lumbosacral injuries. The injuries of the thoracic and lumbosacral areas are diagnosed in 5 (20%) patients. The analysis of MRI scans showed that in both main subgroups, T2 and short tau inversion recovery (STIR) sequence hyperintensity foci along the spinal cord length of 60–80 mm (Table 2, Fig. 2).

3.2.2 NR2 Antibody Results

The NR2 Ab values in groups of healthy subjects (range, 0.5–2.22 ng/mL) and radiculopathy (range, 0.57–2.58 ng/mL) did not differ (Fig. 2a). At the same time, NR2 Ab amounts in serum from IS patient (range, 0.61–9.37 ng/mL) and patients with SC infarction/PTM were significantly increased compared to that for patients with radiculopathy ($p = 0.0001$) and healthy subjects ($p = 0.013$). There was no difference in NR2 Ab levels between IS patients and those with SC infarction (Table 3, Fig. 2a).

A comparison of NR2 Ab distributions shown in Fig. 2b for patients with PTM (range, 0.27–3.50 ng/mL) and SC infarction (range, 0.84–8.12 ng/mL) showed a significant difference ($p = 0.008$). In SC infarction group, a positive moderate correlation between the concentration of NR2 Ab and the size of lesions in spinal cord defined by MRI ($r = 0.49$, $p < 0.05$) was found. The latter might indicate an increased permeability of the vessels located in the area of SC ischemic injury and release into the bloodstream (Fig. 3).

In addition, a negative moderate correlation of NR2 Ab levels with the duration of the disease ($r = -0.45$, $p < 0.05$) was found in

Table 3
Evaluation of NR2 antibodies in serum of patients

| Group | N | Gender | | NR2 Ab ng/mL | | | |
		M	F	Median [Q25; Q75]	Average	Maximum	p^a
SC infarction	27	17	10	1.92 [1.60; 2.45]	2.30	8.12	–
PTM	25	15	10	1.51 [1.29; 1.98]	1.63	3.50	0.45
IS	30	17	13	1.46 [0.99; 2.42]	1.96	9.37	0.18
Rad	32	18	14	1.25 [0.87; 1.59]	1.29	2.58	0.0001
Healthy subjects	15	8	7	1.30 [1.01; 1.53]	1.31	2.22	0.013

^{a}Kruskal–Wallis criteria in comparison with group SC infarction

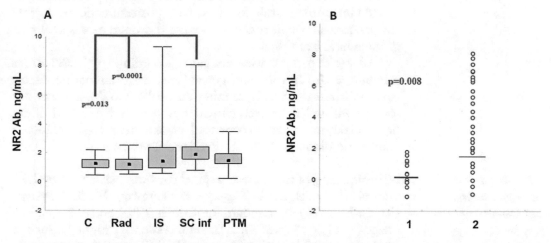

Fig. 3 NR2 Ab levels in serum of subject (**a**) healthy (C), radiculopathy (Rad), IS, SC infarction, and post-traumatic myelopathy (PTM), Kruskal–Wallis and post hoc tests, Mann–Whitney U test, median [25; 75], (**b**) patients with traumatic myelopathy (1) and SC infarction (2), Mann Whitney U test

a subset of patients with SC infarction, attesting a NR2 Ab longevity in these patients and their gradual degradation in the peripheral bloodstream.

4   Methods

This was a blinded, prospective study implying a single blood draw on patients with suspected spinal cord infarction on admission.

4.1   Patients Enrollment

Adults presenting with clinical symptoms of vascular myelopathy ($n = 40$, >50 years) admitted within 2015–2017 to Neurology Clinic at Pavlov First St. Petersburg State Med Univ (St. Petersburg, Russia) and Burdenko Natl Sci and Practical Centre for Neurosurgery (Moscow, Russia) were included in the study.

Patients were excluded if they had a history of recent surgeries (within 12 months), cerebral infarction, cardiovascular disorders, pulmonary embolism, brain and SC tumors, acute and chronic infection diseases (TB, AIDS), pregnancy, used neuroprotective medication (within 6 months before recruitment), did not sign informed consent form, and had contraindications for MRI. A total of 27 patients with SC infarction are assigned to subgroup Ia. Patients with post-traumatic myelopathy (PTM) comprised subgroup Ib ($n = 25$). A group of healthy persons ($n = 15$), patients with radiculopathy ($n = 32$), and acute ischemic stroke ($n = 30$) comprised the comparison group.

4.2   Procedures

4.2.1   Clinical Evaluation

All patients had a standard clinical neurological evaluation. The muscular strength by Medical Research Council Scale (MRS) for patients with SC infarction and American Spinal Injury Association (ASIA) impairment scale for those with posttraumatic myelopathy (http://asia-spinalinjury.org) to assess the degree of spinal cord damage have been used.

All stroke patients were evaluated according to NIHSS scores, standard stroke history, and general medical examination. Stroke patients underwent CT scanning and 90% had MRI scans performed within 72 h. Each patient was seen and followed during hospitalization by an experienced stroke neurologist. Hospital course and discharge examination data were also noted.

4.2.2   Magnetic Resonance Imaging and Analysis

Neuroimaging of the spine and spinal cord was performed by 1.5 T sagittal and axial MRI (Signa, GE Hangweti Medical Systems Co. Ltd., USA) in standard modalities (T1- and T2-weighted images [WI]). "Water and fat suppression" programs such as fluid-attenuated inversion recovery (FLAIR) and short tau inversion recovery (STIR), capable of visualizing an increased amount of extracellular fluid are used in the study as well. Physiological curves of the spine, the height of intervertebral discs and intensity of the MR signal, the presence of protrusions or prolapses, the lumen of the spinal canal, the size and the presence of the spinal cord lesions in the form of modified MR signal have been evaluated.

Intraspinal hemorrhages are better visualized by the T2-gradient-echo (T2 GRE) sequence. Spinal cord edema is seen as a hyperintense signal on T2-WI against the background of normal nervous tissue. Additionally, edema is better visible in STIR images. T1-WI provides information about the spinal cord anatomical structure involved and usually performed at the first stage of MRI study. Anterior spinal artery infarctions in T2-WI sagittal images are usually represented by a "pencil-like" hyperintense lesion in the central area of more than two segments. A bilateral hyperintense signal in the anterior horns as "snake eyes" or "owl eye" patterns on axial T2-WI could be distinguished due to gray matter higher susceptibility to the ischemia [12]. The "flow void" phenomenon is revealed for patients with spinal dura arteriovenous fistula (SDAVF) due to detecting dilated serpentine perimedullary veins. It is seen as pointed hypointensive formations in sagittal section in T2-WI. Such formations are mainly located on the dorsal surface of the spinal cord. Spinal angiography (SA) remains the definitive method of diagnosing spinal dural arteriovenous fistulas (SDAVF), since it allows detecting the exact location of fistulas and was performed in two patients. It is necessary to note that SA modality remains an invasive and a potentially dangerous method with selective catheterization of spinal cord arteries. Therefore, spinal MR-angiography, which is fast and more economical, is considered more acceptable [11]. Spinal cord lesions were measured in the sagittal and axial views. Before the analysis, the MR data was checked for artifacts. Spine CT scanning with CT-myelography was performed in patients with traumatic myelopathy who underwent stabilizing surgical treatment of the spine with the installation of metal fixators before the study. Patients with acute brain ischemia underwent a brain CT scan on a Bright Speed Excel scanner (GE Hangwei Medical Systems Co. Ltd., USA) according to stroke protocol. The structure of cerebral hemispheres, subcortical structures, the brainstem, the presence and the size of hypodense zones in the brain, the midline shift, subarachnoid space, and the brain ventricular system were analyzed.

4.2.3 NR2 Antibodies Detections

A researcher blinded to the clinical and neuroimaging data performed the assays. An ELISA (CIS-test; CIS Biotech, Inc.) was used for detection of NR2 Ab in the serum samples, according to the manufacturer's instruction manual. Briefly, diluted blood sera (1:50) and set of calibrators/standards are added to NR2 peptide-coated wells of microplates (MTP) and incubated the plates for 30 min at 37 °C. After the wells are washed with buffer, horseradish peroxidase-labeled Protein A-HRP is added to the wells and MTP is incubated for another 30 min. The reaction is revealed by o-phenylenediamine (detection solution) after additional washing. The color reaction is developed for 10 min at room temperature, addition of Stop reagent to each well, and monitored at dual wave 450/630 nm on a microplate reader. The NR2 Ab concentrations in serum are determined using the calibration curve: absorbance units vs. NR2 Ab concentrations.

4.3 Statistical Methods

Statistical analysis is performed using SAS 9.4 statistical package (https://www.sas.com/en_us/software/sas9.html).   Standard descriptive statistics are calculated using Pearson chi-square statistics (χ^2) and Fisher' exact test in reporting patient characteristics. To compare more than two groups Kruskal–Wallis and post hoc tests are used while two independent variables are assessed by Mann–Whitney U test. The Spearman rank correlation test is employed to analyze the dependence within two variables.

Quantitative results are presented in the form of dotted graphs, charts "boxes and whiskers," and tables. The values are given as median ± mean standard deviation. A two-sided p value <0.05 was considered significant.

5 Discussion

There is a critical need for noninvasive neurovascular biomarkers that could aid in differential assessment of SC infarction. It is hypothesized that NR2 antibodies would be helpful in evaluation of SC lesions as well as cerebral infarctions that might be related to ascending/descending fibers in the brain and spinal cord grey matter. It has been perceived that volume of cortical layers is much more sizable compared to spinal cord grey matter capacity. Therefore, humoral immune response to arterial occlusion in spinal cord is expected somewhat lower than that in cerebral infarction.

Indeed, humoral immune response to NR subtype of NMDA receptor due to SC infarction is sensitive and prolonged during edema and lesion formation defined by MRI. NR2 antibodies were found in patients with chronic IS, SC infarction and post-traumatic myelopathy but not in radiculopathy and controls confirming the primary neurovascular cause of the disease involving gray matter supplied by the central artery. At the same time, NR2Ab amounts in SC infarction and PTM moderately correlated (Fig. 4) with lesion volumes ($r = 0.49$) while in cerebral infarction this correlation was shown to be more significant ($r = 0.79$) [58, 70].

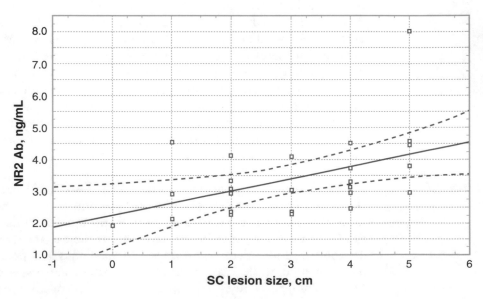

Fig. 4 The correlation curve of NR2 Ab and spinal cord lesion size ($r = 0.49$, Spearman rank correlation test)

Amounts of NR2 antibodies associated with lesions defined by CT/MRI are located in the cervical ($n = 18$), thoracic ($n = 13$), cervical/thoracic ($n = 3$), thoracic/lumbar ($n = 8$) regions of the spinal cord. It is necessary to note that lumbar ($=3$) and lumbosacral ($n = 7$) regions of impairments did not affect NR antibodies concentrations presumably because of descending terminals of nervous fibers surrounded by white matter.

The attempt at measuring S100B antibodies in samples from all study participants did not show significant differences between the groups suggesting low clinical value of the biomarker in differentiating SC impairments from IS and controls [70].

Further optimization of neuroimaging of the SC infarction combined with the implementation of neurovascular biomarkers of humoral immune response could be useful for differential diagnosis of disease, rendering it treatable and potentially preventing unfavorable consequences.

References

1. Fanning JP, Wong AA, Fraser JF (2014) The epidemiology of silent brain infarction: a systematic review of population-based cohorts. BMC Med 12:119. https://doi.org/10.1186/s12916-014-0119-0

2. Nasr DM, Rabinstein A (2017) Spinal cord infarcts: risk factors, management, and prognosis. Curr Treat Options Neurol 19:28

3. Barbon A, Fumagalli F, Caracciolo L, Madaschi L, Lesma E, Mora C, Carelli S, Slotkin TA, Racagni G, Di Giulio AM, Gorio A, Barlati S (2010) Acute spinal cord injury persistently reduces R/G RNA editing of AMPA receptors. J Neurochem 114(2):397–407. https://doi.org/10.1111/j.1471-4159.2010.06767.x

4. Hansen KB, Yi F, Perszyk RE, Menniti FS, Traynelis SF (2017) NMDA receptors in the central nervous system. Methods Mol Biol 1677:1–80. https://doi.org/10.1007/978-1-4939-7321-71

5. Robertson CE, Brown RD Jr, Wijdicks EF, Rabinstein AA (2012) Recovery after spinal cord infarcts: long-term outcome in 115 patients. Neurology 78(2):114–121. https://doi.org/10.1212/WNL.0b013e31823efc93

6. Rubin MN, Rabinstein AA (2013) Vascular diseases of the spinal cord. Neurol Clin 31(1):153–181. https://doi.org/10.1016/j.ncl.2012.09.004

7. Heldner MR, Arnold M, Nedeltchev K, Gralla J, Beck J, Fischer U (2012) Vascular diseases of the spinal cord: a review. Curr Treat Options Neurol 14(6):509–520. https://doi.org/10.1007/s11940-012-0190-9

8. Nedeltchev K, Loher TJ, Stepper F, Arnold M, Schroth G, Mattle HP, Sturzenegger M (2004) Long-term outcome of acute spinal cord ischemia syndrome. Stroke 35(2):560–565. https://doi.org/10.1161/01.STR.0000111598.78198.EC

9. Anderson SE, Boesch C, Zimmermann H, Busato A, Hodler J, Bingisser R, Ulbrich EJ, Nidecker A, Buitrago-Téllez CH, Bonel HM, Heini P, Schaeren S, Sturzenegger M (2012) Are there cervical spine findings at MR imaging that are specific to acute symptomatic whiplash injury? A prospective controlled study with four experienced blinded readers. Radiology 262(2):567–575. https://doi.org/10.1148/radiol.11102115

10. Vargas MI, Gariani J, Sztajzel R, Barnaure-Nachbar I, Delattre BM, Lovblad KO, Dietemann JL (2015) Spinal cord ischemia: practical imaging tips, pearls, and pitfalls. AJNR Am J Neuroradiol 36(5):825–830. https://doi.org/10.3174/ajnr.A4118

11. Lien C-Y, Lui CC, Lu CH, Chang WN (2014) Management of a case with misdiagnosed spinal dural arterio-venous fistula. Acta Neurol Taiwanica 23(1):29–35. PMID: 24833213

12. Weidauer S, Nichtweiss M, Hattingen E, Berkefeld J (2015) Spinal cord ischemia: aetiology, clinical syndromes and imaging features. Neuroradiology 57(3):241–257

13. Szwedowski D, Walecki J (2014) Spinal cord injury without radiographic abnormality (SCI-WORA) – clinical and radiological aspects. Pol J Radiol 79:461–464. https://doi.org/10.12659/PJR.890944

14. Weissman JD, Khunteev GA, Heath R, Dambinova SA (2011) NR2 antibodies: risk assessment of transient ischemic attack (TIA)/stroke in patients with history of isolated and multiple cerebrovascular events. J Neurol Sci 300(1–2):97–102

15. Ponomarev GV, Dambinova SA, Skoromets AA (2018) Neurotoxicity in spinal cord impairments. In: Peplow PV, Dambinova SA, Gennarelli TA, Martinez B (eds) Acute brain impairment: scientific discoveries and translational research. Royal Society of Chemistry, London, pp 198–213

16. Kamel H, Iadecola C (2012) Brain-immune interactions and ischemic stroke: clinical implications. Arch Neurol 69(5):576–581. https://doi.org/10.1001/archneurol.2011.3590

17. Amantea D, Micieli G, Tassorelli C, Cuartero MI, Ballesteros I, Certo M, Moro MA, Lizasoain I, Bagetta G (2015) Rational modulation of the innate immune system for neuroprotection in ischemic stroke. Front Neurosci 9:147. https://doi.org/10.3389/fnins.2015.00147

18. Rayasam A, Hsu M, Kijak JA, Kissel L, Hernandez G, Sandor M, Fabry Z (2018) Immune responses in stroke: how the immune system contributes to damage and healing after stroke and how this knowledge could be translated to better cures? Immunology 154 (3):363–376. https://doi.org/10.1111/imm.12918

19. Fanning JP, See Hoe LE, Passmore MR, Barnett AG, Rolfe BE, Millar JE, Wesley AJ, Suen J, Fraser JF (2018) Differential immunological profiles herald magnetic resonance imaging-defined perioperative cerebral infarction. Ther Adv Neurol Disord 11:1756286418759493. https://doi.org/10.1177/1756286418759493

20. Lu K, Cho CL, Liang CL, Chen SD, Liliang PC, Wang SY et al (2007) Inhibition of the MEK/ERK pathway reduces microglial activation and interleukin-1-beta expression in spinal cord ischemia/reperfusion injury in rats. J Thorac Cardiovasc Surg 133:934–941

21. Matsumoto S, Matsumoto M, Yamashita A, Ohtake K, Ishida K, Morimoto Y et al (2003) The temporal profile of the reaction of microglia, astrocytes, and macrophages in the delayed onset paraplegia after transient spinal cord ischemia in rabbits. Anesth Analg 96:1777–1784

22. Lin A, Tran T, Bluml S, Merugumala S, Liao HJ, Ross BD (2012) Guidelines for acquiring and reporting clinical neurospectroscopy. Semin Neurol 32:432–453. https://doi.org/10.1055/s-0032-1331814

23. Granger DN, Kvietys PR (2017) Reperfusion therapy – what's with the obstructed, leaky and broken capillaries? Pathophysiology 24 (4):213–228. https://doi.org/10.1016/j.pathophys.2017.09.003

24. Faghihi R, Zeinali-Rafsanjani B, Mosleh-Shirazi M-A, Saeedi-Moghadam M, Lotfi M, Jalli R, Iravani V (2017) Magnetic resonance spectroscopy and its clinical applications: a review. J Med Imaging Radiation Sci 48:233–253e

25. Lambertsen KL, Biber K, Finsen B (2012) Inflammatory cytokines in experimental and human stroke. JCerebBlood Flow Metab 32:1677–1698. https://doi.org/10.1038/jcbfm.2012.88

26. Ross AJ, Sachdev PS, Wen W et al (2006) Prediction of cognitive decline after stroke using proton magnetic resonance spectroscopy. J Neurol Sci 251(1–2):62–69

27. Lin A-Q, Shou J-X, Li X-Y, Ma L, Zhu X-H (2014) Metabolic changes in acute cerebral infarction: findings from proton magnetic resonance spectroscopic imaging. Exp Ther Med 7(2):451–455

28. Santhakumari R, Reddy IY, Archana R (2014) Effect of type 2 diabetes mellitus on brain metabolites by using proton magnetic resonance spectroscopy-A systematic review. Int J Pharm Bio Sci 5(4):1118

29. Naveen V, Vengamma B, Mohan A, Vanajakshamma V (2015) N-Terminal pro-brain natriuretic peptide levels and short-term prognosis in acute ischemic stroke. Ann Indian Acad Neurol 18(4):435–440. https://doi.org/10.4103/0972-2327.165478

30. Chen X, Zhan X, Chen M, Lei H, Wang Y, Wei D, Jiang X (2012) The prognostic value of combined NT-pro-BNP levels and National Institutes of Health Stroke Scale (NIHSS) scores in patients with acute ischemic stroke. Intern Med 51(20):2887–2892

31. Li J, Gu C, Li D, Chen L, Lu Z, Zhu L, Huang H (2018) Effects of serum N-terminal pro B-type natriuretic peptide and D-dimerlevels on patients with acute ischemic stroke. Pak J Med Sci 34(4):994–998. https://doi.org/10.12669/pjms.344.15432

32. Nambi V, Hoogeveen RC, Chambless L, Hu Y, Bang H, Coresh J, Ni H, Boerwinkle E, Mosley T, Sharrett R, Folsom AR, Ballantyne CM (2009) Lipoprotein-associated

phospholipase A2 and high-sensitivity C-reactive protein improve the stratification of ischemic stroke risk in the Atherosclerosis Risk in Communities (ARIC) study. Stroke 40 (2):376–381. https://doi.org/10.1161/STROKEAHA.107.513259

33. Zhao M, Wang X, He M, Qin X, Tang G, Huo Y, Li J, Fu J, Huang X, Cheng X, Wang B, Hou FF, Sun N, Cai Y (2017) Homocysteine and stroke risk: Modifying effect of methylenetetrahydrofolate reductase C677T polymorphism and folic acid intervention. Stroke 48(5):1183–1190. https://doi.org/10.1161/STROKEAHA.116.015324

34. Kaplan E (2003) Association between homocyst(e)ine levels and risk of vascular events. Drugs Today 39:175–192

35. Roach ES, Bettermann K, Jose Biller J (2010) Toole's cerebrovascular disorders, 6th edn. Cambridge University Press, Cambridge, p 422

36. Laskowitz DT, Kasner SE, Saver J, Remmel KS, Jauch EC (2009) BRAIN Study Group. Clinical usefulness of a biomarker-based diagnostic test for acute stroke: The Biomarker Rapid Assessment in Ischemic Injury (BRAIN) study. Stroke 40:77–85. https://doi.org/10.1161/STROKEAHA.108.516377

37. Ren C, Kobeissy F, Alawieh A, Li N, Li N, Zibara K, Zoltewicz S, Guingab-Cagmat J, Larner SF, Ding Y, Hayes RL, Ji X, Mondello S (2016) Assessment of serum UCH-L1 and GFAP in acute stroke patients. Sci Rep 14 (6):24588. https://doi.org/10.1038/srep24588

38. Bokesch PM, Izykenova GA, Justice JB, Easley KA, Dambinova SA (2006) NMDA receptor antibodies predict adverse neurological outcome after cardiac surgery in high-risk patients. Stroke 37:1432–1436

39. Skitek M, Jerin A (2013) N-methyl-D-aspartate–receptor antibodies, S100B protein, and neuron-specific enolase before and after cardiac surgery: association with ischemic brain injury and erythropoietin prophylaxis. Lab Med 44 (1):56–62

40. Kidher E, Patel VM, Nihoyannopoulos P, Anderson JR, Chukwuemeka A, Francis DP, Ashrafian H, Athanasiou T (2014) Aortic stiffness is related to the ischemic brain injury biomarker N-methyl-D-aspartate receptor antibody levels in aortic valve replacement. Neurol Res Int 2014:970793. https://doi.org/10.1155/2014/970793

41. Hock A, Henning A, Boesiger P, Kollias SS (2013) 1H-MR spectroscopy in the human spinal cord. AJNR Am J Neuroradiol 34:1682–1689. https://doi.org/10.3174/ajnr.A3342

42. Wyss PO, Hock A, Kollias S (2017) The application of human spinal cord magnetic resonance spectroscopy to clinical studies: a review. Semin Ultrasound CT MR 38 (2):153–162. https://doi.org/10.1053/j.sult.2016.07.005

43. Salamon N, Ellingson BM, Nagarajan R, Gebara N, Thomas A, Holly LT (2013) Proton magnetic resonance spectroscopy of human cervical spondylosis at 3 T. Spinal Cord 51:558–563. https://doi.org/10.1038/sc.2013.31

44. Holly LT, Freitas B, McArthur DL, Salamon N (2009) Proton magnetic resonance spectroscopy to evaluate spinal cord axonal injury in cervical spondylotic myelopathy. J Neurosurg Spine 10:194–200. https://doi.org/10.3171/2008

45. Taha Ali TF, Badawy AE (2013) Feasibility of 1H-MR spectroscopy in evaluation of cervical spondylotic myelopathy. Egypt J Radiol Nucl Med 44(1):93–99. https://doi.org/10.1016/j.ejrnm.2012.11.001

46. Ellingson BM, Salamon N, Hardy AJ, Holly LT (2015) Prediction of neurological impairment in cervical spondylotic myelopathy using a combination of diffusion MRI and proton MR spectroscopy. PLoS One 10(10): e0139451. https://doi.org/10.1371/journal.pone.0139451

47. Hasturk A, Atalay B, Calisaneller T, Ozdemir O, Oruckaptan H, Altinors N (2009) Analysis of serum pro-inflammatory cytokine levels after rat spinal cord ischemia/reperfusion injury and correlation with tissue damage. Turk Neurosurg 19:353–359

48. Sivadasan A, Alexander M, Patil AK, Mani S (2013) Spectrum of clinico-radiological findings in spinal cord infarction: report of three cases and review of the literature. Ann Indian Acad Neurol 16(2):190–193

49. Bede P, Finegan E, Hardiman O (2017) From pneumomyelography to cord tractography: historical perspectives on spinal imaging. Future Neurol 12(3):121–124. https://doi.org/10.2217/fnl-2017-0018

50. Guéz M, Hildingsson C, Rosengren L, Karlsson K, Toolanen G (2003) Nervous tissue damage markers in cerebrospinal fluid after cervical spine injuries and whiplash trauma. J Neurotrauma 20(9):853–858

51. Winnerkvist A, Anderson RE, Hansson LO, Rosengren L, Estrera AE, Huynh TT, Porat EE, Safi HJ (2007) Multilevel somatosensory evoked potentials and cerebrospinal proteins:

indicators of spinal cord injury in thoracoabdominal aortic aneurysm surgery. Eur J Cardiothorac Surg 31(4):637–642

52. Stanca DM, Mărginean IC, Soriţău O, Dragoş C, Mărginean M, Mureşanu DF, Vester JC, Rafila A (2015) GFAP and antibodies against NMDA receptor subunit NR2 as biomarkers for acute cerebrovascular diseases. J Cell Mol Med 19(9):2253–2261. https://doi.org/10.1111/jcmm.12614

53. Hergenroeder GW, Redell JB, Choi HA, Schmitt L, Donovan W, Francisco GE, Schmitt K, Moore AN, Dash PK (2018) Increased levels of circulating glial fibrillary acidic protein and collapsin response mediator protein-2 autoantibodies in the acute stage of spinal cord injury predict the subsequent development of neuropathic pain. J Neurotrauma. https://doi.org/10.1089/neu.2018.5675

54. Martirosyan NL, Carotenuto A, Patel AA, Kalani MY, Yagmurlu K, Lemole GM Jr, Preul MC, Theodore N (2016) The role of microRNA markers in the diagnosis, treatment, and outcome prediction of spinal cord injury. Front Surg 8(3):56

55. Voloshyna I, Krivenko V, Voloshyn M, Deynega V (2016) Serum NR2 peptide antibodies and stroke recurrence in high-risk hypertensives. J Hypertens 34(Suppl 1):PS 02–PS 37

56. González-García S, González-Quevedo A, Hernandez-Diaz Z, Alvarez Camino L, Peña-Sanchez M, Cordero-Eiriz A, Brown M, Gaya JA, Betancourt-Losa M, Fernandez-Almirall I, Menendez-Sainz MC, Fernandez-Carriera R (2017) Circulating autoantibodies against the NR2 peptide of the NMDA receptor are associated with subclinical brain damage in hypertensive patients with other pre-existing conditions for vascular risk. J Neurol Sci 375:324–330. https://doi.org/10.1016/j.jns.2017.02.028

57. Dambinova SA, Khounteev GA, Skorometz AA (2002) Multiple panel of biomarkers for TIA/stroke evaluation. Stroke 33 (5):1181–1182

58. Dambinova SA, Khounteev GA, Izykenova GA, Zavolokov IG, Ilyukhina AY, Skoromets AA (2003) Blood test detecting autoantibodies to N-methyl-D-aspartate neuroreceptors for evaluation of patients with transient ischemic attack and stroke. Clin Chem 49 (10):1752–1762

59. Vincent A, Bien CG, Irani SR, Waters P (2011) Autoantibodies associated with diseases of the CNS: new developments and future challenges. Lancet Neurol 10:759–772. https://doi.org/10.1016/S1474-4422(11)70096-5

60. Arvanitakis Z, Brey RL, Rand JH, Schneider JA, Leurgans SE, Yu L, Buchman AS, Arfanakis K, Fleischman DA, Boyle PA, Bennett DA, Levine SR (2013) Antiphospholipid antibodies, brain infarcts, and cognitive and motor decline in aging (ABICMA): design of a community-based, longitudinal, clinical-pathological study. Neuroepidemiology 40 (2):73–84. https://doi.org/10.1159/000342761

61. Hacohen Y, Wright S, Gadian J, Vincent A, Lim M, Wassmer E, Lin JP (2016) N-methyl-d-aspartate (NMDA) receptor antibodies encephalitis mimicking an autistic regression. Dev Med Child Neurol 58:1092–1094. https://doi.org/10.1111/dmcn.13169

62. Di Marco B, Bonaccorso CM, Aloisi E, D'Antoni S, Catania MV (2016) Neuroinflammatory mechanisms in developmental disorders associated with intellectual disability and autism spectrum disorder: a neuroimmune perspective. CNS Neurol Disord Drug Targets 15:448–463. https://doi.org/10.2174/1871527315666160321105039

63. Mehregan H, Najmabadi H, Kahrizi K (2016) Genetic studies in intellectual disability and behavioral impairment. Arch Iran Med 19:363–375. doi:10.0161905/AIM.0012

64. Hoffmann C, Zong S, Mané-Damas M, Molenaar P, Losen M, Martinez-Martinez P (2016) Autoantibodies in neuropsychiatric disorders. Antibodies 5:9. https://doi.org/10.3390/antib5020009

65. Nibber A, Clover L, Pettingill P, Waters P, Elger CE, Bien CG, Vincent A, Lang B (2016) Antibodies to AMPA receptors in Rasmussen's encephalitis. Eur J Paediatr Neurol 20:222–227. https://doi.org/10.1016/j.ejpn.2015.12.011

66. Wang H, Zhang XM, Tomiyoshi G, Nakamura R, Shinmen N, Kuroda H, Kimura R, Mine S, Kamitsukasa I, Wada T, Aotsuka A, Yoshida Y, Kobayashi E, Matsutani T, Iwadate Y, Sugimoto K, Mori M, Uzawa A, Muto M, Kuwabara S, Takemoto M, Kobayashi K, Kawamura H, Ishibashi R, Yokote K, Ohno M, Chen PM, Nishi E, Ono K, Kimura T, Machida T, Takizawa H, Kashiwado K, Shimada H, Ito M, Goto KI, Iwase K, Ashino H, Taira A, Arita E, Takiguchi M, Hiwasa T (2017) Association of serum levels of antibodies against MMP1, CBX1, and CBX5 with transient ischemic attack and cerebral infarction. Oncotarget 9(5):5600–5613. https://doi.org/10.18632/oncotarget.23789

67. Bala MM, Paszek E, Lesniak W, Wloch-Kopec D, Jasinska K, Undas A (2018) Antiplatelet and

anticoagulant agents for primary prevention of thrombosis in individuals with antiphospholipid antibodies. Cochrane Database Syst Rev 7: CD012534. https://doi.org/10.1002/14651858.CD012534.pub2

68. Dambinova SA (2008) Biomarkers for transient ischemic attack (TIA) and ischemic stroke. Clin Lab Int 32:7–11. http://www.clinlabint.com/fileadmin/pdf/digital_issues_archives/CLI_Nov08.pdf

69. Gusev EI, Skvortsova VI (2001) Cerebral ischemia. Meditsina Publishers, Moscow, pp 54–60

70. Ponomarev GV, Lalajan NV, Dambinova SA, Skoromets AA (2018) The neurotoxicity biomarkers as potential indicators of spinal cord ischemia. J Neurol Psychiatry (Russian) 2:28–33. https://doi.org/10.17116/jnevro20181182152-57

Chapter 13

Impaired Retinal Vasoreactivity as an Early Marker of Stroke Risk in Diabetes

Kerstin Bettermann and Kusum Sinha

Abstract

More than 70 million Americans are prediabetic, an early stage in the hyperglycemic continuum which is associated with an increased risk of developing future diabetes and diabetes-associated vascular complications, such as macrovascular complications (coronary artery disease, peripheral arterial disease, and stroke) and microvascular complications (diabetic retinopathy). The retina is a unique site to directly study the human circulation. Retinal blood flow is controlled by autoregulatory metabolic and pressure mechanisms which are impaired in prediabetes and diabetes contributing to retinopathy and vision loss. Understanding the pathophysiologic basis for changed blood vessel responses across the hyperglycemic continuum is important for the discovery of new treatments and preventive strategies during early disease stages.

Retinal vasoreactivity measurements may be a more sensitive non-invasive indicator of early stages of atherosclerosis rather than traditional markers of cardiovascular risk. Prospective studies may determine whether changes in retinal vascular behavior in individuals with prediabetes are a harbinger of future cardiovascular disease or retinopathy. Retinal and cerebral endothelial dysfunctions are already present in prediabetes and early stages of diabetes stressing the importance for early screening and intervention to prevent cerebrovascular disease.

Key words Stroke risk, Diabetes, Retinal imaging, Vessel reactivity, Biomarker

1 Introduction

Stroke is the second leading cause of death and a major cause of disability [1]. In 2016, 5.8 million people died from a stroke, which accounts for about 10% of all deaths globally [2, 3]. One third of stroke victims die within a year, and of those who survive, almost half are no longer able to live independently. Risk factors for stroke include hypertension, atrial fibrillation, hypercholesterolemia and diabetes, among others. Diabetes is an important risk factor for stroke as it causes microvascular and macrovascular disease affecting multiple organ systems including the brain and the eye [4–6].

Diabetic patients are a high risk stroke population as they have a three-fold increased probability of recurrent stroke, a greatly enhanced stroke morbidity, and a three-fold increased stroke

Philip V. Peplow et al. (eds.), *Stroke Biomarkers*, Neuromethods, vol. 147, https://doi.org/10.1007/978-1-4939-9682-7_13,
© Springer Science+Business Media, LLC, part of Springer Nature 2020

mortality compared to euglycemic patients [7, 8]. The global burden of stroke and diabetes is substantial and steadily growing due to the current diabetes epidemic and the aging of the baby boomer generation. The combined effect will result in a dramatic increase in the annual stroke incidence. Therefore, early identification and prevention of cerebrovascular disease are key for this high-risk population. To achieve these goals new low cost, accurate and noninvasive diagnostic tools are needed that allow identification of cerebrovascular disease at subclinical disease stages, and to monitor the efficacy of therapeutic interventions aimed at stroke prevention [9, 10].

Non-invasive imaging of the retinal vasculature may hold promise as a biomarker for stroke risk. Compared to the cerebral microvasculature the retinal circulation can be directly and non-invasively visualized in vivo. Because the retina and brain share embryological, anatomical and physiological similarities, studies of retinal blood vessels may prove to be useful as a surrogate marker for cerebrovascular disease.

The retina is an extension of the diencephalon and possesses a tightly controlled blood–retinal barrier that is analogous to the blood–brain barrier. Like the cerebral vasculature, retinal vessels are important in maintaining adequate tissue perfusion via autoregulation to adapt to rapidly changing metabolic demands.

It is well known that prolonged hyperglycemia can cause microvascular complications, such as retinopathy and lacunar strokes, as well as macrovascular complications such as large vessel strokes. Retinal vascular dysfunction occurs early in diabetes and prediabetes, and contributes to the pathogenesis of retinopathy [9, 11]. High glucose levels cause chronic inflammation and increased oxidative stress which lead to endothelial cell dysfunction, damage to pericytes and astrocytes and advancing atherosclerosis [12]. These changes ultimately cause increased vessel permeability, hyperproliferation and endothelial cell swelling and abnormal vascularization of the retina with resulting vision loss. In previous studies these retinal vessel changes in diabetes paralleled changes in the cerebral vasculature across the diabetic spectrum showing worsening cerebral and retinal autoregulation with disease progression from prediabetes to advanced diabetes [13]. These findings suggest that microvascular changes in the eye parallel the development of cerebrovascular disease, although further confirmatory studies are required.

Epidemiological studies have shown that alterations in the architecture of retinal vessels are associated with higher risk of cardiovascular disease and stroke [6–11]. Wider retinal vein diameters and decreased arteriole-to-venule ratios (AVRs) correlate with overall higher stroke risk, presence of lacunar infarcts and chronic cerebral ischemic matter disease on brain MRI [14, 15]. However, structural changes may not be an ideal way to

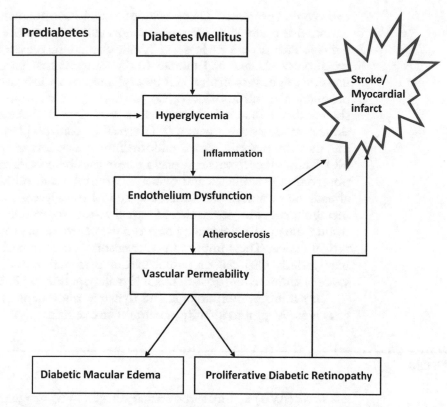

Fig. 1 Relationship between hyperglycemia, retinal, cardiovascular, and cerebrovascular disease. Hyperglycemia causes chronic inflammation, endothelial dysfunction leading to atherosclerosis, and ultimately end organ disease. Dynamic retinal vessel analysis allows detection of early disease stages allowing initiation of early treatment for prevention of vascular disease

assess the efficacy of therapeutic interventions or to identify patients with early changes of the cerebral vasculature, when treatment would be most efficacious, as they detect later stages of vascular disease. Instead measures of endothelial dysfunction which occur at earlier disease stages may be more sensitive to identify patient at stroke risk and to help guide therapeutic interventions and assess their efficacy (*see* Fig. 1).

Dynamic Vascular Assessment (DVA, Imedos Inc., Jena, Germany) of retinal arterioles and venules allows measurement of early endothelial dysfunction in prediabetes and diabetes. The DVA utilizes a flickering light stimulus to induce changes in retinal vessel diameters allowing assessment of retinal vascular function. Flickering light is a well-established metabolic stimulus for the retinal vasculature [2, 12] which causes vasodilation and increased blood flow in healthy individuals [16–18]. In diabetic individuals this retinal vasodilation response is attenuated [14, 18, 19].

Reduced retinal vasodilation in response to flickering light in prediabetes and diabetes may indicate several underlying pathological processes. These include impaired autoregulation and endothelial dysfunction. Vascular abnormalities may cause retinal damage such as pericyte loss which may change the release of local metabolites. Animal and human studies suggest that part of the flickering light vasodilation can be explained by an increase in the production of nitric oxide (NO) [20]. In a recent study, it was shown that retinal vessels in persons with type-1 diabetes have similar responses to exogenous NO as healthy controls [19], implying that the diabetic retinal endothelium is not less sensitive to NO. Thus, other factors may play a role in the altered vasoreactivity observed in prediabetes and diabetes. Arterioles and venules may already be in a maximally dilated state in hyperglycemia to meet metabolic demand, or the altered retinal vasomotor response could result from impaired signaling between the neurosensory retina and retinal vessels. These impaired neurosensory coupling mechanisms may include glial cell or retinal barrier dysfunction and altered vascular endothelium growth factor signaling pathways [7, 21, 22].

In summary, morphologic and dynamic assessment of retinal vessels are helpful to identify patients at stroke risk.

2   Materials

All static (AVR) and dynamic retinal vessel (DVA) imaging is performed using a modified fundus camera (Zeiss FF450, Zeiss Jena, Germany), a recording unit and specialized image analysis software (Imedos Inc., Jena, Germany). The Retinal Vessel Analyzer has been developed for the needs of research institutions and practitioners who do basic research or participate in clinical trials (*see* Fig. 2). The system allows for simple adaptation to almost any kind of clinical hypothesis in the area of vascular research. Special software allows connections to other data sources as well as easy changes to generate different graphical data representation. The system visualizes retinal diameters in real time. Vessel calibers can be analyzed in-time or off-line [23]. Analysis can be performed in a standardized fashion by one trained evaluator using the AVR and DVA software which corrects for any artifacts in the tracings due to spontaneous erroneous measurements.

3   Methods

3.1 Structural Vessel Imaging of the Retinal Arteriole-to-Venule (AVR) Ratios

One challenge of measuring retinal vessels by fundus photography is the calibration of retinal photographs. This is addressed through normalizing artery-to-vein diameters (AVR) or arterial length to diameter ratios (LDR) to obtain dimensionless ratios.

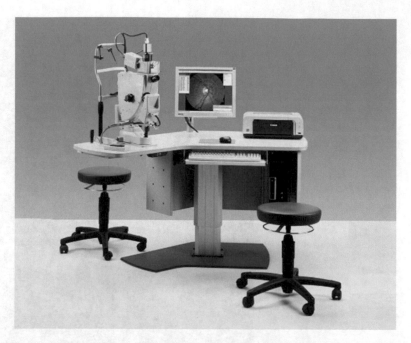

Fig. 2 Modified Zeiss FF450 mydriatic fundus camera (Zeiss Jena, Germany) with video digital high-resolution color CCD camera and PC based imaging software (Imedos Inc., Jena, Germany) allowing static and dynamic retinal vessel measurements and funduscopic photography

3.1.1 Procedure

Before any measurements are taken it is important to talk to the subject about what to expect during the examination. All measurements should be performed in a dark and quiet examination room (*see* **Notes 1–3**). Prior to the funduscopic examination, the pupils are pharmacologically dilated to allow easier and better view of the macula and retinal vessels. Short-acting topical parasympatholytic eye drops are used to paralyze the pupillo-constrictor muscle of the iris. Before obtaining any photographs, the pupils should be maximally dilated to avoid poor image quality.

Eyeglasses, but not contact lenses will be removed. The subject will fixate on a blinking light attached to the fundus camera. The patient fixates only with the eye that is not being examined. The fixation will be adjusted by the examiner so that the optic nerve head is in the center of the fundus monitor. The fundus camera's focus and background light will be adjusted to provide crisp images. All photographs will be taken using a field angle of 50°. Flash light intensity will be adjusted as needed to provide well illuminated images, while avoiding reflection and discomfort of the study subject by overly bright flash intensities. Subjects are encouraged to blink multiple times during the examination (*see* **Notes 4–6**). Multiple images will be taken and stored to obtain optimal photographs for data analysis.

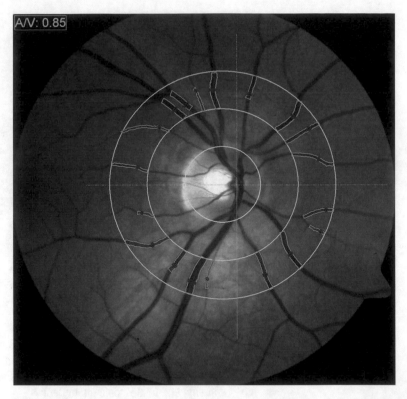

Fig. 3 Morphologic retinal vessel analysis. The *VesselMap software* enables determination of the arteriolar to venular ratio which is a quantitative parameter to determine vascular risk. Concentric rings are positioned over the center of the papilla and measurements are taken within the most outer ring. Arteriolar segments (in red) are marked yielding a sum score of diameters (the arterial vessels artery equivalent, CRAE), and venule segments (marked in blue) yielding the venule vessel equivalent (CRVE). In combination with an individual's medical history and an evaluation of microvascular fundus result, a valuable risk assessment can be made

3.1.2 AVR Measurements   The macula will be centered on the computer screen. Using the Imedos Visualis software, concentric rings will be placed over the macula and measurements of arteriolar and venular diameters will then be obtained within the most outer ring (*see* Fig. 3). All clearly visible retinal vessels will be manually marked and identified as arteriole or venule. Selection criteria for the chosen vessel segments include main vessels (segment diameter > 80 μm), vessels with a clear contrast to fundus background, segments that have no crossings or bifurcations, vessels that have no nearby vessels within one vessel diameter of the chosen segment. The vessel segment diameters are combined automatically by the software and summary indices are derived, including the central retinal arteriolar equivalent (CRAE) and the central retinal venular equivalent (CRVE). From these equivalents, the summary index of all arteriolar and venular diameters will be automatically calculated and expressed as arteriole-to-venule ratio (AVR) taking into consideration vascular branching patterns [24–27]. AVRs are calculated separately for each eye.

Interpretation of measurements: Smaller CRAEs, larger CRVES, and lower AVRs are associated with chronic cerebrovascular and cardiovascular disease [24, 28–30]. The implemented AVR software also automatically generates stroke risk scores based on the Atherosclerosis Risk in Community (ARIC) study [31].

Validation: Reproducibility coefficients for static AVR measurements have been previously reported to be between 0.78 and 0.99 [26].

3.2 Dynamic Retinal Vessel Analysis

Using the same fundus camera and dynamic vessel analysis software (DVA, Imedos Inc.), vascular reactivity of retinal arterioles and venules can be assessed over time following exposure to flickering light [9, 10, 32]. Flickering light provides a strong physiological stimulus to the retina causing vasodilation of arterioles and venules. The DVA module analyzes the brightness profile of retinal blood vessels using two optical pathways, and light reflected by the retina back to the imaging unit. Changes in vasoreactivity can be measured in real-time, and the software automatically generates a detailed time-space profile of changes in vessel diameters of arterioles and venules (*see* Fig. 4).

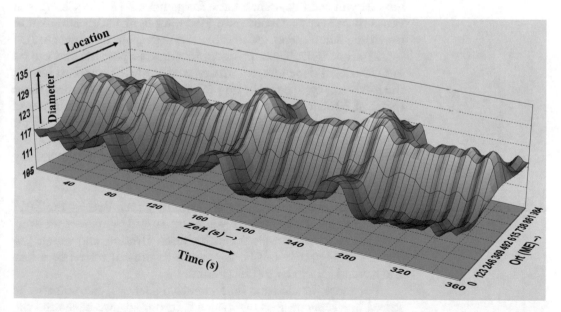

Fig. 4 Dynamic retinal vessel analysis. The DVA software provides online measurements of retinal vessel diameters along vessel segments over time providing functional, local, and time analysis. Various physiological stimuli can be used to assess vessel reactivity (flicker light, pharmacological agents, different breathing gases, etc.)

3.2.1 Procedure

For the DVA measurements the subject will fixate on a needle within the fundus camera with the eye that is being examined. The other eye will be covered with an eye patch to avoid straining during the examination and to improve fixation. The fixation point will be adjusted by the examiner by moving the needle so that the superior temporal vessels are centered in the image. The fundus camera focus and the background green light intensity will be adjusted to provide crisp images. All measurements will be taken using a camera field angle of 30°. The flicker filter will be inserted into the fundus camera. Background light intensity will be adjusted as needed to provide well illuminated images, avoiding strong vascular reflections and discomfort of the study subject which will impair image quality and analysis. Once set, the illumination intensity should not be changed during the examination. Subjects are encouraged to blink multiple times during the examination. Eye-tracking technology in the DVA software will compensate continuously for small eye movements.

At the start of the examination, the superior or inferior temporal retinal arteriole and venule are marked on the computer screen using the mouse curser. A vascular segment length of approximately 1.5 mm in length is chosen within an area of 1–2 optic disc diameters from the optic nerve disc margin. Vessels should have a diameter > 80 μm and a clear contrast to fundus background. The marked vessel segments should not contain any vessel crossings, bifurcations or any other nearby vessels. The chosen region of interest will then be scanned at a frequency of 25 Hz. Changes in vessel diameters at baseline and following light flicker stimulation over time will be reported in dimensionless arbitrary units (AU).

Multiple light flicker protocol settings are possible and can be individualized depending on the type of study. At standard setting the following cycle is used: After 50 s of recording of baseline, three cycles of 20 s flicker light, followed by 80 s pauses are administered and the results are continuously recorded and analyzed. The flicker stimulus is generated by shuttering the observation light of the fundus camera with 12.5 Hz at a contrast ratio of approximately 25:1 (*see* Fig. 5).

3.2.2 DVA Data Analysis

Data, in the form of AUs are continuously recorded by the DVA software. The DVA software will automatically generate an averaged vasoreactivity profile for each chosen arteriole and venule (*see* Fig. 2). This profile will be compared to normal reference values and a standard report can be printed.

For more detailed and individualized analyses, data can also be stored in a separate Excel worksheet. In our lab, we have used the following analysis approach: The resting baseline is averaged from the last 15 s prior to each flicker stimulation. Peak vasodilation period is calculated by averaging the highest diameters achieved during flicker and within approximately the first 3 s after the

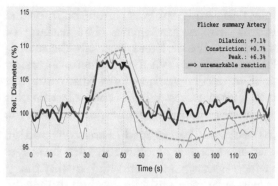
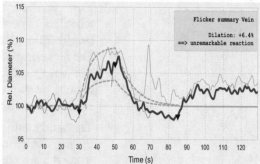

Fig. 5 Diameter changes of arterioles (red) and venules (blue) following flicker light stimulation over time. The flicker light stimulation consists of 3 cycles of 20 s of flickering light at 12.5 Hz followed by 80 s of measurement without stimulus. The light stimulus causes vasodilation in the retinal vessels followed by a recovery period. Multiple elements of this response can also be stored and analyzed offline

stimulus ends to capture maximal vasodilation. Percentage change in vessel diameter is then calculated comparing averaged baseline diameter with averaged peak diameter using the following equation: percentage change in diameter = ((diameter peak-diameter baseline)/diameter baseline) × 100). In addition, maximal vasoconstriction after peak vasodilation (i.e., lowest diameter period) is calculated by averaging the three consecutive lowest diameters (after peak dilation).

Lastly, we calculate the range of diameter change, known as relative amplitude (percentage change in relative amplitude = percentage change in peak vasodilation + percentage change in vasoconstriction) [16]. However, there are multiple alternative ways for data analysis depending on study goals.

Validation: Based on longitudinal measurements in healthy controls the coefficient of variability for measuring dynamic retinal vasoreactivity ranges between 1.5% and 2.8% [9].

Differences in flicker magnitude between different reported studies may reflect the use of different flicker light protocols (duration and Hz) as well as the recent change in classification of prediabetes and diabetes.

4   Notes

1. All studies should be performed in a quiet, darkened and temperature controlled room to control for environmental factors that may affect vascular function.

2. Study subjects should avoid any stimulants that could interfere with vessel reactivity such as caffeine, certain pharmaceutical agents or exercise prior to measurement of retinal vasoreactivity.

3. Prior to the examination, study subjects should be informed what to expect during the measurements. The flashing of bright light while sitting in the dark and especially the strobe-like flickering of light during the DVA measurement can startle the study subject if it occurs unexpectedly. Study subjects should be assured that afterimages of lights and purplish color at the end of the study are temporary, harmless and will resolve.

4. The study subject should not be engaged in any conversation during the examination as people involuntarily turn their eyes toward a speaker which makes fixation and high quality imaging difficult. However, subjects should be coached during the examination to keep their eyes fixed on the chosen fixation point, especially during the light flicker stimulation.

5. Eye blinking during the examination is beneficial as it will result in a better a tearing film of the cornea and improved image quality.

6. Funduscopic based imaging requires practice and persistence. Best results are achieved if the study subject and the examiner are comfortable with all procedures, and the examiner is adept to normal inter-individual variations in retinal vascular anatomy. The techniques should be practiced frequently as funduscopic imaging requires technical skill and experience.

References

1. Johnson W, Onuma O, Owolabi M, Sachdev S (2016) Stroke: a global response is needed. Bull World Health Organ 94(9):634–634a. https://doi.org/10.2471/blt.16.181636

2. Varghese C, Onuma O, Johnson W, Brainin M, Hacke W, Norrving B (2017) Organizational update: World Health Organization. Stroke 48 (12):e341–e342. https://doi.org/10.1161/strokeaha.117.016941

3. Benjamin EJ, Blaha MJ, Chiuve SE, Cushman M, Das SR, Deo R, de Ferranti SD, Floyd J, Fornage M, Gillespie C, Isasi CR, Jiménez MC, Jordan LC, Judd SE, Lackland D, Lichtman JH, Lisabeth L, Liu S, Longenecker CT, Mackey RH, Matsushita K, Mozaffarian D, Mussolino ME, Nasir K, Neumar RW, Palaniappan L, Pandey DK, Thiagarajan RR, Reeves MJ, Ritchey M, Rodriguez CJ, Roth GA, Rosamond WD, Sasson C, Towfighi A, Tsao CW, Turner MB, Virani SS, Voeks JH, Willey JZ, Wilkins JT, Wu JH, Alger HM, Wong SS, Muntner P (2017) Heart Disease and Stroke Statistics—2017 update: a report from the American Heart Association. Circulation 135(10):e146–e603. https://doi.org/10.1161/cir.0000000000000485

4. Bano A, Chaker L, Mattace-Raso FUS, van der Lugt A, Ikram MA, Franco OH, Peeters RP, Kavousi M (2017) Thyroid function and the risk of atherosclerotic cardiovascular morbidity and mortality: the Rotterdam Study. Circ Res 121(12):1392–1400. https://doi.org/10.1161/circresaha.117.311603

5. Chen R, Ovbiagele B, Feng W (2016) Diabetes and stroke: epidemiology, pathophysiology, pharmaceuticals and outcomes. Am J Med Sci 351(4):380–386. https://doi.org/10.1016/j.amjms.2016.01.011

6. Bruno A, Liebeskind D, Hao Q, Raychev R (2010) Diabetes mellitus, acute hyperglycemia, and ischemic stroke. Curr Treat Options Neurol 12(6):492–503. https://doi.org/10.1007/s11940-010-0093-6

7. Titchenell PM, Lin CM, Keil JM, Sundstrom JM, Smith CD, Antonetti DA (2012) Novel atypical PKC inhibitors prevent vascular endothelial growth factor-induced blood-retinal barrier dysfunction. Biochem J 446 (3):455–467. https://doi.org/10.1042/bj20111961

8. Singh A, Brooks DD, Abrams TA, Poorak MD, Gunio D, Kandhal PK, Lakhanpal A,

Sethuraman SN, Bruno A (2017) Pre-stroke glycemia in patients with diabetes. Diabetes Metab Syndr 11(Suppl 2):S891–S893. https://doi.org/10.1016/j.dsx.2017.07.011

9. Bettermann K, Slocomb J, Shivkumar V, Quillen D, Gardner TW, Lott ME (2017) Impaired retinal vasoreactivity: an early marker of stroke risk in diabetes. J Neuroimaging 27 (1):78–84. https://doi.org/10.1111/jon. 12412

10. Lott ME, Slocomb JE, Shivkumar V, Smith B, Quillen D, Gabbay RA, Gardner TW, Bettermann K (2013) Impaired retinal vasodilator responses in prediabetes and type 2 diabetes. Acta Ophthalmol 91(6):e462–e469. https://doi.org/10.1111/aos.12129

11. Hafner J, Karst S, Sacu S, Scholda C, Pablik E, Schmidt-Erfurth U (2018) Correlation between corneal and retinal neurodegenerative changes and their association with microvascular perfusion in type II diabetes. Acta Ophthalmol 10(11). https://doi.org/10.1111/aos. 13938

12. Chen AL, Sun X, Wang W, Liu JF, Zeng X, Qiu JF, Liu XJ, Wang Y (2016) Activation of the hypothalamic-pituitary-adrenal (HPA) axis contributes to the immunosuppression of mice infected with Angiostrongylus cantonensis. J Neuroinflamm 13(1):266. https://doi.org/10.1186/s12974-016-0743-z

13. Pearce I, Simo R, Lovestam-Adrian M, Wong DT, Evans M (2018) Association between diabetic eye disease and other complications of diabetes: implications for care. A systematic review. Diabetes Obes Metab. https://doi.org/10.1111/dom.13550

14. Bettermann K, Slocomb JE, Shivkumar V, Lott ME (2012) Retinal vasoreactivity as a marker for chronic ischemic white matter disease? J Neurol Sci 322(1–2):206–210. https://doi.org/10.1016/j.jns.2012.05.041

15. Orasanu G, Plutzky J (2009) The pathologic continuum of diabetic vascular disease. J Am Coll Cardiol 53(5-Suppl):S35–S42. https://doi.org/10.1016/j.jacc.2008.09.055

16. Lott ME, Slocomb JE, Shivkumar V, Smith B, Gabbay RA, Quillen D, Gardner TW, Bettermann K (2012) Comparison of retinal vasodilator and constrictor responses in type 2 diabetes. Acta Ophthalmol 90(6): e434–e441. https://doi.org/10.1111/j. 1755-3768.2012.02445.x

17. Michelson G, Patzelt A, Harazny J (2002) Flickering light increases retinal blood flow. Retina 22(3):336–343

18. Garhofer G, Zawinka C, Resch H, Kothy P, Schmetterer L, Dorner GT (2004) Reduced response of retinal vessel diameters to flicker stimulation in patients with diabetes. Br J Ophthalmol 88(7):887–891. https://doi.org/10.1136/bjo.2003.033548

19. Pemp B, Weigert G, Karl K, Petzl U, Wolzt M, Schmetterer L, Garhofer G (2009) Correlation of flicker-induced and flow-mediated vasodilatation in patients with endothelial dysfunction and healthy volunteers. Diabetes Care 32 (8):1536–1541. https://doi.org/10.2337/dc08-2130

20. Schmetterer L, Findl O, Strenn K, Graselli U, Kastner J, Eichler HG, Wolzt M (1997) Role of NO in the O2 and CO2 responsiveness of cerebral and ocular circulation in humans. Am J Phys 273(6):R2005–R2012

21. Mendrinos E, Petropoulos IK, Mangioris G, Papadopoulou DN, Stangos AN, Pournaras CJ (2008) Lactate-induced retinal arteriolar vasodilation implicates neuronal nitric oxide synthesis in minipigs. Invest Ophthalmol Vis Sci 49(11):5060–5066. https://doi.org/10.1167/iovs.08-2087

22. Pournaras CJ, Rungger-Brandle E, Riva CE, Hardarson SH, Stefansson E (2008) Regulation of retinal blood flow in health and disease. Prog Retin Eye Res 27(3):284–330. https://doi.org/10.1016/j.preteyeres..02.002

23. Garhofer G, Bek T, Boehm AG, Gherghel D, Grunwald J, Jeppesen P, Kergoat H, Kotliar K, Lanzl I, Lovasik JV, Nagel E, Vilser W, Orgul S, Schmetterer L (2010) Use of the retinal vessel analyzer in ocular blood flow research. Acta Ophthalmol 88(7):717–722. https://doi.org/10.1111/j.1755-3768.2009.01587.x

24. Wong TY, Klein R (2002) Retinal arteriolar emboli: epidemiology and risk of stroke. Curr Opin Ophthalmol 13(3):142–146

25. Ikram MK, Witteman JC, Vingerling JR, Breteler MM, Hofman A, de Jong PT (2006) Retinal vessel diameters and risk of hypertension: the Rotterdam Study. Hypertension 47 (2):189–194. https://doi.org/10.1161/01. hyp.0000199104.61945.33

26. Hubbard LD, Brothers RJ, King WN, Clegg LX, Klein R, Cooper LS, Sharrett AR, Davis MD, Cai J (1999) Methods for evaluation of retinal microvascular abnormalities associated with hypertension/sclerosis in the Atherosclerosis Risk in Communities Study. Ophthalmology 106(12):2269–2280

27. Liew G, Sharrett AR, Wang JJ, Klein R, Klein BE, Mitchell P, Wong TY (2008) Relative importance of systemic determinants of retinal arteriolar and venular caliber: the atherosclerosis risk in communities study. Arch Ophthalmol 126(10):1404–1410. https://doi.org/10. 1001/archopht.126.10.1404

28. Kifley A, Wang JJ, Cugati S, Wong TY, Mitchell P (2008) Retinal vascular caliber and the long-term risk of diabetes and impaired fasting glucose: the Blue Mountains Eye Study. Microcirculation 15(5):373–377. https://doi.org/10.1080/10739680701812220

29. Mandecka A, Dawczynski J, Blum M, Müller N, Kloos C, Wolf G, Vilser W, Hoyer H, Müller UA (2007) Influence of flickering light on the retinal vessels in diabetic patients. Diabetes Care 30(12):3048–3052. https://doi.org/10.2337/dc07-0927

30. Wong TY, Klein R, Sharrett AR, Nieto FJ, Boland LL, Couper DJ, Mosley TH, Klein BE, Hubbard LD, Szklo M (2002) Retinal microvascular abnormalities and cognitive impairment in middle-aged persons: the Atherosclerosis Risk in Communities Study. Stroke 33(6):1487–1149

31. Ikram MK, Ong YT, Cheung CY, Wong TY (2013) Retinal vascular caliber measurements: clinical significance, current knowledge and future perspectives. Ophthalmologica 229 (3):125–136. https://doi.org/10.1159/000342158

32. Lott ME, Slocomb JE, Gao Z, Gabbay RA, Quillen D, Gardner TW, Bettermann K (2015) Impaired coronary and retinal vasomotor function to hyperoxia in Individuals with Type 2 diabetes. Microvasc Res 101:1–7. https://doi.org/10.1016/j.mvr.2015.05.002

Part IV

Clinical Methods

Chapter 14

Imaging Biomarkers: Keys to Decision-Making in Stroke

J. D. Weissman, J. C. Boiser, C. Krebs, and G. V. Ponomarev

Abstract

Imaging biomarkers are medical imaging features useful for the diagnosis, treatment, and assessment of stroke, including etiology, acuity, chronicity, suitability for therapy, therapy options, and prognosis. This chapter is intended for non-clinicians with a scientific background, who are interested in learning about stroke and biomarkers. The discussion is focused on clinical assessment scales and magnetic resonance- and X-ray computed tomography-based methods.

Key words Stroke, Thrombosis, Embolism, Infarction, Brain, Therapy, Diagnosis, Thrombolysis, Anticoagulation, Biomarker

1 Introduction

Stroke is a leading cause of death and disability around the world, accounting for over 6.2 million lives lost per year and 80 million stroke survivors [1]. With an aging population, the incidence of stroke is increasing, doubling for each decade after age 55. Nearly one of four strokes occur in individuals who have previously had a stroke and 87% of strokes are ischemic strokes [2]. Ten to twenty percentage of strokes are hemorrhagic [3] and require different management than nonhemorrhagic strokes. The management of ischemic stroke may vary greatly depending on whether the etiology is thrombotic or embolic, or from the thrombosis of cerebral veins. Advances have also come from discarding earlier, less effective therapies and practices. The use of warfarin in preference to aspirin for secondary prevention of recurrent thrombotic stroke in patients with intracranial stenosis is discouraged by the results of the WASID (Warfarin-Aspirin Symptomatic Intracranial Disease) Trial [4]. Earlier generation clot removal devices have been replaced by more effective ones [5].

The most recent successful clinical trials of endovascular therapy have come to a rapid and definitive conclusion more swiftly than expected due to satisfying interim analysis endpoints that

Philip V. Peplow et al. (eds.), *Stroke Biomarkers*, Neuromethods, vol. 147, https://doi.org/10.1007/978-1-4939-9682-7_14,
© Springer Science+Business Media, LLC, part of Springer Nature 2020

showed a strong positive treatment effect. This occurred before the originally planned target enrollment for the trials had been achieved.

All of these advances are in part due to the use of "imaging biomarkers." The term "biomarker" has an evolving definition and was originally applied on a molecular basis to "chemicals, metabolites, enzymes and other biochemical substances" that reflected a biological or pathological condition [6]. A more expansive and modern definition would be a measurement of "any substance, structure, or process that can be quantitatively measured in the body or its products and influence or predict the incidence of outcome or disease" [7]. Biomarkers contrast to the patients' self-reported symptoms or typical descriptions of clinical signs, neither of which are quantitative and may not be well reproduced between observers. A good biomarker is reproducible between patients and assessors and quantitative and could be used for diagnosis, selecting therapy or measuring the effect of therapy. A biomarker is synonymous with the concept of "surrogate endpoints" in clinical trials. Properly designed biomarkers may facilitate or speed up diagnosis and statistical analysis, and help select populations that are more likely to respond to therapy.

A US Food and Drug Administration pathway for biomarker review has existed since 2007, but is not required for biomarker use in routine clinical use, or in clinical research for FDA approval of new therapies [8]. An online database show that despite the existence of this process for ~11 years, no stroke-related imaging biomarkers are currently FDA-approved (FDA List of Qualified Biomarkers, 2018) [9]. Interestingly, an imaging biomarker for renal size has been established for kidney disease. As a result, imaging biomarkers for stroke reside in the published literature and may vary from study to study.

The emphasis in this chapter is on imaging biomarkers derived from magnetic resonance imaging (MRI) and X-ray computed tomography (CT). The primary focus is the diagnosis of acute and non-acute stroke, the sensitivity and specificity of current CT and MRI methods to distinguish stroke from non-acute stroke, the prognosis of stroke, and the selection of a particular stroke therapy. We also highlight specific imaging findings that are less specific and approaches that make them more specific.

2 Ischemia, TIA, and Stroke

Ischemic stroke results from infarction of central nervous system tissue including the brain, retina, or spinal cord. This results in an inadequate supply of nutrients such as oxygen and glucose and accumulation of the by-products of metabolism such as H^+, CO_2, and lactic acid as aerobic and anaerobic metabolic processes fail.

Stroke damage is further exacerbated by excitotoxicity, which results from the ongoing depolarization of neurons by excitatory neurotransmitters such as glutamate [10–12]. Unremitting excitation and the absence of nutrients increases metabolic demand on neurons and results in energy depletion, failure of ionic pumps, neuronal swelling, neuronal membrane rupture, and neuronal death rather than a lapse of function that would be expected to recover with repletion of oxygen and nutrients. The apoptotic process of programmed cell death is also activated with stroke and causes additional damage.

The time course of stroke ranges from very short term and possibly reversible changes shortly after the disruption of circulation to nonreversible changes that progress over months to years. These changes are reflected in different imaging modalities with varying specificity.

3 Stroke Imaging Modalities

X-ray computed tomography (CT) remains the foundation of stroke imaging. CT is inexpensive in relation to magnetic resonance imaging (MRI) and can be quickly performed without any regard to implanted devices, patient motion, and patient monitoring with EKG and oximetry, and imaging times are shorter with CT scans.

X-ray absorption is determined by the atomic number of the atomic nuclei in the molecules of the imaged tissue. Computed tomographic images are displayed with white being maximal absorption or attenuation of X-rays and black being total passage of X-rays. The degree of absorption is quantified in the Hounsfield scale with air at −1000 and water at 0 U (Table 1). Most pathology is associated with decreased attenuation on CT images. Exceptions to this include hemorrhage, calcification, and freshly clotted blood, all of which are associated with increased attenuation. The typical replacement of clotted blood with serum or cerebrospinal fluid (CSF) over time is associated with decreased attenuation of X-rays.

Digital angiography remains an important tool in stroke diagnosis and management, particularly in the imaging of vascular occlusions and direction of intravascular therapy.

For MRI, interpretation of image intensity is more complicated than for CT. The MRI signal is generated for standard proton MRI by hydrogen nuclei (protons) that are predominantly in water molecules, though fat and other biomolecules also contribute to the signal. The signal is proportional to the concentration of protons and the effect of T1 and T2 relaxation times with the MRI pulse sequence. There are also effects of motion and local magnetic field differences. Magnetic field differences may arise from the magnetic susceptibility of surrounding tissue or air and from local variation in the amount of paramagnetic substances, chiefly hemoglobin and its derivatives but also including other substances.

Table 1
CT Image intensity for various tissues of neurological interest from Prokop and Galanski [116]

Tissue or fluid	X-ray absorption	Color	Hounsfield units
Air	None	Black	−1000
Fat	Slight	Black	−100 to −50
Water/CSF	Minimal	Black	0
Serum	Minimal		13–38
Unclotted blood	<WM < GM	Dark grey	45–68
White matter (WM)	<GM		20–30
Grey matter (GM)			37–45
Clotted blood	>GM	Light grey	40–90
Bone		White	300–1900

From [116]

The MRI imaging process separates the signal from different spatial locations via different encoding techniques, generating an image. MRI images are displayed with the highest signals as whiter and low signals as darker. Relative signal intensities between different tissues and pathological states are more important than absolute signal intensities as there is no universal "Hounsfield-like" scale unit for MR image intensities. It is necessary to look at relative signal changes and the performance of a given MRI scanner as it is configured and operated. Table 2 lists relative signal properties of different tissues in MRI. Most forms of pathology, including ischemic stroke and most non-stroke pathologies, have increased signal on T2 or fluid-attenuated inversion recovery (FLAIR) images and decreased signal on T1-weighted images. One exception to this is pathology involving paramagnetic substances such as hemoglobin by-products.

4   Acute Stroke

Cerebral blood flow is regulated by a complex mixture of autoregulatory and neurovascular coupling mechanisms mediated by various metabolites and excitatory neurotransmitters including H^+, Ca^{2+}, oxygen, adenosine, and excitatory amino acids such as glutamate [13, 14]. These are produced by and act on neurons, astroglia, and blood vessels.

In the first few moments of a stroke decreased blood flow may occur with or without detectable clinical deficits depending on the affected region of the brain. Acute stroke is usually painless, and

Table 2

MR image intensities on different pulse sequences for tissues of neurological interest

Pulse sequence	Relative intensity (low to high)
T1-weighted	CSF < GM < WM < fat Infarcted brain < non-infarcted brain CSF < GM < WM < or = hyperacute HEM CSF < GM < WM; +edema = acute HEM CSF < Acute HEM = GM < WM CSF < GM < WM < early/late subacute HEM CSF < or = chronic_HEM < GM < WM
T2-weighted	WM < GM < CSF Infarcted brain > non-infarcted brain Non-edematous brain < edematous brain Hemosiderin + <hemosiderin− Flowing < non-flowing blood WM < GM < CSF < or = hyperacute HEM <WM < GM < CSF < or = acute/early/late subacute HEM
FLAIR**	CSF < white matter/grey matter Infarcted > infarcted brain CSF < hyperacute HEM CSF < acute HEM
GRE	Hemosiderin < WM < GM < CSF

From Gomori and Grossman [101], Alemany Ripoll et al. [117]
GM grey matter, *WM* white matter, *HEM* hematoma. In current use the relative signal intensities of hematomas appear to be more variable.

even acute motor or sensory deficits may not be noticed by the subject because of impaired perception (termed anosagnosia or neglect). Anosagnosia is a feature of cortical lesions in general. Also, stroke in certain brain areas may not be noticed by the patient. The failure of patients to notice the effects of stroke is one reason that patients may not seek medical attention.

Non-contrast CT imaging of stroke may not detect changes for several hours [15–17] (Fig. 1a). The first CT-detected changes may be a decrease on grey/white contrast (Fig. 1b) resulting from focal tissue edema [18–20]. The ASPECTS (Alberta Stroke Program Early CT Score) score [17] is a useful way to quantify the extent of stroke on CT scans (Fig. 1c, d). The cortical and subcortical areas are divided into ten regions and one point is deducted for any early stroke signs in the corresponding region.

The CT image of acute stroke may contain clues of the underlying pathology including the dense MCA sign correlating with middle cerebral artery occlusion, the insular dot sign [21], or a dense thrombosed basilar artery [22] (Fig. 1g) that may precede or

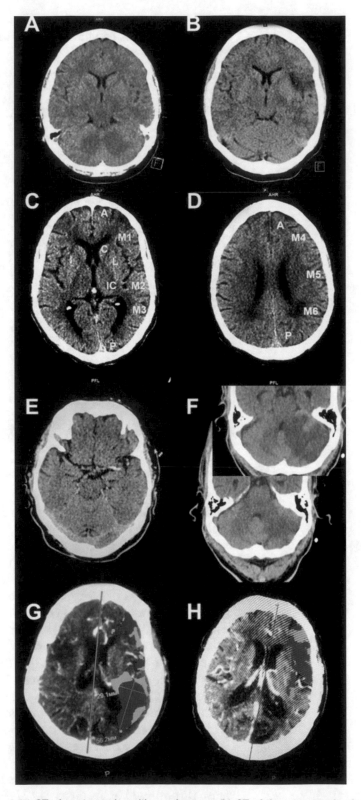

Fig. 1 (a) CT of acute stroke with no changes. (b) CT of the same acute stroke hours later with decreased grey–white contrast. (c) The ASPECTS Scoring System applied to non-contrast CT. The affected hemisphere is divided into

give useful clues to tissue changes. Vessel pathology including calcium emboli, fat emboli, or vascular dissection may also be detected before tissue changes. CT perfusion imaging with contrast is more sensitive and reflects flow differences within minutes (Fig. 1, Panel g, h), though patient handling and other practical matters may delay detection. Rapidly obtained non-contrast CT imaging may be preferable to obtaining an advanced study at a later time [23].

The initial changes from strokes occur after minutes as high signal restricted diffusion on diffusion weighted MRI (DW-MRI) [24] with T2-weighted and FLAIR changes following in hours. In contemporary clinical practice, CT scans are usually performed first, because of time considerations. It is common to see faint T2 and FLAIR signal changes on MRI obtained shortly after stroke onset (Fig. 2a–c). MRI Apparent Diffusion Coefficient (ADC) images verify acute strokes as dark areas whereas some prior strokes can show false positive bright areas on DWI, they will generally not appear dark on ADC. With time T2 and FLAIR changes become more prominent with edema (Fig. 2d, e). An acute stroke is often accompanied by evidence of multiple acute or both acute and prior strokes, which may provide evidence for the underlying stroke mechanism. The most common mechanisms are embolic, thrombotic, or related to focal manifestations of hemodynamic factors (the latter are also known as watershed infarcts (WI)).

Following acute stroke, not all hyperintensities are due to irreversible changes such as necrosis. Sanossian et al. in 2009 found that FLAIR vascular hyperintensities were strongly associated with a high grade of leptomeningeal collateral blood flow on subsequent angiography [25] (Table 3).

5   Embolic and Thrombotic Stroke

Embolic strokes occur when an embolus of blood clot, cholesterol plaque fragment, fat or other non-blood obstruction lodges in a brain blood vessel, blocking flow distally. Most emboli are composed of clotted blood. Emboli originate from the heart, great vessels, carotid or vertebral arteries, or proximal intracranial arteries. An embolic stroke is usually noticed when acute and

Fig. 1 (continued) ten cortical and subcortical regions and one point is deducted for any evidence of acute stroke in each area for a score of 0–10. (**d**) Another brain slice in the same case with ASPECTS regions defined. (**e**) An example of the dense MCA sign in a patient with middle cerebral artery distribution stroke. (**f**) An example of the dense basilar artery sign in a patient with cerebellar stroke. Insert shows another level with large cerebellar infarction. (**g**) CT perfusion-weighted image showing core infarct (red) and ischemic penumbra (green)

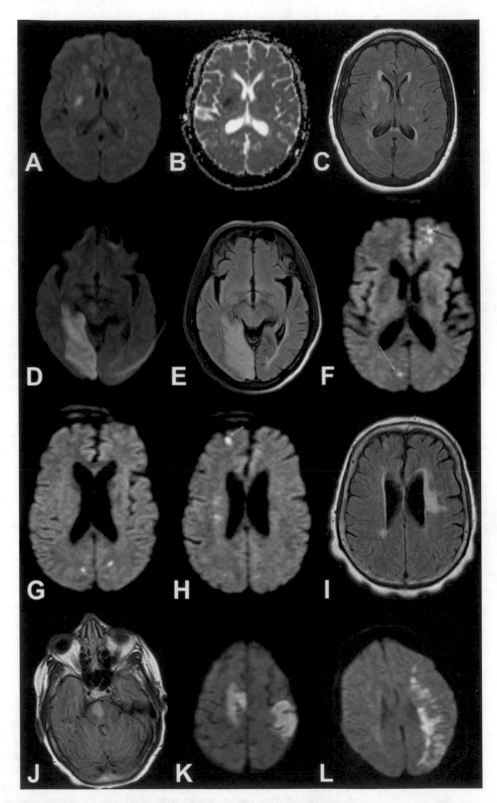

Fig. 2 Acute stroke changes detected by MRI. (**a**) MRI of small acute subcortical basal ganglia region stroke with DWI changes (bright spot). (**b**) Same region showing correlating ADC dark spot. (**c**) Same region showing

Table 3
Imaging markers for acute ischemic stroke

X-ray computed tomography	Magnetic resonance imaging
Decreased attenuation (darker)	Decreased signal (T1-weighted)
	Increased signal (T2-weighted, diffusion and FLAIR[a])
Edema (darker)	Edema (brighter)

[a]Fluid attenuated inversion recovery

there may be multiple emboli. Both large and small cerebral vessels may be occluded, and the size and location of the affected vessels will affect treatment options and determine the size of the stroke and reflect the embolic source. The infarcted brain may often have a distal peripheral cortical wedge shape (Fig. 2k), but emboli may occur in any location. Multiple embolic strokes in multiple vascular distributions are typical for cardioembolic stroke. A large embolus may block a large vessel and fragment into smaller more distal emboli within the same distribution.

Typical embolic risk factors include mechanical and bioprosthetic replacement valves, atrial fibrillation, cardiac ventricular thrombi and valvular vegetations, dilated cardiomyopathy, myocardial infarction, congestive heart failure, ventricular hypokinesis, atrial myxoma, marantic endocarditis, infectious endocarditis, atrial septal aneurysm, patent foramen ovale, friable arterial atherosclerotic plaques, and hypercoagulable states (aka thrombophilia).

Thrombotic strokes occur primarily in smaller cerebral vessels and result from pathological changes in blood vessel walls due to atherosclerosis or lipohyalinosis. Thrombotic infarction occurs in areas of preexisting atherosclerotic disease and this is usually at

Fig. 2 (continued) FLAIR image with subtler bright signal. Note that the FLAIR image is much sharper and with more anatomic detail than the ADC image but is not as sensitive to acute stroke. (**d**) Different case of subacute R occipital stroke showing high signal area in occipital cortex on DWI. (**e**) Corresponding FLAIR image to (**d**) showing more anatomic detail and edema with loss of detail in occipital sulci on R compared to L. (**f**) Another case showing multiple acute cardioembolic strokes in both hemispheres that are bright on DWI. (**g**) As above additional slices showing additional cardioembolic lesions. (**h**) As above additional slice showing additional cardioembolic lesions. (**i**) FLAIR image of same case not showing faint signal from acute emboli but showing large area of probable chronic microvascular change. (**j**) Patient with basilar thrombosis and acute R pontine infarction on FLAIR image. (**k**) Patient with multiple cardioembolic strokes with typical peripheral cortical location. (**l**) Patient with L MCA lesion with internal watershed infarction between deep and peripheral branches of L MCA

branch points in the cerebral vascular tree or in small penetrating vessels in the brainstem, basal ganglia and deep white matter. The process may be acute or gradual and may, in the case of smaller strokes, be asymptomatic. The inciting event for thrombotic infarction can be disruption of the fibrous cap over a cholesterol plaque, exposing the blood within the vessel to thrombogenic materials and inducing clotting. Lipohyalinosis and fibrinoid necrosis of small cerebral vessels can also be associated with thrombotic stroke [26]. Risk factors for thrombotic strokes include hypertension, hyperlipidemia, diabetes, elevated homocysteine, smoking, cocaine use, oral contraceptives, hypercoagulable or thrombophilia states, inflammatory conditions, and vascular trauma.

The distinction between embolic and thrombotic stroke is of importance for secondary prevention of stroke (prevention of additional strokes following an initial hospitalization for stroke). Embolic strokes typically require anticoagulant (warfarin or direct thrombin inhibitor) therapy, whereas thrombotic strokes require antiplatelet therapy (aspirin or clopidogrel). In the present era of revascularization and clot retrieval therapy, large proximal cerebral infarctions may have a mixture of thrombotic and embolic components and the decision to choose between antiplatelet therapies and anticoagulants is not as immediately relevant.

In the case where images show multiple lesions, DWI and FLAIR images can "time stamp" lesions as acute and differentiate them from chronic lesions (Fig. 2f–i). Thrombotic and embolic strokes may overlap in appearance (Fig. 2j, k). The peripheral (often but not exclusively "wedge" shaped) appearance of embolic strokes (Fig. 2k) may be seen with "external" watershed infarcts (WI—see below) between the anterior and middle cerebral artery distributions and the middle and posterior cerebral artery distributions.

CT and MRI images will typically show signal characteristics of the type of obstruction. Typical vessel occlusions of clot origin are hyperdense on CT and hyperintense on MRI and will not show "flow void" effects and will show blockages on MR, CT, or catheter arteriograms (Table 4).

5.1 Watershed Infarctions (WI)

Watershed infarctions occur at the interface between arterial distributions within the brain. These infarctions may present in symmetrical fashion in association with hypotension or in asymmetrical fashion if there is unilateral large vessel occlusion or stenosis. Internal carotid artery stenosis may be associated with asymmetrical watershed infarction. WI are estimated to account for approximately 10% of all cerebral infarctions. The most common feature is an episode of hypotension or low blood flow that preferentially affects the peripheral areas between vascular distributions. In many cases such an episode is suspected but not documented. Thrombotic and embolic factors have also been proposed in the etiology of

Table 4
Markers for embolic vs. thrombotic stroke[a]

Embolic	Thrombotic
Wedge shaped and cortical	Smaller and subcortical

[a]There is considerable overlap

Table 5
Markers for watershed stroke

External or cortical	Internal or subcortical
Wedge-shaped between ACA[a] and MCA[b] distributions	Linear or "Rosary bead" pattern between deep and superficial MCA distributions
Wedge-shaped between PCA[c] and MCA distributions	

[a]Anterior cerebral artery
[b]Middle cerebral artery
[c]Posterior cerebral artery

WI [27]. Either localized patchy or confluent patterns may be encountered, but WI are rarely "solid." Internal WI may have a "rosary bead" linear appearance.

WI are classified according to the pair of vascular distributions that form the watershed and by whether they are external or cortical versus internal or subcortical. External WI occur between the anterior and middle and anterior and posterior cerebral artery distributions. Internal or subcortical WI occur between the cortical branches of the anterior, middle and posterior cerebral arteries and the much deeper lenticulostriate, Heubner, and anterior choroidal arteries (Fig. 2i, j). Purely hemodynamic factors contribute to all types of WI, but internal WI may preferentially derive from microemboli that fail to clear during hypotension [28].

Prior to the interventional therapy era, watershed infarctions were presumably treated with aspirin or other secondary preventative therapies as appropriate. Liu et al. in 2016 [29, 30] reported a retrospective series of 120 cases that earlier carotid stenting of WI was associated with improved outcome (Table 5).

5.1.1   TIA

The definition of transient ischemic attack (TIA) is a sudden, focal neurologic deficit of presumed vascular origin that lasts less than 24 h. Deficits that last more than 24 h but resolve by 48 h were called reversible ischemic neurologic deficits (RINDs). It was noted in the 1980s with the development of clinically useful brain MRI that both TIA and RIND were often associated with imaging findings, most commonly focal hyperintense areas on T2-weighted images. With increasing availability of MRI in the

1990s, including innovations such as T2-FLAIR (Fluid Attenuated Inversion Recovery) and DWI (Diffusion Weighted Imaging), finding cerebral ischemic lesions on diffusion weighted imaging in patients with a clinical diagnosis of TIA became increasingly common. Therefore, the concept that no permanent ischemia occurred because the clinical symptoms had resolved had to be discarded.

Kidwell and colleagues from UCLA in 1999 [31] had the first published series of diffusion MRI in TIA. In their cohort, 48% (20/42) demonstrated positive diffusion abnormalities, with longer TIA symptom duration (7.3 vs. 2.2 h) associated with the presence of DWI lesions. Nearly half of the patients with positive lesions had normal repeat MRI. This finding was reproduced in other studies with the evolution of the concept of "transient symptoms of infarction" or TSI. Crisostomo et al. in 2003 found in a sample of 78 patients with clinical TIA that 21% of cases had DWI findings. Patients who had TIA events with aphasia and motor deficits lasting more than 1 h were nearly 100% likely to have DWI findings [32]. In a small follow-up study, Prabhakaran and coworkers found that TIA patients had higher risk of recurrent stroke and TIA [33]. The information gleaned from MRI imaging appears to help better predict the risk of TIA and stroke recurrence using the RRE-90 [34], an online statistical predictor that incorporates MRI findings. This was compared to the ABCD2 score [35–38] which is based on purely clinical criteria. The MRI and clinical data from which these scores are derived are shown in Table 6. On the other hand, given the correlation of large DWI lesions with ischemic core (see below) and the negative findings in some DWI studies of TIA patients, it is likely that DWI may be insensitive to some transient ischemic events.

Recognizing that finding imaging abnormalities in TIA cases would potentially influence the urgency of diagnostics and treatment, The TIA Working Group was convened in 2000 and released its initial consensus statement in 2002 [39]. After further refinement and discussion, in 2009, the American Heart Association revised the definition for TIA and stroke as follows: "Transient ischemic attack (TIA) is a transient episode of neurological dysfunction caused by focal brain, spinal cord, or retinal ischemia, without acute infarction" [40]. Corollary to this, "An ischemic stroke is defined as an infarction of central nervous system tissue." In terms of imaging features, a "true TIA" should have no imaging findings. A "false positive" TIA will look like an acute or subacute acute stroke (Fig. 2a).

Guidelines for the management of acute stroke recommend against thrombolytic therapy or revascularization in patients that present with stroke and experience complete resolution of their deficits and have negative imaging before therapy can be started. Such cases would be considered TIA by the 2002 TIA Working

Table 6
ABCD2 vs. RRE-90 scales for stroke prediction after TIA

Clinical markers vs. imaging markers	
ABCD2 SCORE[a]	RRE-90 SCORE[b]
Age ≥ 60 years	MRI within 72 h of stroke
BP > 140/90	History of TIA within month
Unilateral weakness	Multiple acute infarcts
Speech disturbance Without weakness?	Simultaneous Infarcts in different distributions
Duration ≥ 60 min	Infarcts of different ages?
Diabetes	Stroke subtype

[a]Johnston et al. [35]
[b]Ay et al. [36]

Group definition. Patients that have certain persistent imaging abnormalities such as perfusion–diffusion mismatch might benefit from therapy on a case-by-case basis. Patients that improve but have moderate or severe residual clinical disability may benefit from thrombolytic therapy or revascularization [41].

5.1.2 The Time Course of Subacute to Chronic Ischemic Stroke

A completed stroke will progress through the various stages of necrosis of brain tissue to an astroglial scar. Initial changes include the formation of "ghost" neurons by rupture and death from excitotoxicity and swelling. There is associated spongiform change that is followed within days by acute inflammation with macrophages and similar cells. Following this chronic inflammation develops with mononuclear cells. Coagulative necrosis follows after several weeks with the appearance of chronic inflammatory changes at around a month including neovascularization and luxury perfusion. After several months, inflammatory changes begin to decrease and there is the emerging presence of cavitation and astrogliosis with Wallerian degeneration [42] (Fig. 3).

Some patients experience hemorrhagic transformation following ischemic stroke. Risk factors include treatment with recombinant tissue plasminogen activator, large stroke volume, especially in cardioembolic strokes, age, hypertension, and hyperglycemia. The typical appearance is of a patchy accumulation of blood seen as a hyperdense area within a previously hypodense CT-defined area of infarction. This is usually less hyperdense that a primary hemorrhagic stroke and the prognosis for the patient is usually better than for a primary hemorrhagic stroke [43] (Table 7).

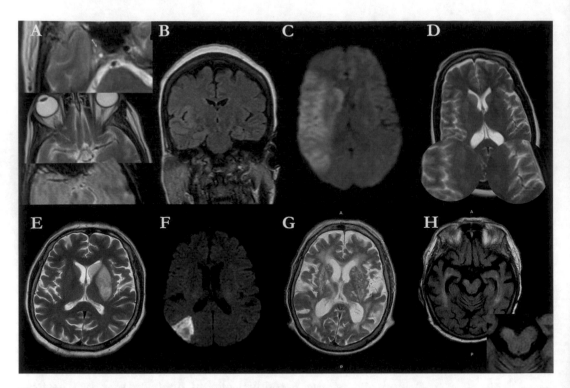

Fig. 3 Time course of ischemic stroke. (**a**) cute Stroke secondary to R internal carotid occlusion (arrow 1) with decreased flow void in R MCA (arrow 2) and susceptibility artifact in R MCA (arrow 3). (**b**) Edema at site of infarction in R insula compared to L insula. (**c**) DWI change compatible with core stroke across MCA distribution. (**d**) T2-weighted image shows lack of flow void in distal R MCA distribution indicating poor collaterals. (**e**) Maximum edema in Day 3 infarction in L caudate and putamen. (**f**) Hemorrhagic transformation in embolic stroke on DWI image. (**g**) Wallerian degeneration in R internal capsule beneath area of encephalomalacia from stroke in R parietal area. (**h**) Atrophy of R cerebral peduncle beneath area of Wallerian degeneration

Table 7
Imaging markers for chronic ischemic and chronic hemorrhagic stroke

Feature	Pathology	X-ray CT	MRI
Core infarct	Atrophy, replaced by CSF	Low attenuation: dark	T1-W/FLAIR: dark T2-W: bright
Peripheral infarct	atrophy, glial cell scar	+/− Low attenuation: dark	T1-W: dark T2-W/FLAIR: bright
Connected distant regions	Atrophy, Wallerian degeneration	+/− Low attenuation: dark	T1-W: dark T2-W/FLAIR: bright
Remaining blood products	Hemosiderin	Isointense	T1-weighted: isointense T2-Weighted/FLAIR/SWI/GRE/DWI: hypointense, "blooming"

CSF cerebrospinal fluid, *T1-W* T1-weighted, *T2-W* T2-weighted, *FLAIR* fluid attenuated inversion recovery, *SWI* susceptibility weighted imaging, *GRE* gradient recalled echo, *DWI* diffusion weighted imaging

6   Stroke Therapeutic Decision-Making and the Role of Imaging

The 2018 American heart Association (AHA) Guidelines for the early management of patients with acute ischemic stroke [41] reemphasize the "Time is Brain" concept of rapid evaluation and treatment. All patients with suspected acute stroke should have a non-contrast CT on arrival to the emergency room, preferably within 20 min. One major purpose of the rapid CT scan is to detect intracerebral hemorrhage and non-stroke entities such as brain tumors. The decision to admit to the hospital is based on a combination of clinical and imaging factors; even patients with symptoms that resolve (i.e., TIA) may be admitted for further evaluation and observation. Some patients with suspected stroke may turn out to have other non-stroke conditions including psychiatric conversion disorders and "stroke mimics" (see below).

The benefits of thrombolysis (IV alteplase or tenecteplase) and mechanical thrombectomy are highly time dependent so imaging, clinical assessment and alteplase reconstitution and administration must be performed promptly. The initial clinical experience with IV thrombolysis with streptokinase was a disappointment to a large part because patients with "wake-up" strokes were included [44]. Subsequent stroke trials have generally relied on the "last known normal" (LKN) time if the stoke onset is not witnessed and patients that wake up with stroke are assumed to have a time of onset as the time of going to bed as a default.

For IV thrombolysis, exclusion of any acute intracranial hemorrhage (ICH) is essential and remains a contraindication to intravenous thrombolysis. Widespread hypoattenuation (over 30% of the MCA distribution) on non-contrast CT is a relative contraindication to the use of thrombolysis [45]. On the other hand, several other imaging signs were felt to be relative contraindications years ago but post hoc analysis of clinical trial data initially showed that there was no benefit for their use as exclusion criteria when IV-rTPA was the only therapeutic option. These include the appearance of <33% MCA distribution CT hypodense areas associated with acute infarctions, loss of gray/white matter demarcation, Alberta Stroke Program Early Computed Tomography Score (ASPECTS), measures for WMH/leukoaraiosis [16, 17], MRI hyperintensities [46] and the CT hyperdense MCA sign [47]. The situation with microbleeds as detected on T2-weighted, diffusion-weighted or gradient echo MRI scans is more complicated with meta-analyses showing some increased risk [48, 49] particularly in cases with a large number of microbleeds. Patients who meet inclusion criteria for TPA but who have large strokes would benefit from TPA, but hemorrhage risk would be greater [50]. It should be noted that the therapy decision process has been evolving with the availability of intravascular therapy options.

6.1 The Role of Intra-arterial Thrombolysis

Current guidelines for management of acute stroke continue to emphasize rapid evaluation of patients for intravenous IV alteplase treatment, emphasizing telestroke consultation where necessary and transfer to facilities where endovascular therapy for stroke is available [41]. Intra-arterial alteplase use played a limited role in endovascular trials but is currently used as a rescue therapy when mechanical endovascular treatments do not achieve enough reperfusion.

6.1.1 Patient Selection for Endovascular Therapy

Current interventional stroke therapy beyond intravenous or intraarterial TPA is based on the 2015 updated AHA/ASA (American Heart Association/American Stroke Association) guidelines for the early management of patients with acute ischemic stroke. A central component is rapidly identifying patients with (1) large vessel occlusions accessible to currently available stent retrievers and who have (2) significant areas of therapeutically reversible ischemia, that is, ischemic penumbra or diffusion–perfusion mismatch. Such patients are the best candidates for revascularization/clot extraction.

The early attempts with intravascular revascularization therapy were successful in improving blood flow through larger arteries but clinical results were disappointing in part because of the performance of earlier devices [5] but also because of including patients that might benefit less because of large areas of unsalvageable "core infarct" or small areas of salvageable "ischemic penumbra [23].

There was accumulated evidence for improved outcome in stroke patients with greater amounts of pial-derived collateral blood supply who underwent intravenous and intra-arterial thrombolysis [51, 52], presumably because pial collaterals could extend viability beyond the 4.5 h window of intravenous therapy. Measures of collateral flow could be used to select appropriate patients.

These observations resurrect the concepts of core stroke and ischemic penumbra. Early non-imaging studies in animals established the ischemic threshold for core stroke as the reduced level of cerebral blood flow (CBF, usually expressed as ml blood flow per minute per 100 g of brain) below which brain tissue is not viable under the conditions of the experimental system. Above this there is a second threshold for *ischemic penumbra* where the brain is functionally impaired (perhaps with loss of measurable EEG activity) and will become necrotic with prolonged ischemia but it will resume normal function if blood flow is raised. Outside this is a region of *benign oligemia* where neurological function is impaired, but function will recover without treatment. Early studies established a core infarct ischemic threshold as <10 ml/min/100 g brain and an ischemic penumbra threshold in the range of 16–18 ml/min/100 g brain with normal flow approximately 50–54 ml/min/100 g [53]. Thresholds vary between different experimental brain models [54].

*6.1.2   Clinical Scales
of Suitability and Success
in Intravascular Treatment*

In 2003 the Technology Assessment Committees of the American Society of Interventional and Therapeutic Neuroradiology and the Society of Interventional Radiology together with neurologists Randall Higashida and Anthony Furlan proposed imaging scores for categorizing the concepts of proximal stroke location, degree of vessel occlusion, degree of collateral supply and also a measure of revascularization that could be used to promote clinical trial research and patient care along this new direction [55]. These are shown in Tables 8, 9, 10 and 11 below and emphasize the need to create clear and measurable biomarkers for these properties. It was and still is clear that available intravascular methods are most effective for proximal lesions of the middle cerebral or internal carotid arteries.

The earliest trials of endovascular therapy relied on perfusion imaging methods to distinguish between core infarction and ischemic penumbra. This approach was initially problematic due to many different early technical methods as shown by the meta-analysis by Bandera et al. in 2006 [56]. With the advent of automated perfusion imaging methods based on timed contrast CT angiography or magnetic resonance angiography (MRA) with automated analysis, progress has been considerable.

Core infarction is associated with the MRI biomarkers of restricted diffusion (DWI) and the later-appearing T2 and FLAIR hyperintensity, and T1 hypointensity. Some studies have suggested that DWI lesion reversal may occur with treatment, though the transient appearance of DWI image changes may also account for improvement [57]. Among the perfusion weighted imaging (PWI) parameter maps, very reduced blood volume correlates well with core stroke volume and "$T_{max} > 6$ s" maps (maps of the region

Table 8
Grading of intravascular occlusion

Name	Description
"T" Occlusion	Embolus to ICA terminus.
Proximal MCA	M1 Trunk occlusion at or proximal to the lenticulostriate arteries
Distal MCA	M1 trunk occlusion distal to the lenticulostriate arteries
M2	Division occlusion beyond the bifurcation of M1
Vertebrobasilar	
Anterior cerebral	
Tandem	Proximal ICA occlusion may be present along with an intracranial embolus and should be separately reported

Table 9

The Thrombolysis in Infarction Score (TICI) perfusion categories [55]

Score	TICI description
0	No perfusion. *No flow beyond the point of occlusion*
1	Penetration with minimal perfusion *The contrast material fails to opacify the entire cerebral bed distal to the obstruction for the duration of the angiographic imaging test.*
2a	Partial perfusion. *Only partial filling (<2/3) of the entire vascular territory is visualized*
2b	Partial perfusion. *Complete filling of all of the expected vascular territory is visualized, but the filling is slower than normal*
3	Complete perfusion. *Antegrade flow into the bed distal to the obstruction occurs as promptly as into the obstruction*

Table 10

Collateral flow grading system (angiographic)

Grade	Description
0	No collaterals visible to the ischemic site
1	Slow collaterals to the periphery of the ischemic site with persistence of some of the defect
2	Rapid collaterals to the periphery of ischemic site with persistence of some of the defect and to only a portion of the ischemic territory
3	Collaterals with slow but complete angiographic blood flow of the ischemic bed by the late venous phase
4	Complete and rapid collateral blood flow to the vascular bed in the entire ischemic territory by retrograde perfusion

From Higashida et al. [55]

Table 11

The modified Arterial Occlusive Lesion (mAOL) score on follow-up CTA to assess recanalization of the occlusive lesion (thrombus) seen on the baseline CTA

mAOL score	Description
0	Primary occlusive lesion remains same
1	Debulking of thrombus without recanalization
2	Partial or complete recanalization of the primary lesion with thrombus/occlusion in the distal vascular tree
3	Collaterals with slow but complete angiographic blood flow of the ischemic bed by the late venous phase
4	Complete recanalization of the primary occlusion with no thrombus in the vascular tree at or beyond the primary occlusive lesion

From Higashida et al. [55]

where the contrast peak is delayed greater than 6 s from the time of onset of injection) correlate with ischemic penumbral volume. Clinical scales such as the NIH stroke scale correlate with core stroke volume, particularly white matter volumes [58] but cannot distinguish between deficits due to the ischemic penumbra and core stroke. Variations in consciousness, seizures and aphasia can complicate clinical assessments but are obviously not going to affect imaging measurements as much.

A high ratio of ischemic penumbra to core stroke volume favors a good result with intravascular therapy as does a core infarct that is not too large. Large core stroke volumes raise the risk of hemorrhagic transformation, especially after thrombolysis [50] and this has been correlated with various PWI parameter maps in different studies using different PWI software: low CBV [59], prolonged delay time [60], very prolonged Tmax [61] and low cerebral blood volume (CBV) [62].

7   Imaging and Clinical Core–Penumbra Mismatch.

There are currently two approaches to patient selection in intravascular therapy trials. The first is imaging-based selection with perfusion–core infarct mismatch based on comparison of images that reflect core infarct with ones that reflect ischemic penumbra. The ASPECTS scoring system for measuring non-contrast based CT stroke extent and the RAPID (RApid processing of PerfusIon and Diffusion software) software [63] for determining PWI maps of core infarct volume and ischemic penumbra are standardized imaging biomarkers for decisions points in current stroke therapy [64, 65]. Perfusion parameter maps of cerebral blood flow (CBF), cerebral blood volume (CBV), mean transit time (MTT), time to peak (TTP) and maximum transit time (MTT) are calculated automatically. This approach has been used in the recent successful interventional stroke trials including the MR CLEAN [66], ESCAPE [67], EXTEND—IA [68], SWIFT—PRIME [69], REVASCAT [70], and THRACE [71, 72]. As in the case of the implementation of r-TPA, attention to the "last known well" or witnessed stroke onset time and rapid workflow including automatic PWI data processing contributed to the success of the recent trials. A typical PWI study demonstrating an acceptable low core infarct volume with acceptable penumbra core volume mismatch ratio and successful clot extraction is shown in Fig. 4.

The second approach is hybrid clinical/imaging based with a mild NIH stroke scale deficit associated with a proximal internal carotid artery (ICA) or middle cerebral artery (MCA) occlusion with a small core infarct: clinical–core infarct mismatch. This was first used in the DAWN trial [64]. For this trial the requirement was

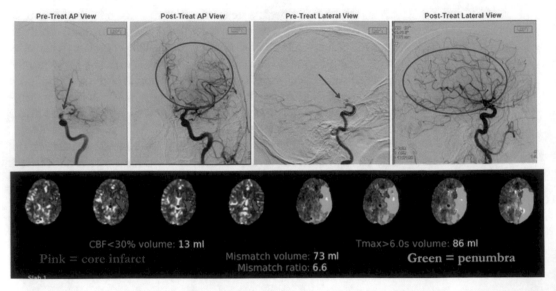

Fig. 4 Endovascular therapy case. Eighty-one-year-old patient with HTN who presented with NIHSS 23 and negative CT for blood. Bottom row: CT-PWI maps from RAPID software showing core infarct (pink) of 13 ml and penumbral volume of 86 ml. This was a mismatch ratio of 6.6:1 and mismatch volume of 73 ml, which, combined with the NIHSS score, made the patient an excellent candidate. Top row: Angiographic views from AP and lateral perspective showing pre- and post-treatment effect of stent retriever. Patient responded well with NIHSS = 3 at discharge. These images are courtesy of Dr. Michael Frankel, Emory University and Grady Memorial Hospital

for a patient with moderate NIH stroke scale deficit small infarct core of 20–30 ml (depending on age) that could optionally be measured with diffusion MRI (a very fast procedure that is widely available and can be done at a facility without intravascular capabilities). Patients with somewhat larger core volumes could enroll with more severe NIH stroke scale deficits. As in many of the endovascular trials, enrolment was stopped early when an interim analysis of the collected data showed that the rate of functional independence at 90 days was 49% in the thrombectomy group compared to 13% in the control group.

The recent set of endovascular trials were largely successful because they identified the population of patients who would benefit from therapy and also those who would not. This means that some patients with minimal clinical deficits were not enrolled, nor were patients with large core infarct volumes and minimal penumbral volumes. It was earlier proposed to measure clot volume as a screening biomarker based on the relative resistance of larger clots to thrombolytic therapy [23], but post hoc analysis of the recent data has shown no effect of clot size on outcome with the new methods [73].. Outcome measures are usually clinical, though surrogate outcome measures on imaging of recanalization or final infarct volume [74] have also been used. Table 12 summarizes biomarker use for patient selection and outcome measurement in selected recent and older trials of intravascular therapy.

Even before the successful revascularization trials, improved outcome has been shown to be associated with improved collateral status, reduced ratio of stroke core to ischemic region, and reduced ASPECTS score [75, 76]. These findings were more impressive than those relating to recanalization of proximal supplying vessels [77–79]. These results and other factors motivated the study of possible intervention after 6 h based on optimal collateral and core infarct status rather than relying on time after know stroke onset. The recently published DAWN [64] and DEFUSE-3 [80] trial results validate these concepts when applied to strokes beyond the 6-h window, to include "wake up" and strokes of unwitnessed onset [39, 79],. Selection criteria for these trials included occlusion at the level of the cervical or intracranial artery or M1 segment of the middle cerebral artery, relatively low core infarct volumes, and low core infarct to under-perfused brain ratios as follows (additional criteria were also used).

In the recent successful trials, patients met enrollment criteria and were not excluded from standard medical treatment with alteplase. Those who underwent interventional care with stent retrievers did better than those with standard medical therapy, including rTPA when applicable with a 70% relative reduction in disability.

Furthermore, the majority (as many as 90%) of patients with stroke who present after 6 h do not meet the abovementioned criteria, usually due to too small or too large a large core infarct volume and so interventional therapy and other available therapies do not have much to offer for these patients [81]. Also, patients with very high initial NIH stroke scales (>17) and patients over 80 years of age did not do as well.

7.1 White Matter Hyperintensities (WMH), Silent Strokes, MRI Hyperintensities of Presumed Vascular Origin, and Enlarged Perivascular Spaces

The presence of focal pathological findings on postmortem examination of the brains of carefully screened normal elderly individuals was noted by Tomlinson et al. [82], mostly in cortical locations. With the new X-ray CT and MRI imaging methods focal findings in asymptomatic normal individuals were frequently found [83]. Pathological correlation is not possible in asymptomatic cases but Fazekas and others did postmortem correlations in patients who had been scanned prior to death and found pathology that differed with the hyperintensity location [84]. Focal white matter hyperintensities remain the most common findings in the scans of asymptomatic patients. These appear as focal bright areas on T2-weighted and FLAIR MRI images or as hypodense areas on CT. Nonetheless, these changes are a source of concern for neurologists and patients alike.

Various scales describing the quality and quantity of white matter hyperintensities have been proposed. Actual clinical use outside of research studies is variable because of variation in the

Table 12

Intravascular Stroke Therapy Trials Patient Selection Criteria

Trial name	Reference	Widow	Intervention	NIHSS	Age	ASPECTS	Core infarct	Penumbra/core	Collateral	Mismatch	Outcome 90 days
Proact II	[118]	6 h	IA pro-UK	4–30[a]	18–85						MRS <2 Positive
MR RESCUE	[119]	8 h	Merci/Penumbra	6–29	18–85		<90 ml	>0.42			MRS Negative
MR CLEAN	[66]	6 h	IA-TPA/UK or stent-R	2–42 [b]14	18-						MRS[c] positive
ESCAPE	[67]	12 h	Stent-R	0–42	18-	6–10	<1/3 MCA by CBV/ASPECTS		>50% MCA pial		MRS[c] positive
EXTEND-IA	[68]	6 h	Solitaire FR	0–42	(68–70)		(19.6)	>0.42			MRS[c] positive[d]
SWIFT PRI ME	[69]	6 h	Solitare FR	≥8 < 30	18–85	> = 6	<1/3 MCA or >1.8 core >50 ml[e]	>1.8 RAPID			MRS[c] positive

Trial	Time	Device		Age		Core		Mismatch
DEFUSE 3 [80]	6–16 h	Trevo/Solitaire/Penumbra	≥6	18–85	>=6	<70 ml	>1.8 RAPID	Mismatch > =15 ml
DAWN [64]	6–24 h	Trevo	≥10	>18		<1/3 MCA (51 ml)		0-30 cc core, NIHSS >10
REVASCAT [70]	8 h	Solitaire	≥6	18–80	≥7 by CT			MRS[c] positive

Exclusion: ICH, abnormal PT, PTT, platelet counts, uncontrolled HTN, non-meningioma tumors, acute CT changes in >1/3 MCA distribution

Inclusion: MRS < 2 baseline (independent) pre-stroke, IV rTPA control, TIMI grade 1 M1/M2 MCA. Outcome measures of MRS, NIHSS, BI at 90 days. POV MRS <2. ITT analysis. Abbreviations. Pro-UK—pro-urokinase

[a]Except for isolated aphasia or hemianopia

[b]Stratified at <14, > =14

[c]Shift analysis

[d]>8 reduction NIHSS, reperfusion by 24 h

[e]>20 ml if age 80–85

Table 13
Fazekas Scale for White Matter Hyperintensities

Abnormality	Degree
Periventricular hypointensity	
0	Absent
1	Punctate or pencil thin
2	Beginning confluence
3	Confluent
Deep white matter hypointensity	
0	Absent
1	Punctate
2	Beginning confluence
3	Confluent
Misc. type of lesion	
Cortical hyperintensity	Present/not present
Basal ganglia	Present/not present
Infarct	Present/not present

From Fazekas et al. [83]

type and number of MRI sequences that are required, including less frequently employed spin density sequences and different classification systems for types of hyperintensities. The Fazekas scale was developed for use in Alzheimer's disease and multi-infarct dementia [83] and has historically been the most commonly used (Table 13), but the 2013 STandards for ReportIng Vascular changes on nEuroimaging (STRIVE) Neuroimaging standards [85] and the 2017 American Heart Association scientific statement: "Prevention of Stroke in Patients With Silent Cerebrovascular Disease" ("2017 AHA Silent Stroke Guidelines") [86] are a consensus approach for dealing with WMHs based on a structured literature review. The STRIVE neuroimaging standards define five categories based on appearance (Table 14) including small subcortical infarcts, lacunes of presumed vascular origin, white matter hyperintensities of presumed vascular origin, prominent perivascular spaces, cerebral microbleeds, and otherwise unexplained brain atrophy. The consensus opinion is that several categories of these findings raise concern for stroke risk, particularly silent brain infarctions and cerebral microbleeds (Fig. 5). In actual clinical practice these scales do not seem to be used very much and various terms have been used to describe WMH including "microvascular changes," "leukoaraiosis," "subcortical ischemic leukoencephalopathy," "age-related findings," and "Binswanger's disease," usually without pathological confirmation.

Table 14
Strive lesion categories

Category	Comment
Recent small subcortical infarct	Consistent with recent infarction of small vessels. <20 mm diameter. May be irregular. Content: not CSF
Lacunes of presumed vascular origin	Usually 3–15 mm in diameter Contents compatible with CSF
White matter hyperintensity of presumed vascular origin	<3 mm, spherical. Hyperintense on FLAIR, T2-weighted MRI. Either periventricular or deep white matter; not brainstem
(Prominent) perivascular spaces	CSF-filled spaces that follow the typical course of a vessel as it passes through grey or white matter. Round/ovoid, <3 mm, often symmetrical
Cerebral microbleed	On T2[a]-weighted, GRE[a], SWI[b] images 2–5 mm dark areas with blooming
Brain atrophy	Reduced brain volume, focal or generalized and not related to trauma, stroke or other conditions

From Wardlaw et al. [85]
[a]Gradient recalled echo
[b]Susceptibility weighted imaging

7.1.1 WMH: Correlation with Stroke and Non-stroke Risk Factors

Small, asymptomatic white matter hyperintensities (WMH) are perhaps the most problematic entity in this category. They are most commonly found in the pons, basal ganglia, or cortical areas and are more frequent with age and in individuals with stroke risk factors, hypertension, and migraine [87]. They tend to be more common in women compared to men [88]. They bear a resemblance to changes seen in multiple sclerosis (MS) but are typically more prevalent in the central pontine, parietal, and frontal deep white matter and less prominent in temporal, periventricular, callosal, and peripheral brainstem areas where MS changes tend to be more prominent. Pathological correlation suggests a microvascular pathology [89] including microembolism, disruption of small penetrating arterioles, disruption of small veins, gliosis, incomplete microscopic infarction [90, 91], hemodynamic changes, and blood–brain barrier disruption [15]. In a very recent prospective cross-sectional study of 125 young adults without cerebrovascular disease, reduced cardiovascular risk factor status was correlated with a lower number of white matter hyperintensities, higher cerebral vessel density, and higher cerebral blood flow on MRI scans [92]. These findings suggest a relationship between modifiable cardiovascular risk factors and biomarkers of cerebrovascular structure and function and white matter hyperintensities in young adults.

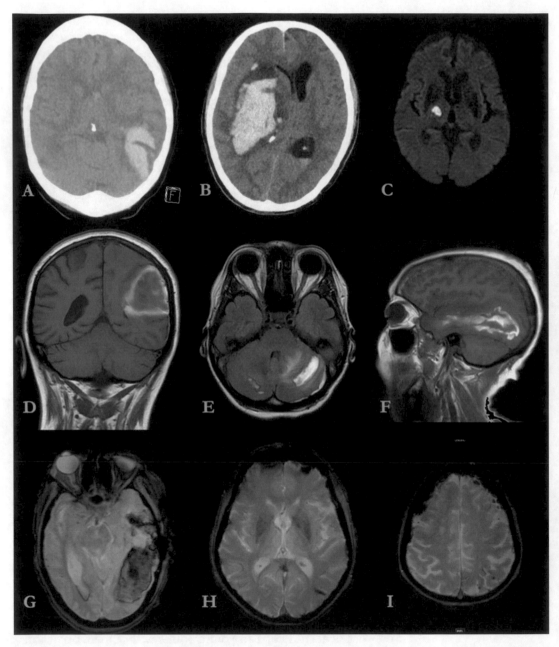

Fig. 5 Intracranial hemorrhage. (**a**) CT of acute intracranial hemorrhage. (**b**) CT of 3-day intracranial hemorrhage showing clot retraction. (**c**) Diffusion MRI of hyperacute thalamic hemorrhage showing high signal. (**d**) T1-weighted MRI showing acute bleed with small amount of methemoglobin at border of hematoma causing increased signal. (**e**) FLAIR MRI of early subacute hematoma showing increased signal effect. (**f**) T1 MRI of late subacute MRI showing rim of methemoglobin. (**g**) GRE MRI of late subacute MRI showing slight rim of hemosiderin that has formed. (**h**) MRI of 82-year-old patient with L parietal-occipital microbleed. (**i**) MRI of same patient showing two additional and other possible microbleeds on next slice

7.1.2 WMH Correlation with Stroke Risk Factors

WMH imaging patterns have been correlated with specific etiologies. A "rosary bead" pattern where impaired hemodynamics with reduced regional cerebral blood flow are seen in association with the hyperintensities [93] but more recent results render these associations less clear [94]. Carotid stenosis is associated with an increased incidence of ipsilateral silent brain infarctions, but not image-classified lacunar infarctions, including downstream silent brain infarcts [95, 96] with MRI being a possibly more sensitive indicator. High-resolution 3 T MRI imaging analysis of the shapes of WMHs has shown that irregularly shaped hyperintensities are more likely with multiple vascular risk factors [90]. Non-stroke inflammatory conditions such as multiple sclerosis (see below) are associated with white matter lesions in a periventricular ovoid lesion or "Dawson's finger" pattern [97, 98].

7.1.3 Asymptomatic or Silent Strokes

Asymptomatic strokes are larger than WMHs and are a step up in severity and clinicopathological relevance from WMHs and are a significant portion of total stroke burden [95, 99] and are diagnosed when a patient has an MRI or occasionally a CT for some other reason and an unsuspected stroke is detected. Like nonspecific white matter changes, silent strokes have increased signal on T2- or FLAIR MRI images, but they differ from nonspecific white matter changes by size and the presence of T1-weighted imaging hypointense signal changes or hypodense areas on non-contrast CT scans [100]. This might be an additional stroke to an acute stroke that was suspected, and it may be acute or subacute, that is, older than the suspected stroke. The lack of symptoms in silent stroke may be because of the impaired perception (anosagnosia) of the neurological deficit or because the asymptomatic or silent stroke is relatively small and/or located in an area of the brain where strokes may occur without readily observable signs. Based on the number of asymptomatic strokes detected when patients present with acute strokes, it is estimated that 50–80% of strokes that occur are clinically silent.

8   Intracerebral Hemorrhage (ICH) and Cerebral Microbleeds (CMB)

Hypertension is the most common risk factor for ICH and these most commonly occur in the basal ganglia (especially the putamen), cerebellum, thalamus, and pons. The use of thrombolytics and anticoagulation are the next most frequent risk factors with amyloid angiopathy, arterial aneurysms, vascular dissection, vasculitis, septic emboli, and venous infarction being less common. Chronic hypertension leads to the formation of Charcot–Bouchard aneurysms of small perforating vessels which are particularly common in the basal ganglia and pons. Many cases of hemorrhage are due to bleeding within a non-hemorrhagic stroke (the so-called hemorrhagic transformation), which may be minimal or massive in extent. These are usually classified separately from ICH.

The appearance of extravasated blood on CT images is dependent on the concentration of hemoglobin, and this is increased by clot retraction (Fig. 5a, b) and decreased by macrophage cleanup of the blood as time passes. On MRI images the sequential chemical changes in the degradation products of hemoglobin greatly affect the appearance of MRI images and are summarized in Tables 7 and 15 and illustrated in Fig. 5a–g according to the scheme of Gomori and Grossman, 1988 [101]. The "dating" of a hematoma from its signal change is approximate and may be affected by ongoing bleeding and specific anatomic location. In practical terms, patients with intracerebral bleeding of any kind are usually managed with serial X-ray CT. Figure 5 shows images of range of different types of intracerebral hemorrhagic events.

8.1   Cerebral Microbleeds

MRI and particularly T2-weighted images reflect the presence of small amounts of paramagnetic blood products with greater sensitivity than CT. This is particularly true of diffusion weighted and gradient echo and susceptibility weighted images which can detect small perivascular hemosiderin accumulations in the brain, otherwise known as cerebral microbleeds (CMB). These were not detectable in the early years of clinically useful MRI due to limitations on magnetic field gradients, lower field strengths, and pulse sequence designs employing spin echoes rather than gradient echoes [102]. CMBs appear as dark spots with somewhat fuzzy edges due to the "blooming" effect of the local magnetic field differences (Fig. 6h, i) on the MRI images. The dark spots are larger than the hemosiderin deposits because of the magnetic field distortions, and

Table 15

Sequence of changes and MRI appearance of hemoglobin and its degradation products

Time	Chemical state	Comments
Normal (no bleed)	Oxyhemoglobin (non-paramagnetic)	RBC intact, moving minimal signal changes
Immediately post bleed	Deoxyhemoglobin (paramagnetic)	RBC intact, stationary No T1 imaging effects Decreased signal on T2 images
Subacute	Methemoglobin (paramagnetic)	RBC disrupted Increased signal on T1 images + −Decreased signal on T2 images
Chronic	Hemosiderin (superparamagnetic)	In macrophage lysosomes Decreased signal on T2 images Decreased signal on T1 images (if concentrated)

From Gomori and Grossman [101]

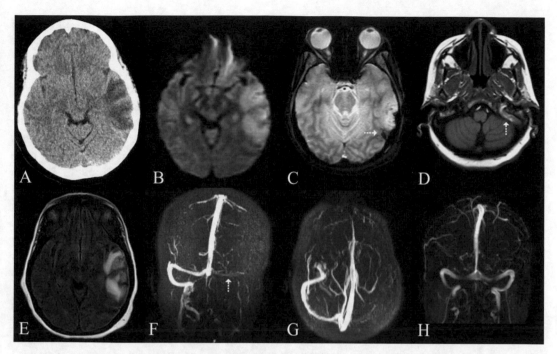

Fig. 6 Cerebral venous thrombosis case (panels **a–g**). (**a**) Axial CT of patient aged 42 years without prior known medical problems presenting with a seizure while driving. (**b**) Axial diffusion MRI showing acute infarction in L temporal lobe. (**c**) Axial gradient echo MRI showing T2∗ area in posterior part of L temporal infarction (arrow) compatible with microbleed. Patient also had red cells in cerebrospinal fluid. (**d**) Axial T1-weighted MRI showing blood clot in L sigmoid venous sinus (arrow). (**e**) Axial FLAIR MRI showing increased signal in area of infarction. (**f**) Coronal projection of MR venogram showing absent signal in L transverse (arrow), sigmoid and internal jugular veins. (**g**) Axial projection of normal MR venogram showing absent signal in L transverse, sigmoid and internal jugular veins. (**h**) Coronal projection of MR venogram of normal subject. This patient was treated with heparin and anticoagulants for the cerebral venous thrombosis and with acyclovir for possible herpes simplex virus (HSV) encephalitis. It turned out that the patient had two previously unknown risk factors for thrombophilia (elevated homocysteine and decreased protein C) and negative polymerase chain reaction tests for HSV DNA

these increase with magnetic field strength. CMBs are distinguished from flow voids in pial blood vessels by their size and by the lack of blooming effect.

CMBs are quite common in the general elderly population, found in 10–25% [103], and correlated with microscopic hemorrhagic foci [104] and have increased frequency in patients who have suffered intracerebral bleeds and are associated with increased risk of hemorrhagic transformation in patients who undergo thrombolytic therapy [49],.

8.2 Cerebral Venous Thrombosis and Infarction

Thromboses of the dural sinuses and or cerebral veins (CVT) are relatively uncommon, accounting for 0.5–1% of all strokes [105] with 78% of cases occurring in patients under 50 years of age and 75% of adult patients are female [106]. Risk factors vary from arterial-side stroke and include thrombophilia, inflammatory bowel disease, cancer, oral contraceptives, pregnancy, the postpartum state, dehydration, and head trauma [107]. The underlying causes of CVT are expressed in the Virchow's triad concept [108]. The modern evolution of this concept is that thrombosis is the result of multiple factors including thrombophilia, venous stasis, or obstruction and inflammation of the venous endothelium (Table 16).

The symptoms of venous infarction of the brain are unlike that of arterial infarction and thrombosis. Headache, particularly prodromal headache, is quite common and more frequent than in arterial stroke. The onset of symptoms can be slower than arterial-side stroke. Other contrasting signs or symptoms include dilation of scalp veins, subacute or chronic onset of signs, bilateral signs, seizures, and hemorrhage, particularly atypical lobar hemorrhage or hemorrhage that crosses typical vascular distributions. CVT explained 5% of cases of intracerebral hemorrhage and was three times as frequent in women as in men [109].

The venous geometry of the brain contrasts greatly with the arterial geometry with significant variation including frequent unilateral absence of the transverse sinus and proximity of the major venous sinuses to the paranasal and other air-filled sinuses. This may in some cases contribute to the risk for CVT. In patients with large middle cerebral artery infarctions consequent to the development of massive brain edema abnormalities in cerebral venous drainage patterns [110] contributed to the risk of CVT.

In contrast with arterial-side stroke resulting from arterial thrombosis, cerebral infarction from venous thrombosis is much more variable. It is estimated that cerebral parenchymal infarction may occur with 50% of detected cerebral venous thromboses. One consequence of venous thrombosis is the impairment of CSF reabsorption which may lead to hydrocephalus. Another is the relative reversibility of FLAIR and T2-weighted imaging changes that may occur with treatment (heparin and/or oral anticoagulants).

Table 16
Venous thrombosis and venous infarction. risk factors

Antithrombin III deficiency
Dehydration
Hyperhomocysteinemia
Pregnancy
Presence of Factor V Leiden
Protein C deficiency
Oral contraceptives
Substance abuse

Occlusion of superficial cerebral veins produces infarctions with characteristic patterns of hyperdense dural sinuses, gyral swelling, and parenchymal hemorrhage that may not be specific to stroke and may also be seen in tumor, trauma, and infection. Infarcts of the deeper cerebral venous structures produce focal characteristic patterns that are usually associated with deteriorating clinical condition.

The typical initial imaging findings of cerebral venous infarction are hyperdense dural sinuses or cortical veins on non-contrast CT although contrast CT and digital subtraction angiography may be more sensitive. Variations in cerebral venous anatomy complicate diagnosis. In the superior sagittal sinus, a triangle of dense thrombus may be seen. MRI is more sensitive for the detection of small cortical vein thromboses and for the delineation of micro-hemorrhages and determining the age of larger hemorrhages. CT imaging methods may be preferred when patients have altered mental status, are moving or may have incompatible devices. The treatment of cerebral venous thrombosis remains anticoagulation. The presence of hemorrhages in association with cerebral venous thrombosis is not an absolute contraindication to anticoagulation [107]. The above considerations are illustrated in images of a case of cerebral venous thrombosis (Fig. 6).

8.3 Stroke Mimics

There are no specific biochemical or imaging biomarkers for stroke mimics. Stroke mimic are patients that initially present with neurological symptoms compatible with stroke, usually in the emergency room, but turn out not to be strokes. Not every patient who is screened for stroke should be considered as a stroke mimic, even if the templated EMR note text includes stroke in an overly broad differential diagnosis. The mimic should present some degree of diagnostic challenge to an experienced neurologist.

Stroke mimics are more frequently encountered in "telestroke" evaluations where emergency room staff and physicians consult stroke neurologists remotely via video or internet links, accounting for 27% of suspected strokes compared to 8–14% of strokes that are evaluated by neurologists directly [111–113]. Such stroke mimics are frequently distinguished by having fewer vascular risk factors, less of an NIH Stroke Scale score, a greater likelihood of having an underlying seizure disorder or having psychiatric problems. These and other stroke mimics are shown in Table 17. Most stroke mimic cases are sorted out within hours, but some may make it through a stroke evaluation and even get alteplase. One practical consideration is that while administering alteplase to a "normal individual" may be relatively safe, it is not cheap.

Stroke mimics that have preexisting focal imaging findings from prior strokes or other conditions such as multiple sclerosis (MS) take longer to sort out. There is also no reason why a patient with MS cannot have a stroke.

Table 17
Stroke mimics

Conditions that may mimic stroke and non-focal or no imaging abnormalities	Conditions that may mimic stroke with focal imaging abnormalities
Psychogenic conversion disorders	Multiple sclerosis
Seizure disorders with post-ictal paralysis	Status epilepticus
Toxic-metabolic encephalopathy	
Migraine	Migraine
	Jacob-Creutzfeld disease

Nonspecific white matter hyperintensities are common in migraine. Migraine can cause a wide variety of focal neurological deficits (motor, sensory, visual, cranial nerve, and cognitive) [114] and hemiplegic migraine is associated with an increased risk for stroke, with mutations of the calcium channel and responds clinically to a calcium channel blocker, verapamil [115]. Often, migraine events are associated with headache, though "acephalgic" events are increasingly common with aging. Migraine events may involve multiple vascular distributions. The sudden clinical deficits from stroke typically have more rapid onset than MS attacks and follow vascular distribution patterns. Other subacute conditions with focal imaging findings may present with acute symptoms of unrelated nature, such as a toxic-metabolic encephalopathy. Figure 7 contains anecdotal examples of stroke mimic images and stroke mimic cases that took a little longer to sort out.

In one case a patient presented with a history of hypertension and complaints of weakness and confusion. The initial DWI MRI showed focal hyperintensity and a diagnosis of possible stroke was in the differential diagnosis list. However, despite an initial improvement, the patient had a progressive course with stimulus-sensitive myoclonus and other developments including a ribbon pattern spread of DWI hyperintensity and was finally diagnosed with a prion disorder (Creutzfeld-Jacob disease, CJD). This case reinforces the observation that CJD occurs in older patients who more frequently have vascular risk factors and might even have unrelated prior strokes.

In a second case, a patient presented with altered mental status and rhythmic movements of the tongue and palate in a pattern similar to palatal myoclonus, a movement disorder common after infarctions in the brainstem in the region of the dentate and red nuclei. Diffusion weighted MRI did show abnormal signal in the dentate nucleus and stroke was mentioned in the differential, though the imaging pattern would be somewhat atypical for stroke, being symmetrical. This was recognized by a movement disorder

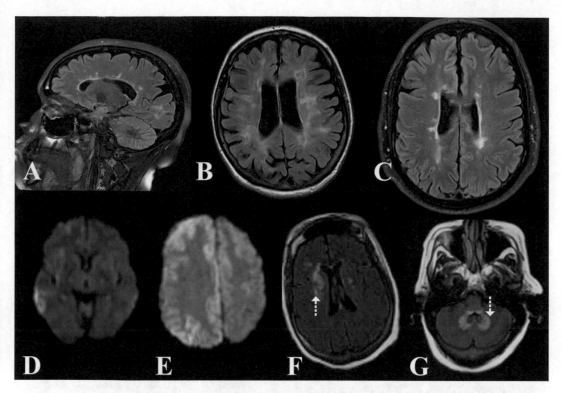

Fig. 7 Stroke mimics. (**a–c**) FLAIR MRI images of MS patient brains showing typical "Dawson's fingers" or ventricular surface based ovoid hyperintensities. (**d**) DWI image of patient showing focal cortical hyperintensity. (**e**) Same patient, weeks later. Follow-up image showing evolution of focal cortical hyperintensity to generalized ribbon pattern of CJD. (**f–g**) FLAIR MRI of patient presenting with abnormal movements showing subcortical and dentate nucleus hyperintensities. The subcortical are likely from subacute or chronic microvascular disease but the dentate lesion is from metronidazole toxicity and not typical of stroke

specialist as characteristic of metronidazole toxicity. Although the patient's family did not report metronidazole as a current or past medication, several empty metronidazole bottles were found in the patient's medicine cabinet.

Most stroke mimics are resolved using purely clinical means with variable effort being the most common discriminator.

9   Summary

History remains the most important part of clinical assessment, but images provide a valuable assist, particularly when the history is unavailable or unreliable. The use of specific criteria in image reporting and interpretation is valuable to the clinician because the list of possible causes of imaging changes is potentially very extensive and a sense of relative probability is necessary.

References

1. WSDC (2018) World stroke day campaign 2018. https://eso-stroke.org/eso/world-stroke-day-29-october-2018/
2. CDC (2018) Stroke facts. CDC, Atlanta, GA
3. An SJ, Kim TJ, Yoon B-W (2017) Epidemiology, risk factors, and clinical features of intracerebral hemorrhage: an update. J Stroke 19:3–10
4. Chimowitz MI et al (2005) Comparison of warfarin and aspirin for symptomatic intracranial arterial stenosis. N Engl J Med 352:1305–1316
5. Broussalis E, Trinka E, Hitzl W, Wallner A, Chroust V (2013) Comparison of stent-retriever devices versus the merci retriever for endovascular treatment of acute stroke. AJNR Am J Neuroradiol 34(2):366–372
6. WHO (1993) WHO international programme on chemical safety. Biomarkers and risk assessment: concepts and principles. WHO, Geneva. http://www.inchem.org/documents/ehc/ehc/ehc155.htm
7. Strimbu K, Tavel JA (2010) What are biomarkers? Curr Opin HIV AIDS 5:463–466
8. Goodsaid F, Frueh F (2007) Biomarker qualification pilot process at the US Food and Drug Administration. AAPS J 9:E105–E108
9. FDA (2018) Biomarker qualification program – list of qualified biomarkers. List of qualified biomarkers. https://www.fda.gov/Drugs/DevelopmentApprovalProcess/DrugDevelopmentToolsQualificationProgram/BiomarkerQualificationProgram/ucm535383.htm
10. Fan J et al (2010) N-Methyl-d-aspartate receptor subunit- and neuronal-type dependence of excitotoxic signaling through postsynaptic density 95. J Neurochem 115:1045–1056
11. Ankarcrona M et al (1995) Glutamate-induced neuronal death: a succession of necrosis or apoptosis depending on mitochondrial function. Neuron 15:961–973
12. Choi DW (1988) Glutamate neurotoxicity and diseases of the nervous system. Neuron 1:623–634
13. Attwell D et al (2010) Glial and neuronal control of brain blood flow. Nature 468:232–243
14. McConnell HL, Kersch CN, Woltjer RL, Neuwelt EA (2017) The translational significance of the neurovascular unit. J Biol Chem 292:762–770
15. Hainsworth AH et al (2017) Neuropathology of white matter lesions, blood-brain barrier dysfunction, and dementia. Stroke 48:2799–2804
16. Dzialowski I et al (2006) Extent of early ischemic changes on computed tomography (CT) before thrombolysis: Prognostic value of the Alberta Stroke Program early CT score in ECASS II. Stroke 37:973–978
17. Demchuk AM et al (2005) Importance of early ischemic computed tomography changes using ASPECTS in NINDS rtPA stroke study. Stroke 36:2110–2115
18. Truwit C, Barkovich A, Gean-Marton A et al (1990) Loss of the insular ribbon: another early CT sign of acute middle cerebral artery infarction. Radiology 176:801–906
19. Kamalian S, Kemmeling A, Borgie R, Morais L (2013) Admission insular infarction >25% is the strongest predictor of large mismatch loss in proximal MCA stroke. Stroke 44:3084–3089
20. Tomura N et al (1988) Early CT finding in cerebral infarction: obscuration of the lentiform nucleus. Radiology 168:463–467
21. Barber PA et al (2001) Hyperdense sylvian fissure MCA "dot" sign. A CT marker of acute ischemia. Stroke 32:84–88
22. Goldmakher GV et al (2009) Hyperdense basilar artery sign on unenhanced CT predicts thrombus and outcome in acute posterior circulation stroke. Stroke 40:134–139
23. Berkhemer OA, Kamalian S, González RG, Majoie CBLM, Yoo AJ (2013) Imaging biomarkers for intra-arterial stroke therapy. Cardiovasc Eng Technol 4:339–351
24. Barber PA et al (1999) Identification of major ischemic change: diffusion-weighted imaging versus computed tomography. Stroke 30:2059–2065
25. Sanossian N et al (2009) Angiography reveals that fluid-attenuated inversion recovery vascular hyperintensities are due to slow flow, not thrombus. Am J Neuroradiol 30:564–568
26. Caplan LR (2015) Lacunar infarction and small vessel disease: pathology and pathophysiology. J Stroke 17:2
27. Torvik A (1984) The pathogenesis of watershed infarcts in the brain. Stroke 15:221–223
28. Momjian-Mayor I, Baron JC (2005) The pathophysiology of watershed infarction in internal carotid artery disease: review of cerebral perfusion studies. Stroke 36:567–577

29. Liu H, Chu J, Zhang L, Yan Z, Zhou S (2016) Early carotid artery stenting for cerebral watershed infarction is safe and effective: a retrospective study. Eur Neurol 76:256–260

30. Liu H et al (2016) Clinical comparison of outcomes of early versus delayed carotid artery stenting for symptomatic cerebral watershed infarction due to stenosis of the proximal internal carotid artery. Biomed Res Int 2016:1–7

31. Kidwell CS et al (1999) Diffusion MRI in patients with transient ischemic attacks. Stroke 30:1174–1180

32. Crisostomo RA, Garcia MM, Tong DC (2003) Detection of diffusion-weighted MRI abnormalities in patients with transient ischemic attack: correlation with clinical characteristics. Stroke 34(4):932–937

33. Prabhakaran S, Chong JY, Sacco RL (2007) Impact of abnormal diffusion-weighted imaging results on short-term outcome following transient ischemic attack. Arch Neurol 64:1105

34. Song B et al (2015) Validation of the RRE-90 scale to predict stroke risk after transient symptoms with infarction: a prospective cohort study. PLoS One 10:1–11

35. Johnston SC et al (2007) Validation and refinement of scores to predict very early stroke risk after transient ischaemic attack. Lancet 369:283–292

36. Ay H et al (2010) A score to predict early risk of recurrence after ischemic stroke. Neurology 74:128–135

37. Sheehan OC et al (2010) Population-based study of ABCD 2 score, carotid stenosis, and atrial fibrillation for early stroke prediction after transient ischemic attack. Stroke 41:844–850

38. Josephson SA, Sidney S, Pham TN, Bernstein AL, Johnston SC (2008) Higher ABCD2-score predicts patients most likely to have true transient ischemic attack. Stroke 39:3096–3098

39. Albers G, Caplan L, Easton J (2002) Transient ischemic attack – proposal for a new definition. N Engl J Med 347:1713–1716

40. Easton JD et al (2009) Definition and evaluation of transient ischemic attack: a scientific statement for healthcare professionals from the American Heart Association/American Stroke Association Stroke Council; Council on Cardiovascular Surgery and Anesthesia; Council on Cardio. Stroke 40:2276–2293

41. Powers WJ et al (2018) 2018 Guidelines for the early management of patients with acute ischemic stroke: a guideline for healthcare professionals from the American Heart Association/American Stroke Association. Stroke 49:e46–e110

42. Mărgăritescu O et al (2008) Histopathological changes in acute ischemic stroke. Rom J Morphol Embryol 50:327–339

43. Paciaroni M et al (2008) Early hemorrhagic transformation of brain infarction: rate, predictive factors, and influence on clinical outcome: results of a prospective multicenter study. Stroke 39:2249–2256

44. Hacke W et al (1995) Intravenous thrombolysis with recombinant tissue plasminogen activator for acute hemispheric stroke. The European Cooperative Acute Stroke Study (ECASS). JAMA 274:1017–1025

45. Wahlgren N et al (2008) Multivariable analysis of outcome predictors and adjustment of main outcome results to baseline data profile in randomized controlled trials: safe implementation of thrombolysis in Stroke-MOnitoring STudy (SITS-MOST). Stroke 39:3316–3322

46. Meisterernst J et al (2017) Focal T2 and FLAIR hyperintensities within the infarcted area: a suitable marker for patient selection for treatment? PLoS One 12:1–10

47. Kalinin MN, Khasanova DR, Ibatullin MM (2017) The hemorrhagic transformation index score: a prediction tool in middle cerebral artery ischemic stroke. BMC Neurol 17:1–16

48. Charidimou A, Shoamanesh A (2016) Clinical relevance of microbleeds in acute stroke thrombolysis: comprehensive meta-analysis. Neurology 87:1534–1541

49. Tsivgoulis G et al (2016) Risk of symptomatic intracerebral hemorrhage after intravenous thrombolysis in patients with acute ischemic stroke and high cerebral microbleed burden ameta-analysis. JAMA Neurol 73:675–683

50. Lansberg MG et al (2007) Risk factors of symptomatic intracerebral hemorrhage after tPA therapy for acute stroke. Stroke 38:2275–2278

51. Christoforidis GA, Mohammad Y, Kehagias D, Avutu B, Slivka AP (2005) Angiographic assessment of pial collaterals as a prognostic indicator following intra-arterial thrombolysis for acute ischemic stroke. AJNR Am J Neuroradiol 26(7):1789–1797

52. Kucinski T et al (2003) Collateral circulation is an independent radiological predictor of outcome after thrombolysis in acute ischaemic stroke. Neuroradiology 45:11–18

53. Astrup J, Siesjö BK, Symon L (1981) Thresholds in cerebral ischemia – the ischemic penumbra. Stroke 12:723–725

54. Symon L (1985) Flow thresholds in brain ischaemia and the effects of drugs. Br J Anaesth 57:34–43

55. Higashida RT, Furlan AJ, Assessment T (2003) Trial design and reporting standards for intra-arterial cerebral thrombolysis for acute ischemic stroke. Stroke 34:2774

56. Bandera E et al (2006) Cerebral blood flow threshold of ischemic penumbra and infarct core in acute ischemic stroke: a systematic review. Stroke 37:1334–1339

57. Tisserand M et al (2016) Does diffusion lesion volume above 70 mL preclude favorable outcome despite post-thrombolysis recanalization? Stroke 47:1005–1011

58. Lyden PD, Zweifler R, Mahdavi Z, Lonzo L (1994) A rapid, reliable, and valid method for measuring infarct and brain compartment volumes from computed tomographic scans. Stroke 25:2421–2428

59. Campbell BCV et al (2013) Advanced imaging improves prediction of hemorrhage after stroke thrombolysis. Ann Neurol 73:510–519

60. Wu B et al (2018) Optimal delay time of CT perfusion for predicting cerebral parenchymal hematoma after intra-arterial tPA treatment. Front Neurol 9:1–8

61. Yassi N et al (2013) Prediction of poststroke hemorrhagic transformation using computed tomography perfusion. Stroke 44:3039–3043

62. Jain AR et al (2013) Association of CT perfusion parameters with hemorrhagic transformation in acute ischemic stroke. Am J Neuroradiol 34:1895–1900

63. Straka M, Albers GW, Bammer R (2010) Real-time diffusion-perfusion mismatch analysis in acute stroke. J Magn Reson Imaging 32:1024–1037

64. Nogueira RG et al (2018) Thrombectomy 6 to 24 hours after stroke with a mismatch between deficit and infarct. N Engl J Med 378 (1):11–21. https://doi.org/10.1056/NEJMoa1706442

65. Haussen DC et al (2016) Automated CT perfusion ischemic core volume and noncontrast CT ASPECTS (Alberta stroke program early CT score): correlation and clinical outcome prediction in large vessel stroke. Stroke 47:2318–2322

66. Berkhemer OA et al (2015) A randomized trial of intraarterial treatment for acute ischemic stroke. N Engl J Med 372:11–20

67. Goyal M et al (2015) Randomized assessment of rapid endovascular treatment of ischemic stroke. N Engl J Med 372:1019–1030

68. Campbell BCV et al (2015) Endovascular therapy for ischemic stroke with perfusion-imaging selection. N Engl J Med 372:1009–1018

69. Saver JL et al (2015) Stent-retriever thrombectomy after intravenous t-PA vs. t-PA alone in stroke. N Engl J Med 372:2285–2295

70. Jovin TG et al (2015) Thrombectomy within 8 hours after symptom onset in ischemic stroke. N Engl J Med 372:2296–2306

71. Achit H et al (2017) Cost-effectiveness of thrombectomy in patients with acute ischemic stroke: the THRACE randomized controlled trial. Stroke 48:2843–2847

72. Goyal M et al (2016) Endovascular therapy in acute ischemic stroke: challenges and transition from trials to bedside. Stroke 47:548–553

73. Borst J et al (2017) Value of thrombus CT characteristics in patients with acute ischemic stroke. Am J Neuroradiol 38:1758–1764

74. Yoo AJ et al (2012) Infarct volume is a pivotal biomarker after intra-arterial stroke therapy. Stroke 43:1323–1330

75. Bang OY et al (2011) Collateral flow predicts response to endovascular therapy for acute ischemic stroke. Stroke 42:693–699

76. Bang OY, Goyal M, Liebeskind DS (2015) Collateral circulation in ischemic stroke: assessment tools and therapeutic strategies. Stroke 46:3302–3309

77. Sylaja PN et al (2007) Acute ischemic lesions of varying ages predict risk of ischemic events in stroke/TIA patients. Neurology 68:415–419

78. Hill MD et al (2014) Alberta stroke program early computed tomography score to select patients for endovascular treatment interventional management of stroke (IMS)-III trial. Stroke 45:444–449

79. Marks MP et al (2018) Endovascular treatment in the DEFUSE 3 study. Stroke 022147:118. https://doi.org/10.1161/STROKEAHA.118.022147

80. Albers GW et al (2018) Thrombectomy for stroke at 6 to 16 hours with selection by perfusion imaging. N Engl J Med 378:708–718

81. Saposnik G, Strbian D (2018) Enlightenment and challenges offered by DAWN trial (DWI or CTP assessment with clinical mismatch in the triage of wake up and late presenting strokes undergoing neurointervention with trevo). Stroke 49:498–500

82. Tomlinson BE, Blessed G, Roth M (1968) Observations on the brains of non-demented old people. J Neurol Sci 7:331–356

83. Fazekas F et al (1987) MR signal abnormalities at 1.5 T in Alzheimer's dementia and normal aging. Am J Roentgenol 149:351–356

84. Fazekas F et al (1993) Pathologic correlates of incidental MRI white matter signal hyperintensities. Neurology 43:1683–1689

85. Wardlaw JM et al (2013) Neuroimaging standards for research into small vessel disease and its contribution to ageing and neurodegeneration. Lancet Neurol 12:822–838

86. Smith EE et al (2017) Prevention of stroke in patients with silent cerebrovascular disease: a scientific statement for healthcare professionals from the American Heart Association/American Stroke Association. Stroke 48:e44–e71

87. Field TS, Benavente OR (2011) Penetrating artery territory pontine infarction. Rev Neurol Dis 8:30–38

88. Kuller LH et al (2004) White matter hyperintensity on cranial magnetic resonance imaging: a predictor of stroke. Stroke 35:1821–1825

89. Pantoni L, Garcia JH (1995) The significance of cerebral white matter abnormalities 100 years after Binswanger's report. A review. Stroke 26:1293–1301

90. De Bresser J et al (2018) White matter hyperintensity shape and location feature analysis on brain MRI; proof of principle study in patients with diabetes. Sci Rep 8:1–10

91. Brun A, Englund E (1986) A white matter disorder in dementia of the Alzheimer type: a pathoanatomical study. Ann Neurol 19:253–262

92. Williamson W et al (2018) Association of cardiovascular risk factors with MRI indices of cerebrovascular structure and function and white matter hyperintensities in young adults. JAMA 320:665

93. O'Sullivan M et al (2002) Patterns of cerebral blood flow reduction in patients with ischemic leukoaraiosis. Neurology 59:321–326

94. Moustafa RR et al (2011) Microembolism versus hemodynamic impairment in rosary-like deep watershed infarcts: a combined positron emission tomography and transcranial doppler study. Stroke 42:3138–3143

95. Gupta A et al (2016) Silent brain infarction and risk of future stroke: a systematic review and meta-analysis. Stroke 47:719–725

96. Baradaran H et al (2016) Silent brain infarction in patients with asymptomatic carotid artery atherosclerotic disease. Stroke 47:1368–1370

97. Filippi M et al (2016) MRI criteria for the diagnosis of multiple sclerosis: MAGNIMS consensus guidelines. Lancet Neurol 15:292–303

98. McDonald WI et al (2001) Recommended diagnostic criteria for multiple sclerosis: guidelines from the International Panel on the Diagnosis of Multiple Sclerosis. Ann Neurol 50:121–127

99. Leary M, Saver J (2001) Incidence of silent stroke in the United States. Stroke 32:P134

100. Weber R et al (2012) Risk of recurrent stroke in patients with silent brain infarction in the Prevention Regimen for Effectively Avoiding Second Strokes (PRoFESS) imaging substudy. Stroke 43:350–355

101. Gomori JM, Grossman RI (1988) Mechanisms responsible for the MR appearance and evolution of intracranial hemorrhage. Radiographics 8:427–440

102. Greenberg SM et al (2009) Cerebral microbleeds: a field guide to their detection and interpretation. Lancet Neurol 8:165–174

103. Vernooij MW et al (2008) Prevalence and risk factors of cerebral microbleeds: the Rotterdam Scan Study. Neurology 70:1208–1214

104. Fazekas F et al (1999) Histopathologic analysis of foci of signal loss on gradient-echo T2*-weighted MR images in patients with spontaneous intracerebral hemorrhage: evidence of microangiopathy-related microbleeds. Am J Neuroradiol 20:637–642

105. Stam J (2005) Thrombosis of the cerebral veins and sinuses. N Engl J Med 352:1791–1798

106. Canhão P et al (2005) Causes and predictors of death in cerebral venous thrombosis. Stroke 36:1720–1725

107. Saposnik G et al (2011) Diagnosis and management of cerebral venous thrombosis: a statement for healthcare professionals from the American Heart Association/American Stroke Association. Stroke 42:1158–1192

108. Kumar DR, Hanlin ER, Glurich I, Mazza JJ, Yale SH (2010) Virchow's contribution to the understanding of thrombosis and cellular biology. Clin Med Res 8:168–172

109. Janghorbani M et al (2008) Cerebral vein and dural sinus thrombosis in adults in Isfahan, Iran: frequency and seasonal variation. Acta Neurol Scand 117:117–121

110. Yu W et al (2009) Hypoplasia or occlusion of the ipsilateral cranial venous drainage is associated with early fatal edema of middle cerebral artery infarction. Stroke 40:3736–3739

111. Ali SF et al (2014) The telestroke mimic (TM)-score: a prediction rule for identifying stroke mimics evaluated in a telestroke network. J Am Heart Assoc 3:14–16

112. Lekoubou A, Sin DI, Smock A, Ozark S (2018) Incidence of stroke mimics and post R-tpa hemorrhage in the era of telemedicine. Stroke 47:AWP335

113. Pavlovic AM et al (2010) Brain imaging in transient ischemic attack—redefining TIA. J Clin Neurosci 17:1105–1110

114. Fisher CM (1980) Late-life migraine accompaniments as a cause of unexplained transient ischemic attacks. Can J Neurol Sci 7:9–17

115. Knierim E et al (2011) Recurrent stroke due to a novel voltage sensor mutation in cav2.1 responds to verapamil. Stroke 42:14–17

116. Prokop M, Galanski M (2001) Spiral and multislice computed tomography of the body. AM J Roentgenol 181(8):1558

117. Alemany Ripoll M, Stenborg A, Sonninen P, Terent A, Raininko R et al (2004) Detection and appearance of intraparenchymal haematomas of the brain at 1.5 T with spin-echo, FLAIR and GE sequences: poor relationship to the age of the haematoma. Neuroradiology 46:435–443

118. Furlan A et al (1999) Intra-arterial prourokinase for acute ischemic stroke. JAMA 282:2003

119. Kidwell CS et al (2013) A trial of imaging selection and endovascular treatment for ischemic stroke. N Engl J Med 368:914–923

Chapter 15

Neuroimaging Methods for Acute Stroke Diagnosis and Treatment

Mathew Elameer and Christopher I. Price

Abstract

In this chapter, we explore the radiological imaging of acute stroke with a focus on the identification of ischemia. We begin by reviewing the role of neuroimaging in emergency stroke management, followed by an overview of a pathophysiological model for ischemic stroke to explain why radiological appearances vary. Finally, we use theory and examples to consider the advantages and disadvantages of the two major neuroimaging modalities: computed tomography (CT) and magnetic resonance imaging (MRI).

Key words Ischemic stroke, Neuroimaging, Neuroradiology, Thrombolysis, Thrombectomy, Computed tomography (CT), Magnetic resonance imaging (MRI)

1 The Role of Neuroimaging in Acute Ischemic Stroke

The two components of stroke diagnosis are an accurate clinical review including physical examination, and appropriate interpretation of radiological imaging. Clinicians have been diagnosing stroke for centuries prior to the advent of radiology by recognizing sudden onset of specific and persistent patterns of neurological impairment but were unable to determine the etiology premortem [1]. Neuroimaging allows us to overcome the limitations of clinical review by enabling us to do the following:

1. To separate genuine stroke patients from those with identical symptoms, that is, "stroke mimics," which represent up to 25% of stroke unit referrals. Common conditions include seizure, syncope, sepsis, brain tumor, and functional syndromes [2]. Some can be excluded by a scan which positively demonstrates changes of acute stroke such as an established infarct or hemorrhage, or confirmed by appearances of a space occupying lesion. However standard neuroimaging may appear normal in a number of stroke mimics such as migraine, so a thorough

Philip V. Peplow et al. (eds.), *Stroke Biomarkers*, Neuromethods, vol. 147, https://doi.org/10.1007/978-1-4939-9682-7_15,
© Springer Science+Business Media, LLC, part of Springer Nature 2020

clinical assessment and additional investigations are also
required for their diagnosis.

2. To distinguish ischemic from hemorrhagic stroke etiology for
determining emergency treatment. Intravenous thrombolytic
treatment with tissue plasminogen activator (tPA) for hyperacute ischemic stroke is a highly time-critical treatment which
triggers the natural fibrinolysis pathway to remove occlusive
thrombus in selected cases of ischemic stroke. It has been
estimated that during ischemic stroke the rate of neuronal
death is 1.9 million per minute [3] and with every minute
that thrombolysis is delayed, almost two days of equivalent
healthy life are lost [4]. Clinical trials suggest that without
advanced imaging, treatment should be given within 4.5 h of
onset to provide a one in seven chance of significant recovery
(against a one in 30 chance of treatment associated hemorrhage
causing deterioration). There are also many clinical reasons for
why tPA cannot be given such as uncontrolled blood pressure.
Intracranial hemorrhage must be ruled out before thrombolysis can begin, and this can be achieved most effectively through
the use of neuroimaging. In hemorrhagic stroke, reversal of
anticoagulation should be undertaken urgently in patients with
abnormal clotting due to medication or illness (e.g., chronic
liver disease). Lowering of blood pressure during the first 12 h
of onset can reduce poorer outcomes by decreasing hematoma
expansion [5]. After emergency assessment, early antiplatelet
medication to prevent stroke recurrence, such as aspirin,
should only be used for ischemic cases [6].

3. To estimate the age of an infarct, as the benefits of thrombolysis
with tPA diminish rapidly with time, and after around 6 h from
stroke onset the risks of treatment (e.g., hemorrhagic transformation) are considered to outweigh the benefits [7]. However,
the precise time of stroke onset is not known for many patients,
for example, 37% of all acute stroke admissions in the UK
between 2016 and 2017. This is often because stroke onset
has occurred during sleep [8]. Neuroimaging is increasingly
being used to identify those patients with unknown stroke
onset time who are may still benefit from thrombolysis [9].

4. To distinguish large from small vessel occlusion, which is now
critically important in the early selection of patients for treatment with mechanical thrombectomy. This can be an extremely
effective minimally invasive surgical treatment for patients with
severe ischemic stroke due to large vessel occlusion (LVO)
[10]. Clinical trials suggest that when onset is known, suitable
patients within 6 h will have a one in three chance of better
recovery from thrombectomy treatment to physically remove a
large arterial thrombus when used in addition to tPA. A smaller
number of additional patients are unsuitable for tPA (e.g., due

to recent major surgery increasing the risk of treatment associated bleeding) but can receive thrombectomy alone. Clinical prediction instruments have been formulated to attempt to predict LVO from clinical signs and symptoms. These are currently limited in either specificity or sensitivity [11], so neuroimaging remains the standard method of discriminating LVO patients from non-LVO ischemic strokes for referral to centers that provide thrombectomy. A further advantage of using neuroimaging to identify LVO stroke is being able to locate the exact position of the thrombus and how accessible it is for removal, which is also important in the selection of patients for thrombectomy.

5. Assess the extent of the infarct core (dead tissue) relative to the surrounding penumbra (living but vulnerable tissue) in ischemic stroke to select patients for emergency treatments. Very early ischemic stroke will present with a relatively large penumbra and small infarct core, reflecting that there is still salvageable brain tissue [12]. If it is not reperfused by tPA or thrombectomy, the volume of the core will grow at the expense of the penumbra: growth in infarct core has been measured in humans over as little as 3–6 h [13]. There is a subset of stroke patients with a good collateral arterial blood supply who maintain their ischemic penumbra for longer [14], and these patients can benefit from thrombectomy even up to 24 h post stroke onset or considered for treatment even when the onset time is not known [15].

Before we explore how the various imaging neuroimaging methods can fulfill these roles, we will first review the pathophysiology of acute ischemic stroke so that the theory underpinning neuroimaging methodology and interpretation can be better understood.

2   Pathophysiology of Acute Ischemic Stroke

Ischemic stroke evolves over time in a heterogeneous fashion according to a number of pathophysiological and anatomical factors, and in clinical practice the radiological assessment of acute stroke can be more of an art than an exact science. It is important to consider clinical information from history and examination when interpreting neuroradiological appearances, which can overlap between pathologies, and we are continuing to learn the limitations of different modalities. A basic model of ischemic stroke pathophysiology is useful to help interpretation of images produced by modern neuroimaging methods.

Ischemic stroke can be imaged in two different ways: directly (by imaging the arterial patency or cerebral perfusion), or indirectly (by imaging changes in the distribution of water within the brain parenchyma). The direct imaging techniques of angiography and perfusion imaging will be described later in this chapter. For now, we will focus on a model which represents the fluid shifts which occur in the post-stroke period, as it is these changes which result in the imaging findings on plain CT and MRI.

2.1 Natural Course of Water-Shift in Ischemic Stroke

There is considerable variance in the definitions used to describe the natural course of stroke evolution. This is partly due to the heterogeneous nature in which ischemic stroke evolves—for example, the infarct core may progress more quickly than the ischemic penumbra, and the rate of progression can vary from patient to patient. Furthermore, different groups of practitioners may use these terms in different ways: a stroke clinician may focus on temporal definitions (e.g., describing all strokes that present within 24 h as "hyperacute"), whereas a radiologist may focus on pathophysiological definitions (e.g., describing all patients in the first pathophysiological phase of stroke evolution as "hyperacute," regardless of the actual time from onset to imaging).

For the sake of clarity, we will consider the typical natural course of stroke evolution in five stages: early hyperacute, late hyperacute, acute, subacute, and chronic [16].

Early hyperacute phase (<6 h): This phase begins with the sudden interruption of intracerebral arterial blood flow usually due to thrombus combined with narrowing of the arterial lumen by atheroma associated with risk factors (e.g., age, hypertension, smoking). The onset of symptoms is extremely rapid: one animal study detected clinical signs within seconds of arterial ligation, usually increasing to maximum severity within 10 min [17].

The interruption of the arterial blood supply arrests cellular respiration, rapidly depleting the intracellular ATP concentration. This results in a failure of the active Na–K pumps that normally maintain the sodium concentration gradient between the intracellular (low sodium) and extracellular (high sodium) spaces [18].

With the power supply to the membrane pumps interrupted, the cell wall is unable to maintain this gradient, and extracellular sodium moves down the concentration gradient across the membrane. Influxing sodium cations result in the passage of chloride anions, and both ions draw water molecules from the extracellular space, causing the intracellular space to swell (Fig. 1). This swelling is known as *cytotoxic edema* [18] and will spread from the core outward [19].

At this point, the intracellular space is overhydrated, but the extracellular space is dehydrated [20], usually resulting in no overall change to the water content of the brain tissue [18]. Cellular swelling impedes the diffusion of water in the extracellular space.

Late hyperacute (6–24 h): There is a variable degree of overlap with the early hyperacute phase, and cytotoxic edema may continue to progress as before. Cytotoxic edema enables *transvascular edema*

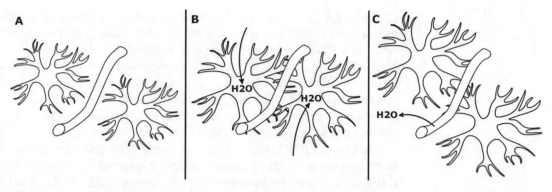

Fig. 1 Cytotoxic and transvascular oedema. (**a**) Normal—two astrocytes and a capillary. (**b**) Cytotoxic oedema: influxing water has caused the cells to swell, impinging on the extracellular space and leading to DWI signal changes on MRI. (**c**) Transvascular oedema—effluxing water from the capillary expands the extracellular space and leading to reduced density on CT

to develop. Transvascular edema consists firstly of *ionic edema*, which is the passive diffusion of sodium and chloride ions, and subsequently water molecules, across the capillary membrane into the extracellular space [18]. As the blood–brain barrier degrades, intravascular proteins also pass into the extracellular space, drawing water by oncotic force, which is known as *vasogenic edema* [18]. The distinction between ionic and vasogenic edema may be theoretically important because blood–brain barrier breakdown may predispose to hemorrhagic transformation causing further deterioration in neurological function [21], but conventional neuroimaging is currently unable to distinguish the etiology of transvascular edema.

Transvascular edema is dependent upon some residual perfusion, so may affect the ischemic penumbra more than the core infarct [21].

By this point, both the intracellular and extracellular spaces are overhydrated, resulting in an increase in the water content of the brain tissue. Diffusion of water in the extracellular space is still restricted by cellular swelling.

Acute (1–7 days): There is considerable overlap between the late hyperacute and acute phases, but the distinction is important for stroke clinicians: patients presenting within the late hyperacute phase may still be eligible for time-dependent thrombolysis or thrombectomy treatment depending upon the residual ischemic penumbra, however once the acute phase has been reached these treatments may no longer be an option.

The water shift during this period is complex and not yet completely understood. Cytotoxic edema has usually peaked within the late hyperacute phase [19] and diminishes throughout the acute phase. Transvascular edema continues to progress, and the overall water content of brain tissue will peak between 2 and 4 days after stroke onset [22]. Despite the resulting reduction in the ratio of intracellular to extracellular space from 24 h onward, the restriction of water diffusion does not begin to relax until 5–7 days post onset [13] which suggests that cellular swelling due to cytotoxic edema cannot be the only cause for diffusion restriction. Alternative mechanisms for diffusion restriction in the acute phase may include a higher concentration of transvascular proteins and debris from apoptotic cells within the extracellular space.

Subacute (1–3 weeks): Many nonviable cells will have undergone apoptosis by the subacute phase. Transvascular edema will continue to resolve in the subacute phase but the extracellular space remains larger than in the pre-stroke brain, partly due to cellular necrosis, and partly due to expansion of the extracellular matrix from reactive astrocytes. The restriction of water diffusion will continue to relax, pseudo-normalizing at 10–15 days, before water diffusion increases above baseline levels due to the expanded extracellular space [16].

In the subacute phase, water in the extracellular space will begin to diffuse more freely even than in healthy brain tissue. The overall water content of the brain tissue remains high due to the expanded overhydrated extracellular space.

Chronic (>3 weeks): By this stage, edema has usually resolved. Parts of the brain tissue may form cystic areas full of CSF (*cystic encephalomalacia*). These areas are separated from surrounding healthy brain tissue by glial scars with an expanded extracellular matrix (*gliosis*).

Now that we have reviewed the pathophysiology of acute ischemic stroke, we can proceed to explore the individual neuroimaging methods and findings most commonly encountered in the emergency setting. We will do this by exploring a typical pathway through the different neuroimaging methods for a patient presenting with acute ischemic stroke, explaining how each method can impact upon the acute management of patients.

3   Neuroimaging Methods in Acute Stroke

Stroke patients may undergo different modes of neuroimaging. The exact methods used can depend upon the clinical circumstances (such as suitability for thrombectomy), in addition to the facilities which are available within the local clinical radiology department (e.g., a patient presenting out-of-hours may not have access to an MRI scanner). The short timescales for effective tPA and thrombectomy treatment also limit what information is often made available to clinicians before a treatment decision must be made. Let us explore a typical pathway (Fig. 2) for a patient being assessed for a hyperacute stroke:

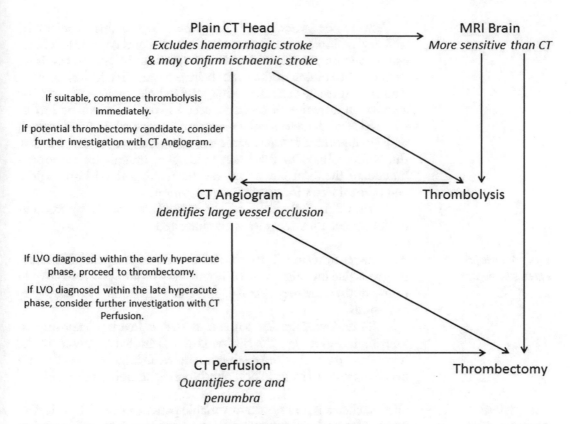

Fig. 2 An example neuroimaging pathway for acute ischaemic stroke. Practice may vary from hospital to hospital. CT Angiogram and CT Perfusion may be replaced by MR Angiogram and MR Perfusion, depending upon local availability. If Perfusion imaging is not used then maximal standard timescales between stroke onset and thrombolysis is 4.5 h, and thrombectomy is 6 h

3.1 Plain Computed Tomography (CT)

3.1.1 Background

Computed Tomography (CT) is universally accepted as the best first choice neuroimaging investigation for acute stroke due to the short procedure time, availability and relatively easy interpretation. It is based upon the principle that physical density is proportional to attenuation of an X-ray beam. Complicated calculations are used to reconstruct a dataset from the X-ray detectors that describes the volume of the brain as a series of small cubes (voxels), each with a brightness that corresponds to the average density of the tissue within each voxel.

The CT scanner consists of a circle-shaped gantry with an X-ray tube source on one side and a detector array on the other side. The gantry rotates around the patient throughout the scan to acquire a series of projections across as many different angles as possible. Older scanners moved the table one step at a time, resulting in the acquisition of separate axial slices through the brain. This was a time-consuming process and resulted in a trade-off either between slices of the patient being skipped by the scanner, or multiple unnecessary irradiations of adjacent slices.

The newer generation CT scanners bypass this problem by utilizing multiple parallel arrays of detectors and continuously and simultaneously moving the table and rotating the gantry, resulting in the entire volume of the brain being scanned in a helical fashion. This has three advantages—first, the whole brain can now be scanned in a matter of seconds; second, the helix can be reconstructed into whatever spatial resolution (aka voxel size) is required, with an important caveat that we will address shortly; third, because the entire volume of the brain is imaged, images are no longer limited to the axial plane and can be reconstructed into sagittal and coronal views for ease of interpretation.

Plain CT is distinguished from other forms of CT by whether or not intravenous contrast is administered.

3.1.2 Advantages and Disadvantages

Advantages of plain CT are that it is relatively inexpensive, universally available including out-of-hours, easy to interpret, effective in ruling out hemorrhagic stroke, and there are no immediate safety exclusions.

The disadvantage for patients is that it involves exposure to ionizing radiation. For the clinician there is limited sensitivity in the hyperacute phase, limited distinction between large and small vessel occlusion and no information about core–penumbra mismatch.

3.1.3 Variable Parameters

Slice thickness is an important variable parameter in plain CT. The same helix can be retrospectively reconstructed into different slice thicknesses even after the patient is off the table, although the necessary datasets are exceptionally large and therefore usually deleted or moved from the scanner's hard drive after 24–48 h. Ideally, both thin and thick slices should be reconstructed and reviewed (Fig. 3).

Thick slices result in a voxel size of around 1 mm (X) × 1 mm (Y) × 3–5 mm (Z). One advantage is that there are fewer images to store, transmit, and view. Furthermore, because more X-ray photons have contributed to the formation of each voxel in a thick rather than a thin slice, there is reduced quantum noise (which is the phenomenon that results in the black-and-white "grainy" random variation in CT images, similar to the static seen on analogue television sets). *This makes thick slices the best choice to assess for parenchymal edema.*

Thin slices result in a voxel size of around 1 mm (X) × 1 mm (Y) × 0.5–1 mm (Z). Their advantage is that they result in a more accurate representation of tissue density, because a shorter length of brain tissue is averaged to form each voxel. *This makes thin slices the best choice for assessing for hyperdense tissue (such as arterial thrombus).*

In-plane spatial resolution can also be improved by reducing the voxel size in the X and Y dimensions to <1mm, or by using a dedicated high resolution reconstruction filter known as a "bone",

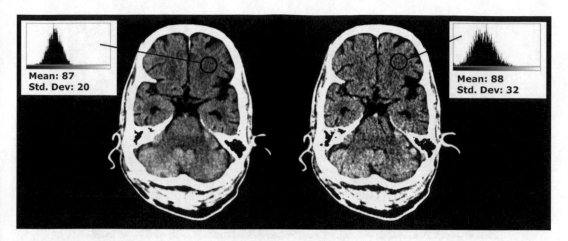

Fig. 3 Thick and thin slices reconstructed from the same plain CT. *Left:* Thick (5 mm) axial slices demonstrate lower quantum noise, as evidence by a smaller standard deviation for a similar mean brightness (the numerical values of pixel brightness here do not reflect the true Hounsfield Unit values because these measurements were retrospectively obtained from JPEG images). *Right:* Thin (1 mm) axial slices provide a more accurate representation of hyperdense structures (the black arrow represents a hyperdense basilar artery in cross-section created because the lumen is full of fresh thrombus, not well visualised on the thick slice)

"hard", or "high resolution" kernel, but both options will increase the image noise. In situations where material contrast is naturally high and spatial resolution is critical (for example, investigation of potential skull fractures) it is better to use a high resolution kernel and smaller voxel size. Stroke patients sometimes present with a history of head trauma either preceding or following the onset of their stroke symptoms, and reconstruction of both high- and low-resolution kernel images may be performed for this reason. *But low resolution or "soft" kernels should always be used to assess the brain parenchyma.*

Image window differs from image kernel and slice thickness in that it can be easily adjusted by the clinician or radiologist after the images have been reconstructed by the radiographer. It is necessary to "window" an image because of the vast range of densities encountered within the body, from around −1000 Hounsfield Units (HU) for air to +1000 HU for cortical bone (Fig. 4). The HU represent X-ray penetration of different tissue types and hence their appearance on images. Water and CSF are around 0–10 HU, white matter is around 20–30 HU, grey matter is around 35–45 HU, acute thrombus is around 50–75 HU, and acute hemorrhage is around 60–100 HU. The human visual system is limited in the range of shades of grey it can discriminate, so in order to discern small changes in density between voxels the range of around 2000 possible values for HU must be condensed into something much smaller. This is done by adjusting the "window width" to magnify the difference between tissues. The center HU

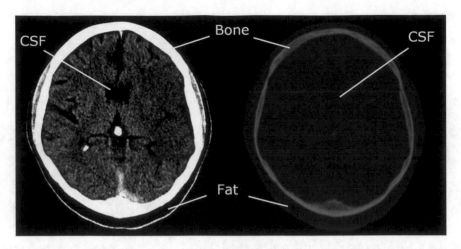

Fig. 4 A demonstration of the power of windowing. *Left:* Narrow stroke windows (window width of 35 HU; window level of 40; range 5–75 HU). CSF and fat both appear black because they are below the range of the window. No internal bone structure is visualised because it is all above the range of the window. *Right:* Very wide windows (width 2000; level 1000; range: −1000 to 3000 HU). No tissue is either black or white because all tissues are within the window range. The relative differences between fat, CSF, white matter and grey matter are all insignificant compared to their difference from bone

value of the window ("window level") can also be adjusted. A very narrow window is used for acute stroke to maximize the visible difference between grey matter, white matter, and edematous tissue—typically, *a window width of 30–40 HU is used*. The window level is usually also centered around 30–40 HU, so stroke windows are able to distinguish between densities within a range of around 0–80 HU. Narrow stroke windows will display bone, calcified atherosclerosis, and early hemorrhage with the same brightness because they are all above the range of the window. As such, brains should also be viewed with a wider window too—a typical general brain window width is around *80–100 HU* and should still be centered around 30–40 HU, to produce a range of around −40 to 120 HU.

3.1.4 Diagnostic Scope and Imaging Findings

Hemorrhagic Stroke

Plain CT is useful in the hyperacute phase of hemorrhagic stroke because it is particularly sensitive to the presence of blood [23]. In order to minimize the time from stroke onset to commencement of therapy, suitable patients may be started on thrombolytic agents as soon as hemorrhagic stroke has been ruled out on plain CT. Acute hemorrhage is easily identified by *markedly increased density within the brain parenchyma* on plain CT (Fig. 5), which if viewed on the appropriate stroke or brain windows, will make it appear exceptionally bright when compared to surrounding normal brain parenchyma. These changes are transient, and the hemorrhage will usually become isodense in the subacute phase before eventually becoming hypodense (and very difficult to distinguish from

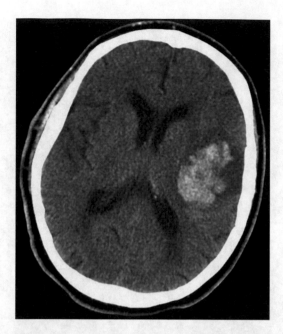

Fig. 5 Plain CT showing acute intracerebral haemorrhage in the left cerebral hemisphere, which is easily identified by a markedly increased density within the brain parenchyma

previous ischemic stroke) in the chronic phase [23]. CT is generally considered to be highly sensitive to hemorrhage within the first 8 days especially if interpreted by a radiologist or experienced stroke clinician [23], with sensitivities reaching 93% in the acute phase but dropping to 17% in the chronic phase [24].

Hyperacute Ischemic Stroke

Changes within the brain parenchyma during the early hyperacute phase are typically not visible on plain CT because the overall water content of the brain parenchyma has not yet increased. However, occasionally fresh arterial thrombus can be identified as a slightly increased density within the arterial lumen often referred to as the *hyperdense artery sign* (Fig. 6c). In this instance, the lack of brain parenchymal change can be diagnostically useful in identifying a potential hyperacute large vessel occlusion, which might help to confirm ischemic stroke as the etiology for the symptoms and highlight the potential for thrombectomy.

The sensitivity of plain CT to diagnose ischemic stroke in the early hyperacute phase has been reported with a wide range in the literature [23], with one study suggesting it may be as low as 47% due to the limit in tissue density changes which can be detected by this single technology [25]. However, if clinicians are confident that their assessment of the patient makes stroke the likely diagnosis, a CT scan without hemorrhage or other pathological changes will be judged as consistent with a clinical impression of ischemic stroke and used to support a tPA or anticoagulation treatment

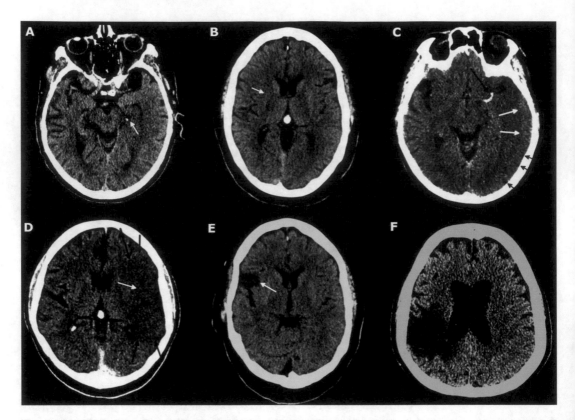

Fig. 6 Plain CT findings throughout the evolution of ischaemic stroke. (**a**) Early hyperacute. Hyperdense left posterior cerebral artery (white arrow). No parenchymal changes. (**b**) Early to late hyperacute. Transvascular oedema results in the subtle loss of grey/white matter differentiation localised to the right putamen (white arrow), indicative of a right MCA stroke. No sulcal effacement. (**c**) Late hyperacute to acute. Left MCA contains hyperdense thrombus (long black arrow). Transvascular oedema results in poorly defined hypodensity within the left MCA territory with sulcal effacement (short black arrows). A small amount of grey matter can still be distinguished (white arrows). (**d**) Acute. Transvascular oedema results in a well-defined region of hypodensity within the left MCA territory (black arrows) with total loss of grey/white matter differentiation and sulcal effacement (white arrow). (**e**) Subacute. Necrosis and early gliosis evidence within a well-defined subset of the right MCA territory (white arrow). Internal contents are still denser than CSF (difficult to discriminate on such narrow stroke windows) indicating large amounts of cellular debris. Oedema within the neighbouring well perfused areas of the brain has resolved. (**f**) Chronic. Large region of brain affected by a stroke 20 years previously is very hypodense due to cystic encephalomalacia. The periphery of this region is chronically gliotic but this cannot be distinguished from cystic encephalomalacia on CT. We will distinguish the two later in this chapter using MRI images subsequently acquired from the same patient

decision. It is not necessary to show evidence of thrombus to initiate tPA as many smaller lesions cannot be identified, but there is still an advantage from treatment.

Late Hyperacute or Acute Ischemic Stroke

Unlike MRI, plain CT is unable to image the compartmental redistribution of water within the brain parenchyma. However, it is able to image changes in the average water content of brain tissue across both intracellular and extracellular compartments, and the increased water content in acute stroke will result in a lower density of the brain parenchyma. This is most easily identified by a region of *hypoattenuation* within the brain parenchyma, usually visible only once the stroke has passed the early hyperacute phase into the late hyperacute or early acute phases. Later acute or subacute changes include a discernible increased volume of brain parenchyma, evidenced by *effacement of the cerebral sulci* (Fig. 6d). Consequently, the sensitivity of CT performed within 7 days of onset is greater than in the early hyperacute phase alone [23] and has been reported between 63–95% [26–28].

3.1.5 Quantitative Plain CT

Plain CT interpretation, like most neuroimaging methods, is normally achieved qualitatively in clinical practice. However, the Alberta Stroke Program Early Computed Tomography Score (ASPECTS) is often used as a quantitative plain CT score. It is a ten-point scale which divides the brain supplied by the anterior circulation into anatomical segments and deducts a point for each region that has infarcted or shows early ischemic changes [29]. A detailed explanation of the score and training to apply it is available at www.aspectsinstroke.com. ASPECTS is gaining favor as an approximate surrogate biomarker for core infarct volume, with one study finding that an ASPECTS of <7 can predict an infarct core of 70 mL or larger on diffusion MRI with 74% sensitivity and 86% specificity [30]. ASPECTS has been used in the patient selection process for large thrombectomy trials in the early exclusion of patients with large established infarct cores without the need for advanced imaging [31, 32], and its role in thrombectomy selection has been acknowledged by the international stroke community [33, 34].

3.1.6 Impact of Plain CT upon the Acute Stroke Treatment Pathway

Plain CT remains the first line neuroimaging method in the setting of acute stroke because it is effective at filtering out patients with hemorrhagic stroke [23], enabling early administration of thrombolytic agents. It can also provide a crude way of separating chronic from acute stroke; however, it is limited in its ability to identify patients suitable for treatment with thrombectomy: patients who are still potentially eligible for treatment with thrombectomy following a plain CT should undergo further investigation with CT Angiography.

3.2 CT Angiography

3.2.1 Background

CT Angiography (CTA) fundamentally differs from plain CT in its use of intravenous contrast agent administration. Iodine is favored almost universally because it is relatively biologically inert, yet physically much denser than soft tissue. A good quality intracranial angiogram will increase the density of the arteries ideally above 250 HU, almost an order of magnitude higher than the density of the brain parenchyma. The increase in the density of opacified arteries relative to the brain parenchyma results in the ability to directly visualize the patency of the intracranial arteries and those in the neck. It is increasingly being performed routinely during the assessment of patients arriving in hospital within a timescale that may be suitable for thrombectomy, which requires direct visualization of a potentially accessible thrombus before removal is attempted.

3.2.2 Advantages and Disadvantages

The advantages of CTA are its ability to distinguish large from small vessel occlusion, localize the thrombus within the arterial tree, and visualize the collateral circulation to the region of stroke. It is readily available quickly and out-of-hours within clinical services.

The disadvantages for patients are an exposure to ionizing radiation and the use of intravenous contrast medium which carries a risk of renal injury. For clinicians there is limited use in assessing the core–penumbra mismatch and it is not suitable for assessing brain parenchyma.

3.2.3 Variable Parameters (Further to Those Described for Plain CT)

The field-of-view for CTA depends upon the purpose of the imaging. When investigating for intracranial arterial aneurysms, only the brain (from the circle of Willis to the cranial vertex) needs to be imaged, whereas identification of carotid stenosis (e.g., as part of the workup for treatment of a transient ischemic attack) requires only the carotid arteries to be imaged. If investigating for large vessel occlusion, both the carotid arteries and the intracranial arteries need to be imaged. This field-of-view is sometimes described as *arch to vertex.*

Maximum Intensity Projections (MIPs) describe how multiple slices can be stacked together during postprocessing in such a way that a given voxel will take the brightness of the brightest voxel in the same location from any of the individual slices in the stack. In practice, this means that tortuous opacified arteries which normally move in and out of plane on a single slice (and thus normally appear fragmented and difficult to follow) will appear continuous and therefore easier to follow on a single image (Fig. 7). This is attractive because it can help speed up image interpretation, in addition to emphasizing any asymmetry in the cerebral vasculature. However, this process can also mask subtle arterial pathology such as mild stenosis or dissection, so MIPs are best used as an adjunct to standard thin slices rather than a replacement.

The *Contrast Phase* refers to the time delay between injecting the contrast and performing the scan. The contrast bolus has to travel from a cannulated peripheral vein, through the right heart, into the lungs, into the left heart and through the aortic arch before it can reach the carotid and vertebral arteries that ultimately supply the brain. Historically CTAs were performed by scanning with a fixed delay of around 20 s after bolus injection so that the contrast medium had time to reach the cerebral circulation, however due to heterogeneity in the length of time taken for the bolus to travel

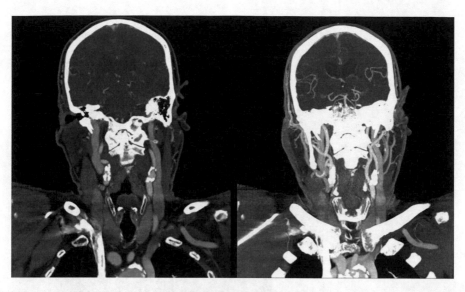

Fig. 7 Standard coronal thick slices compared to MIPs from a CT Angiogram. *Left*: 5 mm coronal thick slices. The intracranial arteries are moving in and out of plane, therefore appear fragmented and it is necessary to look at multiple slices or a computer constructed 3D view. *Right*: 20 mm coronal MIP formed from four adjacent 5 mm slices. The path of the artery from each constituent slice is mapped on the MIP image enabling the arterial tree to be followed with greater ease

from vein to brain, suboptimal opacification of the intracranial arteries used to be a frequent problem. Modern scanners use bolus tracking technology to overcome this. More recently, the technique of *multiphase CTA* has been developed. This involves rescanning the brain after a normal bolus-tracked arch to vertex CTA, usually an additional two times, with all acquisitions separated by around 8 s [35]. This enables the temporal resolution of arterial blood flow, which can overcome the shortcomings of conventional CTA in assessing the collateral circulation [35, 36].

3.2.4 Diagnostic Scope and Imaging Findings

Identifying Large Vessel Occlusion (LVO)

Like plain CT, CT Angiography is highly specific (96%) but not very sensitive (47%) in the overall diagnosis of hyperacute ischemic stroke [37, 38]. It is however extremely accurate in the localization and *diagnosis of large vessel occlusion*, with both sensitivities and specificities of 98–100% [38–40]. The whole of the perfused arterial tree (proximal to the thrombus) is mapped, which means that the thrombus can be described in terms of which artery it is obstructing, and how proximal it lies (Fig. 8). The location of the LVO is important: all of the recent major randomized controlled trials investigating mechanical thrombectomy have only included patients with occlusion of the intracranial internal carotid artery (ICA), the proximal middle cerebral artery (MCA), or the proximal anterior cerebral artery (ACA) (collectively, the proximal anterior circulation) [10]. Large vessel occlusions in the posterior circulation (e.g., the basilar artery) have been successfully treated with

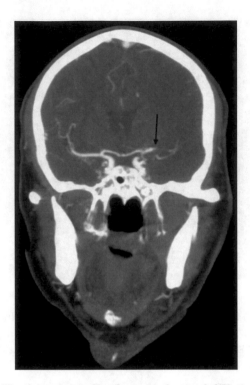

Fig. 8 Five millimeter coronal slice from an abnormal CTA demonstrates an occlusion of the left MCA (black arrow). This is an anterior proximal large vessel occlusion that may be treatable with thrombectomy

thrombectomy in case reports and guidelines suggest these patients should be considered for thrombectomy [33] until the results from ongoing randomized controlled trials become available to provide more specific guidance for management of patients with basilar artery occlusion [41].

Neurointervention Planning

Aside from its use in treatment decisions and stratification of patients, knowing the location of the LVO can help neurointerventionists plan their intervention (e.g., where they should expect to place the arterial catheter device). Similarly, the assessment of the common and cervical internal carotid arteries is invaluable to the neurointerventionist: sometimes patients can present with tandem occlusion, consisting of an intracranial LVO in addition to extracranial internal carotid arterial obstruction, and these patients may require additional management such as carotid stenting prior to accessing the intracranial circulation [42].

Assessing the Collateral Circulation

Collateral scoring using multiphase CTA has been used successfully to identify patients with different outcomes in the process of selection for thrombectomy in one clinical trial. Patients with a "better" score had a better outcome, reflecting the increased blood flow to the penumbra from the additional collateral vessels [31].

Multiphase CTA provides an attractive compromise between providing information that can help select the ideal candidates for thrombectomy, but with relatively little modification to conventional CTA protocols: it could therefore be adopted relatively easily by all departments currently offering single phase CTA. Despite the promising early results and potential transferability of this neuroimaging method, it is still in its relative infancy and further validation from future studies would be useful.

3.2.5 Impact of CTA upon Acute Stroke Management Pathway

The optimum neuroimaging methods for patient selection for thrombectomy have not yet been established, and the decision to treat is still made on a case-by-case basis following urgent multidisciplinary discussion involving a neurointerventionist and a stroke physician [33]. The importance of CTA in making these decisions cannot be underplayed, as the only other noninvasive neuroimaging method that can diagnose and localize LVO effectively is magnetic resonance angiography (which is more expensive and far less available than CTA). For patients presenting in the early hyperacute phase with a proximal large vessel occlusion on CTA and evidence of a small infarct core on plain CT, the decision to proceed to thrombectomy can sometimes be made at this stage with no further neuroimaging until the procedure itself [33]. Sometimes the decision to treat is more complicated, for example if patients present in the late hyperacute phase: some of these patients may benefit from thrombectomy up to 24 h post-onset, dependent upon the collateral circulation, volume of salvageable tissue and size of the core infarct [15]. The distinction between infarct core and ischemic penumbra can be investigated further with CT Perfusion.

3.3 CT Perfusion

3.3.1 Background

CT Perfusion (CTP) is typically used to determine how much of the brain tissue is viable: typically, ischemic strokes will consist of a central infarct core which cannot be salvaged, surrounded by an ischemic penumbra which may be salvaged by reperfusion therapy. Other methods of neuroimaging are able to quantify the infarct core volume alone (ASPECTS or diffusion MRI), but perfusion imaging is able to quantify both the infarct core and, uniquely, the ischemic penumbra by directly imaging cerebral blood flow.

This is achieved by selecting a slice of brain tissue which includes a section through the infarct (typically at the level of the basal ganglia), administering contrast agent, and performing serial scans of the same slice approximately every 1–3 s for 60–90 s [43]. The change in density over time within each voxel can be represented by the time enhancement curve and several key measurements can be extracted [44]:

- Mean transit time (MTT), which reflects the length of time that a red blood cell spends within a given volume of brain tissue (unit: s).

- Cerebral blood volume (CBV), which reflects the total volume of blood within the tissue, per unit mass of brain tissue (unit: mL/100 g).
- Cerebral blood flow (CBF), which reflects the rate at which the tissue fills with blood, per unit mass of brain tissue (unit: mL/s/ 100 g).

This information is usually extracted from the CTP images using deconvolution methods, which are the most complicated of the various proposed options but also the most accurate because they do not rely upon assumptions about cerebral physiology that are not always valid [44]. The three values are related by the central volume principle which states that CBF = CBV/MTT [44].

3.3.2   Advantages and Disadvantages

The main advantage of CT Perfusion is its ability to accurately assess core–penumbra mismatch.

The disadvantage for the patient remains the exposure to ionizing radiation and use of intravenous contrast medium which carries a risk of renal injury. From a clinical viewpoint, it requires the use of dedicated CTP expertise and software not routinely available in all emergency imaging departments and lengthens the radiological assessment time before treatment. Therefore, it should only be used when the additional information would change the clinical decision.

3.3.3   Variable Parameters

The *length of scan* is crucially important in obtaining useful information because of CT perfusion's reliance upon dynamic measures. Scanning for less than 60 s may result in incomplete measurement of the time enhancement curve in patients with poor cardiac output as the contrast will fill and wash out from the brain tissue more slowly in these patients. For this reason, scan times of up to 90 s have become standard [43]. The trade-off is that a longer scanning time (with the same frequency of sampling) will result in a larger radiation dose.

Greater *temporal frequency* of measurement is important for good quality images but was historically limited by scanner technology. However, modern scanners are capable of imaging a slice in under a second, enabling much higher temporal frequency. The trade-off is now between sampling at a higher frequency (e.g., 1 s intervals), resulting in a higher radiation dose, and sampling at a lower frequency (e.g., 4 s intervals), resulting in a lower radiation dose but potentially missing the peak of the time enhancement curve due to under-sampling. Many modern CTP protocols now optimize radiation dose without compromising sampling of the time enhancement curve by varying the temporal frequency throughout the scan so that for the first 40 s (when the peak is expected to occur) the sampling frequency is every second, but for the remainder of the scan the sampling frequency reduces to once every 2–3 s [44].

Field-of-view is the other important variable parameter in CTP. Older generations of CT scanners were unable to sample large volumes of brain tissue quickly enough to facilitate a sufficient temporal frequency, so sampling was limited to a narrow volume of brain tissue (e.g., a single slice at the level of the basal ganglia). This can be problematic if the stroke extends far beyond the selected region [45]. To ensure the stroke is included within the field-of-view some CTP protocols may now involve scanning a larger volume of brain tissue [46], although this will usually limit the maximum temporal sampling frequency, in addition to significantly increasing the radiation dose.

3.3.4 Diagnostic Scope and Imaging Findings

Diagnosis of Hyperacute Ischemic Stroke

As it can show changes reflecting reduce blood flow before irreversible tissue injury has occurred, CTP performs better in the diagnosis of hyperacute ischemic stroke than plain CT, with sensitivities ranging from 75% to 88% and specificities from 89% to 99% [47].

Distinguishing Ischemic Penumbra from Infarct Core

The greatest unique value of perfusion imaging lies in its ability to distinguish ischemic penumbra from infarct core: patients with a relatively large volume of ischemic penumbra relative to the infarct core may be eligible for thrombectomy or thrombolysis treatment, even if they are presenting outside of the recommended time frames or with an unknown onset time [48].

Infarct core findings (Fig. 9) are due to a reduced capillary perfusion pressure with no or little physiological compensation:

- MTT is elevated in the infarct core.
- CBV will be markedly reduced in the infarct core.
- CBF will be markedly reduced in the infarct core.

Infarct core is normally defined as the CBV lesion volume [44].

Ischemic penumbra findings are also due to a reduced capillary perfusion pressure, but there is associated reactive vasodilation resulting in a greater capacitance of the brain capillary bed:

- MTT is elevated in the ischemic penumbra.
- CBV is normal or may even be raised.
- CBF will be moderately reduced.

Ischemic penumbra is normally defined as either the MTT or CBF lesion volume [44].

3.3.5 Impact of CTP upon Acute Stroke Management Pathway

There is still poor consensus on the exact CTP threshold values to determine suitability for thrombectomy or thrombolysis but examples from thrombectomy and thrombolysis trials specify an infarct

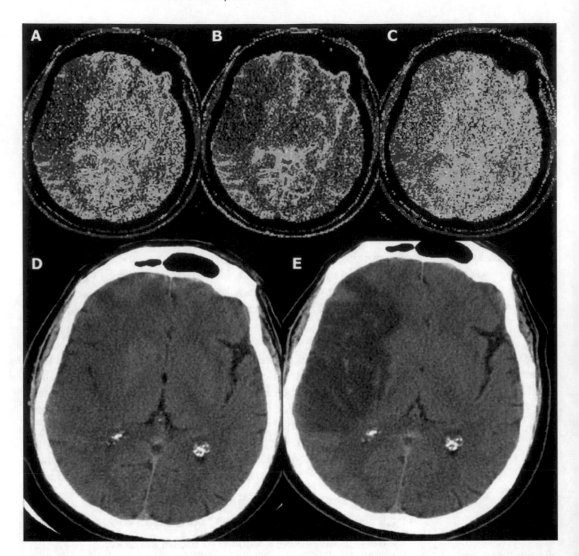

Fig. 9 CT images in a patient with acute ischaemic stroke causing left hemiparesis who was not treated with thrombolysis or thrombectomy. Admission CTP images show (**a**) cerebral blood flow (CBF), (**b**) cerebral blood volume (CBV), and (**c**) mean transit time (MTT). The presence of a matched CBF/CBV perfusion deficit suggests irreversibly ischemic infarct "core", likely to correlate with DWI findings, and not a target for reperfusion therapies; (**d**) shows the admission non-contrast CT, and (**e**) shows the final infarct volume on the follow-up CT, which closely matches the admission CBF and CBV lesions (**a** and **b**)

core volume <70 mL; a ratio of penumbra to core of greater than 1.8; and a penumbra volume of 15 mL or higher, as being associated with more favorable outcomes with treatment [49, 50]. In contrast to CTA and plain CT which are readily available out-of-hours at most hospitals with Emergency departments, CTP is generally restricted to specialist neuroimaging centers or clinical trials due to its complexity and the need for specialist software.

Perfusion imaging might be the only neuroimaging method which can quantify ischemic penumbra, but it is not the only method to measure the infarct core: as previously discussed, the ASPECTS plain CT score of <7 can predict an infarct core of 70 mL or larger with 74% sensitivity and 86% specificity [30]. Furthermore, CTP has been found to be significantly worse than diffusion MRI at estimating the size of the infarct core. Diffusion MRI remains the gold standard neuroimaging method for assessing the infarct core volume [51].

3.4 Magnetic Resonance Imaging

3.4.1 Background

The introduction of MRI into clinical practice has been hugely beneficial for when there is uncertainty about the etiology of stroke symptoms and the timing of ischemic stroke onset. This chapter will describe simplified key principles which explain the major differences between the appearances of the MRI sequences used during the imaging of acute stroke.

3.4.2 Source of MRI Signal

In contrast to X-ray-based imaging methods, MRI does not measure physical (atomic mass and electron) density. Rather, it takes its measurements from protons: scanners can be tuned to image a variety of different protons, but hydrogen protons are used almost universally in clinical MRI because of their abundance and distribution within the body.

Whereas CT measures the interaction (attenuation) of an externally generated X-ray beam with tissue as it passes from source to detector, the detected signal in MRI arises from the protons within the tissue itself. The section below provides a basic explanation of how hydrogen protons within brain tissue generate a detectable MRI signal which allows sensitive differentiation between different tissue types and states.

Spin

Protons have a positive electrical charge. Along with many other particles, they also demonstrate a property called "spin," that is, rotation about an axis. From Maxwell's laws of electromagnetism, a rotating electrical charge will induce a perpendicular magnetic field. We can now imagine each individual proton to possess a tiny magnetic field, like a little compass needle: outside of the MRI scanner, all these tiny compass needles will be randomly aligned and therefore humans do not usually have a net magnetic field. However, when placed inside an exceptionally strong static magnetic field (typically 1.5–3 T in modern MRI scanners), many of the protons in the body will align parallel to the magnetic field along the Z-axis (mostly pointing "up" or in the direction of the magnetic field, with a minority pointing "down" or against it). There is now a net magnetic field arising from our voxel which reflects the sum of billions of spinning protons within this cube of brain tissue. It is, however, a static magnetic field that cannot be easily detected: this static magnetic field must be turned into an oscillating one before we can produce a detectable signal.

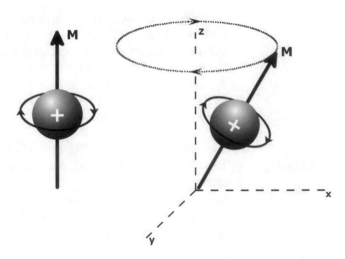

Fig. 10 Diagrammatic representations of proton spin and precession. *Left*: A proton in the absence of a static magnetic field will 'spin' about its axis (solid circular arrow), generating a magnetic dipole M. *Right*: A proton placed in a strong magnetic field parallel with the Z-axis will also precess about the Z-axis (dashed circular arrow) in addition to spinning about its axis (solid circular arrow)

Precession

Protons are always spinning about their axis, but this is not the only rotational movement they are capable of: rather than picturing them as rotating spheres, let us picture them instead as rotating spinning tops. Now we can picture a second form of "tilting" rotation known as precession (Fig. 10). By touching a spinning top, we apply an external force which will "tip" it to be more horizontal rather than vertical, and the axis itself will now wobble (precess) while simultaneously spinning around its axis as it did before it was tipped. The source of the force (torque) that facilitates precession in a spinning top analogy is gravity; however, in MRI the source of this force is the exceptionally strong static magnetic field.

Phase and Excitation

The precessing magnetic field summed from all the protons within the tissue can be represented as a vector, M, with two distinct components: a vertical component (that aligns "up," or along the Z axis) which is called Mz, and a horizontal component (that aligns along the transverse X, Y plane) which is called Mxy.

In the relaxed energy state, Mz will be at its maximum relative value of $+1$: Mz is positive due to the majority of "spin up" protons when placed in a strong static magnetic field, but its absolute value will depend upon the total number of protons within a given voxel. Mxy, however, will have a net value of 0 because the protons are not precessing in sync ("phase") with each other—for example, at any given moment in time, just as many protons will be tilted toward the "east" direction ($x =$ positive) as there are protons tilted toward the "west" direction ($x =$ negative).

In order to produce a detectable coherent oscillating magnetic field (i.e., in order to increase Mxy from 0 to a relative value of +1), we must synchronize the precessions of the protons within our voxel. This is achieved through the transmission of a radiofrequency (RF) pulse at a frequency specific to hydrogen protons. The transmission of the RF pulse also has the secondary effect of "exciting" or "flipping" some spinning protons into the higher (spin down) energy state, diminishing the proportion of excess "spin up" protons and therefore the magnitude of Mz.

Despite the loss of Mz and growth of Mxy happening by different mechanisms, they occur simultaneously and proportionally in response to the same resonant RF pulse. This cannot be explained by classical mechanics alone, so our spinning top analogy is limited in conceptualizing the magnetic consequences of exciting protons. In truth, magnetic resonance can only be fully explained using quantum mechanical models of particle interactions—the true quantum model of MRI is much wilder and more fascinating than compass needles and spinning tops but is far beyond the scope of this chapter.

Thankfully, in practice we do not need to understand the quantum interaction between longitudinal and transverse magnetization in order to understand the differences between stroke MRI sequences. As such, it is convenient from here on to just simply imagine that the RF pulse acts to "tilt" the magnetic vector M from Mz to Mxy. Therefore, when Mz reaches 0, Mxy will reach +1, and we can say that the RF pulse has tipped M through 90' (Fig. 11).

Signal Induction

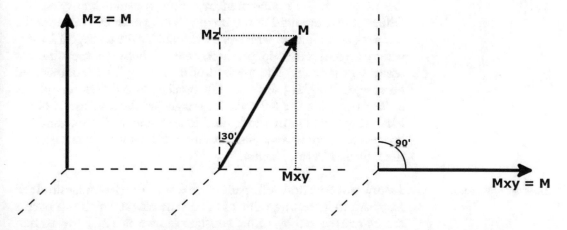

Fig. 11 The excitation of a large number of protons within a voxel, represented by the net magnetic vector M. *Left*: Pre-excitation, M is aligned with the Z-axis. It is positive due to the excess number of 'spin up' protons relative to 'spin down' protons. *Centre*: Application of a radiofrequency pulse 'tips' M from the Z axis to the X, Y plane. *Right*: The pulse will continue until all of the magnetisation has been tipped into the X, Y plane. The magnitude of Mxy after a 90' excitation is the same as the magnitude of Mz was in the resting state

After excitation and precession synchronization, Mxy will now be oscillating with a value ranging from -1 to $+1$ as the net magnetic vector precesses around its origin. Returning to Maxwell's laws of electromagnetism, we know that an oscillating magnetic field can induce an electric current in a coil, and the same coils that were used to transmit the excitatory RF pulse can be used to detect the oscillation in Mxy.

If we were able to measure the peak amplitude of the current induced in the coil at the exact moment the RF pulse was switched off, we would find that its value would vary from voxel to voxel proportionally to the number of hydrogen protons within each voxel. This method of imaging (Free Induction Decay or FID) was used in very early MRI experiments but in clinical practice has been superseded entirely by spin-echo and gradient-echo techniques, which we will explore shortly.

3.4.3 Relaxation Processes

What happens if we wait a period of time after stopping the excitatory pulse before detecting the induced signal is far more interesting and useful in neuroimaging than simply counting protons. The net magnetic vector M will regain its Mz component via "spin–lattice" relaxation and lose its Mxy component via transverse "spin–spin" dephasing. It is crucial to understand that these two processes occur simultaneously, but spin–lattice relaxation occurs independently from spin–spin dephasing.

Longitudinal Relaxation

Longitudinal "spin–lattice" relaxation results from the transfer of energy from excited (spin down) protons to the lattice as they realign with the strong static magnetic field. This will result in a recovery of Mz back to its resting value of 1, and the rate at which this occurs can be represented by a time constant known as T1. Different tissues found in the brain will have different values of T1 and producing images which are weighted by the average T1 value within a voxel can help produce contrast between the different tissues in the brain. The method of measuring T1 is complicated because, unlike Mxy, Mz does not oscillate and therefore will not induce a signal in the RF coils. T1 images are the least useful of the MRI neuroimaging methods used in assessment of stroke and the images are of equivalent clinical value to CT, so we will not explore them further in this chapter.

Transverse Relaxation

Transverse relaxation will result in a decay of the peak magnitude of Mxy back to its resting value of 0, and the rate at which this occurs can be represented by a time constant known as T2. T2 relaxation almost always occurs more rapidly than T1 relaxation because the protons contributing to Mxy will all eventually transfer their excess energy to the lattice. Therefore, T2 can be considered the combined effect of spin–lattice and spin–spin relaxation processes (the latter being the component of T2 that distinguishes it from T1).

If we measure how quickly Mxy decays from the FID, we will find that it actually decays at a rate much quicker than would be expected due to T2 relaxation alone. This quicker rate of Mxy decay is represented by the time constant T2*.

The difference between T2* and T2 relaxation arises because T2 relaxation is not the only process which results in the decay of Mxy, although its contribution is usually the most useful because the rate of T2 relaxation is an inherent property of the nuclear environment of a given proton (e.g., how well a given proton is coupled to its neighbors varies from tissue to tissue). The non-T2 contributions to Mxy decay are due to magnetic field inhomogeneity; either due to fluctuations in the magnetic fields generated by the MRI scanner, or due to the presence of compounds such as iron-containing blood products disrupting the local magnetic field (which allows T2* to be a useful imaging mode to detect small amounts of chronic hemorrhage).

Before we can explain how we distinguish between the T2 and non-T2 contributions to Mxy decay, we have to first consider what they have in common: they all result in the decay of Mxy via dephasing of precessing protons within a voxel.

Dephasing

If we take a bird's-eye view of a precessing proton, we can imagine the magnetization vector spinning around like the minute hand on a clock (Fig. 12). Immediately after switching off the excitatory 90′ RF pulse, all the individual protons will be synchronized or "in phase" with each other as they precess (that is, their minute hands will all be showing the same time). Small fluctuations in the magnetic field within the voxel will result in some protons precessing at either a faster or slower speed than the average; this means that M, the sum of all the proton magnetization vectors, will lose magnitude of its Mxy component more rapidly than the individual protons within the voxel because they are no longer precessing "in phase" with each other.

Gradient Coils

In order to understand why different MRI sequences are useful for stroke assessment it is important to now briefly consider how the anatomical spatial localization of a voxel is determined. The localization of a voxel to a particular coordinate in space is achieved through different methods for the X, Y, and Z directions. It is the localization of a voxel in the X direction that is relevant here, so let us assume we have already selected a single slice (Z), and a single row of voxels on our slice (Y), and now we only need to discern where in the row a particular voxel lies.

The distinction between voxels in a given row is normally achieved by the application of a gradient coil which will serve to modulate the magnetic field from left to right. It is critical to understand that the direction of the magnetic field does not change—it is still longitudinal (i.e., parallel to the Z-axis); its

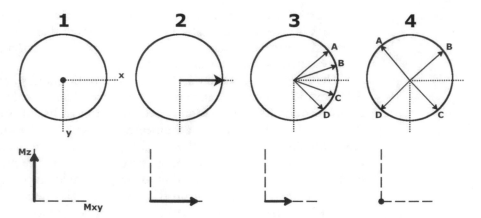

Fig. 12 Transverse relaxation due to dephasing. (1) In the resting state, there is no coherent transverse magnetisation. (2) Following a 90′ excitation, *M* is tipped from the *Z* axis to the *X, Y* plane and spins are aligned in phase with each other. Precession of *Mxy* will occur in a clockwise direction, and these oscillations can induce a signal in the detector coil. (3) After a period of time, subtle differences in magnetic field strength will cause different protons (A–D) to precess at different frequencies (for example, A is slower than B, and D is quicker than C). Dephasing results in a reduction of net transverse magnetisation and therefore a loss of MR signal. (4) After a further period of time, all coherence is lost and the magnitude of *Mxy* reduces to 0. This usually happens much more rapidly than recovery of the longitudinal magnetisation *Mz*

strength however will be different for each voxel in the row. The strength of the magnetic field around a precessing proton will alter the frequency at which it precesses and therefore the frequency of the signal it induces in the RF coil. Each voxel in the row will now have a characteristic frequency (like the different instruments in an orchestra), and the amplitude of the net *Mxy* signal from each individual voxel can be isolated from the detected signal via a mathematical process known as the Fourier Transform (in the same way that a symphony can be split into its constituent frequencies by a spectrum analyzer on a hi-fi system).

The problem with using a gradient coil in this way is that by increasing the frequency for some voxels relative to others, we will cause them to dephase relative to protons in voxels subjected to a weaker part of the gradient: the lack of coherent phase means that the contributions from different voxels will destructively interfere with each other and our detected signal will scramble. This is one factor that limits the information available from MRI.

3.4.4   Gradient-Echo Imaging (GRE)

In order to counter this effect for specific purposes, the use of an additional rephrasing gradient is employed: by reversing the polarity of the gradient coil and applying it for the same length of time as the initial (dephasing) gradient, the protons in voxels that had slowed down during dephasing will speed up by the same amount during rephasing. This kind of pulse sequence is referred to as a *gradient-echo* or GRE sequence. While this has adjusted for the field inhomogeneity due to the applied magnetic gradients, it has not adjusted for the effects of variances in local susceptibility or the effect of the nuclear environment of the spins. Gradient echo sequences are therefore T2* weighted, representing magnetic susceptibility in combination with T2 of the brain tissue.

Although some contrast is lost between tissues, GRE is the most commonly used MRI method of detecting acute hemorrhage due to the impact of blood degradation products upon magnetic susceptibility, and therefore can play an important role in stroke neuroimaging using MRI. However, in order to diagnose acute ischemic stroke on MRI it is essential to separate the true T2 component of *Mxy* relaxation from T2*.

3.4.5   Spin-Echo Imaging (SE)

T2-weighted images can be produced in a variety of ways, but by far the most common is known as the *spin-echo* sequence. This differs from the GRE approach in that after the initial dephasing gradient, instead of rephasing by using a second (reversed) gradient, it rephases protons by the use of a second excitatory radiofrequency pulse.

In order to understand how this works, let us return to the magnetization vector *M*, which so far we had tipped into the *Mxy* plane by the application of a single 90′ pulse. A further 90′ pulse would actually tip the magnetization vector back into the *Mz* plane, only now with a reversed (or downward pointing) magnitude. A third 90′ pulse would tip the vector back into the *Mxy* plane, again with an opposite phase to when it first passed into the *Mxy* plane (i.e., if the magnitude was 1 after 90′, it would be −1 after a further 180′ pulse).

SE sequences employ a second 180′ pulse after the initial 90′ pulse to rephase the protons: those protons which have precessed more quickly than their peers will still be precessing more quickly after the 180′ rephasing pulse, but they will be speeding back toward the average phase rather than away from it. Between the 90′ pulse and the 180′ pulse, the induced signal will be reducing due to dephasing of the spins, but the induced signal will then increase back up to its maximum after the same period of time has passed after the 180′ pulse.

Spin echo images will therefore correct for all the dephasing that had occurred due to magnetic field inhomogeneity and will only detect the dephasing due to spin–spin interactions (therefore, pure T2 decay rather than T2*). After the peak of the rephasing moment has passed, dephasing will occur again and the *Mxy* decay will return to following T2* rather than T2.

3.4.6   T2-Weighted Spin-Echo Imaging (T2w-SE)

The length of time from the 90′ pulse to the maximum induced signal (or echo) is known as TE or "echo time" and has a crucial impact upon the contrast between tissues in an image. In an ideal world, we would measure the entire relaxation curve for each voxel

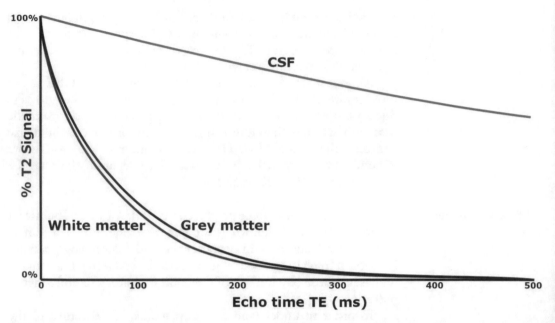

Fig. 13 A diagram representing the T2 decay of different tissues within the brain. An echo time of approximately 100 ms will produce good contrast between grey matter, white matter, and CSF

in order to quantitatively retrieve the average T2 value for the voxel. In practice, we can only acquire signal from a spin-echo sequence at a particular point along the relaxation curve: the center of the echo. The magnitude of the signal detected from each voxel will therefore be proportional to the total number of initially excited protons in the voxel (i.e., the proton density) multiplied by the proportion of residual transverse magnetization, as defined by the T2 relaxation curve at the time TE.

As we can *see* from Fig. 13, which demonstrates the T2 curves for grey matter, white matter, and CSF, a relatively long TE of around 100 ms will result in the greatest difference in detected signal between three voxels comprised of these different tissues. *Spin-echo sequences with a long TE will therefore demonstrate T2 weighting.*

For TEs shorter than this, the relative differences between residual T2 values for these different tissues will decrease and thus the contrast will be less dependent upon T2. It will however still depend upon the total number of initially excited protons. *Spin-echo sequences with a short TE will therefore usually demonstrate proton-density weighting.*

3.4.7 T2-FLAIR

Fluid Attenuation Inversion Recovery, or FLAIR, is another essential sequence used in the diagnosis of acute stroke. Before we explain how FLAIR works, it is important to understand that the more free a hydrogen proton is, the more poorly it couples with its molecular environment, or with other freely moving hydrogen protons. For this reason, rapidly moving free water molecules (such as those found in the CSF spaces in the brain) will take a long time to lose their excess energy via either longitudinal or transverse relaxation (in other words, CSF will have extremely long T1 and T2 times, up to 3–4 s). Water molecules that exist within other environments (such as within a cell) will still have relatively long T1 and T2 relaxation times, but not as long as free water.

FLAIR works by selectively nulling a signal based on its T1 relaxation time. The most commonly used FLAIR sequences can be explained as a modified spin-echo sequence; the modification being the addition of a pre-excitation, or "inversion recovery" phase before the first 90' pulse of a standard spin echo sequence. The inversion recovery pulse is 180', which will tip all of the initial Mz straight through Mxy back into the Mz plane again, although now with a magnitude of −1 rather than +1. Because all our magnetization is now back in the Mz plane there is no Mxy component of Mz, and protons will lose their energy via longitudinal T1 relaxation only. As they lose their energy, the magnitude of Mz will rise from −1 to 0, and eventually from 0 back to +1. Different tissues will regain their positive Mz at different rates, and we can therefore predict at exactly what time a given tissue (e.g., free water) will have an Mz of exactly 0. If we chose this exact moment to start our regular spin-echo sequence, all tissues apart from water will have positive Mz to tip 90' into Mxy as in a regular spin-echo sequence. This is because free water will have had no net Mz recovered to tip into Mxy and will therefore not be able to induce any signal in the detector coil. Signal from free water will be nulled and therefore appear black on all FLAIR images. If we perform the remaining spin-echo sequence with a long TR and a short TE, the rest of the image will appear T2 weighted (including showing bright for edematous tissue).

The time from the initial inversion 180' pulse to the initial spin-echo 90' pulse is known as the inversion time, or TI (not to be confused with T1!) and can actually be varied to null different tissues, for example STIR sequences utilize a shorter TI to null fat signal, but the only inversion sequence commonly used in stroke imaging is FLAIR.

3.4.8   Diffusion-Weighted MRI (DWI and ADC)

FLAIR sequences could be considered to provide limited information about water diffusion, in that water which is totally free (and therefore highly diffusing) will be nulled, but there is no information about the degree of diffusion in the remaining water within the brain (such as the water in the extracellular space).

Diffusion-weighted MRI (or DWI) was developed in order to quantify the degree of diffusion of water within a given voxel, which is a sensitive indicator of early ischemia and an extremely valuable assessment in clinical practice when the diagnosis of stroke is uncertain. Once again, we can consider DWI sequences to be a modification of the standard spin-echo sequence. The modification in DWI is to apply two extremely strong magnetic gradients, one either side of the 180′ rephasing pulse. The first of these pulses serves to completely dephase all the protons along the axis of the gradient. The second serves to completely rephase them, but this is dependent upon no molecules having moved along the axis of the gradient (because they will need to be in exposed to exactly the same strength of magnetic field after the 180′ pulse as they were before in order to rephase). Therefore, water molecules which are diffusing along the direction of the applied diffusion gradients will not rephase, and the voxels they end up in during the echo will demonstrate a lower intensity of signal as a result.

This process will only be sensitive to diffusion in the direction that the diffusion gradient is applied in; it is typically repeated with the diffusion gradient applied across (at least) three different directions and the mean value of diffusion restriction calculated for each voxel—this average is known as the *isotropic diffusion-weighted image*, or sometimes the *trace map*.

Ideally, these images would not have any contrast other than diffusion contrast, to be a true map of water diffusion. However, because the diffusion gradient has to be applied for a finite length of time, TE will always be prolonged, resulting in T2-weighted images. This can be problematic if there is a lesion in the brain that is heavily T2 weighted; it can be difficult to distinguish this T2 *shine-through* from true diffusion restriction.

In order to remove the effect of T2 shine-through, two sets of images are acquired; one is the isotropic diffusion-weighted image performed exactly as described above, and the other is a spin-echo image acquired in exactly the same way with the same resolution as the diffusion-weighted image, but crucially the diffusion gradient coils are not switched on. This other image will contain the same degree of T2 weighting as the isotropic diffusion-weighted image, but without any measurement of the diffusion. The two series are distinguished by the letter "b" (indicating strength of the magnetic field of the diffusion gradient); thus, there will normally be a "$b = 0$" T2-weighted image, in addition to a "$b = 1000$" isotropic diffusion-weighted image. The T2 weighting can therefore be removed from the isotropic diffusion-weighted image in order to map a pure representative of water diffusion; this is known as the apparent diffusion coefficient, or *ADC map*. The brightness of a voxel on the ADC map is proportional to the degree of diffusion (so regions with greater diffusion will appear bright on ADCs, but dark on trace maps).

*3.4.9 Advantages
and Disadvantages*

A key advantage of MRI is the high sensitivity for acute ischemic stroke, especially in the hyperacute phase, when it can also aid estimation of time-from-onset. It shows evidence of chronic hemorrhage which is not detectable by CT, and provides high-quality angiographic and perfusion imaging to enable very accurate anatomical representation of the arterial system and estimation of infarct core volume.

There are several contraindications including older metal implants and cardiac devices such as pacemakers, which prevent some patients from undergoing MRI. The main challenge is the lack of availability of technology and image interpretation out-of-hours, but it is also relatively expensive and time consuming.

*3.4.10 Variable
Parameters*

We have only covered a few of the many possible pulse sequences in this book chapter, and the range of variable parameters for MRI is still expanding. We have described some of the most common pulse sequences, but it is important to be aware that different manufacturers may use variations of these sequences on their own devices. Furthermore, pulse sequences can be modified or even designed from scratch within an imaging department.

Key variables common to almost all MRI pulse sequences are the repetition time (TR) and echo time (TE) because they control the amount of T1 and T2 weighting in the image. A further variable specific to inversion recovery sequences is the inversion time (TI) which dictates the tissue that will be nulled.

*3.4.11 Diagnostic Scope
and Imaging Findings*

MRI is considered to be the gold standard for imaging diagnosis of acute ischemic stroke. It can provide almost all of the information that the CT methods we have reviewed can provide (including specific sequences for angiography and perfusion imaging). Furthermore, MRI has the advantages over CT of being able to estimate the pathophysiological phase of an infarct more accurately than CT [13, 16], in addition to being more accurate than any other neuroimaging method in accurately assessing infarct core volume [51]. The sensitivity of MRI in diagnosing hyperacute and acute stroke also exceeds that of any of the other neuroimaging methods we have reviewed [52, 53]. The major exception of this rule is that MRI is possibly worse than plain CT in the diagnosis of hemorrhagic stroke, with one study reporting a sensitivity of only 46% in the acute phase compared to 93% for plain CT. Hence CT still remains the first mode of imaging for most emergency stroke admissions. However, the sensitivity of MRI is far greater than that for plain CT in detecting chronic hemorrhagic stroke (93% vs. 17% for plain CT) [24].

The barriers to MRI are often more practical than functional, in that it is considerably more expensive than CT and far less readily available, particularly out-of-hours. Scanning times are also longer than in CT, which could potentially delay time-critical stroke therapy, and it is often not safe for patients with metal or electronic implants to be exposed to the powerful magnetic field during MRI.

Examples of ischemic stroke in the hyperacute, acute, subacute and chronic stages are demonstrated in Fig. 14 with an explanation of how the underlying pathophysiology relates to the imaging findings at all stages. The useful diagnostic features of MRI are outlined in Fig. 15 (which provides a time-line of signal changes across the sequences commonly encountered in stroke MRI in relation to CT changes).

3.4.12 Impact of MRI upon Acute Stroke Management Pathway

How MRI changes management of acute stroke patients is dependent upon the clinical context. It is frequently used in clinical trials as a more accurate alternative to CT Perfusion in the measurement of the infarct core volume, but this is still unusual in routine practice.

In current clinical practice, its role can be considered to be a problem-solving tool, appropriate at any stage of the stroke patient pathway where it is likely to change management. For example; in a patient who presents within the time window for thrombolysis, but there is a negative CT scan and clinical concern that there could be a stroke mimic that may contraindicate thrombolysis, MRI can be used to try and positively confirm the diagnosis of stroke prior to commencing thrombolysis by seeking DWI changes.

Like CT imaging, MRI has additional uses following the administration of an intravenous contrast agent, this time containing Gadolinium rather than Iodine. In a similar manner to CTP, MRI can provide very detailed maps of core and penumbral volume to assist in decisions for thrombolysis or thrombectomy treatment. MR angiography provides detailed arterial maps useful for determining the location and size of stenosis, aneurysms, and arteriovenous malformations. It is also possible to create a quick MRI view of arterial anatomy without contrast medium using a technique known as "time of flight" angiography, although this can be influenced by a number of variables such as arterial perfusion pressures and the results require more cautious interpretation than a contrast enhanced scan.

4   Conclusion

In this chapter we have explained how and why CT and MRI brain imaging modalities produce valuable neuroradiological biomarker data during the assessment of stroke patients. The approach used in practice reflects the type of clinical decision being made. The speed and simplicity of CT imaging is usually sufficient for standard stroke cases, whereas the wealth of additional information generated by MRI is of great value for a smaller number of patients. Future

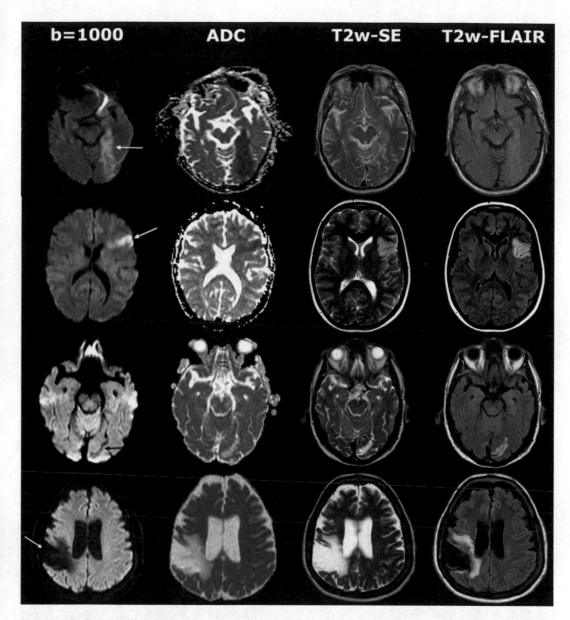

Fig. 14 MRI findings throughout the evolution of ischaemic stroke. Note that each row is from a different patient with a different area of the brain involved. *Row 1*: Early hyperacute. Cytotoxic oedema within the left PCA territory causes diffusion restriction (bright on the $b = 1000$ trace DWI and dark on the ADC). The T2 SE and FLAIR findings are equivocal indicating minimal transvascular oedema. *Row 2*: Acute. Diffusion is still restricted (bright trace signal) but this has started to ease (dark/iso on ADC). Significant transvascular oedema demonstrated by bright T2 SE and FLAIR signal. *Row 3*: Subacute. Water is diffusing freely (dark trace and bright ADC signal). Resolving transvascular oedema and evolving gliosis result in bright T2 SE and FLAIR signal. *Row 4*: Chronic. Large CSF-filled cavity represents cystic encephalomalacia, distinguished from chronic gliosis on the FLAIR signal by nulled CSF signal centrally

Time	Early hyperacute		Acute		Chronic
		Late hyperacute		Subacute	
Plain CT			Hyperdense artery		
				Hypodense brain parenchyma	
			Loss of grey-white matter differentiation		
				Sulcal effacement	
					Gliosis
					Cystic encephalomalacia
CTA			Filling defect		
CTP		Core / penumbra volume mismatch			
MRI DWI		High signal			
				Low signal	
MRI ADC		Low signal			
				High signal	
MRI T2			High signal		
MRI FLAIR			High signal		
					Low signal

Fig. 15 An approximate outline of the timing of findings on the different neuroimaging methods commonly used in the diagnosis and management of ischaemic stroke

combinations of blood and neuroradiological biomarkers will progress our understanding of pathological mechanisms during stroke and may improve the accuracy of clinical diagnosis.

References

1. Paciaroni M, Bogousslavsky J (2008) The history of stroke and cerebrovascular disease. In: Fisher M (ed) Handbook of clinical neurology, vol 92. Elsevier, Amsterdam, pp 3–28. https://doi.org/10.1016/S0072-9752(08)01901-5

2. Fernandes PM, Whiteley WN, Hart SR, Salman RA (2013) Strokes: mimics and chameleons. Pract Neurol 13(1):21–28. https://doi.org/10.1136/practneurol-2012-000465

3. Saver JL (2006) Time is brain – quantified. Stroke 37(1):263–266. https://doi.org/10.1161/01.STR.0000196957.55928.ab

4. Meretoja A, Keshtkaran M, Saver JL, Tatlisumak T, Parsons MW, Kaste M, Davis SM, Donnan GA, Churilov L (2014) Stroke thrombolysis: save a minute, save a day. Stroke 45(4):1053–1058. https://doi.org/10.1161/STROKEAHA.113.002910

5. Anderson CS, Heeley E, Huang Y, Wang J, Stapf C, Delcourt C, Lindley R, Robinson T, Lavados P, Neal B, Hata J, Hisatomi A, Parsons M, Li Y, Wang J, Heritier S, Li Q, Woodward M, Simes RJ, Davis SM, Chalmers J (2013) Rapid blood-pressure lowering in patients with acute intracerebral haemorrhage. N Engl J Med 368(25):2355–2365. https://doi.org/10.1056/NEJMoa1214609

6. Sandercock PA, Counsell C, Tseng MC, Cecconi E (2014) Oral antiplatelet therapy for acute ischaemic stroke. Cochrane Database Syst Rev 3:CD000029. https://doi.org/10.1002/14651858.CD000029.pub3

7. Wardlaw JM, Murray V, Berge E, del Zoppo GJ (2014) Thrombolysis for acute ischaemic stroke (Review). Cochrane Database Syst Rev 7:CD000213. https://doi.org/10.1002/14651858.CD000213.pub3

8. Bragg S, Paley L, Kavanagh M, McCurran V, Hoffman A, Rudd A (2017) Sentinel Stroke National Audit Programme (SSNAP), clinical audit, April 2017 – July 2017, public report. In: National results. Sentinel Stroke National Audit Programme. https://www.strokeaudit.org/Documents/National/Clinical/AprJul2017/AprJul2017-PublicReport.aspx. Accessed 31 Dec 2018

9. Wu O, Schwamm LH, Sorensen GA (2011) Imaging stroke patients with unclear onset times. Neuroimaging Clin N Am 21(2):327–344. https://doi.org/10.1016/j.nic.2011.02.008

10. Goyal M, Menon BK, van Zwam WH, Dippel DWJ, Mitchell PJ, Demchuk AM, Davalos A, Majoie CBLM, van der Lugt A, de Miquel MA,

Donnan GA, Roos YBWEM, Bonafe A, Jahan R, Diener HC, van den Berg LA, Levy EI, Berkhemer OA, Pereira VM, Rempel J, Millan M, Beumer D, Stouch B, Brown S, Campbell BCV, van Oostenbrugge RJ, Saver JL, Hill MD, Jovin TG (2016) Endovascular thrombectomy after large-vessel ischaemic stroke: a meta-analysis of individual patient data from five randomised trials. Lancet 387 (10029):1723–1731. https://doi.org/10.1016/S0140-6736(16)00163-X

11. Smith EE, Kent DM, Bulsara KR, Leung LY, Lichtman JH, Reeves MJ, Towfighi A, Whiteley WN, Zahuranec DB (2018) Accuracy of prediction instruments for diagnosing large vessel occlusion in individuals with suspected stroke: a systematic review for the 2018 guidelines for the early management of patients with acute ischemic stroke. Stroke 49(3): e111–e122. https://doi.org/10.1161/STR.0000000000000160

12. Kaufmann AM, Firlik AD, Fukui MB, Wechsler LR, Jungries CA, Yonas H (1999) Ischemic core and penumbra in human stroke. Stroke 30(1):93–99. https://doi.org/10.1161/01.STR.30.1.93

13. Beaulieu C, de Crespigny A, Tong DC, Moseley ME, Albers GW, Marks MP (1999) Longitudinal magnetic resonance imaging study of perfusion and diffusion in stroke: evolution of lesion volume and correlation with clinical outcome. Ann Neurol 46(4):568–578. https://doi.org/10.1002/1531-8249(199910)46:4<568::AID-ANA4>3.0.CO;2-R

14. Bang OY, Saver JL, Buck BH, Alger JR, Starkman S, Ovbiagele B, Kim D, Jahan R, Duckwiler GR, Yoon SR, Vinuela F, Liebeskind DS (2008) Impact of collateral flow on tissue fate in acute ischaemic stroke. J Neurol Neurosurg Psychiatry 79(6):625–629. https://doi.org/10.1136/jnnp.2007.132100

15. Nogueira RG, Jadhav AP, Haussen DC, Bonafe A, Budzik RF, Bhuva P, Yavagal DR, Ribo M, Cognard C, Hanel RA, Sila CA, Hassan AE, Millan M, Levy EI, Mitchell P, Chen M, English JD, Shah QA, Silver FL, Pereira VM, Mehta BP, Baxter BW, Abraham MG, Cardona P, Veznedaroglu E, Hellinger FR, Feng L, Kirmani JF, Lopes DK, Jankowitz BT, Frankel MR, Costalat V, Vora NA, Yoo AJ, Malik AM, Furlan AJ, Rubiera M, Aghaebrahim A, Olivot JM, Tekle WG, Shields R, Graves T, Lewis RJ, Smith WS, Liebeskind DS, Saver JL, Jovin TG (2018) Thrombectomy 6 to 24 hours after stroke with a mismatch between deficit and infarct. N Engl J Med 378(1):11–21. https://doi.org/10.1056/NEJMoa1706442

16. Allen LM, Hasso AN, Handwerker J, Farid H (2012) Sequence-specific MR imaging findings that are useful in dating ischemic stroke. Radiographics 32(5):1285–1297. https://doi.org/10.1148/rg.325115760

17. Jones TH, Morawetz RB, Crowell RM, Marcoux FW, FitzGibbon SJ, deGirolami U, Ojemann RG (1981) Thresholds of focal cerebral ischemia in awake monkeys. J Neurosurg 54 (6):773–782. https://doi.org/10.3171/jns.1981.54.6.0773

18. Liang D, Bhatta S, Volodymyr G, Simard JM (2007) Cytotoxic edema: mechanisms of pathological cell swelling. Neurosurg Focus 22(5): E2

19. Loubinoux I, Volk A, Borredon J, Guirimand S, Tiffon B, Seylaz J, Meric P (1997) Spreading of vasogenic edema and cytotoxic edema assessed by quantitative diffusion and T2 magnetic resonance imaging. Stroke 28(2):419–427. https://doi.org/10.1161/01.STR.28.2.419

20. Matsuoka Y, Hossmann KA (1982) Cortical impedance and extracellular volume changes following middle cerebral artery occlusion in cats. J Cereb Blood Flow Metab 2 (4):466–474. https://doi.org/10.1038/jcbfm.1982.53

21. Simard JM, Kent TA, Chen M, Tarasov KV, Gerzanich V (2007) Brain oedema in focal ischaemia: molecular pathophysiology and theoretical implications. Lancet Neurol 6 (3):258–268. https://doi.org/10.1016/S1474-4422(07)70055-8

22. Thrane AS, Thrane VR, Nedergaard M (2014) Drowning stars: reassessing the role of astrocytes in brain edema. Trends Neurosci 37 (11):620–628. https://doi.org/10.1016/j.tins.2014.08.010

23. Wardlaw JM, Keir SL, Seymour J, Lewis S, Sandercock PA, Dennis MS, Cairns J (2004) What is the best imaging strategy for acute stroke? Health Technol Assess 8(1):1–180

24. Steinbrich W, Gross-Fengels W, Krestin GP, Heindel W, Schreier G (1990) Intracranial hemorrhages in the magnetic resonance tomogram. Studies on sensitivity, on the development of hematomas and on determination of the cause of the hemorrhage. Fortschr Röntgenstr 152(5):534–543. https://doi.org/10.1055/s-2008-1046917

25. Buttner T, Uffmann M, Gunes N, Koster O (1997) Early CCT signs of supratentorial brain infarction: clinic-radiological correlations. Acta Neurol Scand 96(5):317–323. https://doi.org/10.1111/j.1600-0404.1997.tb00290.x

26. Wardlaw JM, Lewis SC, Dennis MS, Counsell C, McDowall M (1998) Is visible infarction on computed tomography associated with an adverse prognosis in acute ischemic stroke? Stroke 29(7):1315–1319. https://doi.org/10.1161/01.STR.29.7.1315

27. Mohr JP, Biller J, Hilal SK, Yuh WT, Tatemichi TK, Hedges S, Tali E, Nguyen H, Mun I, Adams HP Jr, Grimsman K, Marler JR (1995) Magnetic resonance versus computed tomographic imaging in acute stroke. Stroke 26(5):807–812. https://doi.org/10.1161/01.STR.26.5.807

28. Sandercock P, Molyneux A, Warlow C (1985) Value of computed tomography in patients with stroke: Oxfordshire Community Stroke Project. Br Med J (Clin Res Ed) 290(6463):193–197. https://doi.org/10.1136/bmj.290.6463.193

29. Pexman JHW, Barber PA, Hill MD, Sevick RJ, Demchuk AM, Hudon ME, Hu WY, Buchan AM (2001) Use of the Alberta Stroke Program Early CT Score (ASPECTS) for assessing CT scans in patients with acute stroke. Am J Neuroradiol 22(8):1534–1542

30. Demeestere J, Garcia-Esperon C, Garcia-Bermejo P, Ombelet F, McElduff P, Bivard A, Parsons M, Levi C (2017) Evaluation of hyperacute infarct volume using ASPECTS and brain CT perfusion core volume. Neurology 88(24):2248–2253. https://doi.org/10.1212/WNL.0000000000004028

31. Goyal M, Demchuk AM, Menon BK, Muneer E, Rempel JL, Thornton J, Roy D, Jovin TG, Willinsky RA, Sapkota BL, Dowlatshahi D, Frei DF, Kamal NR, Montanera WJ, Poppe AY, Ryckborst KJ, Silver FL, Shuaib A, Tampieri D, Williams D, Bang OY, Baxter BW, Burns PA, Choe H, Heo JH, Holmstedt CA, Jankowitz B, Kelly M, Linares G, Mandzia JL, Shankar J, Sohn SI, Swartz RH, Barber PA, Coutts SB, Smith EE, Morrish WF, Weill A, Subramaniam S, Mitha AP, Wong JH, Lowerison MW, Sajobi TT, Hill MD (2015) Randomized assessment of rapid endovascular treatment of ischemic stroke. N Engl J Med 372(11):1019–1030. https://doi.org/10.1056/NEJMoa1414905

32. Jovin TG, Chamorro A, Cobo E, de Miquel MA, Molina CA, Rovira A, Roman LS, Serena J, Abilleira S, Ribo M, Millan M, Urra X, Cardona P, Lopez-Cancio E, Tomasello A, Castano C, Blasco J, Aja L, Dorado L, Quesada H, Rubiera M, Hernandez-Perez M, Goyal M, Demchuk AM, von Kummer R, Gallofre M, Davalos A (2015) Thrombectomy within 8 hours of symptom onset in ischemic stroke. N Engl J Med 372(24):2296–2306. https://doi.org/10.1056/NEJMoa1503780

33. Wahlgren N, Moreira T, Michel P, Steiner T, Jansen O, Cognard C, Matte HP, van Zwam W, Holmin S, Tatlisumak T, Petersson J, Caso V, Hacke W, Mazighi M, Arnold M, Fischer U, Szikora I, Pierot L, Fiehler J, Gralla J, Fazekas F, Lees KR, ESO-KSU, ESO, ESMINT, ESNR, EAN (2016) Mechanical thrombectomy in acute ischemic stroke: consensus statement by ESO-Karolinska Stroke update 2014/2015, supported by ESO, ESMINT, ESNR and EAN. Int J Stroke 11(1):134–137. https://doi.org/10.1177/1747493015609778

34. Powers WJ, Derdeyn CP, Biller J, Coffey CS, Hoh BL, Jauch EC, Johnston KC, Johnston SC, Khalessi AA, Kidwell CS, Meschia JF, Ovbiagele B, Yavagal DR, American Heart Association Stroke Council (2015) 2015 American Heart Association/American Stroke Association focused update of the 2013 guidelines for the early management of patients with acute ischemic stroke regarding endovascular treatment. Stroke 46(10):3020–3035. https://doi.org/10.1161/STR.0000000000000074

35. Menon BK, d'Esterre CD, Qazi EM, Almekhlafi M, Hahn L, Demchuk AM, Goyal M (2015) Multiphase CT Angiography: A new tool for the imaging triage of patients with acute ischemic stroke. Radiology 275(2):510–520. https://doi.org/10.1148/radiol.15142256

36. Menon BK, Smith EE, Modi J, Patel SK, Bhatia R, Watson TW, Hill MD, Demchuk AM, Goyal M (2011) Regional leptomeningeal score on CT angiography predicts clinical and imaging outcomes in patients with acute anterior circulation occlusions. Am J Neuroradiol 32(9):1640–1645. https://doi.org/10.3174/ajnr.A2564

37. Ritter MA, Poeplau T, Schaefer A, Kloska SP, Dziewas R, Ringelstein EB, Heindel W, Nabavi DG (2006) CT Angiography in acute stroke – does it provide additional information on occurrence of infarction and functional outcome after 3 months? Cerebrovasc Dis 22(5-6):362–367. https://doi.org/10.1159/000094852

38. Mortimer AM, Simpson E, Bradley MD, Renowden SA (2013) Computed tomography angiography in hyperacute ischemic stroke: prognostic implications and role in decision-making. Stroke 44(5):1480–1488. https://doi.org/10.1161/STROKEAHA.111.679522

39. Lev MH, Farkas J, Rodriguez VR, Schwamm LH, Hunter GJ, Putman CM, Rordorf GA, Buonanno FS, Budzik R, Koroshetz WJ, Gonzalez RG (2001) CT angiography in the rapid triage of patients with hyperacute stroke to intraarterial thrombolysis: accuracy in the detection of large vessel thrombus. J Comput Assist Tomogr 25(4):520–528. https://doi.org/10.1097/00004728-200107000-00003

40. Bash S, Villablanca JP, Jahan R, Duckwiler G, Tillis M, Kidwell C, Saver J, Sayre J (2005) Intracranial vascular stenosis and occlusive disease: evaluation with CT angiography, MR angiography, and digital subtraction angiography. Am J Neuroradiol 26(5):1012–1021

41. Van der Hoeven EJ, Schonewille WJ, Vos JA, Algra A, Audebert HJ, Berge E, Ciccone A, Mazighi M, Michel P, Muir KW, Obach V, Puetz V, Wijman CA, Zini A, Kappelle JL, BASICS Study Group (2013) The basilar artery international cooperation study (BASICS): study protocol for a randomised controlled trial. Trials 14(1):200. https://doi.org/10.1186/1745-6215-14-200

42. Li W, Chen Z, Dai Z, Liu R, Yin Q, Wang H, Hao Y, Han Y, Qiu Z, Xiong Y, Sun W, Zi W, Xu G, Liu X (2018) Management of acute tandem occlusions: stent-retriever thrombectomy with emergency stenting or angioplasty. J Int Med Res 46(7):2578–2586. https://doi.org/10.1177/0300060518765310

43. Konstas AA, Goldmakher GV, Lee TY, Lev MH (2009) Theoretic basis and technical implementations of CT perfusion in acute ischemic stroke, part 2: technical implementations. Am J Neuroradiol 30(5):885–892. https://doi.org/10.3174/ajnr.A1492

44. Konstas AA, Goldmakher GV, Lee TY, Lev MH (2009) Theoretic basis and technical implementations of CT perfusion in acute ischemic stroke, part 1: theoretic basis. Am J Neuroradiol 30(4):662–668. https://doi.org/10.3174/ajnr.A1487

45. Page M, Nandurkar D, Crossett MP, Stuckey SL, Lau KP, Kenning N, Troupis JM (2010) Comparison of 4 cm z-axis and 16 cm z-axis multidetector CT perfusion. Eur Radiol 20(6):1508–1514. https://doi.org/10.1007/s00330-009-1688-8

46. Ukmar M, Degrassi F, Mucelli RAP, Neri F, Mucelli FP, Cova MA (2017) Perfusion CT in acute stroke: effectiveness of automatically-generated colour maps. Br J Radiol 90(1072):20150472. https://doi.org/10.1259/bjr.20150472

47. Shen J, Li X, Li Y, Wu B (2017) Comparative accuracy of CT perfusion in diagnosing acute ischemic stroke: a systematic review of 27 trials. PLoS One 12(5):e0176622. https://doi.org/10.1371/journal.pone.0176622

48. Lui YW, Tang ER, Allmendinger AM, Spektor V (2010) Evaluation of CT perfusion in the setting of cerebral ischemia: patterns and pitfalls. Am J Neuroradiol 31(9):1552–1563. https://doi.org/10.3174/ajnr.A2026

49. Bivard A, Levi C, Krishnamurthy V, McElduff P, Miteff F, Spratt NJ, Bateman G, Donnan G, Davis S, Parsons M (2015) Perfusion computed tomography to assist decision making for stroke thrombolysis. Brain 138(7):1919–1931. https://doi.org/10.1093/brain/awv071

50. Wannamaker R, Guinand T, Menon BK, Demchuk A, Goyal M, Frei D, Bharatha A, Jovin TG, Shankar J, Krings T, Baxter B, Holmstedt C, Swartz R, Dowlatshahi D, Chan R, Tampieri D, Choe H, Burns P, Gentile N, Rempel J, Shuaib A, Buck B, Bivard A, Hill M, Butcher K (2018) Computed tomographic perfusion predicts poor outcomes in a randomised trial of endovascular therapy. Stroke 49(6):1426–1433. https://doi.org/10.1161/STROKEAHA.117.019806

51. Copen WA, Yoo AJ, Rost NS, Morais LT, Schaefer PW, Gonzalez RG, Wu O (2017) In patients with suspected acute stroke, CT perfusion-based cerebral blood flow maps cannot substitute for DWI in measuring the ischemic core. PLoS One 12(11):e0188891. https://doi.org/10.1371/journal.pone.0188891

52. Jauch EC, Saver JL, Adams HP Jr, Bruno A, Connors JJ, Demaerschalk BM, Khatri P, McMullan PW Jr, Qureshi AI, Rosenfield K, Scott PA, Summers DR, Wang DZ, Wintermark M, Yonas H, American Heart Association Stroke Council; Council on Cardiovascular Nursing; Council on Peripheral Vascular Disease; Council on Clinical Cardiology (2013) Guidelines for the early management of patients with acute ischemic stroke: a guideline for healthcare professionals from the American Heart Association/American Stroke Association. Stroke 44(3):870–947. https://doi.org/10.1161/STR.0b013e318284056a

53. Simonsen CZ, Madsen MH, Schmitz ML, Mikkelsen IK, Fisher M, Anderson G (2015) Sensitivity of diffusion- and perfusion-weighted imaging for diagnosing acute ischemic stroke is 97.5%. Stroke 46(1):98–101. https://doi.org/10.1161/STROKEAHA.114.007107

Chapter 16

Ultrasound Assessments of Risk for TIA and Stroke in Vascular Surgery

Melvinder Basra and Robert E. Brightwell

Abstract

Stroke is a significant cause of mortality and morbidity worldwide. The molecular biology of atherosclerotic disease is well understood, and increasing evidence is emerging to permit the interpretation of plaque evolution and the resultant changes in flow dynamics which predict the risk of stroke. This chapter elucidates the new modalities in carotid imaging that may be used for surveillance of carotid artery disease or as part of a risk model to identify patients that may benefit from carotid intervention in the future.

Key words Carotid artery disease, Carotid ultrasound, Atherosclerosis, Stroke

1 Background of Carotid Artery Disease

Across Europe, stroke is the second most common cause of death, causing 1.1 million deaths per year [1]. In a European population of 715 million, approximately 1.4 million strokes occur per annum [2]. In addition to being a significant cause of mortality, stroke results in considerable morbidity, at great cost to the individual and to society. Over half of stroke survivors remain dependent on others for some aspect of daily living [3] resulting in an enormous financial burden with annual stroke costs across Europe reaching approximately 40 billion Euros [1].

Twenty-five percent of ischemic carotid territory strokes are caused by thromboembolism from the internal carotid artery (ICA) or middle cerebral artery (MCA). 25% are caused by small vessel intracranial disease, 20% by cardiac embolism, 5% by specified rarer causes, and 25% from unknown causes despite investigation [4]. Overall, 10–15% of all strokes follow thromboembolism from a previously asymptomatic ICA stenosis of >50% [5].

Philip V. Peplow et al. (eds.), *Stroke Biomarkers*, Neuromethods, vol. 147, https://doi.org/10.1007/978-1-4939-9682-7_16,
© Springer Science+Business Media, LLC, part of Springer Nature 2020

336 Melvinder Basra and Robert E. Brightwell

2 Investigation of Carotid Artery Disease

There are two main methods for measuring the degree of ICA stenosis and these were outlined in the European Carotid Surgery trial (ESCT) [6] and the North American Symptomatic Carotid Endarterectomy trial (NASCET) [7]. Both methods use the residual ICA luminal diameter as the numerator. In the ECST method, the denominator is the estimated total vessel diameter at the level that the residual luminal diameter was measured (often the carotid bulb). In contrast, the NASCET measures the diameter of a healthy portion of ICA cephalad to the stenosis, where the vessel walls are parallel and uses this as the denominator. The severity of stenosis measured will differ with each of these methods and the clinician needs to be aware of the method used in their center in order to offer consistent decision-making regarding intervention.

In the context of large volume plaques within a dilated carotid bulb there may have been secondary enlargement of the bulb. The ECST method offers an important advantage when calculating the severity of stenosis in this setting as the residual luminal diameter may only be slightly less than the distal ICA, and thus the severity of the disease may be underestimated by the NASCET method. In this scenario, recently symptomatic patients should have their degree of ICA stenosis calculated using the ESCT, and offered intervention accordingly.

3 Ultrasound for Assessment of Carotid Artery Disease

Noninvasive assessment of carotid artery disease is highly desirable, particularly in patients that may require serial measurements of plaque burden and disease progression, or surveillance after intervention (such as carotid artery stenting). Formerly, in both the ECST and NASCET patients routinely underwent intra-arterial angiography. However, the Asymptomatic Carotid Atherosclerosis Study [8] found a 30-day death/stroke rate of 2.3% after carotid endarterectomy (CEA) and just over half of these (1.2%) were angiography related, so this practice was discontinued.

Duplex combines brightness mode ultrasound imaging with color Doppler; permitting simultaneous assessment of both the structural characteristics of the vessel and plaque along with real-time measurement of blood flow. A sample image is shown in Fig. 1.

For most individuals, carotid anatomy permits ultrasound assessment of the carotid bifurcation and the proximal 2 cm of the internal carotid artery. Assessment of this region yields information of considerable preoperative importance regarding the

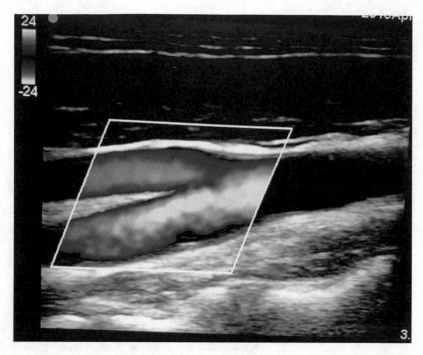

Fig. 1 Duplex Doppler image of the carotid bifurcation

Table 1
Information gleaned from preoperative ultrasound

Anatomical factors
- Configuration of carotid vessels
- Level of carotid bifurcation
- Tortuosity of ICA

Plaque factors
- Degree of stenosis
- Plaque type

Physiological factors
- Flow velocities
- Integrity of circle of Willis
  – Contralateral ICA
  – Vertebral arteries

artery, the atherosclerotic plaque and other anatomical considerations (*see* Table 1).

This information assists preoperative planning and provides some insight with regards to collateral blood supply, which may in turn influence decision-making surrounding the use of a carotid shunt. Duplex allows for the degree of ICA stenosis to be calculated using flow velocities. Luminal narrowing secondary to carotid

Table 2
Duplex ultrasound criteria for defining stenosis

% Stenosis NASCET	Peak systolic velocity (PSV, cm/s)	PSV_{ICA}/PSV_{CCA} ratio
<50%	<125	<2
50–69%	≥125	2–4
60–69%		
70–79%	≥230	≥4
80–89%		
>90%	≥400	≥5
Near occlusion	High, low-string flow	Variable
Occlusion	No flow	Not applicable

atherosclerosis results in velocity changes, which can be detected using duplex and converted into degrees of stenosis (*see* Table 2).

Limitations common to all ultrasound assessments for carotid artery disease include anatomical factors such as a short neck and/or a high carotid bifurcation that obscures the carotid bifurcation deep to the mandible and plaque factors such as highly calcified plaques that result in acoustic shadowing [9]. In this context, an additional imaging modality is indicated.

4   Developments in Ultrasound Technology

At the present time, the percentage diameter reduction in the ICA is used as the main predictor for stroke risk. New research into the mechanisms of plaque rupture and atheroembolic stroke suggests that the degree of narrowing is an imperfect predictor of stroke risk. It is often observed that many patients with high-grade carotid stenosis remain asymptomatic for many years, while others with moderate stenosis develop neurologic symptoms sooner [10].

It has been proposed that other factors such as plaque composition and hemodynamic forces can play a role in determining the risk of stroke. Magnetic resonance image (MRI) studies of stroke patients with only mild carotid artery stenosis have shown that these plaques exhibit other features that suggest vulnerability [11]. Surrogate markers of plaque vulnerability include plaque volume, lipid necrotic core size, surface ulceration, intraplaque hemorrhage, and hemodynamic effects around the plaque. Advances in ultrasound technology have enabled interrogation of the factors that predispose to plaque vulnerability, permitting identification of plaques at high-risk of imminent embolization, potentially resulting in stroke.

Animal studies indicate that enlarging lipid cores, increasing intraplaque hemorrhage and thinning of the fibrous cap all predispose plaques to rupture; however these factors are on a continuum and, unlike flow velocity measurements, at the present time discrete thresholds do not exist that categorize the risk of stroke. Furthermore, one-time imaging does not provide an insight into the long-term biological evolution of a carotid plaque and application of new techniques would necessitate serial measurements alongside clinical correlation before these markers were validated for the stratification of plaque vulnerability [12].

5 Brightness Mode Imaging

Brightness mode (B-mode) imaging is a form of two-dimensional imaging that displays anatomic wall features. Different tissues reflect ultrasound waves to varying degrees, and this leads to structures that appear either heterogeneous, or relatively hypoechoic or hyperechoic. Gray-Weale devised a classification to describe the appearance of different carotid plaques and it was observed that plaques with hyperechoic signals were more likely to be found in patients with neurological symptoms [13]. This method, however, had two major limitations; that the classification depended on subjective visual estimation and that the technique for image acquisition also introduced considerable variability, leading to high levels of interobserver and intraobserver variation.

El-Barghouty et al. [14] introduced standardization for image acquisition and brightness using the blood column and carotid adventitia. Carotid plaques were outlined and the median brightness of pixels in a single longitudinal image of the plaque was measured and expressed as a gray-scale median (GSM) value.

One thousand two hundred and twenty-one patients with 50–99% carotid stenosis were followed for 6–96 months. GSM $\leq$ 40, longitudinal plaque area (hazard ratio = 1.92; 95% confidence interval, 1.50–2.46; $P \leq 0.001$) and the size of discrete white areas in the plaque image (hazard ratio = 2.10; 95% confidence interval, 1.32–3.35; $P \leq 0.002$) were all found to be strong predictors for future stroke [15].

Supporting the theory that rupture of the fibrous cap that overlies the lipid core of an atherosclerotic plaque leads to thromboembolic events, histological studies have corroborated that the necrotic core of unstable plaques is located closer to the lumen than in asymptomatic plaques [16].

On B-mode imaging, this necrotic core appears as a hypoechoic, juxtaluminal black area (*see* Fig. 2). The Asymptomatic Carotid Stenosis and Risk of Stroke (ACSRS) trial of 1121 patients with 50–99% asymptomatic stenosis of the ICA reported that patients with a larger juxtaluminal black area of >10 mm^2 had a

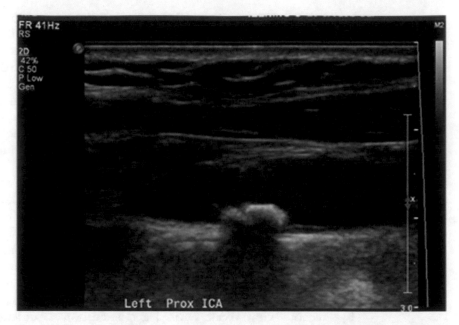

Fig. 2 Juxtaluminal black area (Reproduced from Kakkos et al., 2013 with permission from Elsevier) [17]

higher annualized stroke rate of 5.0% in contrast to patients with a smaller juxtaluminal black area of 8–10 mm^2 who had an annual stroke rate of 3.2% [17].

6   Pixel Distribution Analysis

Our current understanding of atherosclerosis purports that atherosclerotic plaques begin with deposition of lipid to form fatty streaks that gradually coalesce to form a lipid core. Over time, a fibroatheroma forms as fibrous tissue accumulates over the lipid core to form a fibrous cap.

Much of our understanding of atherosclerotic disease is influenced by study of the coronary arteries. When studying coronary artery disease, it emerged that the risk of coronary events was not simply determined by the degree of coronary artery stenosis. Indeed, not all coronary culprit lesions were high-grade stenoses. Emerging data suggests that atherosclerotic plaques are rendered unstable/vulnerable through an enlarging lipid core, intraplaque hemorrhage, fibrous cap thinning/rupture, and finally ulceration. These histological changes in plaque morphology have been observed in carotid endarterectomy specimens retrieved from patients with symptomatic (stroke/TIA) carotid artery disease [12, 18, 19].

The potential translational benefits of improved ultrasound carotid imaging are wide-ranging. In the modern era, when patients may be considered more or less suitable for either surgical carotid endarterectomy or endovascular carotid artery stenting, it is conceivable that identification of plaque stability may influence the choice of treatment modality. For instance, patients with large juxtaluminal necrotic cores may be more vulnerable to endovascular atheroembolization when disturbed by a guide wire, balloon or manipulation of a stent. Improved, noninvasive ultrasound imaging could therefore stratify risk of treatment in asymptomatic patients; it could also quantify plaque responses to medical management.

Detection of these segmental areas of plaque instability was traditionally not possible using Gray-Weale or GSM as they were missed. Ideal noninvasive morphological assessment would not only detect these areas of plaque instability but also elucidate their distribution using a technique with a low intraobserver and inter-observer variability.

Lal et al. utilized the fact that ultrasound is reflected to varying degrees by different tissues, producing B-mode images of varying pixel intensities to devise an image segmentation algorithm. The different components of a carotid plaque (e.g., calcium, fibrous tissue, muscle, lipid, and blood) and their characteristic pixel density were mapped to normalized longitudinal images of carotid plaques giving detailed information on the percentage tissue composition within plaques (*see* Fig. 3). PDA (pixel density analysis) was used to quantify intraplaque hemorrhage, lipid, fibromuscular tissue, and calcium of carotid plaques within 45 carotid arteries prior to undergoing CEA. Statistically significant differences were seen in plaque composition between symptomatic and asymptomatic plaques. Eighteen symptomatic plaques demonstrated larger

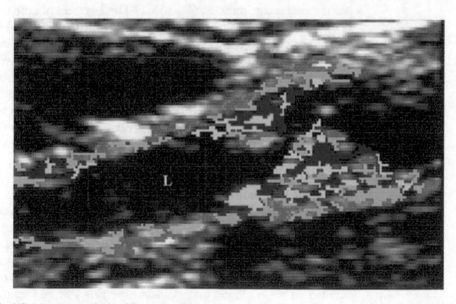

Fig. 3 Pixel distribution analysis of B-mode ultrasound in an atherosclerotic carotid artery plaque (Reproduced from Lal et al., 2002 with permission from Elsevier) [18]

quantities of intraplaque hemorrhage ($P < 0.001$) and lipid ($P = 0.002$), and larger lipid cores ($P = 0.005$) that were closer to the flow lumen ($P = 0.01$). The 27 asymptomatic plaques evaluated in the study demonstrated smaller amounts of calcium ($P < 0.001$) [18]. With the ability to elucidate tissue composition and morphology of plaque within the ICA, pixel density analysis may have the potential to increase the sensitivity and specificity of ultrasound in detecting lesions at increased risk of stroke [20].

7 Three-Dimensional Ultrasound

A carotid plaque is a three-dimensional structure. Enlarging carotid plaques grow both in longitudinal section and cross section, and plaque area grows faster than plaque thickness. Carotid artery imaging that only assesses cross-sectional morphology (lumen-diameter reducing stenosis) only assesses progression of the carotid plaque in two dimensions. Research has shown that the longitudinal-sectional area of a carotid plaque relates to cardiovascular outcomes. Spence et al. followed 1686 patients from an atherosclerosis prevention clinic annually for 5 years and divided them into quartiles based on their carotid plaque areas: 0.00–0.11 cm^2 ($n = 422$), 0.12–0.45 cm^2 ($n = 424$), 0.46–1.18 cm^2 ($n = 421$), and 1.19–6.73 cm^2 ($n = 419$). The 5-year risk of stroke, myocardial infarction, and vascular death increased by quartile of carotid plaque area: 5.6%, 10.7%, 13.9%, and 19.5%, respectively ($P = 0.001$) after adjustment for all baseline patient characteristics [21]. 1085 patients underwent repeat carotid plaque area measurement. 685 (63.1%) had carotid plaque progression, 306 (28.2%) had plaque regression, and 176 (16.2%) had no change in carotid plaque area over the period of follow-up. The 5-year adjusted risk of combined outcome was 9.4%, 7.6%, and 15.7% for patients with carotid plaque area regression, no change, and progression, respectively ($P = 0.003$) [21]. This study concluded that plaque area and progression could be used to identify high-risk patients; thus influencing treatment choices. Plaque area and volume (rather than purely lumen diameter stenosis) may thus be better predictors of future stroke.

With the use of a special probe, three-dimensional ultrasound uses computer software to collate two-dimensional ultrasound cross-sectional slices and reconstruct them into a three-dimensional volume. This reproducible reconstruction eliminates some of the operator variability inherent in ultrasound assessment and can be used to quantify plaque volume. AlMuhanna et al. reported a slightly greater sensitivity of three-dimensional ultrasound to detect small changes in plaque progression over time, when compared to traditional, two-dimensional ultrasound. Using a semiautomatic algorithm for segmentation, they quoted 10–15 min to complete quantification using a three-dimensional system, increasing the clinical utility of such a tool [22]. Other authors such as Wannarong et al. used three-dimensional ultrasound to follow-up 349 patients annually for 5 years. Cox regression analysis demonstrated that progression of total plaque volume was a predictor for stroke, death, and TIA (Kaplan–Meier logrank $P = 0.001$). Intima–media thickness did not predict these events, suggesting that three-dimensional ultrasound may have additional sensitivity when identifying vulnerable plaques over time [23].

Appreciating plaque morphology in three dimensions allows for detailed assessment of the plaque surface. Surface features that may indicate plaque vulnerability such as ulceration (distinct discontinuity in an atherosclerotic plaque with a volume ≥ 1 mm^3) have been found to predict the risk of stroke, myocardial infarction and death. Kuk et al. evaluated 313 patients with asymptomatic carotid artery stenosis and found that patients with a total ulcer volume of ≥ 5 mm^3 experienced a significantly higher risk of developing stroke, TIA, or death (Kaplan–Meier analysis $P = 0.009$). Lower ulcer volumes did not predict events [24].

Madani et al. studied 253 patients with asymptomatic carotid artery stenosis ($>60\%$ on Doppler) identifying carotid plaque ulceration using three-dimensional ultrasound and found that patients with ≥ 3 ulcers were more likely to have a stroke or death during 3 years of follow-up (18% vs. 2% $P = 0.03$). The authors argued that three-dimensional identification of plaque ulceration may assist in the identification of patients with high-grade, asymptomatic carotid artery stenosis that would benefit from carotid intervention [25].

The ability of three-dimensional ultrasound to delineate tissues may also permit the identification of plaque histology to quantify intraplaque hemorrhage or enlarging lipid cores. Clinical three dimensional ultrasound protocols have been developed to quantify plaque composition (hemorrhage, lipid, calcium, fibromuscular tissue) and one small, study of ten patients with asymptomatic $\geq 50\%$ carotid stenosis found that this technique permitted the identification of subtle (approximately 12%) changes in total plaque volume. The authors suggested that this may assist in the stratification of patients with carotid artery disease and permit monitoring of response to pharmacological therapy and selection for carotid intervention. In addition to plaque volume, a larger study of 298 patients with carotid artery disease found that plaque texture change had a hazard ratio of 1.4 for stroke and TIA, whilst greater plaque volume had a hazard ratio of 1.5 [26].

8 Contrast-Enhanced Ultrasound

Contrast-enhanced ultrasound (CEUS) uses B-mode ultrasound augmented by an administration of intravenous microbubbles of an inert gas. The most commonly used agent contains sulfur hexafluoride inside a phospholipid shell [27]. CEUS is performed with the same probe used for conventional ultrasound (5–10 MHz linear probe) but the ultrasound machine is used in a contrast specific mode like pulse inversion or amplitude modulation [27, 28]. After the conventional ultrasound is complete, the microbubbles are injected via a venous cannula in the antecubital fossa. Just before the microbubbles are injected, the probe is placed over the region of most interest (the stenosed segment of ICA) and luminal enhancement begins after 10–30 s and lasts for up to 2–5 min [9]. CEUS offers some key advantages over CT angiogram (CTA). Unlike CTA, CEUS does not rely upon a nephrotoxic contrast medium and does not deliver a radiation dose. Anaphylaxis to the microbubbles is also rare (<1 in 100,000). A typical CEUS image shows the lumen and adventitia as enhanced whereas the carotid plaque and intima–media complex appear hypoechoic (Fig. 4).

By virtue of contrast, CEUS improves the intrareader and interreader visualization of luminal irregularities such as plaque surface ulceration when compared to standard B-mode or color Doppler ultrasound; with a sensitivity and negative predictive value of 87% and 88% respectively [28, 29]. This greater sensitivity to luminal surface change heightens the ability of CEUS to detect subclinical carotid artery disease [30]. Another key advantage of CEUS may lie in its potential to differentiate inflamed carotid plaques, a phenomenon that relates to "late-phase enhancement." It is proposed that contrast microbubbles are ingested by monocytes and these remain adherent to inflamed endothelium; retaining the contrast for a longer period of time [31]. Carotid plaques that retain enhancement for longer, may therefore contain more

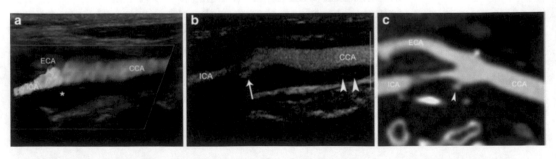

Fig. 4 A symptomatic ulcerated right internal carotid artery. (**a**) Colour Doppler image (**b**) CEUS revealed plaque ulceration (arrow) and neovessels (arrowheads) (**c**) CT angiogram demonstrating plaque ulceration (arrowhead) (Reproduced from Rafailidis et al. 2018, with permission from Springer Nature) [29]

inflammatory cells than those that clear contrast earlier [32], and these "inflamed plaques" may be predisposed to rupture [33, 34]. Late-phase enhancement is assessed by performing CEUS imaging of the carotid plaque 6 min after contrast administration; and may serve as a marker of plaque vulnerability. This imaging modality has biological plausibility and holds promise; it requires ongoing longitudinal study before being accepted for mainstream clinical application.

9   Plaque-Strain Measurement

In addition to the histological characteristics of the plaque, the biomechanical forces involved may also be related to the risk of atheroembolization. These biomechanical forces are an interplay between blood flow (hemodynamic forces) and the nature of the plaque itself (viscoelastic properties).

Strain-imaging (ultrasound elastography) quantifies the degree of deformation that tissues undergo in response to an internal/external force. Whilst this has been validated in a host of other human tissues, carotid plaque motion measurement represents a challenge due to the small size of the target lesion and its highly dynamic location that relates to the cardiac and ventilatory cycle. However, it has been demonstrated that plaques with high strain are more echolucent, suggesting that soft plaques may be identified as having higher strain measurements [35]. Zhang et al. tested this hypothesis with CEUS; and found that plaques with greater neovascularization tended to be softer, more elastic and heterogeneous. The authors suggested that a low GSM (echolucent plaque) with higher elasticity may be indicative of an unstable plaque [36]. This modality is undergoing further assessment within larger, longitudinal studies such as the CREST-2.

10   Conclusion

This chapter outlines the current advances within the field of carotid imaging. The future potential for these new modalities is exciting, but at present no single method is confirmed as having clinical superiority [37]. At this stage a combined methodology may help identify at-risk, unstable plaques in those with carotid disease in whom intervention is planned. If this proves a successful strategy then a role for carotid screening may emerge in the near future, with the associated wide-reaching benefits in stroke prevention.

References

1. Nichols M, Townsend N, Luengo-Fernandez-R, et al. European Cardiovascular Disease Statistics (2012) Sophia antipolis. European Heart Network, European Society of Cardiology, Brussels. www.escardio.org/static_file/.../EU-Cardiovascular-disease-statistics-2012.pdf. Accessed 16 Apr 2018

2. Truelsen B, Piechowski-Jozwiak T, Bonita R et al (2006) Stroke incidence and prevalence in Europe. Eur Neurol 13:581–598

3. Royal College of Physicians National Sentinel Stroke Clinical Audit 2010 Round 7. Public report for England, Wales and Northern Ireland. Prepared on behalf of the Intercollegiate Stroke Working Party May 2011, p 43

4. Ay H, Arsava EM, Andsberg G, Benner T, Brown RD, Chapman SN et al (2014) Pathogenic ischemic stroke phenotypes in the NINDS-stroke genetics network. Stroke 45:3589e96

5. Naylor AR (2015) Why is the management of asymptomatic carotid disease so controversial? Surgeon 13:34e43

6. Randomised trial of endarterectomy for recently symptomatic carotid stenosis: final results of the MRC European Carotid Surgery Trial (1998). Lancet 351(9113):1379–1387

7. Ferguson G, Eliasziw M, Barr HWK et al (1999) The North American symptomatic carotid endarterectomy trial. Stroke 30:1751–1758

8. Executive Committee for the Asymptomatic Carotid Atherosclerosis Study (1995) Endarterectomy for asymptomatic carotid artery stenosis. JAMA 273:1421–1428

9. Rafailidis V, Charitanti A, Tegos T et al (2017) Contrast-enhanced ultrasound of the carotid system: a review of the current literature. J Ultrasound 20:97–109

10. Naylor AR, Rothwell PM, Bell PR (2003) Overview of the principal results and secondary analyses from the European and North American randomised trials of endarterectomy for symptomatic carotid stenosis. Eur J Vasc Endovasc Surg 26:115–129. https://doi.org/10.1053/ejvs.2002.1946

11. Freilinger TM, Schindler A, Schmidt C et al (2012) Prevalence of nonstenosing, complicated atherosclerotic plaques in cryptogenic stroke. J Am Coll Cardiol Img 5:397–405. https://doi.org/10.1016/j.jcmg.2012.01.012

12. Redgrave JN, Lovett JK, Gallagher PJ et al (2006) Histological assessment of 526 symptomatic carotid plaques in relation to the nature and timing of ischaemic symptoms: the Oxford Plaque Study. Circulation 113:2320–2328

13. Gray-Weale AC, Graham JC, Burnett JR et al (1988) Carotid artery atheroma: comparison of pre-operative B-mode ultrasound appearance with carotid endarterectomy specimen pathology. J Cardiovasc Surg 29:676–681

14. El-Barghouty N, Nicolaides A, Bahal V et al (1996) The identification of the high-risk carotid plaque. Eur J Endovasc Surg 11:470–478

15. Nicolaides AN, Kakkos SK, Kyriacou E et al (2010) Asymptomatic internal carotid artery stenosis and cerebrovascular risk stratification. J Vasc Surg 52:1486–1496. e5

16. Bassiouny HS, Sakaguchi Y, Mikucki SA et al (1997) Juxtalumenal location of plaque necrosis and neoformation in symptomatic carotid stenosis. J Vasc Surg 26:585–594

17. Kakkos SK, Griffin MB, Nicolaides AN et al (2013) The size of Juxtaluminal black area in ultrasonic images of asymptomatic carotid plaques predicts the occurrence of stroke. J Vasc Surg 55(Suppl 1):84–85

18. Lal BK, Hobson RW, Pappas PJ et al (2002) Non-invasive identification of the unstable carotid plaque. Pixel distribution analysis of B-mode ultrasound scan images predicts histological features of atherosclerotic carotid plaques. J Vasc Surg 35:1210–1217

19. Lal BK, Hobson RW, Hameed M et al (2006) Noninvasive identification of the unstable carotid plaque. Ann Vasc Surg 20:167–174

20. Hobson RW, Lal BK, Chakhtoura E et al (2003) Carotid artery stenting: analysis of data for 105 patients at high risk. J Vasc Surg 37:1234–1239

21. Spence JD, Eliasziw M, DiCicco M et al (2002) Carotid plaque area: a tool for targeting and evaluating vascular preventive therapy. Stroke 33:2916–2922

22. AlMuhanna K, Hossain MM, Zhao L et al (2015) Carotid plaque morphometric assessment with three-dimensional ultrasound imaging. J Vasc Surg 61(3):690–697

23. Wannarong T, Parraga G, Buchanan D et al (2013) Progression of carotid plaque volume predicts cardiovascular events. Stroke 44:1859–1865

24. Kuk M, Wannarong T, Beletsky V et al (2014) Volume of carotid artery ulceration as a predictor of cardiovascular events. Stroke 45:1437–1441

25. Madani A, Beletsky V, Tamayo A et al (2011) High-risk asymptomatic carotid stenosis ulceration on 3D ultrasound vs TCD microemboli. Neurology 77:744–750

26. De Bruijne M (2014) Three-dimensional carotid ultrasound plaque texture predicts vascular events. Stroke 45:2695–2701

27. Piscaglia F, Nolsoe C, Dietrich CF et al (2012) The EFSUMB guidelines and recommendations on the clinical practice of contrast enhanced ultrasound (CEUS): update 2011 on nonhepatic applications. Ultraschall Med 33:33–59. https://doi.org/10.1055/s0031-1281676

28. Ten Kate GL, Van Dijk AC, Van den Oord SC et al (2013) Usefulness of contrast-enhanced ultrasound for detection of carotid plaque ulceration in patients with symptomatic carotid atherosclerosis. Am J Cardiol 112:292–298

29. Rafailidis V, Chryssogonidis I, Xerras C et al (2018) A comparative study of Colour Doppler imaging and contrast-enhanced ultrasound for the detection of ulceration in patients with carotid atherosclerotic disease. Eur Radiol. https://doi.org/10.1007/s00330-018-5773-8

30. Van den Oord SCH, ten Kate GL, Sijbrands EJG et al (2013) Effect of carotid plaque screening using contrast-enhanced ultrasound on cardiovascular risk-stratification. Am J Cardiol 111:754–759

31. Yanagisawa K, Moriyasu F, Miyahara T et al (2007) Phagocytosis of ultrasound contrast agent microbubbles by Kupffer cells. Ultrasound Med Biol 33:318–325

32. Owen DR, Shalhoub J, Miller S et al (2010) Inflammation within carotid atherosclerotic plaque: assessment with late-phase contrast-enhanced ultrasound. Radiology 255:638–644

33. Virmani R, Burke AP, Willerson JT et al (2007) The pathology of vulnerable plaque. In: Narula J, Leon MB (eds) The vulnerable atherosclerotic plaque: strategies for diagnosis and management. Blackwell Futura, Malden, MA, pp 19–36

34. Moreno PR, Falk E, Palacios IF et al (1994) Macrophage infiltration in acute coronary syndromes. Implications for plaque rupture. Circulation 90:775–778

35. Widman E, Caidahl K, Heyde B (2015) Ultrasound speckle tracking strain estimation of in vivo carotid artery plaque with in vitro sonomicrometry validation. Ultrasound Med Biol 41:77–88

36. Zhang Q, Li C, Zhou M et al (2015) Quantification of carotid plaque elasticity and intraplaque neovascularisation using contrast-enhanced ultrasound and image-registration elastography. Ultrasonics 62:253–262

37. Cires-Drouet RS, Mozafarian M, Ali A, Sikdar S, Lal BK (2017) Imaging of high-risk carotid plaques: ultrasound. Semin Vasc Surg 30:44–53

Melvinder Basra, MBBS, BSc, MSc, MRCS, is a registrar in vascular surgery at the Norfolk & Norwich University NHS Foundation Trust. He developed an interest in surgical research during his time as an undergraduate at Imperial College London.

Robert E. Brightwell, MD, FRCS, is a consultant vascular surgeon at the Norfolk & Norwich University NHS Foundation Trust and honorary clinical lecturer at Imperial College London. He developed a research interest in carotid imaging and biomarkers of stroke during his time conducting a postgraduate doctorate at Imperial College London.

Chapter 17

Preoperative and Intraoperative Markers of Cerebral Ischemia

V. A. Lukshin, D. Yu Usachev, A. V. Shmigelsky, A. A. Shulgina, and A. A. Ogurtsova

Abstract

Different modalities of instrumental neuromonitoring techniques of the risk of cerebral ischemia during carotid endarterectomy (CEA) are described. To assess the rate of ischemic complications, and provide possible solutions to decrease risk factors for carotid clamping (temporary interruption of blood flow through carotid artery during its reconstruction), the selective indwelling shunt and the use of correct intraoperative biomarkers of cerebral ischemia are especially critical. The transcranial Doppler ultrasound (TDUS) examination, cerebral oximetry (CO), and electroencephalography (EEG, evoked potentials) techniques with detailed analyses of principles, methodology, and ischemia thresholds are compared with currently available data. These monitoring modalities during awaked carotid endarterectomy in 148 patients with carotid clamping are evaluated, and potential benefits and limitations are considered for recommendations of the optimal multimodal neuromonitoring strategy based on more sensitive and specific instrumental biomarkers of cerebral ischemia.

Key words Carotid endarterectomy, Cerebral ischemia, Transcranial Doppler ultrasound (TDUS), Near-infrared spectroscopy (NIRS), Electroencephalography (EEG), Thresholds

1 Introduction

Steno-occlusive lesions of the brachiocephalic artery are one of the leading causes of cerebral circulatory disorders which account about 25–30% of all ischemic strokes [1]. Multicenter randomized studies (North American Symptomatic Carotid Endarterectomy Trial—NASCET, European Carotid Surgery Trial—ECST) have proved the benefits of surgical treatment of carotid stenosis more than 70% with a permissible rate of perioperative complications of <5% in symptomatic and 3% in asymptomatic patients [2–4]. Currently, surgical treatment for this pathology includes carotid endarterectomy and endovascular angioplasty and stenting [1, 5].

Carotid endarterectomy (CEA) is an open surgery performed under general or local anesthesia [6, 7]. In case of local anesthesia,

Philip V. Peplow et al. (eds.), *Stroke Biomarkers*, Neuromethods, vol. 147, https://doi.org/10.1007/978-1-4939-9682-7_17,

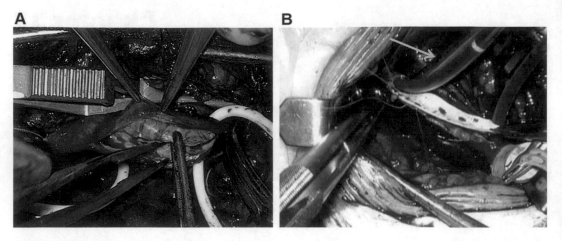

Fig. 1 Intraluminal placed atherosclerotic plaque during open carotid endarterectomy (**a**); Carotid endarterectomy with temporary indwelling shunt (arrow) (**b**)

the contact with the patient during the procedure is maintained that allows one to continuously assess the clinical status of the patient. The procedure implies a surgical dissection of the carotid arteries followed by opening the of the artery lumen by longitudinal or transverse incision of its wall, then removal of an atherosclerotic plaque (ASP) and the altered artery layers with resulting suturing of the arteriotomy incision (Fig. 1).

Recently, in addition to open surgery for brachiocephalic artery stenosis, endovascular treatment is being actively introduced. Endovascular surgery includes intravascular access to the stenosis area via puncture of the femoral artery with subsequent expansion of the stenosis area (angioplasty) and stent placement. The comparison of open and endovascular surgeries in multicenter clinical studies indicates no significant differences in outcome of both procedures in the general patient population [8–10]. However, the carotid endarterectomy is preferable procedure for patients of advanced age (>70 years) and stenosis, caused by calcified plaques, especially, in severely deformed carotid arteries [8, 11, 12]. The advantages of the endovascular procedure for patients with increased risk of CEA (post-radiation stenosis, restenosis, and high stenosis at the C2 vertebra level) are demonstrated [8, 10, 13].

The temporarily clamping of the carotid artery is a common feature of both reconstructive surgeries causing the alteration of cerebral hemodynamics through collateral redistribution of blood supply to the brain tissue [14–16]. The inadequacy of the collateral circulation leads to development of acute cerebral blood flow (CBF) insufficiency that could result in the cerebral infarction [15]. It was shown that at the time of carotid artery clamping, hypoperfusion of the brain tissue registered in 5–7% cases causing an intraoperative ischemic stroke with severe disability or death

[15, 17]. This accounts for up to 25% of all specific complications of carotid arteries' surgery [18].

Currently, there is no consensus in utilizing a particular neuro-protection during carotid revascularization. The feasibility of routine [14, 19, 20] or no shunts [21–23] for all interventions and the selective use of temporary intraluminal shunt (TIS) during carotid arteries reconstruction depending on the indications [13, 24–26] are under the consideration.

For patients with uncertain tolerance for carotid clamping the use of a TIS is a safe method, that allows for minimizing the risk of possible hypoperfusion complications (Fig. 1b) [19, 27, 28]. Additionally, it does not limit the time of arteries reconstruction allowing meticulous atherosclerotic plaque removal and performing more delicate vascular suture, particularly with a shunt inserted into the lumen of the artery [14, 19, 28].

Nevertheless, most patients with isolated carotid lesions (up to 95%) safely tolerate carotid arteries clamping and do not need temporary intravascular insertion of shunt [14, 15, 26]. The routine use of TIS in these cases is associated with technical difficulties and a higher risk for adverse events—up to 3.5% [27]. The main causes of a higher number of complications associated with TIS use are as follows [14, 15, 22]:

1. More complex technical conditions to work in the area of ASP that could be associated with limited mobility and poor visualization of the arterial lumen in the area of the distal and proximal margin of the arteriotomy incision. That makes it difficult to separate the ASP and to control the residues of intact intima that may cause dissection of intima residues leading to arterial thrombosis.

2. Failure of the TIS due to its thrombosis, torsion, incorrect position in the vessel lumen, and its clamping with tourniquets.

3. Increased risk of cerebral thromboembolism.

4. The risk of vascular intima damage by a shunt distally to the arteriotomy incision that may cause the intima dissection and early postoperative thrombosis of the reconstructed artery.

However, opponents of the routine use of a TIS suggest that the incidence of CBF decompensation is comparable and sometimes even less than the incidence of complications associated with the use of a shunt. Thus, performing the main stage of reconstruction without the use of TIS allows achieving a lower number of intraoperative ischemic complications compared to routine use of shunts [22].

At the same time, it is obvious, that for a certain group of patients, a complete denial of TIS leads to a higher rate of ischemic complications. For patients with occlusion of the contralateral internal carotid artery, ischemic complications of carotid endarterectomy increase up to 9% with extraordinary morbidity and mortality rates [20, 29, 30]. Therefore, the most optimal tactics of neuroprotection is the timely and correct use of the TIS based on the identification of preoperative risk factors and intraoperative markers of cerebral ischemia [19, 24, 26, 31, 32].

2 Instrumental Markers of Cerebral Ischemia

In order to prevent adverse events after reconstructive surgery of the carotid arteries it is necessary to consider preoperative and intraoperative risk factors affecting the development of cerebral ischemia.

Preoperative risk factors for cerebral ischemia. Initially the tactics of surgical treatment are determined based on a preoperative study of collateral circulation, the state of the circle of Willis, the presence of stenosis or occlusion of the contralateral carotid artery and arteries of vertebrobasilar system. That plays a significant role when it is impossible to determine obvious markers of cerebral ischemia according to intraoperative neurophysiological monitoring, which is typical in the case of CBF subcompensation.

Intraoperative markers of cerebral ischemia. Currently the intraoperative markers of ischemia are based on the neurophysiological intraoperative monitoring and the evaluation of blood circulation, brain function, and metabolism that are increasingly used in surgical interventions with temporal clamping of the carotid arteries:

1. *Methods for assessing cerebral circulation.* Transcranial Doppler ultrasonography (TDUS) is the most general method that allows to locate the linear velocity of blood flow in the middle cerebral artery (MCA) on the repair side and measure the retrograde pressure. Recently the use of the latter technique is significantly limited due to the risk of cerebral embolism because of invasive introduction of a pressure sensor, qualitative character of blood flow assessment, and the low specificity of the method.

2. *Methods for assessing brain metabolism* by transcranial cerebral oximetry (CO).

3. *Methods for assessing brain function*, which include EEG and somatosensory evoked potentials.

Markers of cerebral ischemia are developed for each instrumental modality, that performance characteristics (sensitivity and specificity) are dependent on the degree of CBF decompensation. Within this paradigm of the decision making for brain protection approach there are preoperative risk factors, intraoperative markers of ischemia, and neurodynamic monitoring based levels of diagnostics.

Neurodynamic intraoperative monitoring. During awake carotid endarterectomy under local anesthesia, the direct contact with the patient is maintained to assess the neurological status continuously. Signs of depression, confusion, the appearance of speech defects, or pyramidal symptoms are regarded as *an ischemic incident* and are considered as indications for temporary indwelling shunt use. This type of intraoperative monitoring of the patient's clinical condition is referred to neurodynamic monitoring and is considered an alternative of neurophysiological intraoperative monitoring. The neurodynamic monitoring is allowing surgeon time to select measures (install TIS, increase O_2 pressure, shorten the procedure, etc.) to protect the brain against an ischemic incident. As mentioned earlier, the rapid installation of a temporary shunt and adequate anesthesiologic support might suppress the emerging clinical symptoms, but increase the risk of the "ischemic incident" into "acute ischemic stroke" transformation.

In this regard, the diagnostic criteria for cerebral ischemia according to various modalities have a special importance especially in cases of direct clinical evaluation of the patient during carotid endarterectomy with local anesthesia when the patient remains conscious.

2.1 Preoperative Assessment of Collateral Flow

Presence of natural cerebral collateral pathways plays an important role in prognosis of carotid occlusion outcome [30]. According to radiological studies, confirmed by autopsies, different variants of incomplete circle of Willis occur in more than 40% of cases, in 8–15% cases with severe collateral flow insufficiency through its anterior and posterior parts [30, 33]. Such patients suffer from increased risks of cerebral ischemic complications, even after temporary carotid occlusion. According to Sundt's classification, performing the carotid endarterectomy for such a group of patients is accompanied by higher complication rates up to 6% despite obligatory use of TIS [23]. Denial of cerebral protection in this group can lead to perioperative ischemic stroke in more than 80% of cases [34]. Another angiographic marker of carotid clamping intolerance is hemodynamic failure of contralateral carotid arteries. According to a meta-analysis of more than 26,000 CEA procedures in 21 series the diagnosis of occlusion or severe stenosis of contralateral internal carotid artery raises the morbidity and mortality rates of carotid surgery up to 1.5 times [29].

Nevertheless, preoperative assessment of individual features of collateral flow could not be considered as a unique method for predicting cerebral ischemia during carotid endarterectomy, with sensitivity of 74%, specificity of 57%, and positive predictive value of 79% [35]. Thus, angiographic markers of possible ischemic complications must be evaluated together with intraoperative instrumental neuromonitoring techniques.

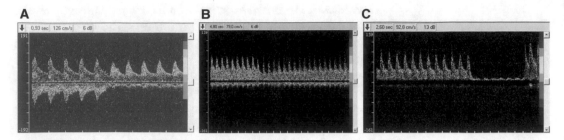

Fig. 2 Compensation (**a**), subcompensation (**b**) and decompensation (**c**) stages of cerebral blood flow by TDUS ipsilateral to carotid endarterectomy

2.2 Transcranial Doppler Ultrasonography

For the last four decades transcranial Doppler ultrasonography (TDUS) has been used as an effective, highly sensitive, and specific method for monitoring the cerebral hemodynamics during the entire course of reconstructive surgical procedure (Fig. 2) and the course of early postoperative period [32, 36, 37]. Additionally, this is a noninvasive method for continuous monitoring of cerebral hemodynamics during reconstructive interventions on the carotid arteries and in the early postoperative period.

TDUS technique is based on monitoring the difference in the frequency between waves emitted and reflected by the Doppler probe from moving blood cells in vessels. The frequency is recorded as the linear velocity of blood flow (LVB) in the controlled cerebral vessels. TDUS application to the cerebral vessels allow measurement with specificity of 90–95% [32]. The LVB measurements to assess blood flow in the major arteries are ascertained by significant correlation of volumetric and linear velocities in the initial segments of the middle cerebral artery (MCA) and irrelevant alteration of cerebral vessels diameter due to changes of vascular wall tonus [38].

The main TDUS marker of cerebral ischemia during temporary carotid clamping is the value of linear blood flow decrease in ipsilateral MCA after carotid clamping. For patients with cerebral blood flow compensation after carotid occlusion the characteristic changes of LVB do not exceed 40% of initial value (Fig. 2a).

TDUS signs of CBF decompensation are presented by a decrease of the mean LVB of more than 85%—these changes are considered as carotid clamping intolerance and an objective indication for using TIS (Fig. 2c). At the same time the final values of the mean LVB at clamping should not exceed 15–20 cm/s—the value of the LVB accompanied with the changes on the electroencephalography (EEG). This threshold level of LVB is confirmed by somatosensory evoked potentials (SSEP) monitoring data and measurements of retrograde pressure [39–41].

Decrease of LVB in the ipsilateral MCA after carotid clamping between 40% and 85% is considered as CBF subcompensation form (Fig. 2b). The tolerance to temporary carotid clamping and the need of TIS at this stage is currently the most controversial issue.

Some researchers suggest the use of TIS on the subcompensation stage routinely as on the decompensation stage. Using this threshold for cerebral ischemia TDUS method demonstrates high sensitivity for ischemic events with low specificity—only in 63% cases TDUS signs of cerebral ischemia could be confirmed by EEG changes [16]. In these cases other modalities of neuromonitoring are helpful for improving TDUS patency for ischemia detection. It is well known that compensation of cerebral circulation occurs by the circle of Willis and is primarily determined by TDUS and by means of cortical arterial collaterals. To assess the distal blood flow electrophysiological monitoring and cerebral oximetry are considered as more reliable monitoring techniques [31, 42, 43].

Another important feature of the TDUS is the ability to detect episodes of cerebral embolism [37]. Migration of emboli from the region of carotid artery reconstruction accounts for up to 50% of all specific complications after carotid endarterectomy [15, 44]. Depending on the structure, there are air and solid microemboli. Air microemboli are less dangerous and not manifested by the development of neurological symptoms or an asymptomatic radiological image of brain structures damage. The registration of solid microemboli that might be fragments of a decaying atherosclerotic plaque, as well as thrombotic deposits on its uneven surface, bear serious symptom of ischemic complications development [37, 44]. Intraoperative registration of more than 10 microemboli per minute might cause "silent" cerebral ischemia registered by magnetic resonance imaging (MRI) in the reconstructed internal carotid artery (ICA) territory and the development of neurological symptoms in 4% patients [37].

There are some limitations and disadvantages of TDUS as follows:

- 9–12% cases might have no an acoustic window to perform ultrasound location in the MCA.

- Need to monitor the location of the sensor and the depth of artery location to avoid wrong detection of the posterior cerebral artery position (PCA). This situation can occur when the depth of the artery's position is increased or patient's head is smaller than usual resulting in PCA branches at a depth of 55–60 mm. There are often no changes in blood flow during temporary clamping of the carotid arteries in these cases. However, a paradoxical increase in blood flow is observed when the PCA is correctly located in certain cases that are regarded as signs of compensation when patient underwent clamping of the carotid arteries without complications.

- False-positive results reducing the specificity of TDUS in the diagnosis of cerebral ischemia when the location of the internal carotid artery is proximal to the PCA. In such cases, a critical decrease in the linear velocity of blood flow may be recorded despite good compensation of the blood flow in the Willis circle.

- Individual characteristics of the patient and technical problems due to vessel location might affect low values of blood flow in the MCA (less than 50 cm/s) decreasing the informativeness of TDUS.

2.3 Cerebral Oximetry

Cerebral oximetry (CO) is a noninvasive method of neuromonitoring that estimates cerebral hemoglobin (Hb) with oxygen saturation (rSO2) in the area of the sensor by near-infrared spectroscopy (NIRS) [43]. Measurements are conducted at different depths to assess the balance between cerebral oxygen delivery and consumption. Adhesive pads attach probes to the patient's scalp. Probes are most commonly applied to the scalp overlying the frontal lobe (Fig. 3a). Cerebral oximeters consist of a monitor that is connected to oximeter probes (Fig. 3b). The simultaneous decrease in rSO2 levels on both sides of the frontal lobe is observed if there are systemic problems, mainly during the decrease of blood pressure. The fall in rSO2 levels on either side while stable contralateral values indicates substantial cerebral oxygen desaturation during the procedure. The latter may point to adverse perioperative outcome (cerebral ischemic event) with better specificity and precede EEG and TDUS recordings [43, 45, 46].

The cerebral oximetry data are extremely dependable on the systemic arterial pressure, presence of pulmonary pathology, and individual patient characteristics. The rSO2 level threshold of 60% can be considered as a marker of intraoperative ischemia with sensitivity of 100% and specificity within 43–57% when more than

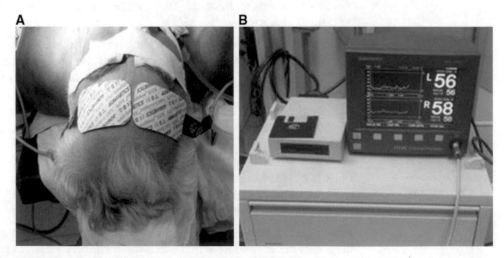

Fig. 3 Cerebral oximetry monitoring of brain metabolism during CEA by NIRS: (**a**) typical placement of CO sensors; (**b**) monitoring terminal

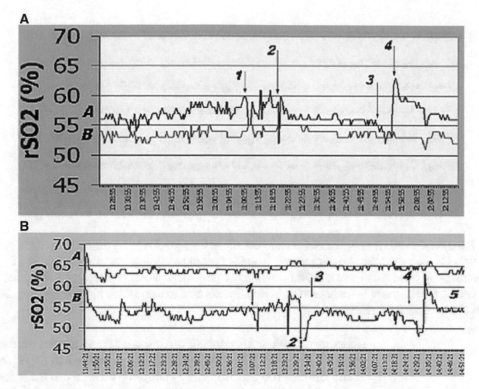

Fig. 4 (a) The pattern of blood flow compensation during carotid occlusion: 1—test clamp; 2—main carotid clamping, start of carotid endarterectomy. rSO2 level decrease by 6–8% below averaged initial values; 3—release of carotid clamps; 4—mild signs if hyperemia after CEA. Upper curve—ipsilateral, lower curve—contralateral hemisphere to CEA. **(b)** The pattern of blood flow decompensation during carotid occlusion: 1—test clamp; 2—main carotid clamping, start of carotid endarterectomy. rSO2 level decrease by 20–25% below averaged initial values; 3—insertion of TIS with rSO2 compensation; 4—release of carotid clamps; 4—mild signs of hyperemia after reperfusion with quick return to initial CO values. Upper curve—contralateral hemisphere to CEA, lower curve—ipsilateral

13% patients developed the "ischemic incident" compared to the initial saturation level (Fig. 4a) [31]. The trade off in rSO2 value to 25% increases specificity up to 97% but accompanied with lower sensitivity of 51%. According to Samra et al. [47], an intermediate rSO2 threshold value of >20% brings the sensitivity and specificity of cerebral oximetry to 72% and 69% correspondently. These rSO2 values correspond to cerebral blood flow decompensation with the decision to install a temporary intraluminal shunt (Fig. 4b). The relative disadvantages of cerebral oximetry include high variability of absolute values and a slower response to changes in blood flow compared to TDUS.

2.4 Brain Function Monitoring

EEG represents a method of choice to determine the tolerance to clamping of the carotid arteries and the need to install a TIS [16, 24, 40]. The success of this reliable technique is facilitated by established correlation of EEG recordings with the development of

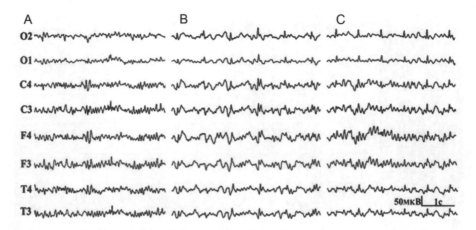

Fig. 5 The dynamics of EEG during carotid endarterectomy on the right side in the patient with critical stenosis of the right ICA. Compensation. A—Background, B—after applying a vascular clamp to the right ICA, C—after the restoration of blood flow. There is a diffuse slowing of the cortical rhythm, a partial reduction of the alpha rhythm and a mild decrease in the level of brain biopotentials after turning off the ICA with restoring the EEG pattern after removing the vascular clamp

ischemic complications during carotid endarterectomies performed without TIS [40]. The development of a pathological EEG pattern at the time of the carotid artery clamping varies from 8.5% to 31% and weakly relates to preoperative angiography data [16, 24].

There were no pronounced changes in the EEG characterized with mild diffused disturbances in the form of a smoothed down increased disorganization, a reduction in the cortical rhythmic amplitude, and the appearance of diffused frequency forms in patients with good CBF compensation during the clamping of the carotid arteries without adverse effects being observed (Fig. 5). Typical bioelectric markers of intraoperative ischemia are represented by (1) a decrease in bioelectric activity, (2) slowdown of the basic rhythm with the predominance of slow oscillations in different periods of increased amplitude on the side of the clamping of the internal carotid artery (ICA), and (3) the appearance of irritation signs in the subcortical-diencephalic and brainstem areas. Local changes in the form of pronounced interhemispheric asymmetry and slow focal waves are also noted (Figs. 5 and 6). The marked impairments of brain biopotentials largely regressed after the restoration of blood flow; however, the EEG pattern, as a rule, did not fully recover to the initial background level.

The presence of severe changes in the EEG is a criterion for determining the risk of ischemia with high specificity that reaches 92–100% and the sensitivity about 50% [16, 40]. Additionally, the incorporation of moderate EEG changes in the intolerance to clamping of the carotid artery allowed to increase the sensitivity of the technique up to 80% [19]. At the same time, routine use of TIS during reconstruction procedure with mild to moderate EEG

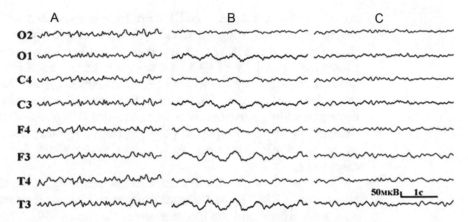

Fig. 6 EEG dynamics during carotid endarterectomy on the left side in a patient with bilateral lesions of the ICA: stenosis of the left ICA (65%), thrombosis of the right. Decompensation. A—Background, B—clamping of the ICA, installation of a temporary shunt, C—after the restoration of blood flow. There is a sharp decrease in the EEG amplitude after temporal clamping of the ICA, the reduction of frequent oscillations, the predominance of pathological slow waves of various periods of increased amplitude, accentuated on the side of the operation. When the blood flow is restored, these changes regress, irregular fluctuations of the alpha and beta range appear with smooth increase of the cortical rhythm amplitude

changes is often not justified due to the high frequency of false-positive results. Indeed, severe EEG changes were registered in 22% and moderate in 19% cases out of 176 uncomplicated CEA without TIS [16]. In patients who showed a severe decompensation on EEG in the postoperative period, ischemic strokes were seen in 9% of cases wherein no ischemic complications were observed in groups without or with mild EEG changes [40].

Currently, digital multichannel encephalographs are used, allowing for real-time processing presented in recorded signals time and frequency modes to conduct more complex frequency analyses increasing the quality of the monitoring [16].

Nevertheless, intraoperative EEG registration has a number of limitations. The first, the results interpretation is complicated and requires highly qualified specialists. Additionally, it has a substantial frequency of false-positive results in assessment of the risk for cerebral ischemia and less informative in diagnosis of cerebral embolism and hyperperfusion syndrome. Besides, EEG monitoring assesses the cortical activity (cortical anastomoses) and is less helpful in evaluation of perforating branches of the large arteries supplying basal ganglia. The adequate EEG recording is limited by the anesthesia type and neuroprotective agents [16], for example, the thiopental use for general suppression of cortical activity [40].

Somatosensory evoked potentials (SSEP) are a valuable alternative to EEG technique that assess adequate collateral circulation and the need for shunt insertion. The perioperative SSEP monitoring criterion for the cerebral ischemia detection is more than a 50% decrease in N20-P25 amplitude of cortical waveforms [48, 49]. The complete disappearance of SSEP amplitude is associated with persistent severe neurological deficit with the sensitivity of 100%, regardless of the TIS use. A number of studies have noted the substantial specificity of 94% in assessment of the need for shunt use additional to high sensitivity of SSEP monitoring [48]. According to bimodal criteria of ischemia in SSEP the selective use of TIS allowed avoiding persistent ischemic complications in 9% of cases in the early postoperative period [49]. At the same time, the occurrence in SSEP false-negative results is described for 1–2% cases [42].

Compared to the traditional EEG monitoring, a significant advantage of SSEP is the resistance to pharmacological agents, the anesthesia depth, and the blood pressure dynamics [49]. There are controversial data on EEG and SSEP efficacy for indication of TIS use. According to the first point of view the EEG monitoring is considered as the "gold standard" in determining indications for use of TIS [40]. The authors' opinion is that SSEP showed low sensitivity in comparison with EEG because of an increase in central holding time in 47% of cases is accompanied with a 50% decrease in SSEP amplitude only in one case. At the same time, another study showed a higher specificity of SSEP in comparison to EEG [49].

Currently the advantage of simultaneous use of two techniques is demonstrated, where the results of SSEP complemented the EEG monitoring, thereby reducing the frequency of false-positive and false-negative results [49]. SSEP has a special importance in the diagnosis of slowly developing cerebral ischemia that is not manifested on the EEG. The SSEP downside is insensitivity to cerebral embolism and hyper reperfusion. A general drawback of both electrophysiological methods is a drastic decrease in sensitivity of cerebral hypoperfusion detection in the presence of extensive ischemic *foci* in the monitored hemisphere.

3   Materials and Methods

About 2000 reconstructive carotid surgeries were performed with intraoperative neuromonitoring in 78% of cases in 2000–2018 at Burdenko National Scientific and Practical Centre for Neurosurgery, Moscow, Russia. About 200 (10%) procedures were performed with use of regional anesthesia. To identify angiographic risk factors for cerebral ischemia, 148 (7.4%) patients underwent the preoperative CT angiography evaluating the state of the opposite internal carotid and vertebral arteries as well the cerebral arterial circle. The main intraoperative markers of cerebral ischemia were assessed by TDUS, cerebral oximetry, and EEG monitoring.

Preoperative CT angiography was performed in all cases to assess the anatomic variant of the circle of Willis. For this we preferred noninvasive high-resolution contrast multislice CT angiography

(CTA), performed in arterial phase of cerebral circulation. For better visualization of cerebral arteries and its differentiation from other vascular structures on the skull base, we used multiplanar reconstruction (MPR) mode. For patients with contraindication for CTA (contrast intolerance) we used non-contrast magnetic resonance angiography (MRA) in time-of-flight vascular mode. This method is a reliable replacement for contrast angiographic examinations [50].

Assessment of main cerebral collateral pathways included separate measurements of anterior and posterior parts of the circle of Willis. According to angiographic findings, we distinguished the following preoperative risk factors for cerebral ischemia: aplasia (absence) or hypoplasia (diameter less than 1 mm) of A1 segments of anterior cerebral arteries (ACA) or anterior communicating artery (ACommA) with aplasia/hypoplasia of posterior collateral pathways (P1 segment of posterior cerebral artery or posterior communicating artery) on the side of carotid endarterectomy. Signs of good collateral flow either through anterior or posterior parts of Willis circle indicated a benign prognosis of carotid clamping tolerance.

TDUS registration of blood flow in the MCA was performed at a depth of 45–55 mm through the temporal acoustic window in special helmet used for fixing the sensor during the operation. Specific markers of cerebral ischemia using LVB after a short-term clamping of the carotid arteries (5–10 s) immediately after their dissection were determined (Table 1).

Registration of microemboli is performed according to the following Doppler characteristics in 19% of cases: (1) an embolic sign is a short signal with a duration of less than 0.3 s depending on the size of the emboli, (2) the smallest signal of 7–8 dB and maximal signal of 18–20 dB from the solid embolus are recorded, (3) the embolic signal has a unidirectional representation within the spectrum during bidirectional Doppler detection, (4) the distinctive sound (short click) is registered.

Table 1
TDUS parameters according to ICA clamping tolerance during awake carotid endarterectomy ($p < 0.05$)

TDUS parameters	Without deficit	"Ischemic incident"
Blood flow before clamping (cm/sec)	67.5 ± 17.2	69.6 ± 21.5
Blood flow after clamping	47.3 ± 11.2	13.7 ± 9.6
% of LVB decrease	27.9 ± 14.3	74.7 ± 18.2
Total (148)	129	19

Cerebral oximetry for intraoperative assessment of metabolism during carotid endarterectomy is performed by use of two sensors installed as depicted on Fig. 3 under the Doppler helmet. The bilateral rSO2 values of hemoglobin oxygen saturation preoperatively, during ICA clamping, and perioperatively—30 min after restoration of the blood flow—are evaluated. The difference between the average value of rSO2 levels (minimum and maximum background) before and after ICA compression is estimated to determine the degree of compensation. The maximal rSO2value immediately after the restoration of blood flow through the reconstructed artery is used to assess hyperperfusion syndrome in the early postoperative period.

3.1 Results

An "ischemic incident" in 19 (12.8%) out of 148 patients was recorded with three patients out of 17 (11.4%) cases of controlled hypertension/barbiturate protection and TIS installation having an outcome on modified Rankin scale (mRS) from no significant (mRS = 1) to moderate disability (mRS = 2–3) 2 weeks after the procedure.

Depending on the blood flow decrease in the MCA during the clamping of the carotid arteries there are three types of compensation observed with different performance characteristics.

The *compensation* stage that characterizes by an average LVB decrease in the MCA within 40% while the systolic blood flow velocity remains more than 50 cm/s. This type of compensation was recorded by TDUS in most of our procedures ($n = 89$) without pathological changes on the EEG and considered as favorable (not accompanied with perioperative cerebral ischemia). Then LVB marker below 40% may be considered as a sensitive one for cerebral ischemia. However, this threshold showed a specificity of 32.2% and then could not be a reliable criterion.

The stage of CBF *decompensation* characterized by the average LVB decrease in the MCA over 85% or with the absolute value of the blood flow less than 30 cm/s. This type of decompensation is accompanied with changes on EEG in 65% of cases and the development of "ischemic incident" in 9 out of 11 cases. TIS installations in these cases allowed avoiding persistent ischemic complications. This approach of CBF decompensation detection yielded a specificity of 81.8%, but a low sensitivity of 47.3%.

The *subcompensation* is characterized by an average LVB decrease in the range of 40–85% of the background value while systolic blood flow was more than 30 cm/s. This category of TDUS changes is the most difficult to detect cerebral ischemia that in 10 (20.8%) out of 48 cases is observed and required the installation of a TIS. The TDUS specificity of 52.6% to detect cerebral ischemia on the subcompensation stage was established.

Cerebral microembolic episodes in 16 cases (10.8%) of reconstructive interventions with TDUS monitoring were recorded. The

Table 2
Parameters of ICA clamping tolerance during awake carotid endarterectomy ($p < 0.05$)

Parameter	Without incident	"Ischemic incident"
rSO2 (%)	67.4 ± 6.7	52.4 ± 5.5
% rSO2 decrease	7.7 ± 3.5	22.5 ± 3.6
Total	129	19

Table 3
Influence of the anatomical pattern of the circle of Willis on the frequency of circulatory decompensation in the territory of the clamped ICA

Circle of Willis	With use of TIS (decompensation)	Without use of TIS (compensation)	Total
Completed	10.5%	89.5%	48.7%
Incomplete with absence of PCoA	28%	72%	18.9%
Incomplete with absence of ACoA	75%	25%	9%
Totally incomplete	86%	14%	12.6%
No data	0	100%	10.8%

characteristic parameters of cerebral oximetry in patients who developed cerebral ischemia during the operation are presented in Table 2. In "ischemic incident" cases, an rSO2 level decrease below 60% or more than 13% compared with the initial saturation level is registered. Less drastic dynamic of cerebral oximetry parameters indicated a satisfactory CBF compensation that did not require the use of a TIS.

In case of contradictory data from modalities used, as well as in the absence of clearly expressed signs of decompensation possible risk factors for the development of perioperative ischemia are taken into account (presented in Table 3). As soon as circle of Willis is completed with the anterior and posterior parts of the cerebral arterial circle, there is a trend in 89.5% patients to show a compensation of cerebral circulation. However, in 10.5% of cases the results of multimodal neuromonitoring are interpreted as decompensation of collateral blood flow that required the installation of a TIS contrary to preoperative diagnostic data.

In case of incomplete dissociation of the arterial circle of the brain, the stage of CBF decompensation is determined with absence of anterior communicating artery (ACoA) in 75% of cases and absence of posterior communicating artery (PCoA) in 28%.

These data seem to indicate a greater significance of ACoA compared with the PCoA in the compensation of CBF during the clamping of the common carotid artery (CCA). The cerebral circulation decompensation in 86% of cases detected when the circle of Willis was completely dissociated being a significant risk factor for the development of intraoperative ischemia during the CCA clamping.

3.2 Discussion

Selective use of a TIS according to the monitoring data of blood circulation parameters, metabolism and cerebral function is considered as an optimal approach to protect the brain during the reconstructive interventions on the carotid arteries. Among the variety of instrumental techniques used in the neuromonitoring of the endarterectomy, the advantage has been given to transcranial Doppler ultrasound, cerebral oximetry and EEG. The most common issue of these modalities is the trade off between compensation and decompensation within each modality with the difficulty in identifying the limits of decompensation. From one side, a decrease in decompensation while identifying intolerance to ICA clamping reduces the sensitivity of the methods assessing the risk of developing ischemic adverse events. On the other hand, expanding the limits of decompensation causes a decrease in specificity that results in the excessively frequent use of a TIS and an increased risk of cerebral embolism. To increase the diagnostic power of instrumental techniques it would be necessary to use a combination of available modalities of neuromonitoring (Table 4).

Here the analysis of results from three neuromonitoring techniques in addition to the assessment of compensation and decompensation allowed one to identify the CEA subcompensation stage. If the TDUS monitoring depicted the compensation during the clamping, the procedure is performed without the use of a TIS

Table 4
The performance characteristics of various modalities and their combinations in detecting intraoperative ischemia

Method	Sensitivity	Specificity
Transcranial Doppler ultrasound (TDUS) (decrease >85%)	47%	82%
Transcranial doppler ultrasound (decrease >40%)	100%	32%
Cerebral oximetry (CO) (decrease of rSO2 > 20%)	**72%**	**69%**
Cerebral oximetry (decrease of rSO2 > 13%)	100%	47%
EEG	56%	51%
TDUS + CO	**86%**	**82%**
TDUS + CO + EEG	**82%**	**72%**

Bold type highlights the most balanced options for the detection of intraoperative ischemia

while a decompensation is detected according to at least one of the techniques, the TIS is installed on clamping. In a subcompensated state of cerebral function, the final decision of using TIS is made according to combined monitoring modalities results as well as the presence of concomitant risk factors: disunity of the circle of Willis and hemodynamically significant lesions of the contralateral carotid arteries.

EEG data in most cases correlated with changes in the TDUS parameters and CO. However, in some cases, while marked CBF disturbances were registered on TDUS, there were no significant changes in the pattern of brain potentials on EEG. This might be explained by a delay in EEG changes after clamping the artery. Besides, medication and individual patient characteristics (sharp fluctuations in blood pressure, increased reactivity of the sino-carotid zone, and increased sensitivity to certain anesthesia) affect the difficulties in interpretation of the EEG patterns.

Cerebral oximetry unlike the TDUS cannot serve as a fast criterion for assessing collateral blood flow because it measures metabolic changes in the vascular flow that requires much more time than changes in the LBF of brachiocephalic arteries.

Considering the delay of the cerebral metabolism and bioelectrical activity, the EEG and cerebral oximetry are less useful in estimating the degree of compensation during the test clamping. At the same time, these methods respond to changes in cerebral hemodynamics associated with intolerance to clamping of the CCA, aid in assessment of defects in the TIS and the vascular anastomosis zone after restoring the blood flow in the arteries. These techniques made it possible to promptly correct the anesthesia and surgical tactics at the main stage of reconstruction.

Based on above presented data an algorithm for the combined interpretation of the results for multimodal neuromonitoring is developed (Fig. 7). Before the main stage of reconstructive intervention, a test clamping of the CCA should be conducted under the control of the TDUS. Detection of signs of CBF decompensation, regardless of the data of other modalities, is considered as an indication for the use of a TIS. The CBF compensation does not require a TIS. The use of this tactic allowed avoiding the development of ischemic complications with a frequency of using a TIS in 10.7% of cases.

A group with subcompensation of cerebral circulation included patients who had a stage of subcompensation according to TDUS, or when TDUS had low informativity due to technical problems, or when individual characteristics of the patient affected the state. Then cerebral oximetry would be a key technique to use when any sign of decompensation is considered as a likely marker of cerebral ischemia and a decision was made to install a TIS. The absence or weakness of retrograde blood flow from the ICA also indicates a likely intolerance to the clamping of the carotid arteries.

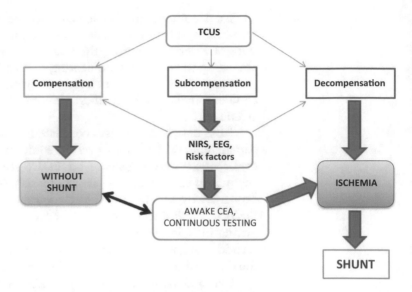

Fig. 7 Algorithm of detecting markers of intraoperative ischemia and determining indications for a temporary intraluminal shunt

4   Conclusion

During the 60 years of carotid endarterectomy use, its technique has undergone only minor changes and modifications. However, the indications for the surgery and new methods for neuroprotection are constantly updated to avoid adverse neurological events like cerebral ischemia. Currently various instrumental methods for the detection of perioperative ischemia have been developed and introduced into practice; used algorithms are proposed to increase performance characteristics for the developing ischemic stroke. At the same time, capabilities of described methods are limited for detailed study of borderline changes in cerebral perfusion and characteristics of the CBF subcompensation stage. The field of cerebral circulation pathophysiology is still evolving and has a significant impact on the clinical outcome of the treatment with steno-occlusive pathology of brachiocephalic arteries that could be further advanced by the progress in neuroimaging and modern neurobiology.

References

1. Powers WJ, Derdeyn CP, Biller J, Coffey CS, Hoh BL, Jauch EC, Johnston KC, Johnston SC, Khalessi AA, Kidwell CS, Meschia JF, Ovbiagele B, Yavagal DR, American Heart Association Stroke C (2015) 2015 American Heart Association/American Stroke Association Focused Update of the 2013 guidelines for the early management of patients with acute ischemic stroke regarding endovascular treatment: a guideline for healthcare professionals from the American Heart Association/American Stroke Association.

Stroke 46(10):3020–3035. https://doi.org/10.1161/STR.0000000000000074

2. MRC European Carotid Surgery Trial (ECST) (1998) Randomised trial of endarterectomy for recently symptomatic carotid stenosis: final results of the MRC European Carotid Surgery Trial (ECST). Lancet 351(9113):1379–1387

3. Abbott AL, Paraskevas KI, Kakkos SK, Golledge J, Eckstein HH, Diaz-Sandoval LJ, Cao L, Fu Q, Wijeratne T, Leung TW, Montero-Baker M, Lee BC, Pircher S, Bosch M, Dennekamp M, Ringleb P (2015) Systematic review of guidelines for the management of asymptomatic and symptomatic carotid stenosis. Stroke 46(11):3288–3301. https://doi.org/10.1161/STROKEAHA.115.003390

4. North American Symptomatic Carotid Endarterectomy Trial C, Barnett HJM, Taylor DW, Haynes RB, Sackett DL, Peerless SJ, Ferguson GG, Fox AJ, Rankin RN, Hachinski VC, Wiebers DO, Eliasziw M (1991) Beneficial effect of carotid endarterectomy in symptomatic patients with high-grade carotid stenosis. N Engl J Med 325(7):445–453. https://doi.org/10.1056/NEJM199108153250701

5. Eckstein HH (2018) European Society for Vascular Surgery guidelines on the management of atherosclerotic carotid and vertebral artery disease. Eur J Vasc Endovasc Surg 55(1):1–2. https://doi.org/10.1016/j.ejvs.2017.06.026

6. Guay J, Kopp S (2013) Cerebral monitors versus regional anesthesia to detect cerebral ischemia in patients undergoing carotid endarterectomy: a meta-analysis. Can J Anaesth 60(3):266–279. https://doi.org/10.1007/s12630-012-9876-4

7. Hajibandeh S, Hajibandeh S, Antoniou SA, Torella F, Antoniou GA (2018) Meta-analysis and trial sequential analysis of local vs. general anaesthesia for carotid endarterectomy. Anaesthesia 73(10):1280–1289. https://doi.org/10.1111/anae.14320

8. Jones DW, Brott TG, Schermerhorn ML (2018) Trials and frontiers in carotid endarterectomy and stenting. Stroke 49(7):1776–1783. https://doi.org/10.1161/STROKEAHA.117.019496

9. Liu ZJ, Fu WG, Guo ZY, Shen LG, Shi ZY, Li JH (2012) Updated systematic review and meta-analysis of randomized clinical trials comparing carotid artery stenting and carotid endarterectomy in the treatment of carotid stenosis. Ann Vasc Surg 26(4):576–590. https://doi.org/10.1016/j.avsg.2011.09.009

10. Noiphithak R, Liengudom A (2017) Recent update on carotid endarterectomy versus carotid artery stenting. Cerebrovasc Dis 43

(1–2):68–75. https://doi.org/10.1159/000453282

11. Benes V, Bradac O (2018) Carotid endarterectomy and carotid artery stenting in the light of ICSS and CREST studies. Acta Neurochir Suppl 129:95–99. https://doi.org/10.1007/978-3-319-73739-3_14

12. Heo SH, Bushnell CD (2017) Factors influencing decision making for carotid endarterectomy versus stenting in the very elderly. Front Neurol 8:220. https://doi.org/10.3389/fneur.2017.00220

13. De Haro J, Michel I, Bleda S, Canibano C, Acin F (2017) Carotid stenting in patients with high risk versus standard risk for open carotid endarterectomy (REAL-1 trial). Am J Cardiol 120(2):322–326. https://doi.org/10.1016/j.amjcard.2017.04.023

14. Aburahma AF, Mousa AY, Stone PA (2011) Shunting during carotid endarterectomy. J Vasc Surg 54(5):1502–1510. https://doi.org/10.1016/j.jvs.2011.06.020

15. Khattar NK, Friedlander RM, Chaer RA, Avgerinos ED, Kretz ES, Balzer JR, Crammond DJ, Habeych MH, Thirumala PD (2016) Perioperative stroke after carotid endarterectomy: etiology and implications. Acta Neurochir 158(12):2377–2383. https://doi.org/10.1007/s00701-016-2966-2

16. Thirumala PD, Thiagarajan K, Gedela S, Crammond DJ, Balzer JR (2016) Diagnostic accuracy of EEG changes during carotid endarterectomy in predicting perioperative strokes. J Clin Neurosci 25:1–9. https://doi.org/10.1016/j.jocn.2015.08.014

17. Li J, Shalabi A, Ji F, Meng L (2016) Monitoring cerebral ischemia during carotid endarterectomy and stenting. J Biomed Res 31. https://doi.org/10.7555/JBR.31.20150171

18. Lareyre F, Raffort J, Weill C, Marse C, Suissa L, Chikande J, Hassen-Khodja R, Jean-Baptiste E (2017) Patterns of acute ischemic strokes after carotid endarterectomy and therapeutic implications. Vasc Endovasc Surg 51(7):485–490. https://doi.org/10.1177/1538574417723482

19. Chongruksut W, Vaniyapong T, Rerkasem K (2014) Routine or selective carotid artery shunting for carotid endarterectomy (and different methods of monitoring in selective shunting). Cochrane Database Syst Rev 6:CD000190. https://doi.org/10.1002/14651858.CD000190.pub3

20. Perini P, Bonifati DM, Tasselli S, Sogaro F (2017) Routine shunting during carotid endarterectomy in patients with acute watershed stroke. Vasc Endovasc Surg 51(5):288–294.

https://doi.org/10.1177/1538574417708130

21. Baram A, Majeed G, Subhi Abdel-Majeed A (2018) Carotid endarterectomy: neither shunting nor patching technique. Asian Cardiovasc Thorac Ann 26(6):446–450. https://doi.org/10.1177/0218492318788777

22. Ben Ahmed S, Daniel G, Benezit M, Ribal JP, Rosset E (2017) Eversion carotid endarterectomy without shunt: concerning 1385 consecutive cases. J Cardiovasc Surg 58(4):543–550. https://doi.org/10.23736/S0021-9509.16.08495-0

23. Bendok BR, Naidech AM, Walker MT (2012) Hemorrhagic and ischemic stroke: medical, imaging, surgical and interventional approaches. Thieme Medical Publishers, Noida

24. Lee J, Lee S, Kim SW, Chang JW (2018) Selective shunting based on dual monitoring with electroencephalography and stump pressure for carotid endarterectomy. Vasc Spec Int 34(3):72–76. https://doi.org/10.5758/vsi.2018.34.3.72

25. Loftus CM (2006) Carotid endarterectomy: principles and technique. Taylor & Francis, Abingdon

26. Wiske C, Arhuidese I, Malas M, Patterson R (2018) Comparing the efficacy of shunting approaches and cerebral monitoring during carotid endarterectomy using a national database. J Vasc Surg 68(2):416–425. https://doi.org/10.1016/j.jvs.2017.11.077

27. Bennett KM, Scarborough JE, Cox MW, Shortell CK (2015) The impact of intraoperative shunting on early neurologic outcomes after carotid endarterectomy. J Vasc Surg 61(1):96–102. https://doi.org/10.1016/j.jvs.2014.06.105

28. Loftus CM (2015) Tips, tricks, subtleties, and superiority of carotid artery surgery. World Neurosurg 83(5):758–761. https://doi.org/10.1016/j.wneu.2013.07.091

29. Antoniou GA, Kuhan G, Sfyroeras GS, Georgiadis GS, Antoniou SA, Murray D, Serracino-Inglott F (2013) Contralateral occlusion of the internal carotid artery increases the risk of patients undergoing carotid endarterectomy. J Vasc Surg 57(4):1134–1145. https://doi.org/10.1016/j.jvs.2012.12.028

30. Banga PV, Varga A, Csobay-Novak C, Kolossvary M, Szanto E, Oderich GS, Entz L, Sotonyi P (2018) Incomplete circle of Willis is associated with a higher incidence of neurologic events during carotid eversion endarterectomy without shunting. J Vasc Surg. https://doi.org/10.1016/j.jvs.2018.03.429

31. Pennekamp CW, Immink RV, den Ruijter HM, Kappelle LJ, Bots ML, Buhre WF, Moll FL, de Borst GJ (2013) Near-infrared spectroscopy to indicate selective shunt use during carotid endarterectomy. Eur J Vasc Endovasc Surgery 46(4):397–403. https://doi.org/10.1016/j.ejvs.2013.07.007

32. Udesh R, Natarajan P, Thiagarajan K, Wechsler LR, Crammond DJ, Balzer JR, Thirumala PD (2017) Transcranial doppler monitoring in carotid endarterectomy: a systematic review and meta-analysis. J Ultrasound Med 36(3):621–630. https://doi.org/10.7863/ultra.16.02077

33. Wang BH, Leung A, Lownie SP (2016) Circle of Willis collateral during temporary internal carotid artery occlusion II: observations from computed tomography angiography. Can J Neurol Sci 43(4):538–542. https://doi.org/10.1017/cjn.2016.10

34. Kim GE, Cho YP, Lim SM (2002) The anatomy of the circle of Willis as a predictive factor for intra-operative cerebral ischemia (shunt need) during carotid endarterectomy. Neurol Res 24(3):237–240. https://doi.org/10.1179/016164102101199846

35. Jaffer U, Normahani P, Harrop-Griffiths W, Standfield NJ (2015) Pre-operative methods to predict need for shunting during carotid endarterectomy. Int J Surg 23(Pt A):5–11. https://doi.org/10.1016/j.ijsu.2015.09.007

36. Sef D, Skopljanac-Macina A, Milosevic M, Skrtic A, Vidjak V (2018) Cerebral neuromonitoring during carotid endarterectomy and impact of contralateral internal carotid occlusion. J Stroke Cerebrovasc Dis 27(5):1395–1402. https://doi.org/10.1016/j.jstrokecerebrovasdis.2017.12.030

37. Spence JD (2017) Transcranial Doppler monitoring for microemboli: a marker of a high-risk carotid plaque. Semin Vasc Surg 30(1):62–66. https://doi.org/10.1053/j.semvascsurg.2017.04.011

38. D'Andrea A, Conte M, Scarafile R, Riegler L, Cocchia R, Pezzullo E, Cavallaro M, Carbone A, Natale F, Russo MG, Gregorio G, Calabro R (2016) Transcranial Doppler ultrasound: physical principles and principal applications in neurocritical care unit. J Cardiovasc Echogr 26(2):28–41. https://doi.org/10.4103/2211-4122.183746

39. Shahidi S, Owen-Falkenberg A, Gottschalksen B (2017) Clinical validation of 40-mmHg carotid stump pressure for patients undergoing carotid endarterectomy under general anesthesia. J Cardiovasc Surg 58(3):431–438. https://doi.org/10.23736/S0021-9509.16.08173-8

40. Tan TW, Garcia-Toca M, Marcaccio EJ Jr, Carney WI Jr, Machan JT, Slaiby JM (2009) Predictors of shunt during carotid endarterectomy with routine electroencephalography monitoring. J Vasc Surg 49(6):1374–1378. https://doi.org/10.1016/j.jvs.2009.02.206

41. Yun WS (2017) Cerebral monitoring during carotid endarterectomy by transcranial Doppler ultrasonography. Ann Surg Treat Res 92 (2):105–109. https://doi.org/10.4174/astr.2017.92.2.105

42. Arnold M, Sturzenegger M, Schaffler L, Seiler RW (1997) Continuous intraoperative monitoring of middle cerebral artery blood flow velocities and electroencephalography during carotid endarterectomy. A comparison of the two methods to detect cerebral ischemia. Stroke 28(7):1345–1350

43. Radak D, Sotirovic V, Obradovic M, Isenovic ER (2014) Practical use of near-infrared spectroscopy in carotid surgery. Angiology 65 (9):769–772. https://doi.org/10.1177/0003319713508642

44. Huibers A, Calvet D, Kennedy F, Czuriga-Kovacs KR, Featherstone RL, Moll FL, Brown MM, Richards T, de Borst GJ (2015) Mechanism of procedural stroke following carotid endarterectomy or carotid artery stenting within the international carotid stenting study (ICSS) randomised trial. Eur J Vasc Endovasc Surg 50(3):281–288. https://doi.org/10.1016/j.ejvs.2015.05.017

45. Pedrini L, Magnoni F, Sensi L, Pisano E, Ballestrazzi MS, Cirelli MR, Pilato A (2012) Is near-infrared spectroscopy a reliable method to evaluate clamping ischemia during carotid surgery? Stroke Res Treat 2012:156975. https://doi.org/10.1155/2012/156975

46. Pennekamp CW, Bots ML, Kappelle LJ, Moll FL, de Borst GJ (2009) The value of near-infrared spectroscopy measured cerebral oximetry during carotid endarterectomy in perioperative stroke prevention. A review. Eur J Vasc Endovasc Surg 38(5):539–545. https://doi.org/10.1016/j.ejvs.2009.07.008

47. Samra SK, Dy EA, Welch K, Dorje P, Zelenock GB, Stanley JC (2000) Evaluation of a cerebral oximeter as a monitor of cerebral ischemia during carotid endarterectomy. Anesthesiology 93 (4):964–970

48. Nwachuku EL, Balzer JR, Yabes JG, Habeych ME, Crammond DJ, Thirumala PD (2015) Diagnostic value of somatosensory evoked potential changes during carotid endarterectomy: a systematic review and meta-analysis. JAMA Neurol 72(1):73–80. https://doi.org/10.1001/jamaneurol.2014.3071

49. Thirumala PD, Natarajan P, Thiagarajan K, Crammond DJ, Habeych ME, Chaer RA, Avgerinos ED, Friedlander R, Balzer JR (2016) Diagnostic accuracy of somatosensory evoked potential and electroencephalography during carotid endarterectomy. Neurol Res 38 (8):698–705. https://doi.org/10.1080/01616412.2016.1200707

50. Ito K, Sasaki M, Kobayashi M, Ogasawara K, Nishihara T, Takahashi T, Natori T, Uwano I, Yamashita F, Kudo K (2014) Noninvasive evaluation of collateral blood flow through circle of Willis in cervical carotid stenosis using selective magnetic resonance angiography. J Stroke Cerebrovasc Dis 23(5):1019–1023. https://doi.org/10.1016/j.jstrokecerebrovasdis.2013.08.018

Chapter 18

Time is Brain: The Prehospital Phase and the Mobile Stroke Unit

Shrey Mathur and Klaus Fassbender

Abstract

In the clinical management of acute ischemic stroke, time is brain. In routine stroke care, valuable time is lost in transporting the patient to the hospital prior to necessary imaging, in-hospital delays and secondary transfer to a comprehensive stroke center. The Mobile Stroke Unit (MSU) is an innovation in the prehospital phase which addresses these issues by facilitating rapid, onsite diagnosis and treatment of acute stroke or similar emergencies. The MSU ambulance is equipped with a specialized CT scanner, a point-of-care laboratory, and stroke medication. Via telemedicine, CT images and real-time videos can be transmitted between the hospital and MSU allowing for expert advice. In this chapter, we discuss the development and configuration of the MSU, its role in the prehospital phase and triage, current evidence and challenges, and future directions for research.

Key words Prehospital stroke, Mobile stroke unit, Telemedicine, Telestroke, Teleradiology, Stroke imaging

1 Introduction

Stroke is one of the most frequent causes of disability and death [1, 2]. Acute ischemic stroke has enormous financial and societal costs due to rehabilitation, long-term care, and lost productivity [3]. With advances over the last decades, there are now safe and effective treatments for stroke [4, 5]. However, the treatment of ischemic stroke is extremely time-dependent as an estimated two million neurons are lost every minute [6]. Leading therapeutic modalities, intravenous thrombolysis, and mechanical thrombectomy are extremely time-dependent [7–9]. International guidelines now support the use of thrombolysis up to 4.5 h after symptom onset with mechanical thrombectomy as a viable option within 6 h of symptom onset or within 24 h of last known normal for select patients [4, 5]. Furthermore, for an excellent outcome of intravenous thrombolysis, the number needed to treat (NNT) is also time-dependent, ranging from approximately five in the first 90 min after

Philip V. Peplow et al. (eds.), *Stroke Biomarkers*, Neuromethods, vol. 147, https://doi.org/10.1007/978-1-4939-9682-7_18,
© Springer Science+Business Media, LLC, part of Springer Nature 2020

symptoms onset to nine with treatment between 91 and 180 min to more than 14 with treatment between 181 and 270 min [10]. In the case of mechanical thrombectomy, for each 30 min delay before reperfusion, the relative likelihood of a good clinical outcome decreases by approximately 15% [11, 12].

With this in mind, concerted efforts have been made to improve in-hospital management of stroke by minimizing delays and optimizing protocols and personnel. However, despite substantial efforts to streamline care, only 3.4–9.1% of patients receive thrombolysis and even fewer receive mechanical thrombectomy [13, 14]. These low rates of treatment are largely due to the fact that patients do not reach the hospital in time for assessment and treatment within the narrow therapeutic windows. In fact, only 15–60% of acute stroke patients arrive at the hospital within 3 h after onset of symptoms [15, 16].

Consequently, further time saving measures are required before the patient reaches the hospital—in the prehospital phase. Prehospital care is multifaceted. Public awareness campaigns help patients identify symptoms early and alert emergency medical services (EMS) [17, 18]. Trained dispatchers are able to rapidly and correctly identify stroke patients [19–22]. Paramedics and other healthcare personnel are able to initiate management and notify the target hospital [23–25]. While optimization of prehospital mechanisms reduces the time from symptoms to the hospital, in conventional stroke management, therapy can only be administered after the patient is brought to the hospital, often outside therapeutic time windows.

The Mobile Stroke Unit is an innovation in the prehospital phase which aims to provide onsite healthcare professionals with the necessary tools to rapidly and effectively evaluate and treat stroke by bringing the hospital to the patient. The concept is based on a specialized ambulance (Mobile Stroke Unit, MSU). The MSU concept was first proposed and actualized in Homburg, Germany (Fig. 1) [26, 27]. In addition to standard ambulance equipment, the MSU is equipped with the necessary tools for diagnosis and treatment of acute stroke or similar emergencies. This includes a specialized CT scanner, a point-of-care laboratory, and stroke medication. Via telemedicine, CT images and real-time videos can be bidirectionally transmitted between the hospital and MSU. These images can be integrated into the hospital's medical records. With the point-of-care laboratory, stroke-specific lab tests (thrombocytes, erythrocytes, leukocytes, hemoglobin, hematocrit, INR, PTT, gamma-GT, p-amylase, creatinine, glucose, etc.) can be analyzed within minutes in the Mobile Stroke Unit [17, 26–33].

In this chapter, we summarize up-to-date evidence for the prehospital acute ischemic stroke care. We highlight challenges and solutions in prehospital stroke care and focus on innovations with the Mobile Stroke Unit.

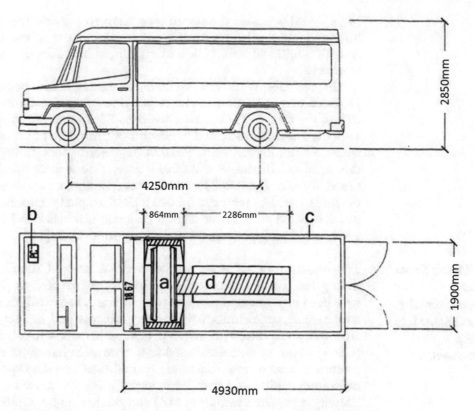

Fig. 1 Early schematics of a "mobile stroke unit" first developed in Homburg. Dimensions indicated in millimeters. (**a**) Integrated small CT, (**b**) operation console, (**c**) isolation against radiation produced by the CT, (**d**) metal free stretcher as CT table

2   The Prehospital Phase

2.1   Engaging the Stroke Rescue Chain

The prehospital phase refers to the stroke diagnosis and management before the patient arrives at the hospital. It starts with symptom onset and includes the activation of EMS, onsite management and transport to the target hospital. Though various structural models exist, effective prehospital care systems must have certain core elements. Important components of the prehospital stroke rescue chain are early identification of stroke symptoms, early activation of EMS, recognition by dispatchers and paramedics, effective onsite diagnosis and management, and effective prenotification of the target hospital. Prehospital management is augmented by mobile telemedicine for remote clinical examination and imaging, integration of CT scanners and point-of-care laboratories in ambulances [34].

2.2   Prenotification

Prenotification of the target hospital regarding an en route stroke patient is an evidence-based measure to accelerate in-hospital management and is recommended by international guidelines

[4, 5]. MSUs enable the prehospital team to provide the target hospital with detailed information about the nature of the stroke and any additional information to expedite subsequent specialized treatment.

In the Get With The Guidelines registry, EMS personnel provided prenotification to the target emergency department for 67% of the total 371,988 stroke patients. EMS prenotification was associated with an increased likelihood of thrombolytic treatment within 3 h (82.8% vs. 79.2%, $P < 0.0001$), shorter door-to-imaging times (26 vs. 31 min, $P < 0.0001$), shorter door-to-needle times (78 vs. 80 min, $P < 0.0001$), and shorter symptom onset-to-needle times (141 vs. 145 min, $P < 0.0001$). EMS hospital prenotification was associated with better clinical evaluation, timelier treatment, and more eligible patients treated with thrombolysis [35].

2.3 Primary Stroke Centers and Comprehensive Stroke Centers: Drip and Ship and Mothership

The organization of acute stroke care has evolved significantly during the past few decades [36]. Primary stroke centers (PSCs) have been implemented to improve stroke care. PSCs include acute stroke teams, stroke units, written care protocols, and an integrated emergency response system [37]. Comprehensive stroke centers (CSCs) integrate specialized services for the management of severe cerebrovascular disease. These are typically staffed with experts in neurointervention and vascular neurology, have advanced neuro-imaging capabilities including MRI and cerebral angiography, specialize in surgical and endovascular techniques (including clipping and coiling of intracranial aneurysms, carotid endarterectomy, and intra-arterial thrombolytic therapy), and have specific infrastructure such as an intensive care unit [38].

As of 2017, comprehensive stroke centers accounted for roughly one-third of all stroke centers in the USA (327 of 1148) and in France (37 of 132) [39–41]. In 2011, 66% of Americans were within a 60-min ground transfer to a primary stroke center and 81% to an intra-arterial therapy (IAT) capable hospital. However, only 56% of Americans had 60 min ground transfer proximity to a CSC [39]. As a result, in some settings it is difficult to admit patients directly to CSC, especially with standard ambulance services. As a proposed solution, patients could receive intravenous therapy in a hospital before being transferred to a CSC for mechanical thrombectomy. This is known as the drip-and-ship paradigm, in contrast to direct admission to a CSC, referred to as mothership.

Along with the development of specialized stroke units, there have been significant advances in reperfusion techniques. Intravenous thrombolysis has been the leading reperfusion method with its efficacy window having been extended to 4.5 h after symptom onset [42–44]. Previously, mechanical thrombectomy was limited to patients with basilar artery occlusions and those with contraindications to intravenous thrombolysis [45–48]. Recently, there has been growing evidence supporting the use of mechanical thrombectomy in a greater proportion of patients—including those with occlusion of the proximal anterior circulation—as time windows for efficacy grow up to 24 h since last known normal [9, 49–51]. With an increasing proportion of patients becoming eligible for mechanical thrombectomy, it becomes important to optimize patient transport from the emergency site either by drip-and-ship or mothership. Further to these is bridging therapy whereby intravenous therapy within 4.5 h is followed by mechanical thrombectomy within 6 h of symptom onset [39]. There is evidence from clinical trials supporting bridging therapy over intravenous thrombolysis for patients with large vessel occlusion (LVO) of the anterior circulation [52–57]. These approaches to patient transfer have important implications for the optimization of prehospital care.

3 The Mobile Stroke Unit

3.1 The Mobile Stroke Unit Concept

It is clear that effective and safe treatment for acute ischemic stroke is time-dependent. However, because thrombolysis for ischemic stroke can only be safely performed after exclusion of possible hemorrhage by computed tomography (CT), crucial time is lost as the patient is brought to the hospital for imaging. The Mobile Stroke Unit (MSU) concept is an innovation which brings the hospital to the patient (Fig. 2). The MSU, as a specialized mobile emergency room, contains multimodal imaging, a point-of-care laboratory, and a telemedicine connection with a hospital, in addition to appropriate medication and assessment tools [26]. The aim is to deliver timely, state-of-the-art prehospital diagnosis and treatment, as well as diagnosis-based triage of the patient to the most appropriate target hospital.

The MSU strategy allows for the administration of treatment directly at the emergency site. Treatments include thrombolysis for

Fig. 2 Generations of mobile stroke units used in Homburg

acute cerebral ischemia, anticoagulant reversal for acute intracranial hemorrhage, management of physiological variables for ischemic or hemorrhagic stroke, and management of further emergencies [31]. The MSU extends specialized stroke care to the prehospital phase of stroke management.

Acute stroke management involves multidisciplinary cooperation. Healthcare professionals (including paramedics, nurses, radiographers, technicians, physicians) are required to quickly collect medical history, perform clinical examinations, and handover information. In a hospital setting, these professionals may be working in different locations and join the patient care pathway at different times. The resulting handovers and multiple interfaces can contribute to errors and delays. With the MSU concept, crucial time is saved by substantially reducing those interfaces. In the MSU, a single, specialized, interdisciplinary team, consisting of paramedics, physicians, nurses, and technicians, can perform a complete diagnostic workup and acute treatment. The MSU acts synergistically with hospital stroke units to close the existing treatment gaps.

3.2 The Mobile Stroke Unit Ambulance

The Mobile Stroke Unit is an ambulance that contains equipment necessary for initial acute stroke treatment, and standard emergency care equipment (Fig. 3) [33]. In the Homburg MSU model, the ambulance is equipped with a multimodal imaging-capable, accumulator-driven, and lead-shielded CT scanner

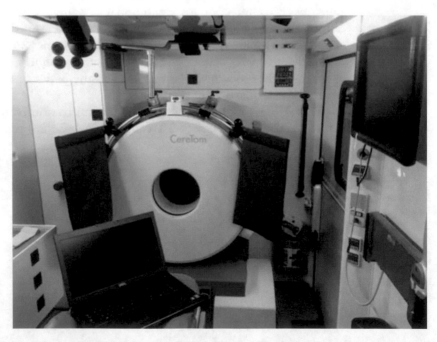

Fig. 3 Homburg Mobile Stroke Unit interior with CT scanner. The Mobile Stroke Unit is an ambulance which contains a multimodal CT scanner, a point-of-care laboratory, as well as a telemedicine system, which allows for transfer of CT images and videos of patient examination to the clinical information system as well as a bidirectional communication

(CereTom, NeuroLogica/Samsung, Danvers, USA). Further to the scanner, a telemedicine system (MEYTEC, Werneuchen, Germany) allows for transmission of digital imaging data and video of clinical examination to the hospital. A point-of-care laboratory system on the Homburg MSU, enables the measurement of platelet count, leukocyte count, erythrocyte count, hemoglobin level, and hematocrit (PocH 100i, Sysmex, Hamburg, Germany), international normalized ratio (INR) and activated partial thromboplastin time (Hemochron Jr., ITC, Edison, USA), and gamma-glutamyltransferase activity, p-amylase activity, and glucose concentration (Reflotron Plus, Roche Diagnostics, Mannheim, Germany) [33, 58].

Subsequent iterations of the MSU have adapted to their respective settings. Larger vehicles have the advantage of carrying larger scanners, robustness in rural off-road conditions, and allowing relatives to accompany patients to provide history and procedural consent [59]. Smaller vehicles may have greater access to narrow roads and lower cost. With this in mind, vehicle models should be selected according to the specific needs of the region and healthcare setting [31].

3.3 Prehospital Brain Imaging

Imaging is important for triage and management of acute stroke. Broadly, in acute events, ischemic stroke (IS) must be reliably distinguished from hemorrhagic stroke (HS) [4, 5]. This distinction is necessary for initiation of thrombolysis and for transport decisions [30].

Mobile Stroke Units benefit from advances in portability and image quality of mobile CT scanners. Multimodal imaging (non-contrast CT, CT angiography, and CT perfusion) has been integrated into existing MSU programs [31, 33]. There are several portable CT scanners on the market, including the CereTom (Neurologica), Tomoscan M (Philips, Eindhoven, The Netherlands), and Somatom Scope (Siemens). Images from most portable CT scanners are of sufficient quality for assessment of the parenchyma and for CT angiography of the intracranial circulation including the circle of Willis. The Somatom Scope, currently in use in Memphis, allows assessment of proximal neck vessels and the aortic arch [31].

The CereTom scanner is in use on the Homburg MSU. The CereTom is an 8-slice small-bore portable CT scanner that allows for non-contrast, angiography, and contrast perfusion. As a space saving measure, the scanner's gantry—rather than the patient—moves along the craniocaudal axis. As a result, a dedicated translating exam table is not required [60]. Furthermore, CereTom operates with an on-board 120 V rechargeable battery pack and can be plugged into a standard 110 V outlet obviating the need for a high-voltage power supply [61].

Both size and weight are factors in selecting a suitable scanner. For example, the CereTom scanner measures $1.5 \times 0.7 \times 1.3$ m and weighs 362 kg (about 800 lbs). At present, more detailed scanners exceed size and weight limitations for roadworthy ambulances. It is likely that with continued innovation, more detailed and advanced scanners may be suitable for MSUs in the future as scanner size decreases and vehicle size increases.

Endovascular intervention is becoming more suitable for an increasing population of patients [50, 62]. However, not all primary stroke centers (PSCs) are equipped and staffed for endovascular neurointervention. This poses a challenge for prehospital triage. Preferably, patients with large vessel occlusion (LVO) should be sent to endovascular-capable centers directly. Congruently, patients who are unlikely to have an LVO or undergo thrombectomy should be transported to PSCs closest to their homes.

Despite continued progress, it remains difficult in a prehospital setting to reliably identify LVO clinically [63–65]. Advances in imaging in the prehospital phase provide an opportunity for screening patients for LVO. CT-Angiogram (CTA) enables prehospital diagnosis of severe strokes with LVO and can aid in guiding transport decisions for the patient [66, 67]. For non-contrast CT (NCCT), the Alberta Stroke Program Early CT Score (ASPECTS) is a systematic approach used to assess early ischemic changes within the middle cerebral artery territory [68]. However, limitations due to interrater variability and scan quality impact its utility [53]. Automated assessment by artificial intelligence (AI) based software, such as e-ASPECTS (Brainomix, Oxford, UK), may facilitate rapid and remote assessment of images [69]. CT Perfusion may have clinical utility in the prehospital selection of candidates for endovascular therapy as it may have greater sensitivity than NCCT in identifying the ischemic changes [70].

3.4 Prehospital Point-of-Care Testing

Laboratory testing is an important component of stroke care. However, in many cases, it is more important to not delay the administration of thrombolysis due to laboratory testing. According to leading guidelines, blood glucose is the only assessment which must precede the initiation of thrombolysis in all patients [5]. If there is strong suspicion of coagulopathy, tests such as international normalized ratio (INR), activated partial thromboplastin time (aPTT), and platelet count, may be necessary. Other recommended tests are leukocyte and erythrocyte counts, hemoglobin levels, gamma-glutamyltransferase (GGT) and pancreatic amylase activity [31]. Creatinine level is helpful for concerns regarding renal function [30]. Still, thrombolysis should not be delayed while waiting for hematological or coagulation testing if there is no reason to suspect an abnormal test considering the low risk of coagulopathy in the general population. In the same vein, baseline troponin assessment is recommended but should not delay initiation of thrombolysis [5].

Point-of-care testing (POCT) allows for faster results than use of the hospital laboratory system [71–73]. In a German study, the use of a point-of-care laboratory has been shown to decrease time from door-to-therapy decision (end of all diagnostic procedures) from 84 min (SD 26) to 40 min (SD 24, $P < 0.0001$) compared with use of a centralized hospital laboratory [28]. With that in mind, further studies are needed to clarify the relevance of point-of-care results on treatment decisions. Further advantages of POCT are portability, simple measurement procedures, use of low sample volumes, and automated data processing. POCT devices require extensive testing, validation, and conformity with legislative directives and health standards before they can be used in a clinical setting [31].

POCT devices are valuable and adaptable to the MSU setting. In the Homburg trial, the laboratory system allowed for measurement of platelet count, leukocyte count, erythrocyte count, hemoglobin, and hematocrit (PocH 100i, Sysmex, Hamburg, Germany), international normalized ratio and activated partial thromboplastin time (Hemochron Jr., ITC, Edison, NY, USA), and gamma-glutamyltransferase, p-amylase, and glucose (Reflotron plus, Roche Diagnostics, Mannheim, Germany) [26]. Similar POCT has been used in MSU trials in the USA [74].

3.5 Telemedicine: Communicating with the Ambulance

Telecommunication approaches between standard ambulances and the stroke center via systems can provide real-time remote specialist advice. Telemedicine includes telestroke assessment (real-time bidirectional videoconferencing and high-speed transmission of videos) and teleradiology (transmission of high-quality images) [31]. Commercially available systems allow MSUs to transmit digital imaging and communication data to hospital records. MSUs provide a valuable resource to rural and remote settings where patient may not have easy access to in-hospital stroke care [75].

However, there are limitations to the management tasks which can be carried out remotely. The treatment of acute stroke in an MSU is a complex exercise involving multiple parallel tasks being carried out expediently by several healthcare professionals within the confined space of an MSU. This includes neurological assessment, monitoring of vital signs, patient positioning, management of patient comfort and possible restraint, CT scanning, point-of-care laboratory testing, and medication preparation and administration [74]. Furthermore, the clinical decision on whether to administer thrombolysis requires training, experience, and careful clinical judgment.

Nevertheless, limitations due to EMS staffing, cost-effectiveness and geographic accessibility necessitate telemedicine. EMS response varies between the majority of European and US settings. In the USA, ambulances are typically not staffed by physicians. As a result, in developing an MSU program, a decision has to be made as to whether to include an on-board physician. Early experience from Houston and Germany suggests the ratio of MSU alerts (from EMS dispatch) to tPA treatments is at least 10 to 1, making it impractical to have a vascular neurologist aboard the MSU for all calls [74].

The Pre-hospital Utility of Rapid Stroke Evaluation Using In-Ambulance Telemedicine (PURSUIT) study assessed the feasibility and accuracy of telemedicine assessment of actors simulating patients with stroke in ambulances [76]. Remote, real-time assessments of National Institutes of Health Stroke Scale (NIHSS) showed absolute agreement for intraclass correlation (ICC) of 0.997 (95% CI 0.992–0.999) and matching within two points of NIHSS occurred for 88% of scenarios.

As part of the BEST-MSU study, 174 patients were simultaneously, independently assessed by an on-board vascular neurologist and a telemedicine vascular neurologist to evaluate interrater variability. There was 98% satisfactory connectivity and 88% agreement on the thrombolysis decision [77]. This level of agreement is comparable to two vascular neurologists evaluating the same patients face-to-face in the emergency department [78].

Technical innovation in the transmission of data between the hospital and MSU plays an important role. Early studies encountered difficulties in telecommunication in part due to suboptimal 3G public network availability [76, 79–82]. Fortunately, with improved technology and 4G mobile systems, telecommunication is becoming more reliable [31]. Telemedicine encounters between the MSU and hospital has been shown to be successfully completed for 99% of patients with 4G connectivity [83].

In the iTREAT study, a low-cost, tablet-based platform and commercial cellular networks (4G/LTE) were used to reliably perform prehospital neurologic assessments (NIHSS) of actors in both rural (central Virginia) and urban settings (San Francisco Bay Area) via videoconferencing [84]. With innovation, it is important to ensure that bidirectional telecommunication is encrypted and secure, and meets the standards for transmission of protected health information.

3.6 Staffing the MSU

Emergency Medical Services (EMS) staffing varies greatly across different countries [85, 86]. In many European settings, EMS services are staffed by a physician. In contrast, in many other countries, including most US settings, ambulances are staffed by paramedics with no physician present [17, 31]. MSU staffing, accordingly, must be adapted to disparity in local legislation. Additionally, to date most MSUs operate in addition to conventional emergency medical services due to legislative limitations [31]. Staffing conventions also impact the cost-effectiveness and sustainability of an MSU practice. In the USA, an MSU with a physician present may not be cost-effective [87]. However, remote physician assessment through the use of telemedicine may serve as a safe and reliable workaround [77]. Most research projects to date have included a vascular neurologist on board the ambulance [33]. However, with increasingly reliable telecommunication, other staffing configurations with different combinations of paramedics, nurses, radiographers, neurologists, and neuroradiologists are possible via telemedicine [31].

In the Homburg model, the MSU team includes a paramedic, a stroke physician, and a neuroradiologist. Additionally, conventional EMS which includes an emergency physician for critically ill patients accompanies the MSU [26]. In Houston, the MSU is staffed by a paramedic, a certified CT technician, a vascular neurologist, and a registered nurse experienced in both clinical research and acute stroke management [74].

In Norway, anesthesiologists trained in the prehospital clinical assessment of acute stroke patients and the interpretation of CT imaging have reliably been able to identify radiological contraindications for thrombolysis and also identify subarachnoid hemorrhage [88–90].

3.7 Dispatching the MSU

Early symptom recognition is essential for timely care in stroke. Once engaged, EMS dispatchers have an important role in identifying and prioritizing stroke patients for the most efficient use of this specialized vehicle. Education programs for EMS dispatchers increase the number of stroke patients treated in timely manner [17]. Shortening times from call-to-dispatch, dispatch-to-ambulance arrival, on-scene time, and scene-to-hospital transport are all important. In a study of 184,179 cases in the USA, median time from call to ED arrival is 36 min (IQR 28.7–48.0) with the longest component being on-scene time (15 min). Strikingly, dispatch identified only 52% of cases as stroke [91].

Identification of potential stroke patients by telephone is a challenging task undertaken by EMS dispatchers [17, 92]. Several prehospital stroke screening scales have been developed and employed in an effort to assist dispatchers in identifying stroke patients with the greatest specificity and sensitivity. Recognition of Stroke in the Emergency Room (ROSIER) scale (sensitivity 93%, specificity 83%) and Dispatcher Identification Algorithm of Stroke Emergency (DIASE) (sensitivity 53%, specificity 97%) have been employed in this regard [93–95]. Furthermore, there are several scales which assist EMS personnel in the early identification of stroke, namely the Cincinnati Pre-Hospital Stroke Scale (CPSS), Los Angeles Pre-Hospital Stroke Screen (LAPSS), Melbourne Ambulance Stroke Screen (MASS), Medic Prehospital Assessment for Code Stroke (Med PACS), Ontario Prehospital Stroke Screening Tool (OPSS), and Face Arm Speech Test (FAST) [94].

4   Mobiles Stroke Units for Time Saving

4.1   MSU Trial Evidence

The first randomized, single-center trial was conducted by Saarland University Medical Centre in Homburg, Germany in 2008 (Table 1) [26]. In this study, patients received either prehospital stroke treatment in the MSU or optimized conventional hospital-based stroke treatment (control group) with a 7 day follow-up. Included patients were aged 18–80 years with onset of one or more stroke symptoms started within the previous 2.5 h. Patients in the MSU group had significantly faster times both time from alarm (emergency call) and from symptom onset to therapy decision, intravenous thrombolysis, intra-arterial thrombectomy, and end of CT to end of laboratory analysis.

Positive results from the Homburg trial have been confirmed in the PHANTOM-S (Pre-Hospital Acute Neurological Therapy and Optimization of Medical Care in Stroke) study [96], in an observational study in Houston [97] and in a case series in Cleveland [98]. Encouragingly, time-to-treatment from these studies were consistently faster than other interventions aiming to reduce delays in the emergency department [17].

Table 1
Results from the first MSU trial conducted in Homburg, Germany demonstrating time savings across all metrics

	MSU group (n = 53)	Control group (n = 47)	P value	Difference (95% CI)
Therapy decision				
Alarm to therapy decision (min)	35 (31–39)	76 (63–94)	<0.0001	41 (36–48)
Symptom onset to therapy decision (min)	56 (43–103)	104 (80–156)	<0.0001	43 (30–58)
Intravenous thrombolysis				
Alarm to intravenous thrombolysis (min)	38 (34–42)	73 (60–93)	<0.0001	34 (23–54)
Symptom onset to intravenous thrombolysis (min)	72 (53–108)	153 (136–198)	0.0011	80 (40–115)
Intravenous thrombolysis or IAT				
Alarm to intravenous thrombolysis or intra-arterial recanalization (min)	38 (34–42)	78 (61–110)	<0.0001	44 (27–73)
Symptom onset to intravenous thrombolysis or intra-arterial recanalization (min)	72 (53–108)	152 (135–209)	<0.0001	80 (46–115)
End of CT				
Alarm to end of CT (min)	34 (30–38)	71 (62–87)	<0.0001	38 (33–43)
Symptom onset to end of CT (min)	56 (43–103)	97 (74–156)	<0.0001	39 (26–52)
End of laboratory analysis				
Alarm to end of laboratory analysis (min)	28 (26–34)	69 (55–81)	<0.0001	38 (32–44)
Symptom onset to end of laboratory analysis (min)	51 (40–95)	99 (70–140)	<0.0001	39 (26–56)

MSU studies perform favorably when compared to stroke management observed in clinical practice. In the Safe Implementation of Thrombolysis in Stroke–Monitoring Study (SITS–MOST), 6853 patients treated at 285 European centers between 2002 and 2006 reported median time from symptom onset to treatment was 140 min (IQR 110–165) [99]. The Get With The Guidelines-Stroke Program, involved 58,353 patients treated in 1395 US hospitals between 2003 and 2012 with a symptom onset to treatment time of 144 min (IQR 115–170) [7].

4.2 Racing Against Time: the Golden Hour

The term "golden hour" has been attributed to the trauma surgeon R Adams Cowley, who recognized that the sooner trauma patients receive definitive care—particularly within the first hour after trauma—the better their chance of survival [100].

> There is a golden hour between life and death. If you are critically injured you have less than 60 minutes to survive. You might not die right then; it may be three days or two weeks later—but something has happened in your body that is irreparable.—R Adams Cowley

The concept of "golden hour" has broadened to hyperacute conditions in which early treatment is more effective than later intervention. These include trauma, myocardial ischemia, septic shock, cardiopulmonary resuscitation, and stroke [101]. In acute ischemic stroke, the benefit of intravenous thrombolytic therapy is strongly time dependent [6–8, 10, 12]. Patients who present to the hospital within the first 60 min of onset have the greatest opportunity to benefit from recanalization therapy [31].

Despite substantial efforts to improve stroke management over the past decade, only a small proportion of patients receive treatment during the golden hour. In the National Institute of Neurological Disorders and Stroke (NINDS) study, less than 1% of patients were randomly assigned to a study group within 60 min [42]. In the Safe Implementation of Treatments in Stroke–International Stroke Thrombolysis Registry (SITS-ISTR) observational study, only 166 (1.4%) of 11,429 patients were treated within 60 min [102]. In the Get With The Guidelines-Stroke registry, 750 of 58,353 (1.3%) patients were treated within this time [7]. In a Finnish study, in-hospital delays were minimized leading to door-to-needle times as short as 20 min [103]. However, no more than 10% of patients were treated within 70 min of symptom onset and the median onset-to-treatment time was 119 min (IQR 80–176). This suggests that there may be a ceiling to the benefit of reducing in-hospital delays, and that the prehospital phase should be a focus of optimization.

MSUs optimize the prehospital phase and enable healthcare professionals to act during the golden hour. In the Homburg trial, patients in the MSU group had a median time from symptom to therapy decision of 56 min (IQR 43–103) versus 104 min

(80–156) with a substantial median difference of 43 min (95% CI 30–58 min) [26]. In the PHANTOM-S sub-study, the median onset-to-treatment time was 24.5 min shorter with the MSU group (80.5 min, IQR 54–126) with 6 times as many patients receiving thrombolysis in the golden hour compared with conventional care (105.0 min, IQR 82–146) [104]. These results have been confirmed in other studies [31]. To date, patients treated in this early time frame have significantly better clinical outcomes [7, 105]. However, further studies are needed to confirm the extent of this benefit as many of the studies to date have had small treatment cohorts.

5   Mobile Stroke Units as a Tool for Triage

Accurate triage and selection of the appropriate target hospital avoids the transfer of patients with large-vessel occlusion to hospitals without endovascular treatment services and prevents the overloading of comprehensive stroke centers by selecting thrombectomy-eligible patients. It is estimated that every minute of delay in transfer reduces the probability that patients will receive intra-arterial treatment by 2.5% [106].

Clinical scales have been evaluated for improving triage and differentiating patients with or without large-vessel occlusion [107]. Several scales have been evaluated both prospectively and retrospectively for this purpose. A prospective evaluation of the Rapid Arterial Occlusion Evaluation scale reported that a score of 5 or higher predicts LVO with a sensitivity of 88% and a specificity of 68% [65]. In another prospective study, the Field Assessment Stroke Triage for Emergency Destination scale, based on the NIHSS with a cutoff value of 2 or higher, reported a sensitivity of 60% and a specificity of 89% for LVO [108]. A retrospective study of the Los Angeles Motor Scale showed that a score of 4 or higher predicts LVO with a sensitivity of 81% and a specificity of 89% [109]. Lastly, the Prehospital Acute Stroke Severity scale was studied retrospectively, with scores of 2 or higher having a sensitivity of 66% and a specificity of 83% for LVO. Importantly, with all of these scales, a large number of LVOs are missed [110, 111]. Furthermore, these scales remain to be tested and validated in the prehospital environment which may be appropriate for MSU studies in the future.

The MSU strategy has proven to be effective at triaging stroke patients (Fig. 4). CT-Angiography has been used in MSUs to assess for LVO at the emergency site [26]. Even use of non-contrast CT, has been associated with reduction of delay before intra-arterial treatment [66]. For patients with hemorrhagic stroke, MSU-based triage allows for transport to hospitals with neurosurgery services, bypassing hospitals without such capabilities [32]. In

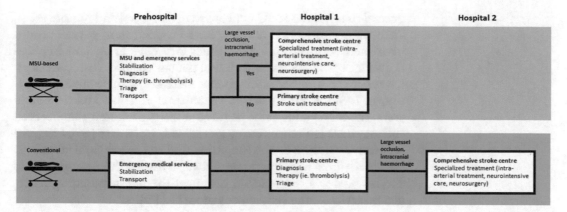

Fig. 4 MSU-based stroke management compared with conventional stroke management. In MSU-based approach, patients are diagnosed at the site of emergency, allowing for case-specific treatment and triage to the most appropriate stroke center, thus avoiding secondary transfers. In conventional stroke management, due to insufficient information about the cause of symptoms, patients are transported to the primary stroke center and eventually, by secondary transfer, brought to a comprehensive stroke center

the PHANTOM-S study, patients with hemorrhage delivered to hospitals without neurosurgery services decreased from 43% in the conventional treatment group to 11% in the MSU group [104, 112]. MSUs are also valuable for investigating other time-sensitive cerebral conditions such as traumatic brain injury or status epilepticus [113].

6   Future Directions

6.1   Integrating the MSU into Existing Emergency Services

As strides in prehospital care research continue to be made, it is important to consider their integration into existing healthcare systems with differing configurations of emergency care, legislative requirements, financial limitations, geography and demographics. Fortunately, the more than 35 ongoing and planned MSU projects worldwide are being conducted in varied settings and accordingly may provide much-needed insight.

The MSU approach is not intended to replace continuous efforts to improve the quality of prehospital and in-hospital stroke care [17]. The crucial time from symptom onset to emergency call is beyond the influence the MSU. Here, public education and awareness campaigns continue to be vital. The MSU, as a rolling billboard in the community, may in itself promote public awareness [31].

Further research is also required to help optimize the relation between MSU and primary and comprehensive stroke centers in various healthcare systems. In settings where there is limited availability of thrombolytic agents, improving basic access and improving primary stroke care takes priority. In addition, more research is needed to tailor the MSU concept to urban, rural and remote areas. With widespread MSU use, a greater proportion of patients could be treated with thrombolysis which may have important implications for organizational and clinical management in stroke services.

Since the number of dispatches increases with population density, an MSU is particularly advantageous in urban areas. In rural areas, the MSU adds value as these settings are often underserved and may have limited access to specialized stroke care [87, 114]. An innovative solution to the long driving distances in rural areas is to set a predefined meeting point midway between the emergency site and hospital where the MSU can meet the first responding emergency services—the rendezvous model [115].

6.2 Safety

As with all innovations in stroke care, patient safety is paramount. With the help of advanced equipment and onsite and remote multidisciplinary expertise, patient management and safety in an MSU is comparable to that in a hospital [33]. At present, technical failure rates are similar to those in routine emergency ambulances and corresponding hospital equipment [31]. A concern is that early, rapid evaluation may result in thrombolytic treatment for an increased number of stroke mimics. This is a common concern with time-saving interventions in acute stroke care. To date, MSU studies have found no significant differences with regard to indicators of safety, such as hemorrhagic complications or mortality [26, 83, 96, 116, 117]. However, safety is still a relevant and important area for future research.

6.3 Long-Term Outcome

Based on the "time is brain" principle, it would be expected that expedient treatment would result in a clear clinical benefit. However, direct evidence from controlled trials comparing similar patients managed by the MSU or by standard emergency medical care is still needed. The first Homburg trial found no differences in 7 day modified Rankin scores and NIHSS scores likely due to insufficient power of the trial to detect differences in these endpoints [26]. The PHANTOM-S trial found that patients in the MSU group were more likely to be discharged home but did not find significant improvements in short-term outcomes [96]. In a large registry study in Berlin, patients treated with the mobile stroke unit were compared to patients receiving conventional care between 2011 and 2015. Patients between the MSU and control group did not differ significantly in the primary outcome of a 90 day modified Rankin score of 0 or 1 (53% vs. 47%, $P = 0.14$), even though 37% of patients treated in the MSU group received thrombolysis within the golden hour [118]. However, when the analysis was adjusted for baseline differences between the non-randomized groups, results significantly favored the MSU cohort. The Benefits of Stroke Treatment Delivered Using a Mobile Stroke Unit (BEST-MSU) trial in Houston is expected to provide important data about clinical outcome and cost-effectiveness of the MSU approach through prospective, direct comparison to standard prehospital triage with a large expected enrolment of 1200 patients [97].

6.4 Cost-Effectiveness

The MSU strategy is an innovative approach to optimizing prehospital acute stroke management. However, valid concerns remain regarding the costs of the MSU, its staffing, and operation. Two independent preliminary analyses of cost-effectiveness have reported encouraging results.

Based on the Homburg trial, the incremental direct costs of the MSU were compared using a one-year cost–benefit analysis across a number of scenarios. The cost-benefit was 1.96 in the baseline experimental setting and with two on-board physicians. In the model, gradually reducing staff and increasing population density can substantially improve the cost-benefit ratio up to 6.85. Cost modeling showed that the MSU strategy is cost-efficient starting from an operating distance of 15.98 km (9.99 miles) or from a population density of 79 inhabitants/km^2 (202 inhabitants per square mile) [87].

A group in Berlin analyzed costs from the PHANTOM-S study evaluating treatment effects of intravenous thrombolysis, intermediate outcomes and gains in quality-adjusted life-years (QALYs). The net annual cost of the MSU program in Berlin was €963,954. In the analysis, there was an expected avoidance of 18 cases of disability—equaling 29.7 QALYs—due to the higher frequency of thrombolysis administrations per year and higher proportions of patients treated in the early time interval. Incremental cost-effectiveness ratio was reported as €32,456 per QALY [119].

Further to this, cost-effectiveness can be improved by substituting onboard physicians for telemedicine-linked remote experts [77], increased usage rates and refining stroke identification algorithms for dispatchers [87]. It must be noted that MSU configuration and associated costs depend on the logistical, financial and legislative constraints of the setting. Optimizing costs of the MSU strategy is important for its long-term viability. Future prospective research is required in defining costs for establishing and maintaining MSUs and costs for both acute and long-term care of patients managed both on MSUs and by standard emergency services. Future services may be expanded to include other cerebral emergencies and treatment modalities. It is important that this financial cost is weighed and considered with the perspective of important stakeholders in acute stroke care.

7 Active MSU Programs

The number of active prehospital mobile stroke unit programs continues to grow with an ever expanding MSU World Map (Fig. 5).

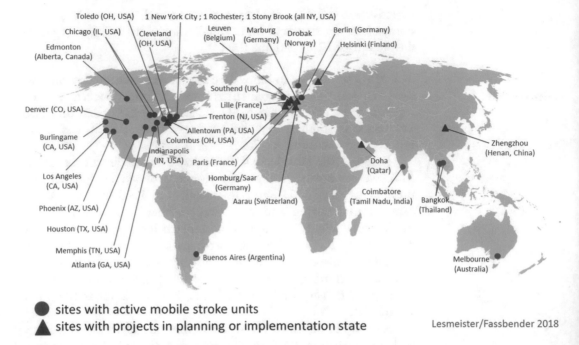

Fig. 5 MSU world map. A world map of active and planned MSU projects worldwide. Blue circles represent centers with active programs, red triangles are centers with MSU in planning

The Mobile Stroke Unit concept was first proposed [27] and actualized [33] by Fassbender and colleagues at the Saarland University Medical Centre in Homburg, Germany with its first design in 2000. Since having demonstrated the significant benefit of the MSU concept, more recent studies have focused on further technical innovations and adaptability to various healthcare settings. In an ongoing trial (NCT02465346), the group aims to evaluate the feasibility, safety and clinical benefit of diagnosis and treatment directly at the emergency site and its role in transfer decision to appropriate target institution in rural and remote regions [30]. A group in Berlin has been studying the MSU concept in a large, urban population [104]. The group has demonstrated the feasibility, safety, time gain, greater treatment rates, and improved 90-day functional outcome of the concept [117, 118]. More recent projects are focused on city-wide coverage with several ambulances in close collaboration with the centralized ambulance service of the Berlin Fire Brigade [30]. Berlin-Prehospital Or Usual Care Delivery (B_PROUD) is a prospective quasi-randomized study (NCT02869386) evaluating functional outcome after 3 months measured by the modified Rankin Scale over the entire range [120]. In France, the Acute Stroke: Prehospital versus in-Hospital initiation of recanalization Therapy (ASPHALT) study aims to assess the cost-utility of an MSU for stroke management in the era of bridging therapy [30]. The group in Norway is investigating the role of on-board anesthesiologists and the applicability of the

MSU concept to air ambulances to adapt the strategy to the Norwegian geographic context [88, 90].

The first MSU in the USA was implemented in Houston, Texas. Along with demonstrating feasibility in American healthcare setting, the group has been able to evaluate technical solutions such as telemedicine and demonstrated radiation safety [77, 116, 121]. BEST-MSU is a large multicenter randomized trial to further investigate clinical benefit and cost-effectiveness [97]. The Cleveland Pre-Hospital Acute Stroke Treatment Group has investigated the operational feasibility of an MSU in the absence of on-board physician. The MSU is staffed with paramedics and technicians and guided by remote physician via telemedicine [66, 83]. In Memphis, Tennessee, a high-resolution scanner that allows improved imaging of head and neck vessels and the aortic arch is being actively investigated. Further, ultrasound and sonothrombolysis are being evaluated as diagnostic and therapeutic options in the MSU [30]. The program at the University of Alberta in Edmonton, Canada covers a large rural region, utilizing a rendezvous model whereby the MSU travels out to meet the incoming ambulance to initiate treatment [30]. MSU programs are also active in many settings including at the Rush Hospital and Northwestern Hospital in Chicago, New York Presbyterian Hospital system, UCLA Health in Los Angeles and Colorado University Hospital in Aurora [30].

PRESTO (PRE-hospital Stroke Treatment Organization) provides a platform for cooperation of global MSU projects. This international consortium was formed in 2016 and aims to provide a forum for collaboration, advocacy, and education [30]. The first research collaborations aim to develop a list of common data elements to be collected by all Mobile Stroke Unit programs and entered into a common research database.

8 Conclusion

In the management of acute ischemic stroke, time is brain. Instead of losing valuable time while waiting for the patient to arrive to the hospital, the MSU strategy helps to optimize prehospital care to start the right treatment at the right time. MSU studies consistently report a reduction in delays before thrombolysis and effective triage to the correct target hospital (primary vs. comprehensive stroke center).

In the future, mobile stroke units may serve as a valuable research platform for the hyperacute management of stroke [31]. MSUs can be used in research investigating diagnostic markers of cerebral damage [122], transcranial duplex ultrasonography [123] and automated imaging decision support tools [58], improved clot-dissolving medications and sonothrombolysis [124]. As MSU programs continue to be researched and implemented across the world, safety, clinical efficacy, best setting, and cost-effectiveness need to be addressed.

References

1. Norrving B, Kissela B (2013) The global burden of stroke and need for a continuum of care. Neurology 80:S5–S12. https://doi.org/10.1212/WNL.0b013e3182762397

2. Feigin VL et al (2014) Global and regional burden of stroke during 1990–2010: findings from the Global Burden of Disease Study 2010. Lancet 383:245–254

3. Ovbiagele B et al (2013) Forecasting the future of stroke in the United States: a policy statement from the American Heart Association and American Stroke Association. Stroke 44:2361–2375. https://doi.org/10.1161/STR.0b013e31829734f2

4. Kobayashi A et al (2018) European Academy of Neurology and European Stroke Organization consensus statement and practical guidance for pre-hospital management of stroke. Eur J Neurol 25:425–433. https://doi.org/10.1111/ene.13539

5. Powers WJ et al (2018) 2018 Guidelines for the early management of patients with acute ischemic stroke: a guideline for healthcare professionals from the American Heart Association/American Stroke Association. Stroke 49:e46–e110. https://doi.org/10.1161/STR.0000000000000158

6. Saver JL (2006) Time is brain—quantified. Stroke 37:263–266. https://doi.org/10.1161/01.STR.0000196957.55928.ab

7. Saver JL et al (2013) Time to treatment with intravenous tissue plasminogen activator and outcome from acute ischemic stroke. JAMA 309:2480–2488. https://doi.org/10.1001/jama.2013.6959

8. Strbian D et al (2010) Ultraearly thrombolysis in acute ischemic stroke is associated with better outcome and lower mortality. Stroke 41:712–716. https://doi.org/10.1161/STROKEAHA.109.571976

9. Goyal M et al (2016) Endovascular thrombectomy after large-vessel ischaemic stroke: a meta-analysis of individual patient data from five randomised trials. Lancet 387:1723–1731. https://doi.org/10.1016/S0140-6736(16)00163-X

10. Lees KR et al (2010) Time to treatment with intravenous alteplase and outcome in stroke: an updated pooled analysis of ECASS, ATLANTIS, NINDS, and EPITHET trials. Lancet 375:1695–1703. https://doi.org/10.1016/S0140-6736(10)60491-6

11. Khatri P et al (2014) Time to angiographic reperfusion and clinical outcome after acute ischaemic stroke: an analysis of data from the Interventional Management of Stroke (IMS III) phase 3 trial. Lancet Neurol 13:567–574. https://doi.org/10.1016/S1474-4422(14)70066-3

12. Sun CH et al (2014) Door-to-puncture: a practical metric for capturing and enhancing system processes associated with endovascular stroke care, preliminary results from the rapid reperfusion registry. J Am Heart Assoc 3:e000859

13. Adeoye O, Hornung R, Khatri P, Kleindorfer D (2011) Recombinant tissue-type plasminogen activator use for ischemic stroke in the United States: a doubling of treatment rates over the course of 5 years. Stroke 42:1952–1955. https://doi.org/10.1161/STROKEAHA.110.612358

14. Schwamm LH et al (2013) Temporal trends in patient characteristics and treatment with intravenous thrombolysis among acute ischemic stroke patients at Get With The Guidelines-Stroke hospitals. Circ Cardiovasc Qual Outcomes 6:543–549. https://doi.org/10.1161/CIRCOUTCOMES.111.000303

15. Agyeman O et al (2006) Time to admission in acute ischemic stroke and transient ischemic attack. Stroke 37:963–966. https://doi.org/10.1161/01.STR.0000206546.76860.6b

16. Evenson KR, Foraker RE, Morris DL, Rosamond WD (2009) A comprehensive review of prehospital and in-hospital delay times in acute stroke care. Int J Stroke 4:187–199. https://doi.org/10.1111/j.1747-4949.2009.00276.x

17. Fassbender K, Balucani C, Walter S, Levine SR, Haass A, Grotta J (2013) Streamlining of prehospital stroke management: the golden hour. Lancet Neurol 12:585–596. https://doi.org/10.1016/S1474-4422(13)70100-5

18. Bouckaert M, Lemmens R, Thijs V (2009) Reducing prehospital delay in acute stroke. Nat Rev Neurol 5:477–483. https://doi.org/10.1038/nrneurol.2009.116

19. Kothari R, Barsan W, Brott T, Broderick J, Ashbrock S (1995) Frequency and accuracy of prehospital diagnosis of acute stroke. Stroke 26:937–941

20. Caceres JA et al (2013) Diagnosis of stroke by emergency medical dispatchers and its impact on the prehospital care of patients. J Stroke Cerebrovasc Dis 22:e610–e614. https://doi.org/10.1016/j.jstrokecerebrovasdis.2013.07.039

21. Buck BH et al (2009) Dispatcher recognition of stroke using the National Academy Medical Priority Dispatch System. Stroke 40:2027–2030. https://doi.org/10.1161/STROKEAHA.108.545574

22. Jones SP et al (2013) The identification of acute stroke: an analysis of emergency calls. Int J Stroke 8:408–412. https://doi.org/10.1111/j.1747-4949.2011.00749.x

23. Kim SK, Lee SY, Bae HJ, Lee YS, Kim SY, Kang MJ, Cha JK (2009) Pre-hospital notification reduced the door-to-needle time for iv t-PA in acute ischaemic stroke. Eur J Neurol 16:1331–1335. https://doi.org/10.1111/j.1468-1331.2009.02762.x

24. O'Brien W, Crimmins D, Donaldson W, Risti R, Clarke TA, Whyte S, Sturm J (2012) FASTER (face, arm, speech, time, emergency response): experience of central coast stroke services implementation of a pre-hospital notification system for expedient management of acute stroke. J Clin Neurosci 19:241–245. https://doi.org/10.1016/j.jocn.2011.06.009

25. Casolla B et al (2013) Intra-hospital delays in stroke patients treated with rt-PA: impact of preadmission notification. J Neurol 260:635–639. https://doi.org/10.1007/s00415-012-6693-1

26. Walter S et al (2012) Diagnosis and treatment of patients with stroke in a mobile stroke unit versus in hospital: a randomised controlled trial. Lancet Neurol 11:397–404. https://doi.org/10.1016/S1474-4422(12)70057-1

27. Fassbender K, Walter S, Liu Y, Muehlhauser F, Ragoschke A, Kuehl S, Mielke O (2003) "Mobile stroke unit" for hyperacute stroke treatment. Stroke 34:e44. https://doi.org/10.1161/01.STR.0000075573.22885.3B

28. Walter S et al (2011) Point-of-care laboratory halves door-to-therapy-decision time in acute stroke. Ann Neurol 69:581–586. https://doi.org/10.1002/ana.22355

29. Kettner M et al (2017) Prehospital computed tomography angiography in acute stroke management. Cerebrovasc Dis 44:338–343. https://doi.org/10.1159/000484097

30. Audebert H et al (2017) The PRE-hospital stroke treatment organization. Int J Stroke 12:932–940. https://doi.org/10.1177/1747493017729268

31. Fassbender K, Grotta JC, Walter S, Grunwald IQ, Ragoschke-Schumm A, Saver JL (2017) Mobile stroke units for prehospital thrombolysis, triage, and beyond: benefits and challenges. Lancet Neurol 16:227–237. https://doi.org/10.1016/S1474-4422(17)30008-X

32. Kostopoulos P et al (2012) Mobile stroke unit for diagnosis-based triage of persons with suspected stroke. Neurology 78:1849–1852. https://doi.org/10.1212/WNL.0b013e318258f773

33. Walter S et al (2010) Bringing the hospital to the patient: first treatment of stroke patients at the emergency site. PLoS One 5:e13758. https://doi.org/10.1371/journal.pone.0013758

34. Audebert HJ, Saver JL, Starkman S, Lees KR, Endres M (2013) Prehospital stroke care: new prospects for treatment and clinical research. Neurology 81:501–508. https://doi.org/10.1212/WNL.0b013e31829e0fdd

35. Lin CB et al (2012) Emergency medical service hospital prenotification is associated with improved evaluation and treatment of acute ischemic stroke. Circ Cardiovasc Qual Outcomes 5:514–522. https://doi.org/10.1161/CIRCOUTCOMES.112.965210

36. Gorelick PB (2013) Primary and comprehensive stroke centers: history, value and certification criteria. J Stroke 15:78–89. https://doi.org/10.5853/jos.2013.15.2.78

37. Alberts MJ et al (2000) Recommendations for the establishment of primary stroke centers. Brain Attack Coalition. JAMA 283:3102–3109

38. Alberts MJ et al (2005) Recommendations for comprehensive stroke centers: a consensus statement from the Brain Attack Coalition. Stroke 36:1597–1616. https://doi.org/10.1161/01.STR.0000170622.07210.b4

39. Gerschenfeld G, Muresan IP, Blanc R, Obadia M, Abrivard M, Piotin M, Alamowitch S (2017) Two paradigms for endovascular thrombectomy after intravenous thrombolysis for acute ischemic stroke. JAMA Neurol 74:549–556. https://doi.org/10.1001/jamaneurol.2016.5823

40. Badhiwala JH et al (2015) Endovascular thrombectomy for acute ischemic stroke: a meta-analysis. JAMA 314:1832–1843. https://doi.org/10.1001/jama.2015.13767

41. Zaidat OO et al (2013) Recommendations on angiographic revascularization grading standards for acute ischemic stroke: a consensus statement. Stroke 44:2650–2663. https://doi.org/10.1161/STROKEAHA.113.001972

42. National Institute of Neurological D, Stroke rt PASSG (1995) Tissue plasminogen activator for acute ischemic stroke. N Engl J Med 333:1581–1587. https://doi.org/10.1056/NEJM199512143332401

43. Hacke W et al (2008) Thrombolysis with alteplase 3 to 4.5 hours after acute ischemic

stroke. N Engl J Med 359:1317–1329. https://doi.org/10.1056/NEJMoa0804656

44. Lees KR et al (2016) Effects of alteplase for acute stroke on the distribution of functional outcomes: a pooled analysis of 9 trials. Stroke 47:2373–2379. https://doi.org/10.1161/STROKEAHA.116.013644

45. Furlan A et al (1999) Intra-arterial prourokinase for acute ischemic stroke. The PROACT II study: a randomized controlled trial. Prolyse in acute cerebral thromboembolism. JAMA 282:2003–2011

46. Investigators IMSS (2004) Combined intravenous and intra-arterial recanalization for acute ischemic stroke: the Interventional Management of Stroke Study. Stroke 35:904–911. https://doi.org/10.1161/01.STR.0000121641.77121.98

47. Investigators IIT (2007) The Interventional Management of Stroke (IMS) II Study. Stroke 38:2127–2135. https://doi.org/10.1161/STROKEAHA.107.483131

48. Broderick JP et al (2013) Endovascular therapy after intravenous t-PA versus t-PA alone for stroke. N Engl J Med 368:893–903. https://doi.org/10.1056/NEJMoa1214300

49. Saver JL et al (2016) Time to treatment with endovascular thrombectomy and outcomes from ischemic stroke: a meta-analysis. JAMA 316:1279–1288. https://doi.org/10.1001/jama.2016.13647

50. Nogueira RG et al (2018) Thrombectomy 6 to 24 hours after stroke with a mismatch between deficit and infarct. N Engl J Med 378:11–21. https://doi.org/10.1056/NEJMoa1706442

51. Albers GW, Marks MP, Lansberg MG (2018) Thrombectomy for stroke with selection by perfusion imaging. N Engl J Med 378:1849–1850. https://doi.org/10.1056/NEJMc1803856

52. Berkhemer OA et al (2015) A randomized trial of intraarterial treatment for acute ischemic stroke. N Engl J Med 372:11–20. https://doi.org/10.1056/NEJMoa1411587

53. Goyal M et al (2015) Randomized assessment of rapid endovascular treatment of ischemic stroke. N Engl J Med 372:1019–1030. https://doi.org/10.1056/NEJMoa1414905

54. Campbell BC et al (2015) Endovascular therapy for ischemic stroke with perfusion-imaging selection. N Engl J Med 372:1009–1018. https://doi.org/10.1056/NEJMoa1414792

55. Saver JL et al (2015) Stent-retriever thrombectomy after intravenous t-PA vs. t-PA alone

in stroke. N Engl J Med 372:2285–2295. https://doi.org/10.1056/NEJMoa1415061

56. Jovin TG et al (2015) Thrombectomy within 8 hours after symptom onset in ischemic stroke. N Engl J Med 372:2296–2306. https://doi.org/10.1056/NEJMoa1503780

57. Bracard S et al (2016) Mechanical thrombectomy after intravenous alteplase versus alteplase alone after stroke (THRACE): a randomised controlled trial. Lancet Neurol 15:1138–1147. https://doi.org/10.1016/S1474-4422(16)30177-6

58. Grunwald IQ et al (2016) First automated stroke imaging evaluation via electronic alberta stroke program early CT score in a Mobile Stroke Unit. Cerebrovasc Dis 42:332–338. https://doi.org/10.1159/000446861

59. Ashkenazi L, Toledano R, Novack V, EI E, Abu-Salamae I, Ifergane G (2015) Emergency department companions of stroke patients: implications on quality of care. Medicine 94:e520. https://doi.org/10.1097/MD.0000000000000520

60. Rumboldt Z, Huda W, All JW (2009) Review of portable CT with assessment of a dedicated head CT scanner. AJNR Am J Neuroradiol 30:1630–1636. https://doi.org/10.3174/ajnr.A1603

61. John S et al (2016) Brain imaging using mobile CT: current status and future prospects. J Neuroimaging 26:5–15. https://doi.org/10.1111/jon.12319

62. Albers GW et al (2018) Thrombectomy for stroke at 6 to 16 hours with selection by perfusion imaging. N Engl J Med 378:708–718. https://doi.org/10.1056/NEJMoa1713973

63. Hastrup S, Damgaard D, Johnsen SP, Andersen G (2016) Prehospital acute stroke severity scale to predict large artery occlusion: design and comparison with other scales. Stroke 47:1772–1776. https://doi.org/10.1161/STROKEAHA.115.012482

64. Katz BS, McMullan JT, Sucharew H, Adeoye O, Broderick JP (2015) Design and validation of a prehospital scale to predict stroke severity: Cincinnati Prehospital Stroke Severity Scale. Stroke 46:1508–1512. https://doi.org/10.1161/STROKEAHA.115.008804

65. Perez de la Ossa N et al (2014) Design and validation of a prehospital stroke scale to predict large arterial occlusion: the rapid arterial occlusion evaluation scale. Stroke 45:87–91. https://doi.org/10.1161/STROKEAHA.113.003071

66. Cerejo R et al (2015) A mobile stroke treatment unit for field triage of patients for intraarterial revascularization therapy. J Neuroimaging 25:940–945. https://doi.org/10.1111/jon.12276

67. John S et al (2016) Performance of CT angiography on a mobile stroke treatment unit: implications for triangle. J Neuroimaging 26:391–394. https://doi.org/10.1111/jon.12346

68. Barber PA, Demchuk AM, Zhang J, Buchan AM (2000) Validity and reliability of a quantitative computed tomography score in predicting outcome of hyperacute stroke before thrombolytic therapy. ASPECTS Study Group Alberta Stroke Programme Early CT Score. Lancet 355:1670–1674

69. Nagel S et al (2017) e-ASPECTS software is non-inferior to neuroradiologists in applying the ASPECT score to computed tomography scans of acute ischemic stroke patients. Int J Stroke 12:615–622. https://doi.org/10.1177/1747493016681020

70. Pepper EM, Parsons MW, Bateman GA, Levi CR (2006) CT perfusion source images improve identification of early ischaemic change in hyperacute stroke. J Clin Neurosci 13:199–205. https://doi.org/10.1016/j.jocn.2005.03.030

71. Zenlander R, von Euler M, Antovic J, Berglund A (2018) Point-of-care versus central laboratory testing of INR in acute stroke. Acta Neurol Scand 137:252–255. https://doi.org/10.1111/ane.12860

72. Harpaz D, Eltzov E, Seet RCS, Marks RS, Tok AIY (2017) Point-of-care-testing in acute stroke management: an unmet need ripe for technological harvest. Biosensors 7. https://doi.org/10.3390/bios7030030

73. Rizos T, Herweh C, Jenetzky E, Lichy C, Ringleb PA, Hacke W, Veltkamp R (2009) Point-of-care international normalized ratio testing accelerates thrombolysis in patients with acute ischemic stroke using oral anticoagulants. Stroke 40:3547–3551. https://doi.org/10.1161/STROKEAHA.109.562769

74. Rajan SS, Baraniuk S, Parker S, Wu TC, Bowry R, Grotta JC (2015) Implementing a mobile stroke unit program in the United States: why, how, and how much? JAMA Neurol 72:229–234. https://doi.org/10.1001/jamaneurol.2014.3618

75. Southerland AM, Brandler ES (2017) The cost-efficiency of mobile stroke units: where the rubber meets the road. Neurology 88:1300–1301. https://doi.org/10.1212/WNL.0000000000003833

76. Wu TC et al (2014) Prehospital utility of rapid stroke evaluation using in-ambulance telemedicine: a pilot feasibility study. Stroke 45:2342–2347. https://doi.org/10.1161/STROKEAHA.114.005193

77. Wu TC et al (2017) Telemedicine can replace the neurologist on a mobile stroke unit. Stroke 48:493–496. https://doi.org/10.1161/STROKEAHA.116.015363

78. Ramadan AR et al (2017) Agreement among stroke faculty and fellows in treating ischemic stroke patients with tissue-type plasminogen activator and thrombectomy. Stroke 48:222–224. https://doi.org/10.1161/STROKEAHA.116.015214

79. Bergrath S et al (2012) Feasibility of prehospital teleconsultation in acute stroke–a pilot study in clinical routine. PLoS One 7:e36796. https://doi.org/10.1371/journal.pone.0036796

80. Eadie L, Regan L, Mort A, Shannon H, Walker J, MacAden A, Wilson P (2015) Telestroke assessment on the move: prehospital streamlining of patient pathways. Stroke 46:e38–e40. https://doi.org/10.1161/STROKEAHA.114.007475

81. Liman TG, Winter B, Waldschmidt C, Zerbe N, Hufnagl P, Audebert HJ, Endres M (2012) Telestroke ambulances in prehospital stroke management: concept and pilot feasibility study. Stroke 43:2086–2090. https://doi.org/10.1161/STROKEAHA.112.657270

82. Van Hooff RJ et al (2013) Prehospital unassisted assessment of stroke severity using telemedicine: a feasibility study. Stroke 44:2907–2909. https://doi.org/10.1161/STROKEAHA.113.002079

83. Itrat A et al (2016) Telemedicine in prehospital stroke evaluation and thrombolysis: taking stroke treatment to the doorstep. JAMA Neurol 73:162–168. https://doi.org/10.1001/jamaneurol.2015.3849

84. Chapman Smith SN et al (2016) A low-cost, tablet-based option for prehospital neurologic assessment: the iTREAT Study. Neurology 87:19–26. https://doi.org/10.1212/WNL.0000000000002799

85. Sikka N, Margolis G (2005) Understanding diversity among prehospital care delivery systems around the world. Emerg Med Clin North Am 23:99–114. https://doi.org/10.1016/j.emc.2004.09.007

86. Dick WF (2003) Anglo-American vs. Franco-German emergency medical services system. Prehosp Disaster Med 18:29–35. discussion 35–27

87. Dietrich M et al (2014) Is prehospital treatment of acute stroke too expensive? An economic evaluation based on the first trial. Cerebrovasc Dis 38:457–463. https://doi.org/10.1159/000371427

88. Hov MR, Nome T, Zakariassen E, Russell D, Roislien J, Lossius HM, Lund CG (2015) Assessment of acute stroke cerebral CT examinations by anaesthesiologists. Acta Anaesthesiol Scand 59:1179–1186. https://doi.org/10.1111/aas.12542

89. Hov MR et al (2017) Pre-hospital ct diagnosis of subarachnoid hemorrhage. Scand J Trauma Resusc Emerg Med 25:21. https://doi.org/10.1186/s13049-017-0365-1

90. Hov MR et al (2018) Interpretation of brain CT scans in the field by critical care physicians in a mobile stroke unit. J Neuroimaging 28:106–111. https://doi.org/10.1111/jon.12458

91. Schwartz J, Dreyer RP, Murugiah K, Ranasinghe I (2016) Contemporary prehospital emergency medical services response times for suspected stroke in the United States. Prehosp Emerg Care 20:560–565. https://doi.org/10.3109/10903127.2016.1139219

92. Puolakka T, Strbian D, Harve H, Kuisma M, Lindsberg PJ (2016) Prehospital phase of the stroke chain of survival: a prospective observational study. J Am Heart Assoc 5. https://doi.org/10.1161/JAHA.115.002808

93. Nor AM et al (2005) The Recognition of Stroke in the Emergency Room (ROSIER) scale: development and validation of a stroke recognition instrument. Lancet Neurol 4:727–734. https://doi.org/10.1016/S1474-4422(05)70201-5

94. Brandler ES, Sharma M, Sinert RH, Levine SR (2014) Prehospital stroke scales in urban environments: a systematic review. Neurology 82:2241–2249. https://doi.org/10.1212/WNL.0000000000000523

95. Krebes S et al (2012) Development and validation of a dispatcher identification algorithm for stroke emergencies. Stroke 43:776–781. https://doi.org/10.1161/STROKEAHA.111.634980

96. Ebinger M et al (2014) Effect of the use of ambulance-based thrombolysis on time to thrombolysis in acute ischemic stroke: a randomized clinical trial. JAMA 311:1622–1631. https://doi.org/10.1001/jama.2014.2850

97. Bowry R et al (2015) Benefits of stroke treatment using a mobile stroke unit compared with standard management: The BEST-MSU study run-in phase. Stroke 46:3370–3374. https://doi.org/10.1161/STROKEAHA.115.011093

98. Taqui A et al (2017) Reduction in time to treatment in prehospital telemedicine evaluation and thrombolysis. Neurology 88:1305–1312. https://doi.org/10.1212/WNL.0000000000003786

99. Wahlgren N et al (2007) Thrombolysis with alteplase for acute ischaemic stroke in the Safe Implementation of Thrombolysis in Stroke-Monitoring Study (SITS-MOST): an observational study. Lancet 369:275–282. https://doi.org/10.1016/S0140-6736(07)60149-4

100. Scalea TM (2015) R Adams Cowley. Trauma 17:77–78

101. Saver JL et al (2010) The "golden hour" and acute brain ischemia: presenting features and lytic therapy in >30,000 patients arriving within 60 minutes of stroke onset. Stroke 41:1431–1439. https://doi.org/10.1161/STROKEAHA.110.583815

102. Wahlgren N et al (2008) Thrombolysis with alteplase 3-4.5 h after acute ischaemic stroke (SITS-ISTR): an observational study. Lancet 372:1303–1309. https://doi.org/10.1016/S0140-6736(08)61339-2

103. Meretoja A, Strbian D, Mustanoja S, Tatlisumak T, Lindsberg PJ, Kaste M (2012) Reducing in-hospital delay to 20 minutes in stroke thrombolysis. Neurology 79:306–313. https://doi.org/10.1212/WNL.0b013e31825d6011

104. Ebinger M et al (2015) Effects of golden hour thrombolysis: a Prehospital Acute Neurological Treatment and Optimization of Medical Care in Stroke (PHANTOM-S) substudy. JAMA Neurol 72:25–30. https://doi.org/10.1001/jamaneurol.2014.3188

105. Kim JT et al (2017) Treatment with tissue plasminogen activator in the golden hour and the shape of the 4.5-hour time-benefit curve in the National United States Get With The Guidelines-Stroke Population. Circulation 135:128–139. https://doi.org/10.1161/CIRCULATIONAHA.116.023336

106. Prabhakaran S et al (2011) Transfer delay is a major factor limiting the use of intra-arterial treatment in acute ischemic stroke. Stroke 42:1626–1630. https://doi.org/10.1161/STROKEAHA.110.609750

107. Grotta JC, Savitz SI, Persse D (2013) Stroke severity as well as time should determine stroke patient triage. Stroke 44:555–557. https://doi.org/10.1161/STROKEAHA.112.669721

108. Lima FO et al (2016) Field assessment stroke triage for emergency destination: a simple and accurate prehospital scale to detect large vessel occlusion strokes. Stroke 47:1997–2002. https://doi.org/10.1161/STROKEAHA. 116.013301

109. Nazliel B et al (2008) A brief prehospital stroke severity scale identifies ischemic stroke patients harboring persisting large arterial occlusions. Stroke 39:2264–2267. https:// doi.org/10.1161/STROKEAHA.107. 508127

110. Heldner MR et al (2016) Clinical prediction of large vessel occlusion in anterior circulation stroke: mission impossible? J Neurol 263:1633–1640. https://doi.org/10.1007/ s00415-016-8180-6

111. Turc G et al (2016) Clinical scales do not reliably identify acute ischemic stroke patients with large-artery occlusion. Stroke 47:1466–1472. https://doi.org/10.1161/ STROKEAHA.116.013144

112. Ebinger M et al (2012) PHANTOM-S: the prehospital acute neurological therapy and optimization of medical care in stroke patients – study. Int J Stroke 7:348–353. https://doi. org/10.1111/j.1747-4949.2011.00756.x

113. Schwindling L et al (2016) Prehospital imaging-based triage of head trauma with a Mobile Stroke Unit: first evidence and literature review. J Neuroimaging 26:489–493. https://doi.org/10.1111/jon.12355

114. Leira EC, Hess DC, Torner JC, Adams HP Jr (2008) Rural-urban differences in acute stroke management practices: a modifiable disparity. Arch Neurol 65:887–891. https:// doi.org/10.1001/archneur.65.7.887

115. Shuaib A, Khan K, Whittaker T, Amlani S, Crumley P (2010) Introduction of portable computed tomography scanners, in the treatment of acute stroke patients via telemedicine in remote communities. Int J Stroke 5:62–66. https://doi.org/10.1111/j.1747-4949. 2010.00408.x

116. Parker SA et al (2015) Establishing the first mobile stroke unit in the United States. Stroke 46:1384–1391. https://doi.org/10. 1161/STROKEAHA.114.007993

117. Weber JE et al (2013) Prehospital thrombolysis in acute stroke: results of the PHANTOM-S pilot study. Neurology 80:163–168. https://doi.org/10.1212/ WNL.0b013e31827b90e5

118. Kunz A et al (2016) Functional outcomes of pre-hospital thrombolysis in a mobile stroke treatment unit compared with conventional care: an observational registry study. Lancet Neurol 15:1035–1043. https://doi.org/10. 1016/S1474-4422(16)30129-6

119. Gyrd-Hansen D, Olsen KR, Bollweg K, Kronborg C, Ebinger M, Audebert HJ (2015) Cost-effectiveness estimate of prehospital thrombolysis: results of the PHANTOM-S study. Neurology 84:1090–1097. https://doi.org/10.1212/ WNL.0000000000001366

120. Ebinger M, Harmel P, Nolte CH, Grittner U, Siegerink B, Audebert HJ (2017) Berlin prehospital or usual delivery of acute stroke care – study protocol. Int J Stroke 12:653–658. https://doi.org/10.1177/ 1747493017700152

121. Gutierrez JM, Emery RJ, Parker SA, Jackson K, Grotta JC (2016) Radiation monitoring results from the first year of operation of a unique ambulance-based computed tomography unit for the improved diagnosis and treatment of stroke patients. Health Phys 110:S73–S80. https://doi.org/10.1097/ HP.0000000000000502

122. Dvorak F, Haberer I, Sitzer M, Foerch C (2009) Characterisation of the diagnostic window of serum glial fibrillary acidic protein for the differentiation of intracerebral haemorrhage and ischaemic stroke. Cerebrovasc Dis 27:37–41. https://doi.org/10.1159/ 000172632

123. Schlachetzki F et al (2012) Transcranial ultrasound from diagnosis to early stroke treatment: part 2: prehospital neurosonography in patients with acute stroke: the Regensburg stroke mobile project. Cerebrovasc Dis 33:262–271. https://doi.org/10.1159/ 000334667

124. Barlinn K et al (2014) Outcomes following sonothrombolysis in severe acute ischemic stroke: subgroup analysis of the CLOTBUST trial. Int J Stroke 9:1006–1010. https://doi. org/10.1111/ijs.12340

Chapter 19

Acute Clinical Intervention and Chronic Management of Cerebral Vascular Accident

D. M. R. Harker

Abstract

This chapter focuses on the clinical aspects of stroke, specifically the diagnostic and treatment aspects of stroke. Studies are currently aimed at both treatment and management but, most importantly, proper and timely diagnosis, as these greatly influence patient outcome and subsequent quality of life. Here, recent work is highlighted, and also the importance of timely and accurate diagnosis is discussed. Certainly, it is my hope that this chapter will inspire physicians, scientists, engineers, and patients themselves to together formulate more accurate diagnostic tests directed toward proper treatment, all aimed at more effectively and more efficiently saving lives.

   Key words Clinical diagnostics, Experimental stroke recovery, Ischemic stroke, Neuronal proteins, Biomarkers, Radiology, Imaging

1 Introduction

The disease commonly described as stroke has diverse and at times confusing vernacular in the medical realm. Terms including cerebrovascular accident (CVA), brain accident, transient ischemic attack (TIA), and apoplexy have all been used to describe the condition [1]. This jargon is further complicated with sub-descriptive terms such as ischemic or hemorrhagic describing the specific etiology as well as transient vs. persistent describing symptom morphology and progression [2]. Stroke has been understood as a cause of debilitating disease for more than 2400 years with Hippocrates recognizing CVA syndromes considerably before science had progressed to the level of accurately describing specific etiologies [2]. As the mysteries of the brain and the basics of how it functions were difficult to study it would not be until 1620 that Jacob Wepfer began to unravel the cause of cerebrovascular disease [3]. Through autopsy reports of patients who developed a sudden onset of what he described as "apoplexy," the popular term for unexplained paralysis at the time, retrospective anatomical studies

Philip V. Peplow et al. (eds.), *Stroke Biomarkers*, Neuromethods, vol. 147, https://doi.org/10.1007/978-1-4939-9682-7_19,

began to shed light on this complex medical quandary; his study of patients who had succumbed to this illness clarified that apoplexy was related to the blood supply of the brain, mainly the carotid and vertebral arteries. As medical sciences developed it became clear that not only hemorrhage, but also all forms of vascular occlusion produced similar disease presentation, which we now commonly describe as stroke [2, 3].

The constellation of symptoms seen in CVA is diverse and can mimic the presentation of any neurologic lesion ranging from weakness, difficulties with speech, changes in vision or even vestibular/cochlear derangements. For most of the time that formal medicine has been practiced, a CVA-related illness was considered a cause of permanent disability inevitably ending in death. Research directed at this significant cause of morbidity and mortality has now evolved the idea of what a stroke is. It is now considered a condition which is not only preventable and treatable but no longer needs to be an inevitable result of aging. With this shift in clinical opinion, appropriately driven by evidence-based research, there has been a greater focus on primary and secondary preventative approaches. As we have seen in the setting of chronic diseases the recognition of high-risk groups prior to onset of injury is necessary to improve patient outcomes [4]. A strong contribution toward this goal is the development of a targeted approach to lifestyle alteration, prophylactic intervention, and screening strategies. The litany of research has also targeted the development of care focused after injuries occur with development of treatment algorithms in the acute and chronic stages of brain injury [5, 6]. CVA patient load is estimated at 795,000 patients (approximately one CVA every 40 s) annually with 134,000 lethal outcomes. There are approximately 6.4 million survivors of CVA present in the USA as of 2012 which has more than doubled since 1993, making CVA the most prevalent form of adult disability [7]. In this patient pool approximately 77% of presentations are first time CVAs and the other 23% are recurring CVAs with an economic burden of approximately 36.5 billion dollars in the USA [8]. When evaluating any disease, it is important to understand not only the numbers but also the genetic bias of the disease in terms of racially and genetically driven factors. Statistical analyses have shown that Black as well as Hispanic and Latino populations are proportionally at increased risk, when compared to Caucasian patients, with Black patients presenting almost twice as often as their Caucasian counterparts [9]. Research has shown that this increased relative risk is even more drastic in younger populations allowing for experienced clinicians to include CVA syndromes in their differential diagnostic process, providing for better early detection and improved outcomes [10]. Another significant difference in patient presentation appears in the minor difference of gender representation prevailing in CVA populations. Women experience a slightly higher rate of gross presentation percentage of acute CVA syndrome then men and the extrapolation of this is important since female patients have a greater chance of presenting unconventionally with less distinct symptoms, leading to difficulties in diagnosis [4, 8]. Lastly another issue that warrants further evaluation is the contribution of risk factors such as high blood pressure, tobacco use, coagulopathies, and diastolic heart failure as well as possible genetic factors [7].

2 Pathology

As in all forms of medicine it is important to first understand the pathology of the disease in question. Categorization of acute CVAs is specifically important and in most cases must be performed immediately by clinical examination or non-contrast-enhanced CT (NECT) scan imaging. By grouping CVA disorders it allows for an understanding of the disease mechanics at work and directs physicians toward proper treatment. By recognizing the issue at hand; whether it be related to ischemia, hemorrhage, or vasoconstriction, skilled practitioners can extrapolate possible differential diagnoses that may not be initially apparent by contrasting disease with similar etiology. The initial and most important segregation in a patient presenting with CVA symptomology requires recognizing hemorrhage vs. ischemia. Hemorrhage disrupts downstream tissue blood supply as the vasculature involved is compromised allowing for intraluminal contents to flow into the extravascular space. This has many consequences including inflammatory responses as extravascular blood irritates surrounding tissues and as the extravasated fluid is confined within the tissue it invades, it will begin to exert pressure. This can be extraordinarily dangerous when occurring in a closed space such as the skull that is reinforced by bone and considered the canonical closed vault. The recognition of hemorrhage as the source of symptoms directs skilled clinicians toward the direction of processes that affect the endothelial cells of the vasculature. Endothelial cells are subject to many forms of damage, but common culprits involve hypertension, collagenous disease, arteriolar-venous malformations (AVM), malignancy, and trauma. Diagnostic evidence of hemorrhage separates patients from those that suffer an ischemic cause of CVA as initial treatment options will differ significantly due to intracranial hemorrhage treatment options' contemporary limitation. Ischemic strokes on the other hand are secondary to partial or complete intraluminal blockage of blood flow and as such there can be a further classification of ischemic strokes that focuses on the progression of symptoms in the first 24 h. Ischemic CVA has two very telling evolutions that give clinicians valuable information in deciding treatment options. In the first case there are ischemic strokes that show resolution or significant improvement of neurologic symptoms in the first 24 h and these are branded transient ischemic attacks (TIA). TIAs are the most beneficial outcome of ischemia for patients, as they suggest the body was able to correct the cause of ischemia by reperfusion, either through the affected vasculature or via collateral vasculature. Secondly CVAs which do not resolve in the short-term progress to eventual neuronal cell death and are designated appropriately as acute ischemic strokes (AIS). The ambiguity of TIA vs. AIS can be confusing, especially to patients and their families, as it can be unclear if neuronal cell death has occurred in the acute setting [10]. Additionally, just as diagnosis of acute hemorrhage can provide insight into potential causes of disease, finding characteristic signs of ischemic CVA can offer differentials toward the cause of neuronal damage such as atherosclerosis, hypoperfusion injury, constriction of the vasculature, cardiac valvular disease, atrial fibrillation, emboli, or thrombus. While having described the basic differences in hemorrhagic vs. ischemic CVA it is important to note that the frequency of ischemia is considerably greater at approximately 87% of acute CVAs while hemorrhagic causes of disease account for only the subsequent 13% of cases [8]. Due to the relative pervasiveness of ischemic disease in the realm of CVA we will focus on its treatment.

On paper, hemorrhage vs. ischemia may seem straight forward, but these two seemingly different pathologies present clinically similar; moreover, without experienced clinical evaluation and the aid of NECT scans, difficulties in accurate diagnosing arise [11]. In some cases, patients present classically with FAST symptoms which stands for (FACE: numbness, uneven smile, drooping, ARM and LEG: weakness, numbness, SPEECH: slurred, mute, TIME: call the ambulance fast), but other times indistinct symptomology can complicate the matter with generalized symptoms of altered mental status, lightheadedness, or variable changes in sensation [8]. A thorough physical examination is important in the diagnosis of acute CVA, but it cannot be the only significant factor. Stroke protocols related to acute treatment are intrinsically reliant on the initial presentation of symptoms and what this means for practitioners is that without proper documentation of a patient's initial presentation on scene, such as timing and severity, treatment options such as thrombolytics may not be reliably administered as they carry a risk that may not be ignored.

While TIA symptomology may be transient, research within the past 10 years has elucidated that neuronal damage and even cellular death may occur during these short periods of ischemia. A great diagnostic tool in these settings is diffusion weighted MRI scans. Diffusion weighted MRI can determine when acute cellular death is present at a much more sensitive degree than NECT scans and evidence of this is often seen after a patient suffers a TIA. This discovery led to a shift in the vernacular determining the use of AIS when neurologic symptoms persist with known tissue loss, while

TIA is preferred in settings of cellular dysfunction in the brain that is not contingent on a diffusion-weighted MRI showing contiguous brain tissue loss [12].

Neuronal tissue is quite sensitive to disturbances in oxygen supply and as such commands approximately 20% of oxygen that is transported in the blood and 15% of all blood leaving the heart [10]. In both cases, ischemia and hemorrhage, the interruption in oxygen and nutrient transport forces neuronal cells into a state of ATP deficiency but due to the mechanism of injury, lesions seen in ischemia tend to be continuous with the affected vasculature while hemorrhagic CVA leads to dissemination of blood into the intracranial space beneath the skull. This collection of fluid in an enclosed space leads to pressure injuries that can be distant from the vascular cause as tissue is forced against the rigid bones of the neurocranium. ATP is the energy currency of neuronal cells and to continue production of ATP in a hypoxemic state the cell is forced to undertake metabolic concessions which produce lactic acid and disrupt homeostasis. The decrease in pH caused by this lactic acidosis causes imbalances in electrolytes especially calcium. The accumulation of calcium allows unopposed interaction with multiple proteins and inactive enzymes in the cellular milieu activating them. Enzyme activation commences the cascade which inevitably ends in cellular damage and death. If blood flow does not return before membrane damage to the cell and organelles becomes irreversible, holes in the cellular membrane will push the cell toward permanent volume and protein loss and mitochondrial membrane damage will activate the apoptotic pathway via cytochrome C efflux into the cytoplasm of the neuron. Inherently a thought experiment about hypoxemic tissue injuries would make it clear that there will be regions where there is complete loss of blood supply. Just as there are areas that will have complete blood supply loss, some surrounding areas will be fed by collateral blood supply leading to a gradient of metabolic changes and differing cellular injury timelines. Areas that have complete blood supply failure will progress rapidly onward to death [6]. Adjacent areas that have either partial or inadequate collateral blood flow experience levels of ischemia that put the cells at risk for metabolic impairment but still provide enough oxygenation to stave off immediate cellular death. This region of the injury has been described as the "penumbra" and allows for cellular survival with the caveat that oxygenation be returned within an adequate timeline. Thus, the penumbra can be thought of as a zone of tissue that has particular interest when investigating treatment options of tissue resuscitation [13].

The homeostatic changes seen in acute ischemic neuronal injury include shifts in electrolyte levels considerably away from the cellular baseline [14]. These changes lead to a significant influx of reactive oxygen species (ROS) produced from mitochondria, lipoxygenase activation and degradation of cellular products. If

ROS generation is not controlled through normal cellular detoxification, such as antioxidants and scavenging mechanisms of reactive species, widespread damage to almost all aspects of cellular structure is possible [6, 14]. The most important structures commonly affected are the cellular and organelle membranes, whose integrity is integral to the survival of the cell. The cellular membrane allows the cell to create an isolated internal environment that is appropriate for cellular survival and if it is intact, the cell has an ability to recover from insults that disrupt normal homeostasis. Many organelle membranes house proteins and other cellular products that are held separate from the cytoplasm because their action would be devastating if allowed free reign inside the cell proper. Take the mitochondria for instance which houses elements such as cytochrome C that when released inevitably leads to programmed cell death. Lysosomes are the disposal system of the cell and contain various degradative enzymes able to catabolize cellular structures when left unchecked, which is why they are sequestered behind an organelle membrane. In addition to intracellular homeostatic changes and membrane disruption, neuronal ischemia causes dysfunction in a neuron's ability to release neurotransmitters, cellular receptor representation on external membranes and effective reuptake of neurotransmitters. The results of these changes are exemplified well through the main excitotoxic neurotransmitter glutamate. The main mechanism through which glutamate exerts its physiologic effect involves receptor mediated influx of calcium which in turn is a potent activator of enzymatic action [15]. In the setting of increased extracellular concentration of glutamate, as in neuronal cell death secondary to ischemic neuronal tissue injury, this influx of calcium can disadvantageously increase activation of enzymes which cleave intracellular proteins and lipids via hydrolytic action. This cellular apparatus is mediated through receptor binding interaction and as such is a possible avenue for research and treatment. Blockade of these receptors in an acute setting could provide treatment options aimed at prolonging the window of normal cellular repair mechanics prolonging the time until more definitive care can be administered [13].

As the mechanism for brain tissue injury is understood to be hypoxemic tissue environments leading to subsequent cellular dysfunction and cell death, injuries of this type, in both hemorrhagic and ischemic conditions, lead to diminished cerebral blood flow (CBF). Most of the investigation for effective treatments has understandably been focused on increasing CBF in hypoxic tissues [13]. The most successful intervention of this type as well as the most widely used is recombinant tissue plasminogen activator (r-tPA). Currently, standard of care therapy for ischemic CVA includes administration of r-tPA to patients who fall into the stringent risk assessment guidelines for administration, but evaluation of r-tPAs safe use, efficacy, and window for treatment is continuing as understanding of cellular death in ischemic disease and pharmacologic treatment in CVA is evolving. Due to the aforementioned risk assessment protocols for r-tPA administration [16], it is projected that a meager 3–5% of CVA patients meet criteria upon arrival to definitive care facilities. As such, research to safely extend these guidelines to include patients who arrive outside of the current acute treatment protocol is understandably ongoing [17]. An interesting way to address this issue is the use of stroke ambulances bringing definitive care to the field allowing for administration of r-tPA in a more timely fashion [13, 18].

3 Prehospitalization

The majority of acute CVA presentations will involve varying forms of hemiparesis, unilateral weakness or flaccid paralysis, with studies citing numbers as high as 75% of acute strokes occurring with this manifestation. CVA symptomology can evolve quickly leading to deficit exacerbation, alleviation, spontaneous resolution, or persistent impairment both in the setting prior to treatment and post therapy. In the case of chronic disability, which is the end result in approximately 25–50% of acute CVAs, patients require constant care and have significantly decreased quality of life [7]. With such diverse causes, presentations, prognoses, and consequences of acute CVA syndromes, it is paramount for treating physicians as well as those interested in research to be familiar with the diagnosis and treatment strategies of acute CVA syndromes. Resources that allow for definitive diagnostics and patient classification such as the National Institute of Health Stroke Scale (NIHSS) provide a framework for physicians to evaluate CVA patients objectively [19]. The importance of this is critical as treatment strategies differ based on the etiology of each individual patient. Due to the importance of timeline, context and patient parameters, it is imperative to separate patients into descriptive categories such as ischemic or hemorrhagic and transient or persistent prior to building a specific treatment plan [10]. The dichotomy of these patients is important not only in treatment but also expected outcomes. Currently the evaluation is secondary to acute evaluation by a specialist with physical examination, lab studies and imaging studies aiding the diagnosis. There is a popular saying in the emergency department (ED) which conveys the necessity for timely management, "Time is tissue" and there exists a no more striking example than the setting of acute stroke. Although there have been significant advances in the recognition of high-risk groups through screening, improvements of diagnostic workup and changes in the treatment of CVA; acute stroke syndromes continue to hover between the second and third most common cause of mortality with no foreseeable decline through to the year 2020 [20]. The fact that this disease is such a significant cause of chronic disability with physical, emotional and financial strain on patients, their families and national medical resources demands that an effort is made to better understand and address it through primary and secondary intervention tactics. A noteworthy improvement could be found through the reliable recognition of acute vs. subacute CVA presentation of patients initially seen by the public and secondarily by prehospital practitioners with an emphasis on the delineation of ischemic vs. hemorrhagic etiology. Advancements in the public teaching of basic clinical evaluation as well as enhancement of imaging and biological markers could expedite this process but each present caveats in the form of sensitivity, specificity, and, most importantly, availability.

The role of physiologic parameters in respect to cellular death was thought to be understood by pathologists broadly in practice for many years but it was not until 1858 that it was first cited by Virchow [21]. Neuronal cell death leads to decreased neuronal functionality and can occur in different forms, such as apoptosis, autophagy, pyroptosis, or oncosis. The nature of cellular death that predominates in an injured tissue is determined by the etiology of the cellular insult in combination with external and internal cellular mechanisms [6, 14, 22]. Specific types of neuronal cell death are understood to require transcription and translation of proteins integral to the process of recycling cellular products and has been acknowledged to occur as soon as 6 h after injury [23]. A more complete understanding of the forms of cellular injury that occur in CVA may be integral to developing contemporary treatments.

As the clinical opinion of acute CVA has shifted toward it being a treatable disease, it is tempered by the fact that current treatment options are severely limited by time windows and specific patient risk assessment profiles. Due to neuronal cells considerable vascular and metabolic demands they undergo progressive changes toward membrane and organelle degradation quickly [24]. With this is mind, intervention must be initiated as soon as possible and a focus on improving prognosis in this time dependent process. When looking at the treatment of CVA "onset to needle time" and "door to needle time" are important parameters for which we can develop standardizations of care around. "Onset to needle time" is a description of the onset of symptoms to the time that definitive treatment is administered which usually involves the patient passing from a public space to emergency medical services (EMS) and finally to an appropriate stroke facility. "Door to needle time" is a description of the delay in when a patient arrives at the hospital to when they receive treatment. As acute CVA syndrome of any etiology is a time-dependent emergency, the enhancement of patient disease recognition, transportation, and diagnostic evaluation are the most important parameters related to patient outcome after an insult to the brain has occurred. Each of these measurements emphasizes a different link in the chain that is the emergency medical response. Improvements in "onset to needle time" are often centered on the recognition of CVA symptoms by either the patient or the family or a bystander and are critical in the initiation of the treatment process [18]. This is because the longest delays in definitive treatment outside of arrival to the hospital usual take place in the patient's or family's acknowledgement that an acute emergency is taking place. The difficulty in this is reflected in how acute CVA presentation can be very diverse ranging from coma to slight weakness or even just mild changes in sensation. As this is an acute process, the recognition of the acute disease state must begin with the general public and studies have shown that this area is significantly lacking. This has necessitated educational initiatives such as the "facial droop, arm weakness, speech difficulty, time to call emergency services" (FAST) criteria promoted by the Cincinnati Stroke Scale and various other organizations [25, 26]. These resources hope to promote immediate EMS response by informing the public to allow them to be able to recognize critical symptoms [8]. It is doubtful that anyone who has ever been inside a hospital has not seen a poster demonstrating the FAST criteria, but more significant public education is still needed. The second area of possible enhancement in "onset to needle time" is the quick initiation of transport to the hospital, most appropriately via ambulance as CVA patients are classically unstable. This requires situational education, especially in high-risk patients as well as their family members, to encourage the patient to be seen in the emergency department (ED) even when symptoms are not particularly severe. Lastly the actual transport of the patient should be focused on immediate stabilization by trained EMS professionals with an emergent emphasis on minimizing delay of ambulance transport until arrival to definitive care. On the other hand, enhancing the "door to needle time" requires streamlining of the hospital services including a stroke alert from the transporting ambulance, immediate neurology consultation at arrival with initial intravenous (IV) line placement, prompt imaging with radiology consult, and lastly, the decision of most appropriate treatment with involvement of the patient or their family [18].

4 Initial Management Prehospitalization

A case by case understanding of initial patient presentation and symptom progression is critical in the acute management of CVA patients. Anecdotal evidence demonstrates improvement in patient care secondary to thorough bystander accounts and even in some situations corroborated by video evidence recorded on mobile phones. The importance of this correlation is made increasingly significant as CVA patients with neurologic symptoms can have severe difficulties with the interview process as they are not always able to provide their own history secondary to neurologic dysfunction [8]. As such, emergency medical services (EMS) should be advised that bystanders, whether they are family or not, should be either interviewed extensively or advised to provide an account of the onset of symptoms to the physician managing the patient. In the setting of acute CVA, time is the most important factor and subsequently on scene delay should be restricted to stabilization. An emphasis on timely transport is necessary leading to the logical conclusion that bystanders should be urged to provide a firsthand account upon arrival to the emergency department. EMS is a critical link in patient care during acute processes and evolution of stroke management should include promoting EMS providers to understand beyond initial presenting symptoms. EMS protocols facilitating transport to appropriate care facilities, sometimes meaning that the closest possible hospital may not be appropriate to provide the highest level of care thus negatively impacting patient outcomes. It is important to ensure that during the interview process the timeline of the patient's progression is thoroughly understood. Reporting when the patient's symptoms first began, such as a witnessed onset of facial droop and slurred speech when it is available, is the most critical factor in the patient's personal history. Delineating the onset of symptoms as opposed to when the patient was found to have symptoms is paramount. When there is confusion on these points it can be helpful to ask the family when the patient was last seen to be at their normal baseline and this timeline will be used as the onset of symptoms based on current stroke care guidelines. This distinction is true of patients found with symptoms after sleeping as well and can place patients outside of the therapeutic window for thrombolytic treatment [8].

In the modern era it is not outrageous to expect a streamlined and effective first response protocol when time sensitive disease processes are identified in the field. It is critical to identify ways to produce improvements in response time, in EMS stabilization as well as transport to an appropriate facility with pre-stroke alerts conveyed to the hospital accepting the patient. When analyzing these processes one of the largest obstacles to overcome is diversity of transport distances and local resources. It is a problem that fights broad standardization and as such protocols must be flexible to work in diverse environments. The "door to needle" time is highly reflective of hospital protocol efficiency and is dependent on many departments working together closely. At the time the patient arrives in the ED the hospital should have already acknowledged a pre-stroke alert and will have resources ready when the EMS team arrives with the patient. At the time the patient arrives, immediate evaluation by an emergency physician is done to ensure the patient's airways, breathing, and circulatory system are stable. Simultaneously, hospital staff should also strive to ensure the patient has at least two appropriately sized IV lines (sides of body check) and once the patient is thought to be stable by the ED physician a thorough neurologic examination should be performed by either the accepting neurologist, if present, or the treating ED physician. Also, drawing blood labs during this period is important for receiving results promptly. CVA presentation includes a variety of traditional and nontraditional symptoms and as such many different etiologies can mimic an acute CVA. Initial examination must include confirmation of possible acute CVA, excluding the mimics and looking for possible comorbidities. The National Institutes of Health Stroke Scale (NIHSS) is the suggested prognostic tool for evaluation of stroke severity during initial presentation. The NIHSS spans 11 categories on its 15-item neurologic evaluation but more importantly it is efficient to use taking less than 10 min to complete by a trained physician and most importantly it yields reproducible results correlating to the size of the suspected infarct [8, 27]. The American Heart Association/American Stroke Association (AHA/ASA) stroke guidelines hope to form "an organized protocol for the emergency evaluation of patients with suspected stroke" once the patient finds themselves being treated in an appropriate care facility [28]. Their goal is the standardization of care aimed at a decision toward proper treatment within the first 60 min of the patient's entry to the ED. An important hospital administration focus on the creation of a "stroke team" involving all necessary disciplines of necessary care including triage staff, the treating ED physician, on call neurologist, charge nurse, CT, and laboratory staff. The ability of the "stroke team" to quickly recognize and react to the presentation of acute CVA patients prioritizes decreased "door to needle" times regardless of the current state of ED resources.

5   Imaging

Imaging plays an important role in the treatment of acute stroke. Current research-based guidelines for patient's presenting with stroke symptoms advocate for immediate non-contrast-enhanced CT (NECT) scan for multiple reasons. Primarily NECT scans are necessary to rule out the presence of a hemorrhage as well as other pathologic processes which may mimic the symptoms of a CVA [29]. Secondarily a NECT scan is required as part of the initial workup of patients who may be treated with thrombolytics such as r-tPA. While NECT scans are increasingly helpful in the management of acute CVA patients they often are not enough to diagnose by themselves as acute strokes may not have developed to a sufficiently large extent for detection. Importantly though, NECT scans can increase a practitioner's confidence that a patient is eligible for thrombolytics when deciding whether to administer r-tPA. As radiologic findings on NECT scans can be subtle, experienced radiologists are key in the treatment of acute CVA and telemedicine allows for instantaneous CT evaluation no matter the on-site resources of the treating facility. While NECT scans are the gold standard for initial investigation of acute CVA symptoms, studies have shown that diffusion-weighted MRI is superior to NECT cans in the diagnosis of acute ischemia. Unfortunately, diffusion-weighted MRI has not been implemented as a devoted resource in the management of acute stroke symptomology due to decreased accuracy in picking up acute hemorrhage as well as significant hurdles involving cost, lack of rapid availability, length of time needed for imaging, and difficulty in administration (patient comfort, MRI technicians, claustrophobia, metal implants). Expanding protocols to include MRI utilization is a possibility as stroke protocol continues to evolve and necessitating research toward overcoming the inherent difficulties of MRI. As the field of medical imaging has become significantly important in the treatment of stroke, other imaging options are being explored. Vascular imaging has been implied as an appropriate choice when endovascular therapy options are considered, and perfusion studies may allow for earlier and greater precision when mapping areas of ischemia, but it is important to note that stroke treatment is time dependent and implementation of new techniques should never prolong administration of r-tPA to suitable patients.

6   Stroke Biomarkers in Biofluids

Although imaging offers some insight and diagnostic value, it is important to note the drawbacks of these imaging techniques. For example, diagnostically, CTs are only a meager 30% sensitive when it comes to the diagnosis of cerebral ischemia. Additionally, the cost, time and difficulties in timely scheduling, do not make MRI any more attractive, in regards to diagnostics tools. Stroke biomarkers have recently become a utile tool that is rapid, sensitive, and offers accurate diagnostic value [30, 31]. Here, some important novel metabolomic biomarkers are highlighted which have found their utility in the field of stroke diagnostics; medically, the aim would be to identify biomarkers which enable the distinction between hemorrhagic and ischemic stroke. NR2 peptide, for example, offers diagnostic potential in the setting of acute cerebrovascular accidents as a biomarker for cerebral ischemia. Studies examining this potential have shown a direct correlation between the concentration of NR2 peptide and the size of the ischemia zone [31].

In 2015, Lui et al., observed two metabolites which seem to decrease substantially in cognitively impaired stroke patients, namely 3-indolepropionic acid and stearoyl-carnitine [32]. Lysophosphocholine, another clinically beneficial biomarker, has been shown to by itself increase the sensitivity of the ABCD2 score—which is a clinical test used outside of a hospital setting for people at high risk of stroke after a transient ischemic attack [30, 33]. In addition, there appears to be a correlation between elevated levels of lactate, pyruvate, and formate, and concomitant decreases in VLDL, LDL CH3, valine as well as 4-hydroxymethyl acetate in stroke patients when assessing plasma levels [34]. Lastly, and just as importantly, microRNAs (miRNAs) are highly beneficial diagnostic markers of hyperacute cerebral infarction (HACI); the expression of miR-16, for example, has lent itself to being a powerful biomarker for not just diagnosis but also stratification, and prognosis of hyperacute cerebral infarction [35].

7 General Treatment of Acute Ischemic CVA

Current generalized treatment that is recommended in the presentation of acute CVA involves consensus-based guidelines agreed upon by specialists with many years of experience. Generalized treatment can be thought of as the absolute basics necessity after the patient has been stabilized and attempts to correct homeostasis by targeting dehydration, hypoxia, hypertension, and blood glucose abnormalities. Each of these conditions has been shown to contribute toward worsened outcomes, which has led to the consensus-based guidelines that have been implemented in acute stroke care. Dehydration has been shown to complicate patients presenting with acute CVA by increasing blood viscosity, increasing incidence of DVTs, decreasing vascular perfusion, and renal dysfunction. The rectification of dehydration involved infusion of isotonic fluids and has only been shown to be effective in hypovolemic states and should not be employed in euvolemic states except when a patient's blood has been found to be hyperviscous. Oxygen transportation dysfunction is characteristic of ischemia, and as such maintaining red blood cell oxygen saturation above 94% is an obvious goal, and research on CVA mortality has shown improved outcomes with the correction of hypoxia. The treatment of hypertension in the setting of acute CVA is slightly more nuanced depending on whether the patient is a candidate for r-tPA treatment. If the patient does not meet the criteria for r-tPA, then the patient's hypertension can persist as long as their blood pressure does not rise above 220 mmHg systolic or 120 mmHg diastolic. Dramatically lowering a patient's blood pressure is almost never done because of the risk of stroke. Hypertension is a chronic condition, and the body becomes accustomed to the higher-pressure state; therefore, in the setting of stroke treating a hypertensive patient's blood pressure may cause the penumbra to grow. When a patient is being considered for r-tPA therapy there are very strict guidelines surrounding the patient's blood pressure. Often the goal is to decrease the blood pressure with either labetalol or nicardipine in a controlled and gradual way prior to r-tPA treatment [36]. Blood glucose has a direct effect on neurologic dysfunction as neurons are dependent on glucose for energy. Given the brain's requirement of glucose, it is not surprising that stress-induced hyperglycemia is commonly seen in acute stroke. Current consensus-based guidelines suggest that blood sugar level should be held between 140 and 180 mg/dl with an emphasis on avoiding hypoglycemic states. The idea being that glucose-derived energy production is necessary to prolong a neurons chance at recovering after a period of hypoxia is resolved [37].

8 Treatments

When discussing treatments in the area of acute ischemic CVA management it is important to remember that patients will present in a wide array of states. As in the management of any other pathology if the patient is unable to swallow due to weakness or unconsciousness, oral medications should be withheld. Current guidelines recommend that the patient be treated with 325 mg of Aspirin within 24–48 h of onset. The only time that this should not be adhered to is when the patient is treated with r-tPA as the antiplatelet therapy should not be combined with thrombolytics. Studies have shown that the use of Aspirin specifically reduces the reoccurrence of ischemia in the days after an acute stroke [38–40]. While Aspirin should be used ubiquitously on all patients that are not receiving r-tPA or have a history of sensitivity, administration of thrombolytics, and specifically r-tPA therapy, is much more tightly scrutinized. Most commonly, a neurologist will make the call if a patient is deemed appropriate for r-tPA therapy by reviewing the patient's history against a "risk assessment score" or list of relative and absolute contraindications. First the patient's age, clinical presentation, and the time of symptom onset are reviewed, followed by absolute contraindications such as current or previous intracranial hemorrhage, recent neurosurgical procedure, uncontrolled hypertension, presence of any internal bleeding, diagnosed bleeding diathesis, hypoglycemia, history of arteriovenous malformation, cancer, or aneurysm. Lastly the prescribing physician must consider the soft or relative contraindications. These considerations do not exclude patients from being treated but may help guide the physician with the help of previous experience of patients in similar categories. There are numerous considerations here but one of particular note is when a patient has only minor symptoms or is swiftly improving, there should be significant deliberation before administering a drug such as r-tPA as the severity of complications ranges from worsening CVA secondary to intracerebral hemorrhage, hemorrhages elsewhere in the body, significant angioedema of the face and tongue, often requiring intubation, and even death. Given the possibility for unacceptable consequences when r-tPA is administered inappropriately it is worth noting that with the aid of risk assessment scores it is estimated that only 2–10% of patients who present with ischemic symptomology are ruled into treatment [41, 42]. Lastly, when dealing with a medication such as r-tPA, it is critical to understand that even some patients who are deemed acceptable for treatment may experience unexpected bleeding and what to do when this happens. Current guidelines recommend discontinuation of r-tPA with reversal using a coagulation therapy as hemorrhage may be fatal. Lastly after a patient has been administered r-tPA critical observation by an experienced physician is beneficial and has led to the practice of "drip and ship" stroke protocols where patients may be treated with r-tPA at an initial facility but shortly after will be transferred to a specialty stroke center for further management. Aspirin and r-tPA may be medications that we consider synonymous with treatment of acute CVA, but the field is evolving and treatments that were once considered unconventional are being studied more rigorously. Some promising areas of expansion involve treatments such as therapeutic hypothermia which has shown promising ability to slow tissue death in patients who have undergone myocardial infarction. Inducing hypertension which is theorized to increase the chances of reperfusion in ischemic tissue is another treatment that has expounded out of the common practice of allowing hypertensive states to persist in patients who do not meet criteria for r-TPA treatment. Lastly more invasive endovascular therapies that could provide cleaning of the occlusion or stenting of collapsed vascular pathways and even hemicraniectomy to allow a release from rising intracranial pressure that can be seen in CVA are surgical options that are possible strategies that show promise as treatments progress. With studies focused on such specialized treatments, the evolution of stroke management could move into an even more concentrated realm necessitating the use of "drip and ship" protocol allowing for patients' best outcomes.

9   Disposition

It is general consensus that all acute CVA patients need to be admitted for observation and management to a certified stroke unit. It has been shown in practice that specialized stroke units provide patients with enhanced care leading to reductions in length of stay at the hospital and a better chance of returning home after treatment even inpatient populations who are not treated with thrombolytics. When the patient requires specialized care even after the post-acute stroke, units have the experience to facilitate transfer to appropriate care facilities delineating specific medical requirements leading to increased rates of neurologic deficit recovery.

10 Heparin and Warfarin/Intermittent Pneumatic Compression/Elastic Stockings Are Effective in DVT Prophylaxis

Once initial stabilization and management of the patient has occurred, acute CVA patients have varying paths to recovery depending on the severity of the initial infarct and progression of symptoms [7]. The most striking facet about acute ischemic stroke is that the median length of stay in hospital is 4 days. This transition may seem truncated as stroke can be a significant life event but as the management of CVA patients relies on a multidisciplinary approach early transfer to specialized facilities to treat stroke patients can often be the best course of action. As the invention of stroke units in hospitals has promoted better treatment so has the understanding that transferring patients to a setting where all their needs can be met simultaneously has shown to be beneficial in overall outcome. Rehabilitation of stroke patients depends primarily on the severity of their deficits as well as their personal support structure necessitating a variety of different types of care facilities. The range varies from subacute care specialties such as inpatient facilities that concentrate on intensive care which allow for hospital level management under the direction of a treating physician while for other patients it may be most appropriate for them to undergo rehabilitation in skilled nursing facilities or even at their own homes with personalized treatment professionals visiting them where they are most comfortable [43]. Some patients will spontaneously recover during their hospitalization while unfortunately others may retain persistent deficits. The most commonly seen issues involve muscular weakness, difficulties with producing or understanding speech, inability to feed themselves, difficulties swallowing normally, changes in cognitive ability, vision changes or even psychiatric disorders brought on by neurologic dysfunction or the understanding of their deficit. Primarily, the goal of post-acute care should involve developing a treatment plan that will ultimately integrate the patient back into their community with the highest quality of life and autonomy possible. Part of this process is to allow for family members to be as involved as much as the patient would like. This is crucial as rehabilitation often requires patient and family efforts as both physical and financial needs can be enormous. There is a strong correlation between outcomes and family support availed by patients, showing that those who have a strong support structure often having better outcomes [7].

Once the patient's condition is determined to be stable and appropriate treatment has been administered, rehabilitation should be the next immediate goal. Concurrent care should be motivated toward stopping any further complications that could arise in the post-acute setting, with an emphasis on avoiding a recurrent ischemic or hemorrhagic event and secondarily encouraging the patient to be as active as possible; specifically, in the setting of personal care including activities such as grooming, feeding, etc. In the setting of severe deficits, patients should be put through a range of motion exercises with changes in resting position as soon as tolerable, progressing toward personal care activities. The worth of immediate "stroke exercise" movement-based therapies has been well documented and exhibits an important concept. Weakness is the most common quality-of-life deficit reported by stroke survivors and using movement specific measure for deficit improvement has been effective. In a similar light, etiologic specific measures can be implemented when dealing with known diagnoses. Current clinical guidelines support this citing suggested carotid endarterectomy for patients who have majority occlusions of the carotid arteries that measure between 70% and 99% [44]. Other great examples include mandatory anticoagulant prescriptions for patients with a history of atrial fibrillation and standard antiplatelet therapy with Aspirin or Clopidogrel after patients suffer a TIA [45–47]. These prophylactic interventions are important in the discussion of acute CVA syndrome as patients with a history of each of these conditions are at significant risk. There is also significant risk for patients in the post-acute and rehabilitation stages of recovery as patients with deficits often have weakness, ataxia, or depression, leading to an increasingly sedentary lifestyle. On a case-by-case basis, anticoagulation drugs like heparin can be used to prevent deep vein thrombosis. In addition, swallowing studies are able to investigate when patients are able to feed normally and dissuade feeding tube nutrition, educated nursing staffs can prevent unnecessary skin breakdown or pressure ulcers through careful observation, fall prevention measures in patients who are unsteady, antidepressant and therapy protocols in patients facing depression, and management of bowel and bladder function as catheters have complications as well. Many of these complications can continue on in the patient's life after return to home; moreover, education of both the family and patient is a cornerstone of successful integration back into their baseline [7].

The goal of all hospital and post-acute facility measures should be care directed at returning the patient home as soon as feasible. Discharge either to home or another similar community care setting is promoted for all patients who have recovered to a point where adequate function is achievable. As each patient transitions to their appropriate requested environment, continual effort and active participation toward their personal treatment goals are required. This process allows for patients to feel greater participation in their care, improving compliance as well as evidence of improved outcomes in patients who are able to undergo early supported discharge (ESD) treatment plans [43].

Organizing a safe and effective strategy for discharge should be among the actions taken on the first day of admission. This allows the patient and the family to plan for the extensive rehabilitation process which may consume significant time and resources. Effective organization of the discharge process should be coordinated by a single patient care advocate to allow for planning with the appropriate specialists who will be involved [7]. Stroke units have become very popular not only for experienced, dedicated care but also having support staff such as patient care advocates who are intimately familiar with the rehabilitation process moving forward from a hospital setting.

11 Conclusions

It is clear we are just beginning to understand the subtle distinctions between the varying degrees of presentation in stroke. Studies are currently aimed at both treatment and management, but most importantly, proper and timely diagnosis as these greatly influence patient outcome and subsequent quality of life. Here, recent work is highlighted, and also the importance of timely and accurate diagnosis is discussed. Certainly, it is my hope that this chapter will inspire physicians, scientists, engineers, and patients themselves to together formulate more accurate diagnostic tests directed toward proper treatment, all aimed at more effectively and more efficiently saving lives.

References

1. Coupland AP, Thapar A, Qureshi MI, Jenkins H, Davieset AH (2017) The definition of stroke. J R Soc Med 110(1):9–12
2. Adams HP, Biller J (2015) Classification of subtypes of ischemic stroke: history of the trial of org 10172 in acute stroke treatment classification. Stroke 46(5):e114–e117
3. Paciaroni M, Bogousslavsky J (2008) The history of stroke and cerebrovascular disease. Handb Clin Neurol 92:3–28
4. Berglund A, Schenck-Gustafsson K, von Euler M (2017) Sex differences in the presentation of stroke. Maturitas 99:47–50
5. Swain S, Turner C, Tyrrell P (2008) Diagnosis and initial management of acute stroke and transient ischaemic attack: summary of NICE guidance. BMJ 337:a786
6. Ramon R, Fernandez-Gajardo R, Gutierrez R, Matamala JM, Carrasco R, Miranda-Merchak-A, Feuerhake W (2013) Oxidative stress and pathophysiology of ischemic stroke: novel therapeutic opportunities. CNS Neurol Disord Drug Targets 12(5):698–714
7. Ovbiagele B, Goldstein LB, Higashida RT, Howard VJ, Johnston SC, Khavjou OA, Lackland DT, Lichtman JH, Mohl S, Sacco RL, Saver JL, Trogdon JG (2013) Forecasting the future of stroke in the United States. Stroke 44(8):2361–2375
8. Judith E, Tintinalli JSS, John Ma O, Donald Yealy GD, Meckler DM (2015) Cline Tintinalli's emergency medicine: a comprehensive study guide, 8th edn. McGraw-Hill Education, New York, NY

9. Noonan AS, Velasco-Mondragon HE, Wagner FA (2016) Improving the health of African Americans in the USA: an overdue opportunity for social justice. Public Health Rev 37(1):12

10. Pare JR, Kahn JH (2012) Basic neuroanatomy and stroke syndromes. Emerg Med Clin North Am 30(3):601–615

11. Chung-Fen T, Thomas B, Sudlow CLM (2013) Epidemiology of stroke and its subtypes in Chinese vs. white populations. Neurology 81(3):264

12. Easton JD, Saver JL, Albers GW, Alberts MJ, Chaturvedi S, Feldmann E, Hatsukami TS, Higashida RT, Johnston SC, Kidwell CS, Lutsep HL, Miller E, Sacco RL (2009) Definition and evaluation of transient ischemic attack. Stroke 40(6):2276–2293

13. Xing C, Arai K, Lo EH, Hommel M (2012) Pathophysiologic cascades in ischemic stroke. Int J Stroke 7(5):378–385

14. Djordje R, Katsiki N, Resanovic I, Jovanovic A, Sudar-Milovanovic E, Zafirovic S, Mousa S, Isenovic ER (2017) Apoptosis and acute brain ischemia in ischemic stroke. Curr Vasc Pharmacol 15(2):115–122

15. Reiner A, Levitz J (2018) Glutamatergic signaling in the central nervous system: ionotropic and metabotropic receptors in concert. Neuron 98(6):1080–1098

16. Jauch EC, Saver JL, Adams HP Jr, Bruno A, Connors JJ, Demaerschalk BM, Khatri P, McMullan PW Jr, Qureshi AI, Rosenfield K, Scott PA, Summers DR, Wang DZ, Wintermark M, Yonas H (2013) American Heart Association Stroke Council; Council on Cardiovascular Nursing; Council on Peripheral Vascular Disease; Council on Clinical Cardiology. Guidelines for the early management of patients with acute ischemic stroke. Stroke J 44(3):870–947

17. Roth JM (2011) Recombinant tissue plasminogen activator for the treatment of acute ischemic stroke. Proc (Bayl Univ Med Cent) 24(3):257–259

18. Audebert HJ, Saver JL, Starkman S, Lees KR, Endres M (2013) Prehospital stroke care: new prospects for treatment and clinical research. Neurology 81(5):501–508

19. Chen SY, Liu JW, Wang YH, Huang JY, Chen SC, Yang SF, Wang PH (2019) The conditions under which piracetam is used and the factors that can improve National Institute of Health Stroke Scale Score in ischemic stroke patients and the importance of previously unnoticed factors from a hospital-based observational study in Taiwan. J Clin Med 8(1):122

20. Smajlović D (2015) Strokes in young adults: epidemiology and prevention. Vasc Health Risk Manag 11:157–164

21. Maynard S, Keijzers G, Gram M, Desler C, Bendix L, Jørgensen EB, Molbo D, Croteau DL, Osler M, Stevnsner T, Rasmussen LJ, Dela F, Avlund K, Bohr VA (2013) Relationships between human vitality and mitochondrial respiratory parameters, reactive oxygen species production and dNTP levels in peripheral blood mononuclear cells. Aging 5(11):850–864

22. Kroemer G, El-Deiry WS, Golstein P, Peter ME, Vaux D, Vandenabeele P, Zhivotovsky B, Blagosklonny MV, Malorni W, Knight RA, Piacentini M, Nagata S, Melino G (2018) Molecular mechanisms of cell death: recommendations of the Nomenclature Committee on Cell Death. Cell Death Differ 25(3):486–541

23. Lipton P (1999) Ischemic cell death in brain neurons. Physiol Rev 79(4):1431–1568

24. Lee SW, de Rivero JP, Vaccari JS, Truettner W, Dietrich D, Keane RW (2019) The role of microglial inflammasome activation in pyroptotic cell death following penetrating traumatic brain injury. J Neuroinflammation 16(1):27

25. English SW, Rabinstein AA, Mandrekar J, Klaas JP (2018) Rethinking prehospital stroke notification: assessing utility of emergency medical services impression and cincinnati prehospital stroke scale. J Stroke Cerebrovasc Dis 27(4):919–925

26. Brandler ES, Sharma M, Sinert RH, Levine SR (2014) Prehospital stroke scales in urban environments. A systematic review. Neurology 82(24):2241–2249

27. Kwah LK, Diong J (2014) National Institutes of Health Stroke Scale (NIHSS). J Physiother 60(1):61

28. Smith EE, Saver JL, Alexander DN, Furie KL, Hopkins LN, Katzan IL, Mackey JS, Miller EL, Schwamm LH, Williams LS (2014) AHA/ASA Stroke Performance Oversight Committee. Clinical performance measures for adults hospitalized with acute ischemic stroke. Stroke 45(11):3472–3498

29. Aguilar MI, Brott TG (2011) Update in intracerebral hemorrhage. Neurohospitalist 1(3):148–159

30. Jové M, Mauri-Capdevila G, Suárez I, Cambray I, Sanahuja J, Quílez A, Farré J, Benabdelhak I, Pamplona R, Portero-Otín M, Purroy F (2015) Metabolomics predicts stroke recurrence after transient ischemic attack. Neurology 84(1):36–45

31. Dambinova SA, Aliev KT, Bondarenko EV, Ponomarev GV, Skoromets AA, Skoromets AP, Skoromets TA, Smolko DG, Shumilina MV (2019) Biomarkers for cerebral ischemia as a novel method for validating the efficacy of neurocytoprotectors. Neurosci Behav Physiol 49:142–146

32. Liu M, Zhou K, Li H, Dong X, Tan G, Chai Y, Wang W, Bi X (2015) Potential of serum metabolites for diagnosing post-stroke cognitive impairment. Mol BioSyst 11(12):3287–3296

33. Dutta D, Bailey S-J (2016) Validation of ABCD2 scores ascertained by referring clinicians: a retrospective transient ischaemic attack clinic cohort study. Emerg Med J 33 (8):543–547

34. Qureshi MI, Vorkas PA, Coupland AP, Jenkins IH, Holmes E, Davies AH (2017) Lessons from metabonomics on the neurobiology of stroke. Neuroscientist 23(4):374–382

35. Tian C, Li Z, Yang Z, Huang Q, Liu J, Hong B (2016) Plasma microRNA-16 is a biomarker for diagnosis, stratification, and prognosis of hyperacute cerebral infarction. PLoS One 11: e0166688. https://doi.org/10.1371/journal. pone.0166688

36. Hecht JP, Richards PG (2018) Continuous-infusion labetalol vs. nicardipine for hypertension management in stroke patients. J Stroke Cerebrovasc Dis 27(2):460–465

37. Lindsberg PJ, Roine RO (2004) Hyperglycemia in acute stroke. Stroke 35(2):363–364

38. National Institute of Neurological Disorders and Stroke rt-PA Stroke Study Group (1995) Tissue plasminogen activator for acute ischemic stroke. N Engl J Med 333(24):1581–1588

39. Marler JR (2007) NINDS clinical trials in stroke. Stroke 38(12):3302–3307

40. Wardlaw JM, Murray V, Berge E et al (2012) Recombinant tissue plasminogen activator for acute ischaemic stroke: an updated systematic review and meta-analysis. Lancet (London) 379(9834):2364–2372

41. Miller DJ, Simpson JR, Silver B (2011) Safety of thrombolysis in acute ischemic stroke: a review of complications, risk factors, and newer technologies. Neurohospitalist 1 (3):138–147

42. Lahr MHH, van der Zee DJ, Vroomen P, Luijckx GJ, Buskens E (2012) Proportion of patients treated with thrombolysis in a centralized versus a decentralized acute stroke care setting. PLoS One 43(5):1336–1340

43. Winstein CJ, Stein J, Arena R, Bates B, Cherney LR, Cramer SC, Deruyter F, Eng JJ, Fisher B, Harvey RL, Lang CE, MacKay-Lyons M, Ottenbacher KJ, Pugh S, Reeves MJ, Richards LG, Stiers W, Zorowitz RD (2016) Guidelines for adult stroke rehabilitation and recovery. Stroke 47(6):e98–e169

44. Whiten C, Gunning P (2009) Carotid endarterectomy: Intraoperative monitoring of cerebral perfusion. Curr Anaesth Crit Care 20 (1):42–45

45. Pan Y, Jing J, Chen W, Meng X, Li H, Zhao X, Liu L, Wang D, Johnston SC, Wang Y, CHANCE Investigators (2017) Risks and benefits of clopidogrel–aspirin in minor stroke or TIA. Time course analysis of CHANCE. Neurology 88(20):1906–1911

46. Wang Y, Wang Y, Zhao X, Liu L, Wang D, Wang C, Wang C, Li H, Meng X, Cui L, Jia J, Dong Q, Xu A, Zeng J, Li Y, Wang Z, Xia H, Johnston SC, CHANCE Investigators (2013) Clopidogrel with aspirin in acute minor stroke or transient ischemic attack. N Engl J Med 369 (1):11–19

47. Wang X, Zhao X, Johnston SC, Xian Y, Hu B, Wang C, Wang D, Liu L, Li H, Fang J, Meng X, Wang A, Wang Y, Wang Y (2015) Effect of clopidogrel with aspirin on functional outcome in TIA or minor stroke. CHANCE substudy. Neurology 85(7):573–579

Part V

Conclusion

Chapter 20

Trends in Biomarkers Development for Stroke

Philip V. Peplow, Bridget Martinez, and Svetlana A. Dambinova

Abstract

Recent progress in biomarker research, neuromethods development, and clinical practice has contributed to advances in stroke diagnoses and treatment optimization. At present, no "perfect" biomarker(s) exists for stroke. A "perfect" biomarker could assist in risk assessment, predict and distinguish the stroke from mimicking conditions, and measure disease progression. Neurofunctional biomarkers, structural imaging biomarkers, and molecular biologic markers all have the potential to revolutionize the screening, diagnosis, prognosis, and prediction of disease recurrence as well as therapeutic monitoring of stroke and transient ischemic attack (TIA).

Key words Biomarkers development, Stroke, Transient ischemic attack, Neurofunctional biomarkers, Structural biomarkers, Molecular biomarkers, Neuromethods

1 Introduction

A biomarker is defined as "any substance, structure, or process that can be measured in the body or its products that influences or predicts the incidence or outcome of disease" [1, 2]. This includes disease risk and diagnosis of acute and chronic conditions in cardiovascular and cerebrovascular circulation such as heart attack and stroke.

Stroke, also known as cerebrovascular accident (CVA), is now characterized as a "brain attack," which is both the most descriptive and realistic image of stroke. As with a heart attack, the appropriate response to a brain attack is emergency action, both by the person it strikes and the medical community. Delays in diagnosis and medical intervention beyond 4–6 h of stroke onset may contribute to clinical deterioration and disability.

As the population ages, stroke has the potential to become an even larger medical and economic burden on the healthcare system. The public misperception that nothing can be done about stroke has prevailed for too long. Recent progress in biomarker research, neuromethods, and clinical practice has contributed to advances in

Philip V. Peplow et al. (eds.), *Stroke Biomarkers*, Neuromethods, vol. 147, https://doi.org/10.1007/978-1-4939-9682-7_20,
© Springer Science+Business Media, LLC, part of Springer Nature 2020

stroke diagnoses and treatment optimization. Additionally, in spite of the important role that brain biomarkers could have in emergency diagnosis, there currently exists no "perfect" biomarker (s) for stroke. A single blood sample collected by first responders and processed rapidly for biomarker(s) testing might help to stratify patients into main stroke groups even prior to admission to hospital.

A "perfect" molecular biomarker could assist in risk assessment, predict and distinguish the stroke from mimicking conditions, and measure disease progression. Specifically, the "perfect" molecular biomarker testing combined with other markers (functional and structural) should meet certain requirements: (1) reflect an early pathology of brain damage; (2) reveal structural impairment(s) in certain brain areas; (3) allow early assessment in biological fluids (within minutes to hours); and (4) correlate with severity of disease progression.

Functional genomics, proteomics, and molecular imaging represent robust and evolving areas of molecular biomarker research that have the potential to revolutionize the screening, diagnosis, prognosis, and prediction of disease recurrence as well as therapeutic monitoring of stroke and transient ischemic attack (TIA).

Functional markers. The primary tool available for assessing TIA or stroke is patient history and physical examination of neurological symptoms. Physicians can track changes in cerebrovascular disease by localizing the disease to specific vessels using various stroke scales. The NIH Stroke Scale defines the most employed scale in the USA that rates severity of symptoms with stroke diagnosis. The rationale for emergent patient care is to improve diagnostic certainty, obtain objective results within minutes, evaluate risks for future events, monitor the patient for recurrence or worsening symptoms, and institute personal timely needed interventions while including operative critical care.

The second biomarker approach is defining *structural alterations by neuroimaging*. Current clinical neuroradiological methods allow the visualization of cerebral lesions and play an important role in decision making in both diagnosis and thrombolytic treatment. Two primary diagnostic modalities are used to distinguish ischemic from hemorrhagic stroke: computed tomography (CT) and, to a lesser extent, magnetic resonance imaging (MRI). More than 90% of hospitals have access to CT scanners, enabling them to evaluate a hemorrhagic stroke within 15 min. The overall accuracy of recognition of hemorrhagic *vs.* ischemic stroke is 67–85%. A large proportion of hospitals have MRI capability as well. However, MRI scans require a longer time window for image-reading and cannot distinguish ischemic from hemorrhagic stroke within the first 6 h of onset. Comparison of MRI and CT showed sensitivity of 83% *vs.* 26%, respectively, to detect acute ischemia as well as acute and chronic hemorrhage. There still remains an option to improve diagnostic certainty of early stroke onset by combining advanced neuroimaging with key biomarkers measured in biological fluids.

The third approach is *substance or biological markers* and is still being researched. Efforts are being directed to finding molecular markers in diseased biological tissues and fluids, particularly the blood.

1.1 Protein Biomarkers

Studies using protein biomarkers in patients with ischemic cerebrovascular disease have mainly focused on pathophysiology, diagnosis, prognosis, and neuronal death in stroke. The majority of protein biomarkers examined include S-100B, neuron specific enolase (NSE), glial fibrillary acidic protein (GFAP), brain natriuretic peptide (BNP), D-dimer, metalloproteinase 9 (MMP-9), monocyte chemotactic protein-1 (MCP-1), NMDA receptor markers (NR2 peptide and NR2 antibody), anti-cardiolipin (anti-phospholipid) antibodies, and lipoprotein-associated phospholipase A2 (Lp-PLA2), which are still in translational research mode.

1.2 Genetic and Epigenetic Biomarkers

A number of epidemiological studies suggested that stroke has genetic susceptibility, and various genetic factors have been investigated [3]. Genetic risk factors seem to be subtype-sensitive, and differential genetic risk factors have been reported to atherosclerotic, cardioembolic, and lacunar stroke [4]. Changes in the KCNQ1 methylation pattern have been shown to be useful as potential stroke biomarkers [5]. Further studies are needed to determine the role of DNA methylation in the diagnosis and prognosis of TIA and stroke.

1.3 Metabolomic Biomarkers

Metabolomics-associated biomarker profiles of fatty acid, amino acid, or polyamine in the blood or urine have been investigated to determine normal and pathologic states. Furthermore, in situ metabolomics-associated biomarkers evaluated during neurosurgery can be applied to monitoring recovery after the procedure [6]. At present, metabolomic studies using mass spectrometry are not widespread and limited to specialized neurosurgery centers.

The integrative use of structural, functional, and molecular biomarkers would be advantageous during time-critical assessment of patients presenting with stroke-like symptoms at a mobile stroke unit. Although the role of the laboratory blood testing in the diagnosis of neurologic disorders is still relatively modest compared to mobile neuroimaging, the need for rapid blood testing of suitable biomarkers remains in high demand. Promising molecular biomarkers and newest strategies for clinical instrumental methods are presented here, with the potential to drive the development of novel stroke management techniques, particularly in early diagnosis.

In this book, neuromethods assessing integrative biomarkers in terms of conventional, novel, and emerging analytes, techniques, platforms, and clinical applications to stroke emergent service are presented. It covers the theoretical background in conjunction with tested protocols reproducing experimental, clinical laboratory, and instrumental methods. The purpose of the issue is to supply readers—graduate students, clinical chemists, physicians, and medical residents—with the latest advances in stroke neuroscience, biotechnology, neuroimaging, and emergent care. This book presents advances in the medical application of neurofunctional, structural, and molecular biomarkers through innovative neuromethods development.

References

1. World Health Organization (2001) Biomarkers in risk assessment: validity and validation. http://www.inchem.org/documents/ehc/ehc/ehc222. Accessed 12 Jan 2019

2. Biomarkers Definitions Working Group (2001) Biomarkers and surrogate endpoints: preferred definitions and conceptual framework. Clin Pharmacol Ther 69(3):89–95. https://doi.org/10.1067/mcp.2001.113989

3. Dichgans M (2007) Genetics of ischaemic stroke. Lancet Neurol 6(2):149–161. https://doi.org/10.1016/S1474-4422(07)70028-5

4. Traylor M, Farrall M, Holliday EG, Sudlow C, Hopewell JC, Cheng YC, Fornage M, Ikram MA, Malik R, Bevan S, Thorsteinsdottir U, Nalls MA, Longstreth W, Wiggins KL, Yadav S, Parati EA, Destefano AL, Worrall BB, Kittner SJ, Khan MS, Reiner AP, Helgadottir A, Achterberg S, Fernandez-Cadenas I, Abboud S, Schmidt R, Walters M, Chen WM, Ringelstein EB, O'Donnell M, Ho WK, Pera J, Lemmens R, Norrving B, Higgins P, Benn M, Sale M, Kuhlenbäumer G, Doney AS, Vicente AM, Delavaran H, Algra A, Davies G, Oliveira SA, Palmer CN, Deary I, Schmidt H, Pandolfo M, Montaner J, Carty C, de Bakker PI, Kostulas K, Ferro JM, van Zuydam NR, Valdimarsson E, Nordestgaard BG, Lindgren A, Thijs V, Slowik A, Saleheen D, Paré G, Berger K, Thorleifsson G, Australian Stroke Genetics Collaborative, Wellcome Trust Case Control Consortium 2 (WTCCC2), Hofman A, Mosley TH, Mitchell BD, Furie K, Clarke R, Levi C, Seshadri S, Gschwendtner A, Boncoraglio GB, Sharma P, Bis JC, Gretarsdottir S, Psaty BM, Rothwell PM, Rosand J, Meschia JF, Stefansson K, Dichgans M, Markus HS; International Stroke Genetics Consortium (2012) Genetic risk factors for ischaemic stroke and its subtypes (the METASTROKE collaboration): a meta-analysis of genome-wide association studies. Lancet Neurol 11(11):951–962. https://doi.org/10.1016/S1474-4422(12)70234-X

5. Gomez-Uriz AM, Milagro FI, Mansego ML, Cordero P, Abete I, De Arce A, Goyenechea E, Blázquez V, Martínez-Zabaleta M, Martínez JA, López De Munain A, Campión J (2015) Obesity and ischemic stroke modulate the methylation levels of KCNQ1 in white blood cells. Hum Mol Genet 24(5):1432–1440. https://doi.org/10.1093/hmg/ddu559

6. Boyd LA, Hayward KS, Ward NS, Stinear CM, Rosso C, Fisher RJ, Carter AR, Leff AP, Copland DA, Carey LM, Cohen LG, Basso DM, Maguire JM, Cramer SC (2017) Biomarkers of stroke recovery: Consensus-based core recommendations from the stroke recovery and rehabilitation roundtable. Neurorehabil Neural Repair 31(10–11):864–876. https://doi.org/10.1177/1545968317732680

Correction to: Laser-Capture Microdissection for Measurement of Angiogenesis After Stroke

Mark Slevin, Xenia Sawkulycz, Laura Combes, Baoqiang Guo, Wen-Hui Fang, Yasmin Zeinolabediny, Donghui Liu, Glenn Ferris, and Anna Ludlaim

Correction to:
Chapter 7 in: Philip V. Peplow et al. (eds.), *Stroke Biomarkers*, Neuromethods, vol. 147, https://doi.org/10.1007/978-1-4939-9682-7_7

The original version of this chapter was revised. Dr. Anna Ludlaim has been included to the co-authors list. Dr. Anna Ludlaim was contribution was also vital to the book with her involvement in the construction of the manuscript procuring the references and some of the draft writing.

The updated online version of this chapter can be found at:
https://doi.org/10.1007/978-1-4939-9682-7_7

Philip V. Peplow et al. (eds.), *Stroke Biomarkers*, Neuromethods, vol. 147, https://doi.org/10.1007/978-1-4939-9682-7_21,
© Springer Science+Business Media, LLC, part of Springer Nature 2020

INDEX

A

Acute ischemic stroke (AIS) 10, 12–14,
   18, 157–166, 173, 195, 197, 199, 200, 203–205,
   211, 227, 236, 267, 273, 274, 353, 400, 412
Adenosine triphosphate (ATP) 43, 61,
   62, 69–72, 133, 300, 401
Angiogenesis 36, 39,
   40, 43, 44, 47, 51, 113–121
Antibodies 12, 38,
   63, 118, 142, 172, 226, 421
Anticoagulation 158, 285,
   289, 298, 307, 413
Arterial hypertension (HT) 126, 127,
   131–133, 231
Arterial occlusion 130, 238, 384
Atherosclerosis 9, 31, 128,
   139, 140, 158, 163, 200, 203, 205, 206, 217,
   218, 226, 228, 246, 247, 251, 267, 306, 336,
   338, 340, 342, 400

B

Bioenergetics 62, 75
Biomarkers .... xi, 3–5,
   9–18, 23, 29, 35–54, 79–106, 125–147,
   157–166, 171–213, 226, 229, 239, 246,
   259–290, 309, 328, 330, 408–409, 419–422
Blood assay 173, 185, 203, 213
Blood biomarkers 79–106,
   127, 132, 133, 144, 146, 160, 161, 196
Brain .... ix, xi, 3, 11,
   35, 66, 80, 113, 126, 163, 171, 195, 225, 245,
   260, 297, 350, 385, 397, 419
Brain biomarkers 3, 420

C

Cardioembolic strokes 32, 158,
   160, 163, 165, 166, 200, 214, 267, 270
Carotid artery disease 335, 336,
   338, 340, 343, 344
Carotid endarterectomy (CEA) 200,
   201, 336, 340, 341, 349–351, 353–359,
   361–363, 365, 374, 413
Carotid ultrasound 336, 338, 340, 341

Cerebral infarction 52, 132, 195,
   203, 206, 208, 226, 227, 229, 232, 236, 238,
   268, 288, 350
Cerebral ischemia 3, 4,
   11–13, 36, 37, 39–41, 44, 49, 50, 52, 61, 69,
   171–173, 197, 199, 203, 211, 214, 349–365
Cerebral small vessel disease (CSVD) 127–130,
   132, 139, 146
Clinical diagnostics 10, 106,
   213, 270
Computed tomography (CT) 10,
   79, 159, 195, 236, 260, 300, 372, 420

D

Diabetes 31, 44, 127,
   128, 131, 158, 161, 199–201, 203, 205, 206,
   214, 217, 218, 231, 245–254, 268, 271
Diagnosis 4, 10, 79,
   113, 127, 158, 173, 195, 226, 260, 297, 353,
   372, 399, 419

E

Electroencephalography (EEG) 127,
   274, 352, 354, 355, 357–360, 362–365
Embolism 226, 236,
   335, 352, 355, 359, 360, 363
Experimental stroke recovery 35–54

F

Fluorescence 45, 46,
   54, 69, 71, 72, 74, 162, 186
Fluorometry 71, 72

G

Genetic risk scores (GRS) 23–32
Genome-wide association studies (GWAs) 23–28,
   31, 32
Glial proteins 39, 48,
   141, 197, 228, 421
Glutamate (Glu) 43, 171,
   173, 195–213, 226, 227, 261, 262, 402
Glutamate receptor (GluR) 173,
   195–213, 226, 231

Philip V. Peplow et al. (eds.), *Stroke Biomarkers*, Neuromethods, vol. 147, https://doi.org/10.1007/978-1-4939-9682-7,
© Springer Science+Business Media, LLC, part of Springer Nature 2020

H

Hemorrhagic stroke (HS) 4, 27,
    35, 38, 39, 79, 80, 84, 161, 208, 210, 213, 230,
    259, 270, 285, 420
Hemorrhagic transformation (HT) 11, 14,
    15, 38, 39, 80, 95, 96, 98, 199, 201–203, 216,
    217, 229, 232, 270, 272, 277, 285, 287

I

Imaging..ix, 3, 10, 36,
    67, 119, 127, 158, 199, 226, 246, 260, 297, 336,
    373, 399, 420
Immunoassay......................................172, 176, 178, 185
Infarction 4, 13, 41,
    95, 114, 133, 159, 186, 196, 225, 260, 267, 342,
    350, 409
Inflammation 12, 36, 43,
    113, 114, 127, 129–131, 133, 138, 144, 202,
    214, 229, 246, 247, 270, 288
Inflammatory biomarkers 133, 144
Injury 11–13, 16,
    17, 35, 37, 42, 53, 65–67, 72, 75, 130, 135,
    141–143, 145, 206, 209, 213, 217, 226, 227,
    230, 233, 234, 310, 314, 315, 385, 398,
    400–402, 404
Ischemia...4, 11,
    36, 89, 138, 160, 171, 199, 226, 270, 352, 383,
    399, 420
Ischemic stroke (IS) 4, 9, 27,
    36, 79, 114, 133, 158, 180, 196, 226, 259, 349,
    375, 399, 420

L

Lacunar infarcts 127, 129,
    133, 134, 138–142, 144, 145, 214, 246, 285
Lacunar strokes 38, 140,
    141, 143, 202, 203, 215, 216, 227, 246, 421
Large vessel occlusion (LVO) 10,
    195, 207, 213, 268, 274, 298, 299, 307,
    310–313, 375, 378, 384
Laser capture micro-dissection (LCM)............... 113–121

M

Magnetic resonance imaging (MRI)............................. 10,
    36, 39, 40, 42, 47, 50–52, 126, 127, 129, 133,
    134, 136, 138, 139, 141–145, 158, 163, 196,
    203–206, 209–211, 213, 214, 226, 229,
    232–234, 236, 238, 246, 260–262, 265–269,
    273, 275, 278–286, 289–291, 300, 301, 308,
    309, 313, 316–328, 338, 355, 374, 400, 408, 420

Mobile stroke unit (MSU) 371–389, 421
Molecules...11–15,
    40, 61, 62, 67, 96, 113, 129, 131, 132, 142, 143,
    261, 300, 301, 325, 326

N

NAD+ 69
Near-infrared spectroscopy (NIRS) 355, 356
Neuroimaging 4, 36, 44,
    80, 101, 127, 146, 162, 185, 197, 199, 202–204,
    208, 209, 211, 226, 236, 237, 239, 282,
    297–330, 374, 420–422
Neuronal proteins 141
Neuron-specific enolase (NSE) 81, 83,
    96, 142, 145, 197, 421
Neuroplasticity biomarkers...35–54
Neuroradiology 275
Neurotoxicity 43, 171–194, 197, 199
Neurovascular biomarkers 195–213,
    226, 231, 238, 239
Nicotinamide adenine dinucleotide
    (NADH) 61, 62, 69, 70
NMDA receptors (NMDAR).....................................143,
    146, 171–173, 186, 198, 199, 214, 225–239, 421
N-methyl-D-aspartate (NMDA)................................142,
    171–173, 238
NR2 .... 143, 172, 173,
    177–180, 182, 184, 185, 187, 188, 194, 199,
    202, 204, 210, 214–216, 226, 228, 230, 231,
    235, 237, 238, 421
NR2 peptides 89, 146,
    172, 173, 175, 177, 179, 181, 183, 184,
    199–203, 214, 215, 217, 218, 237, 408, 421

O

Oxidative phosphorylation (OXPHOS) 61–63,
    69, 71, 72, 75, 133

P

Prehospital stroke 206, 372, 373, 381, 382

R

Radiology..275, 297, 301, 405
Respiration.. 61, 70–72, 300
Respirometry 69, 71, 72
Retinal imaging 248–250
RNA/DNA 25, 47,
    62, 64, 65, 83, 115, 116, 119–121, 140, 143,
    157–166, 421
RNA expression....................................159, 162, 164–166

S

S100B protein 142, 228
Small vessel occlusion (SVO) 9, 158,
        195–197, 201, 202, 213, 298, 304, 310
Spinal cord.. 37, 199,
        202, 214, 225–239, 260, 270
Stroke
    differentiation..79–106
    etiology 158–160, 163, 164, 166
    imaging 261–262, 325
    mimics.. 4, 80, 83,
        89, 91, 101, 104, 167, 200, 201, 203,
        205–208, 210, 211, 273, 288–291, 297, 328, 386
    risk...31, 32, 135,
        140, 144–146, 157, 245–254, 338

T

Telemedicine 372, 373,
        375–377, 379–381, 387, 389, 408
Teleradiology...379
Telestroke 274, 288, 379
Therapy... 4, 10, 12, 14,
        15, 17, 36, 44, 104, 113, 158, 196, 201, 203,
        213, 259–261, 268–270, 273–275, 277–280,
        305, 313, 327, 343, 372, 374, 375, 378, 382,
        383, 388, 402, 403, 408, 409, 411, 413

Thresholds

Thresholds 11, 14,
        26–28, 45, 48, 274, 315, 339, 354, 355, 362
Thrombectomy 9, 10, 196,
        273, 278, 298, 299, 301, 307, 309–313, 315,
        328, 371, 372, 374, 378, 382, 384
Thrombolysis.. 10, 14,
        15, 97, 161, 201, 273–277, 298, 301, 315, 328,
        371, 372, 374, 375, 377–384, 386, 387, 389
Thrombosis 16, 139, 141,
        164, 196, 259, 267, 287–289, 351, 359, 413
Transcranial Doppler ultrasound (TDUS) 352,
        354–358, 361–365
Transient ischemic attack (TIA) 3, 4,
        39, 80, 83, 89, 101, 104, 105, 139, 145, 146,
        157, 160, 185, 201, 203, 214, 231, 260–261,
        269–271, 273, 335–345, 397, 400, 409, 420, 421

V

Vascular... 4, 10, 27,
        36, 65, 89, 113, 128, 162, 185, 195, 227, 246,
        261, 342, 351, 374, 398
Vessel reactivity..251, 253

W

White matter hyperintensities (WMHs) 127,
        128, 133, 134, 138–143, 145, 146, 202, 273,
        279–285, 290